# AEROSOLS and CLIMATE

**Studies in**
**Geophysical Optics and Remote Sensing**

**Series Editor: Adarsh Deepak**

## Published Volumes and Volumes in Preparation

**Advances in Remote Sensing Retrieval Methods**
Adarsh Deepak, Henry E. Fleming, and Moustafa T. Chahine (Eds.)

**Aerosols and Climate**
Peter V. Hobbs and M. Patrick McCormick (Eds.)

**Aerosols and Their Climatic Effects**
Hermann E. Gerber and Adarsh Deepak (Eds.)

**Applications of Remote Sensing for Rice Production**
Adarsh Deepak and K.R. Rao (Eds.)

**Atmospheric Aerosols: Their Formation, Optical Properties, and Effects**
Adarsh Deepak (Ed.)

**Hygroscopic Aerosols**
Lothar H. Ruhnke and Adarsh Deepak (Eds.)

**IRS '84: Current Problems in Atmospheric Radiation**
Giorgio Fiocco (Ed.)

**Light Absorption by Aerosol Particles**
Hermann E. Gerber and Edward E. Hindman (Eds.)

**Microwave Remote Sensing**
A. Chedin (Ed.)

**Optical Properties of Clouds**
Peter V. Hobbs and Adarsh Deepak (Eds.)

**Proceedings in Atmospheric Electricity**
Lothar H. Ruhnke and John Latham (Eds.)

**Radiative Transfer in Scattering and Absorbing Atmospheres: Standard Computational Procedures**
Jacqueline Lenoble (Ed.)

**RSRM '87: Advances in Remote Sensing Retrieval Methods**
Adarsh Deepak, Henry E. Fleming, and John S. Theon (Eds.)

**Spectral Line Shapes, Volume 4**
R.J. Exton (Ed.)

**Tropical Rainfall Measurements**
John S. Theon and Nobuyoshi Fugono (Eds.)

**Selected papers from the**
Symposium on Aerosols and Climate, which was one of the Symposia organized by the International Association of Meteorology and Atmospheric Physics (IAMAP), at the **XIX General Assembly of the International Union of Geodesy and Geophysics**. The General Assembly was held at the University of British Columbia, Vancouver, Canada, 9–22 August 1987. The Symposium was called by Professor P.V. Hobbs and Dr. M.P. McCormick. Preparations for the meeting were made by the IUGG Local Organizing Committee.

# AEROSOLS and CLIMATE

*Edited by*

Peter V. Hobbs
University of Washington
Seattle, Washington

and

M. Patrick McCormick
National Aeronautics and Space Administration
Langley Research Center
Hampton, Virginia

**A. DEEPAK Publishing 1988**
A Division of Science and Technology Corporation
Hampton, Virginia USA

**A. DEEPAK Publishing**
A Division of Science and Technology Corporation
101 Research Drive
Hampton, Virginia 23666–1340 USA

**Library of Congress Cataloging-in-Publication Data**

Aerosols and climate / edited by Peter V. Hobbs and M. Patrick McCormick.
p. cm. — (Studies in geophysical optics and remote sensing)
Selected papers from the Symposium on Aerosols and Climate organized by the International Association of Meteorology and Atmospheric Physics at the XIX General Assembly of the International Union of Geodesy and Geophysics, held at the University of British Columbia, Vancouver, Canada, 9–22 August 1987.
Bibliography: p.
Includes index.
ISBN 0-937194-11-5 :
1. Aerosols—Congresses. 2. Climatic changes—Congresses. 3. Troposphere—Congresses. 4. Stratosphere—Congresses. I. Hobbs, Peter Victor, 1936– . II. McCormick, M. Patrick. III. International Association of Meteorology and Atmospheric Physics. IV. Symposium on Aerosols and Climate (1987 : University of British Columbia). V. Series.
QC882.A358 1988 88-29946
551.6—dc19 CIP

Printed in the United States of America

## CONTENTS

Page

*Preface* ..... ix

### 1. SOURCES OF ATMOSPHERIC AEROSOL

Laboratory Investigations of the Influence on Marine Aerosol Production of the Interaction of Oceanic Whitecaps and Surface-Active Material ..... 1
*D.K. Woolf and E.C. Monahan*

The Role of Nonprecipitating Cloud Cycles and Gas-to-Particle Conversion in the Maintenance of the Submicron Aerosol Size Distribution Over the Tropical Oceans ... 9
*William A. Hoppel*

Dead Sea Evaporation Rates in Relation to Aerosol Number Density Variations ..... 21
*A. Cohen, Y. Mahrer, M.A. Weiss and M. Litvak-Neumann*

The Primary Biogenic Fraction of the Atmospheric Aerosol ..... 31
*R. Jaenicke and S. Matthias*

### 2. SENSING AND DISTRIBUTION OF ATMOSPHERIC AEROSOL

Remote Sensing of Atmospheric Aerosols Over the Oceans Using Multispectral Radiances Measured With the Advanced Very High Resolution Radiometer Onboard the NOAA Meteorological Satellites ..... 39
*K. Masuda, T. Takashima and C.R. Nagaraja Rao*

Satellite Mapping of Aerosol Loading Over Vegetated Areas ..... 51
*Yoram J. Kaufman and Claudia Sendra*

Development and Application of Aerosol Remote Sensing With AVHRR Data From the NOAA Satellites ..... 69
*C.R. Nagaraja Rao, L.L. Stowe, E.P. McClain and J. Sapper*

A Spaceborne Backscatter Lidar Proposed for a German Space-Shuttle Mission ..... 81
*Heinrich Quenzel, Matthias Wiegner, Martina Kästner, Wolfgang Renger and Martin Endemann*

Optical Remote Sensing of Atmospheric Aerosol, Water Vapor, and Trace Gas $NO_2$ ... 89
*Zhao Bolin, Mao Jietai, Zhang Chengxiang and Wang Qiang*

Inference of Atmospheric Aerosol Size Distribution From Ground Measurements of Solar Radiation ..... 101
*U. Amato, V. Cuomo, F. Fontana, M. Macchiato and C. Serio*

Rocket and Balloon Measurements of the Vertical Distribution of Aerosols in the Indian Tropical Region . . . . . 117
*A. Jayaraman and B.H. Subbaraya*

An Approach to a Global Optical Aerosol Climatology . . . . . 125
*G.A. d'Almeida and P. Koepke*

Long-Term Series of Atmospheric Turbidity, Estimated From Records of Sunshine Duration and Cloud Cover . . . . . 139
*L. Helmes and R. Jaenicke*

Simulation of the Desert Dust Cycle Using an Atmospheric General Circulation Model . . . . . 147
*S. Joussaume and R. Sadourny*

## 3. SOME PROPERTIES AND EFFECTS OF ATMOSPHERIC AEROSOL

Physical Properties of Aerosols in Rural and Forested Areas . . . . . 153
*Josef Podzimek*

On the Scaling of Atmospheric Aerosol Scattering and Extinction With Wavelength . . . 165
*Eric P. Shettle and Eli Trakhovsky*

Aerosol Sizes and Nighttime Cooling Rates in the Harmattan . . . . . 175
*C.O. Oluwafemi*

Impact of Nonstandard Conditions on Visibility Measurements . . . . . 185
*D. Weintraub and V.K. Saxena*

## 4. VOLCANIC EFFECTS

Aerosols From Past and Present Volcanic Emissions* . . . . . 195
*D.J. Hofmann*

Stratospheric Extinction and Mass Variations After a Major Volcanic Eruption, Derived From Lidar Measurements at Northern Midlatitudes . . . . . 215
*H. Jäger, M. Littfass, D.J. Hofmann and J.M. Rosen*

The Abnormally Cold Summers of Central China and Their Relation to Volcanic Eruptions . . . . . 223
*Xu Qun*

Modeling of Optical Properties of Postvolcanic Stratospheric Layer . . . . . 233
*V.E. Zuev, G.M. Krekov and R.F. Rakhimov*

*Invited review paper.

## 5. RADIATIVE EFFECTS OF ATMOSPHERIC AEROSOL

Radiative Effects of Atmospheric Aerosol Particles* .... 241
*Hartmut Grassl*

Effect of Ship-Stack Effluents on the Radiative Properties of Marine Stratocumulus: Implications for Man's Impact on Climate .... 253
*J.A. Coakley, Jr., R.L. Bernstein and P.A. Durkee*

A Desert Aerosol Model for Radiative Transfer Studies .... 261
*David R. Longtin, Eric P. Shettle, John R. Hummel and James D. Pryce*

Radiation Studies in the Arctic .... 271
*Francisco P.J. Valero, Thomas P. Ackerman, Warren J.Y. Gore and Melinda L. Weil*

## 6. STRATOSPHERIC AEROSOLS

Characteristics of Aerosols in the Lower Stratosphere .... 277
*David C. Woods and Raymond L. Chuan*

Nitric Acid Haze Formation in the Lower Stratosphere: A Major Contributing Factor to the Development of the Antarctic "Ozone Hole" .... 287
*P.J. Crutzen, C. Brühl, U. Schmailzl and F. Arnold*

Size Distribution of Stratospheric Aerosols From SAGE II Multiwavelength Extinctions .... 305
*C. Brogniez and J. Lenoble*

Ten Years of $CFCL_3$ Measurements in the Stratosphere: Model Comparisons for the Northern Hemisphere .... 313
*Robert Leifer*

Stratospheric Aerosol Conditions Over Mauna Loa During Recent Quiescent Volcanic Periods .... 325
*Elmer Robinson and Thomas E. DeFoor*

## 7. AEROSOLS AND CLIMATE CHANGE; NUCLEAR WINTER

Aerosols in Climate Modeling* .... 335
*Thomas P. Ackerman*

Aerosol Radiative Forcing and Model Responses* .... 349
*V. Ramaswamy*

The Effects of Arctic Aerosols on a General Circulation Model Simulation: A Sensitivity Study .... 373
*Jean-Pierre Blanchet*

*Invited review paper.

On the Possibility of Climate Instabilities After Violent Volcanic Eruptions .......... 385
*R. Reiter*

Climatic Effects of Smoke and Dust Produced From Nuclear Conflagrations* ........ 395
*A.B. Pittock, J.S. Frederiksen, J.R. Garratt and K. Walsh*

Airborne Measurements on Smokes From Biomass Burning ........................ 411
*L.F. Radke, D.A. Hegg, J.H. Lyons, C.A. Brock, P.V. Hobbs, R. Weiss and R. Rasmussen*

Potential Nucleation Scavenging of Smoke Particles Over Large Fires: A Parametric Study ........ 423
*Leslie L. Edwards and Joyce E. Penner*

Surface Temperature Effects of Forest Fire Smoke Plumes ........................ 435
*Alan Robock*

Persistent Effects of Residual Smoke Layers .................................. 443
*T.P. Ackerman, R.P. Turco and O.B. Toon*

The Development of Low Level Coastal Flow Fields When Solar Radiation Is Blocked by Smoke ........ 459
*C.R. Molenkamp*

The International Aerosol Climatology Project ................................ 471
*Adarsh Deepak and Gabor Vali*

*Subject Index* .......................................................... 481
*Author Index* ........................................................... 485

*Invited review paper.

# PREFACE

Until comparatively recently, the majority of research on atmospheric aerosols was carried out by those interested in the effects of aerosols on air quality and on cloud and precipitation processes. Observations were largely confined to in situ measurements. In the past several years these efforts have been augmented by those interested in the effects of aerosols on the radiation balance of the earth, on climate, and atmospheric chemistry. Fortunately, the growing interest in global aspects of aerosols has coincided with (and, in part, been prompted by) the capability to monitor aerosols on a global scale from satellites.

One of the principal objectives of the *Symposium on Aerosols and Climate,* which was part of the *XIX General Assembly of the International Union of Geodesy and Geophysics* held in Vancouver, Canada, 9–22 August 1987, was to bring together a broad spectrum of researchers interested in atmospheric aerosols. The goals of the symposium were to review our understanding of the properties and effects of atmospheric aerosols, and to discuss means of pursuing some of the outstanding problems in this subject.

The present volume contains a selection of papers from this symposium. Included are five invited papers that review the effects of volcanic emissions on atmospheric aerosols, the radiative effects of atmospheric particles, aerosols in climate modeling, and the potential effects of a nuclear war on the climate of the earth. Thirty-seven contributed papers provide information on the sources, measurement, distribution, properties, and effects of atmospheric aerosols in the troposphere and stratosphere and the role of aerosols in climate change. All papers included in this book were peer-reviewed and revised before final acceptance.

Preparation of this book has involved a great deal of work by many individuals. In particular, we would like to thank both the authors and reviewers of the papers for their efforts under tight time constraints. It is also a great pleasure to acknowledge the diligent work of Diana McQuestion and Marilou Fromme, of A. Deepak Publishing, in bringing this volume to publishable form. If this book plays a role in advancing our understanding of atmospheric aerosols and their effects, the efforts of all these individuals, as well as our own, will have been well repaid.

*Peter V. Hobbs*
*M. Patrick McCormick*
27 May 1988

# LABORATORY INVESTIGATIONS OF THE INFLUENCE ON MARINE AEROSOL PRODUCTION OF THE INTERACTION OF OCEANIC WHITECAPS AND SURFACE-ACTIVE MATERIAL[1]

*D.K. Woolf and E.C. Monahan*
Marine Sciences Institute, University of Connecticut
Groton, Connecticut 06340, USA

ABSTRACT

The production of marine aerosol associated with oceanic whitecaps can be modified by the presence of surface-active material. There is a complex interaction between breaking waves and whitecaps on the one hand, and the surface-active material distributed in the upper ocean and at the sea surface on the other. Some aspects of this interaction are reflected in experiments based on laboratory simulations of breaking waves. Two examples are detailed. In one case a slick apparently enhanced the production of film drops. In the second case, the whitecapping apparently generated a slick that inhibited aerosol production. It is suggested that similar effects will be prevalent on the world's oceans and, consequently, marine aerosol fluxes should be reevaluated.

## 1. INTRODUCTION

A breaking wave entrains air in the form of many bubbles, which are manifested on the surface as a whitecap. When the bubbles burst, two distinct types of drop are produced; these are film drops and jet drops (Blanchard, 1963 and 1983; Woolf et al., 1987). Together these drops form an important component of the atmospheric aerosol. A few workers have simulated oceanic whitecaps in the laboratory, and thereby produced estimates of marine aerosol production (Monahan, 1986; Cipriano et al., 1987). This paper describes experiments using simulated whitecaps in which surface-active material (SAM) in natural seawater appears to have affected the production of aerosol, thus introducing considerable uncertainty into present estimates of marine aerosol production. An ultimate objective of our continuing studies is to assess the impact that SAM, and other material concentrated at natural water surfaces, has and can have on the generation of marine aerosol.

There are several compelling reasons to expect that material on the sea surface, or on the surface of bubbles, will affect the generation of drops by the bubbles. Considerations of the dynamics of bubble bursting and drop production suggest that these processes are critically dependent on properties such as the surface tension and elasticity of the water surfaces (MacIntyre, 1972). Thus any material

[1]The whitecap simulation studies at UCG were made possible by ONR Grant N00014-78-G-0052.



ISBN 0-937194-11-5

that can alter the properties of a water surface potentially can have an impact on drop generation. Some details of the effect of organic materials on drop production have been learnt from laboratory studies. Blanchard (1963) showed that an organic film could eliminate film-drop production by individually bursting bubbles, while bubbles that clustered together before bursting were not affected by similar films. Paterson and Spillane (1969) demonstrated the crucial importance of surface film pressure to film-drop production. Garrett (1968) reported an enhancement of condensation nucleus production (film drops) when "foam-breaking" SAM was applied. Blanchard and Hoffman (1978) found that jet-drop production is slightly inhibited by SAM scavenged by the bubble before surfacing. The effect of these and other phenomenon on aerosol production by oceanic whitecaps is unknown. Laboratory simulations of whitecaps are a useful tool for assessing the impact of SAM on realistic bubble populations.

While SAM affects the behaviour of bubbles, and in extreme cases the development of breaking waves, whitecapping in turn will have a major influence on the distribution of SAM. The turbulence induced by breaking waves is a major mixing agent in the upper ocean. Meanwhile bubbles scavenge SAM from the water column (Sutcliffe et al., 1963). This material may be left as an aggregate in the water column if the bubble dissolves, or may be ejected into the atmosphere or deposited

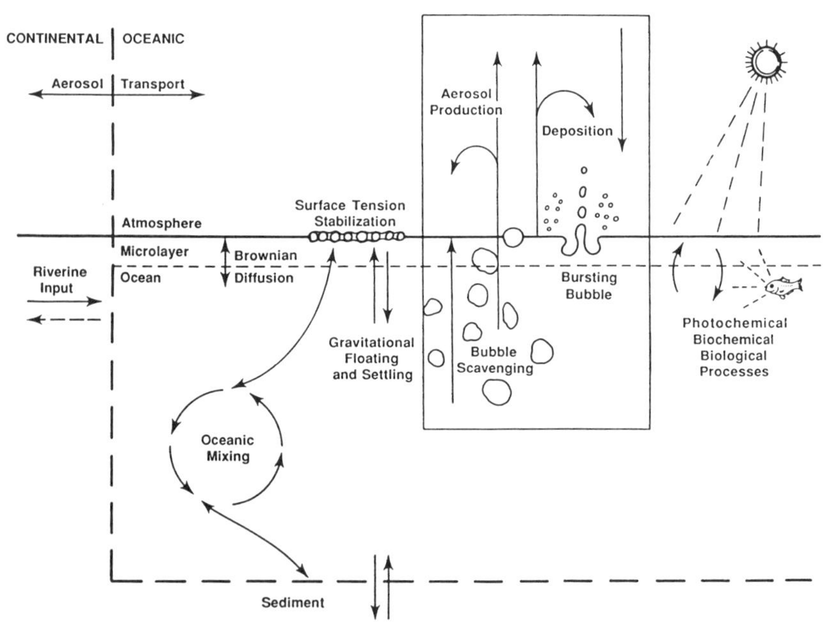


FIGURE 1. Schematic of the cycling of marine surface-active material. The bubble-mediated mechanisms are boxed. Breaking waves will be an important component of "oceanic mixing" at high windspeeds.

on the sea surface if the bubble surfaces. Material already on the water surface can be removed on the aerosol drops generated by bursting bubbles (Blanchard, 1964; Baier, 1972). Other "non-whitecap" mechanisms also transport SAM; the cycle is shown schematically in Fig. 1. In high winds, whitecap-related mechanisms may effectively control the distribution of SAM.

## 2. TEMPERATURE

The University College, Galway Whitecap Simulation Tank (UCG WST), was used to study the temperature dependence of the aerosol production of a standard breaking wave event (see also: Bowyer, 1986; Woolf et al., 1987). The breaking wave events were produced by pumping water to a standard height behind gates at opposite ends of the tank (Fig. 2). The gates were raised simultaneously, allowing water to rush towards the centre from both ends. The collision of the two waves produced a whitecap in the deep central well. The aerosol produced was sampled from the overlying air within an enclosing hood. The hood was continually flushed by clean, filtered air that prevented infiltration and mixed the air in the hood. Detailed descriptions of operational procedures can be found elsewhere (Monahan et al., 1982; Cipriano et al., 1987).

Numerous aerosol measurements were made using several different natural seawater samples. Each water sample was collected locally (Galway Bay, Ireland) when it was needed. Aerosol production varied considerably with temperature, but in addition, the production at a given water temperature varied widely between water samples. In general, the production of small drops was much higher at low temperatures. Woolf et al. (1987) identified the temperature dependence and water-sample dependence of small drop production with the film drops that predominate in this size range (<2 µm radius at 80% relative humidity). Jet-drop production exhibits a different temperature dependence (Bowyer, 1986). There was only limited

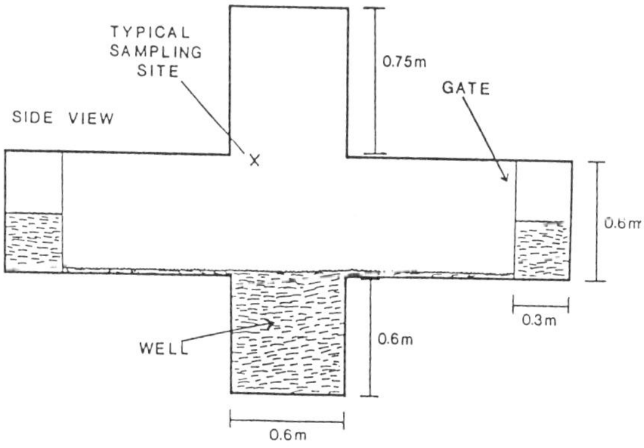


FIGURE 2. The University College, Galway Whitecap Simulation Tank, prepared for a standard breaking wave event (adapted from a figure by Monahan et al., 1982).

evidence that jet-drop production varied with the water sample in these experiments. The production of particles >0.5 µm diameter for one water sample is shown in Fig.3. The total number of particles >0.5 µm diameter was always dominated by film drops, and they were responsible for the temperature dependence of this aerosol. This sample manifested a particularly high level of production at low temperatures (level "e" in Fig. 3). A low level of production, "u", was common to all water samples at a sufficiently high temperature, and corresponds to the normal production measured in earlier experiments in the UCG WST. In order to avoid overcrowding, the results for only one water sample were shown in Fig. 3. The temperature dependence for other water samples are clearly distinguishable. The temperature dependence of at least one water sample changed during its period of use. The physical properties of water vary considerably with temperature (especially viscosity), but the variation of aerosol production between water samples cannot be explained in terms of bulk water properties. A clue to the cause of the erratic small-drop production came from the observation (unfortunately not systematic) that periods of high small-drop

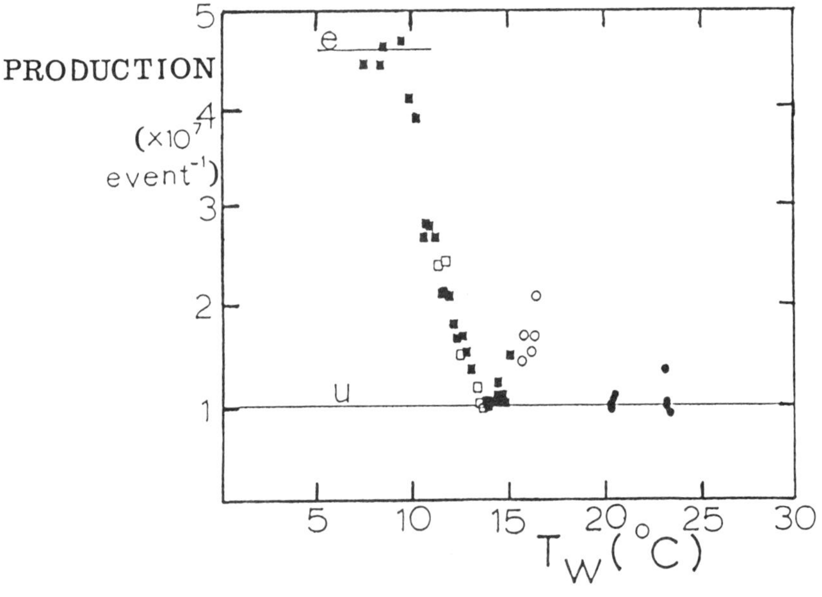


FIGURE 3. Aerosol production plotted against water temperature, for individual standard breaking wave events in the UCG WST using one water sample. All particles larger than 0.5 µm diameter at 80% relative humidity are included. Four different symbols are used, one for each of four distinct measurement sets. The similar trend of the sets marked by open squares and filled squares indicates that the temperature-dependence was consistent as the water warmed slowly on two separate occasions.

production were coincident with periods when a complex slick was visible on the water surface. We hypothesize that film-drop production responded to changes in the state of the water surface. We conjecture that SAM was always present at the water surface, but was modified (perhaps constituents changed phase) when the temperature changed. The variation of aerosol production between water samples is then readily explained by the natural variability of the organic content of seawater.

## 3. TEMPORAL VARIABILITY

During the experiments described above, the rate of whitecap formation and decay was probably insufficient, even under the most intensive use, for the associated bubble scavenging to have a substantial effect on the distribution of SAM. In some other experiments, we used a crude breaking wave simulation in which a steady jet of water fell continuously into a tank of water entraining large quantities of air. In these experiments, we would expect the level of bubble scavenging to be sufficient to produce a major change in the distribution of SAM in the several hours over which the simulation was run. The scavenging bubbles would carry SAM to the surface. Organic material would also be transported on drops to the air above.

The experiments were conducted in the UCG WST, using natural seawater and with the hood steadily flushed with filtered air. Essentially, the aerosol concentration (and thus the signal measured by the particle counter) in the hood, at any one time, will be directly proportional to the instantaneous aerosol production rate. The aerosol concentration was measured throughout the length of "runs" of the simulation; measurements for one run are shown in Fig.4. There was a big drop in the production of large particles over thirty hours, while small particle production was fairly steady. At the same time the water surface became increasingly soiled with material. This led to an increased stability of the bubble rafts, thus increasing the visual size of the whitecap. A similar pattern was repeated, with one exception, during several runs with two distinct water samples. A few samples of water from and beneath the surface were collected. Analysis showed that the surface was enriched in nitrogen and phosphorous compounds at a time when rafting was heavy, thus supporting the visual evidence.

We hypothesize that bubble scavenging steadily built up SAM on the water surface and this was responsible for the decrease in aerosol production. The transport of SAM from the water column to the surface appears to have outweighed the transport of material from the surface to the air above. The steadiness of small-drop production suggests that film-drop production was almost constant. A drop in jet-drop production must be largely responsible for the fall in large-drop production during the course of a run. Interestingly, when standard breaking wave events (see Section 2) were performed in the tank, with one of the thick slicks in place, the associated production of both large and small droplets was much lower than usual.

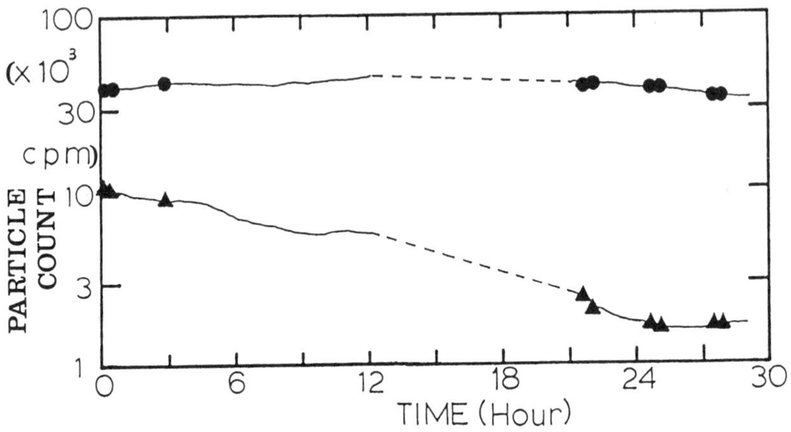


FIGURE 4. The measured particle count (counts per minute) plotted against time for one run of a continuous breaking wave simulation. Circles: 0.5-5 µm diameter (80% relative humidity). Triangles: >5 µm diameter. Symbols mark short digital counts, the full curves show the continuous analogue signal, and the dashed lines correspond to a hiatus in aerosol measurement.

## 4. DISCUSSION

Existing estimates of marine aerosol production (Monahan, 1986) are reliant on the principle that the aerosol production from single simulated whitecaps are representative of the aerosol production from all oceanic whitecaps (with appropriate scaling for the size of the whitecaps). The experiments described in the foregoing sections make it clear that the interference of SAM in aerosol production makes this principle untenable, at least in the broadest usage. Specifically, the results described in Section 2 suggest that film-drop production may be relatively high in some cold, organically-rich waters. The results described in Section 3 suggest that aerosol production can be inhibited in circumstances when high-aerosol production would normally be expected (during high winds or in a surf zone), if the associated bubble scavenging generates a surface slick. Both of the above inferences can only be very tentative at this time.

It is not clear what changes in bubble behaviour were responsible for the changes in aerosol production observed in our experiments. During the experiments described in Section 2, there appeared to be more long-lived bubbles when film-drop production was high. We suspect that SAM increased the stability and clustering of bubbles on the surface on these occasions. Some observations of Blanchard (1963) and Paterson and Spillane (1969) suggest that clustering will increase film-drop production. During the continuous-bubbling experiments (Section 3), the bubble-raft coverage increased during the course of the runs. However, most of the large bubbles (which produce most film drops) burst near the centre of the surfacing bubble plume. This area was probably kept clean by the divergent flow, and so, the steadiness of small-drop production is not very surprising. Small bubbles will surface dispersed over the surface of the tank. We conjecture that the collapse of the cavities of these bubbles can be inhibited by compact SAM on the surface, and this was responsible for the fall in large-drop production during the course of the runs. This hypothesis should be tested.

A more systematic study of the role of SAM requires detailed knowledge of the distribution of SAM, both in the simulations and in the field. In particular, we do not know the composition of the SAM in the water samples we used, and thus, how representative the phenomenon observed are of behaviour in the oceans. A prerequisite of confidently assessing the impact of SAM is to understand the behaviour of all significant classes of SAM. As a first step we have applied measured quantities of a surfactant, oleic acid, to the water surface of some of our simulations. This method of investigation had already been used during studies in the University College, Galway Whitecap Simulation Tank (Monahan et al., 1979). It served as a starting point for our continuing studies. We intend to use other surfactants in our continuing studies. More observations of single-bubble behaviour (bursting individually or in clusters), especially interacting with biochemically complex microlayers (Sieburth, 1983), are desirable. Another important shortfall is limited knowledge of the distribution of SAM in the world's oceans. Also there is contrary evidence on the state of the sea surface microlayer at high windspeeds (when whitecapping will be pronounced). Huhnerfuss et al. (1977) identified an increase in surface tension (and thus film break-up) with increasing windspeed (this seems to be the common perception). However, Demin et al. (1985) have reported an increased SAM concentration in the surface microlayer for windspeeds above 6 $ms^{-1}$. In addition, stable foam often accompanies or follows periods of high windspeed and is common in surf zones. The latter observations suggest that a period of heavy whitecapping will often enrich the sea surface microlayer. In summary, it is likely that existing oceanic distributions of SAM impact significantly on marine aerosol production. Further, pollutants or changes in biota might conceivably result in significant changes in marine aerosol production.

ACKNOWLEDGMENTS

The first studies of the effect of temperature on aerosol production, at University College, Galway, were conducted by Peter Bowyer. The whitecap simulation experiments currently being carried out at the University of Connecticut are supported by ONR Contracts N00014-87K-0069 and N00014-87K-0185.

REFERENCES

Baier, R.E., 1972: Organic Films on Natural Waters: Their Retrieval, Identification and Modes of Elimination, J. Geophys. Res., 77, 5062-5075.

Blanchard, D.C., 1963: The Electrification of the Atmosphere by Particles from Bubbles in the Sea, Prog. Oceanogr., 1, 71-202.

Blanchard, D.C., 1964: Sea-to-Air Transport of Surface-Active Material, Science, 146, 396-397.

Blanchard, D.C., and E.J. Hoffman, 1978: Control of Jet Drop Dynamics by Organic Material in Seawater, J. Geophys. Res., 83, 6187-6191.

Blanchard, D.C., 1983: The Production, Distribution and Bacterial Enrichment of the Sea-Salt Aerosol. In P.S. Liss and W.G.N. Slinn (Eds.), Air-Sea Exchange of Gases and Particles, D. Reidel Publishing Co., Dordrecht, Holland, 407-454.

Bowyer, P.A., 1986: Electrostatic Phenomena Associated with the Marine Aerosol Arising From the Bursting of Whitecap Bubbles, Ph.D. Dissertation, National Univ. Ireland, Dublin, 155 pp.

Cipriano, R.J., E.C. Monahan, P.A. Bowyer and D.K. Woolf, 1987: Marine Condensation Nucleus Generation Inferred from Whitecap Simulation Tank Results, J. Geophys. Res., 92, 6569-6576.

Demin, B.T., S.A. Yermakov, Ye.N. Pelinkovsky, T.G. Talipova and A.I. Sheremet'yeva, 1985: Study of the Elastic Properties of Sea Surface-Active Films, Izvestiya, Atmos. and Oceanic Physics, 21, 312-316.

Garrett, W.D., 1968: The Influence of Monomolecular Surface Films on the Production of Condensation Nuclei from Bubbled Sea Water, J. Geophys. Res., 73, 5145-5150.

Huhnerfuss, H., W. Walter and G. Kruspe, 1977: On the Variability of Surface Tension with Mean Wind Speed, J. Phys. Oceanogr., 7, 567-571.

MacIntyre, F., 1972: Flow Patterns in Breaking Bubbles, J. Geophys. Res., 77, 5211-5228.

Monahan, E.C., 1986: The Ocean as a Source of Atmospheric Particles. In P. Buat-Menard (Ed.), The Role of Air-Sea Exchange in Geochemical Cycling, D. Reidel Pub. Co., Dordrecht, Holland, 129-163.

Monahan, E.C., K.L. Davidson and D.E. Spiel, 1982: Whitecap Aerosol Productivity Deduced from Simulation Tank Measurements, J. Geophys. Res., 87, 8898-8904.

Monahan, E.C., B.D. O'Regan and K.L. Davidson, 1979: Marine Aerosol Production from Whitecaps. Presented at Seventeenth General Assembly of the International Union of Geodesy and Geophysics, 38 pp, preprint.

Paterson, M.P., and K.T. Spillane, 1969: Surface Films and the Production of Sea-Salt Aerosol, Quart. J. R. Met. Soc., 95, 526-534.

Sieburth, J. McN., 1983: Microbiological and Organic-Chemical Processes in the Surface and Mixed Layers. In P.S. Liss and W.G.N. Slinn, (Eds.), Air-Sea Exchange of Gases and Particles, D. Reidel Publishing Co., Dordrecht, Holland, 121-172.

Sutcliffe, W.H., Jr., E.R. Baylor and D.W. Menzel, 1963: Sea Surface Chemistry and Langmuir Circulation, Deep Sea Res., 10, 233-243.

Woolf, D.K., P.A. Bowyer and E.C. Monahan, 1987: Discriminating Between the Film Drops and Jet Drops Produced by a Simulated Whitecap, J. Geophys. Res., 92, 5142-5150.

# THE ROLE OF NONPRECIPITATING CLOUD CYCLES AND GAS-TO-PARTICLE CONVERSION IN THE MAINTENANCE OF THE SUBMICRON AEROSOL SIZE DISTRIBUTION OVER THE TROPICAL OCEANS

*William A. Hoppel*
Naval Research Laboratory
Washington, DC 20375, USA

ABSTRACT

The submicron aerosol size distribution over the remote tropical oceans exhibits a minimum at about 0.06 μm between peaks at about 0.02-and 0.09-μm radius. The minimum at 0.06 μm and the maximum at 0.09 μm are believed to be the result of nonprecipitating cloud cycles, and the particles in the 0.01-μm to 0.03-μm range are thought to result from homogeneous nucleation of new particles at an average rate of the order of 50 particles $cm^{-3}$ $day^{-1}$. In simple terms, the life cycle of an "average" particle is pictured as follows: After formation by homogeneous heteromolecular nucleation it grows into the 0.02-μm range in less than a day and to about 0.04 μm in 2 days; at which time it becomes a cloud condensation nucleus (CCN) and is incorporated into a cloud droplet during cloud cycles. As a cloud droplet it scavenges additional trace material and attains a radius of about 0.09 μm after going through about 10 cloud cycles. After about 20 cloud cycles it is removed by precipitation scavenging.

## 1. INTRODUCTION

Reliable measurements of the submicron aerosol size distribution are a relatively recent accomplishment. Optical size spectrometers are capable of measuring down to sizes of about 0.1 μm. Since the number of particles smaller than 0.1-μm radius exceed the number of larger particles, it is not possible to get a complete picture of the size distribution, to understand the aerosol source mechanisms, or to evaluate the interactions between particles of different sizes solely from data taken with optical particle counters. The advent of the differential mobility analyzer has made measurements down to about 0.005 μm possible (Hoppel, 1978). Previously reported shipboard measurements (Hoppel et al., 1983, 1985) of submicron particles with a differential mobility analyzer gauged the evolution of the size distribution in air masses advecting off the east coast of the U.S.A. These measurements showed the rapid decay of small particles caused by coagulation with larger particles and by growth through condensation of gas-phase reaction products during


A. DEEPAK Publishing
ISBN 0-937194-11-5

the first 12 to 24 hours. In clear weather and after the initial decay of small particles, the size distribution remained remarkably stable over the period of days until the air mass encountered precipitation. Later measurements (Hoppel, et al., 1986) on a transatlantic cruise provided evidence that nonprecipitating clouds have a profound influence on the shape of the aerosol size distribution. In this paper the mechanisms that we believe are responsible for shaping the submicron size distribution in the remote marine environment are discussed.

## 2. MEASUREMENTS

The measurements discussed in this paper were taken during a cruise in 1983 from Charleston, SC to the Canary Islands aboard the USNS LYNCH. After leaving Charleston the ship dipped south so that most of the trip was well within the tropics. Figure 1 shows the size distributions measured at the locations of the ship indicated in Fig. 2.

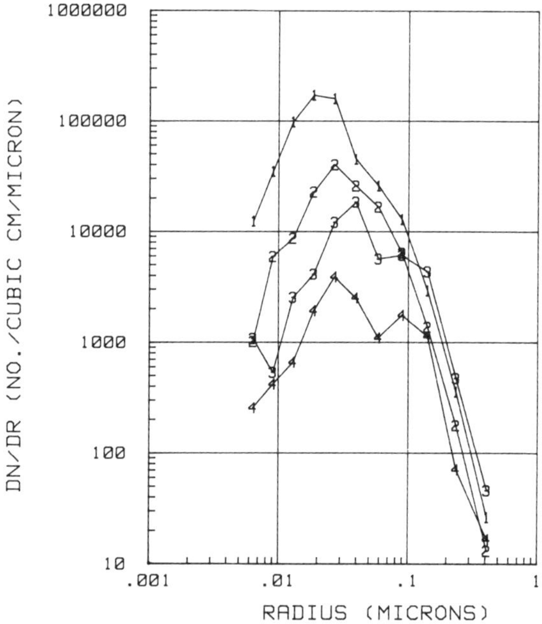


FIGURE 1. Transition of aerosol size distribution from continental to marine.

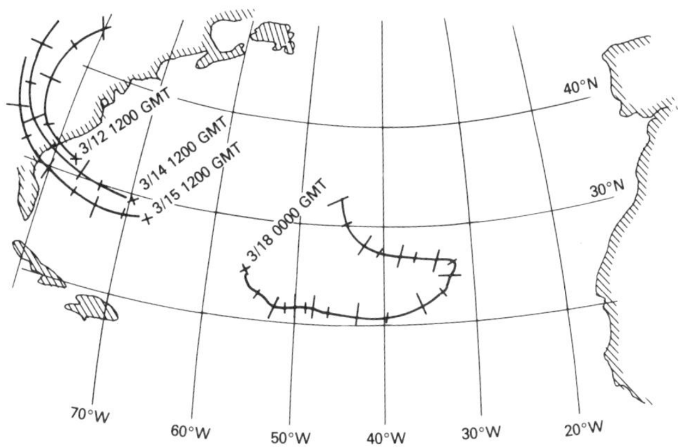


FIGURE 2. Back trajectories for air masses corresponding to the four curves in Fig. 1.

The points on the size distributions are plotted at the center of contiguous channels. Figure 2 also shows the calculated air mass trajectories for these locations where the larger (smaller) tic marks indicate 24- (12-) hour time intervals. [Air trajectories were calculated by NOAA using the method described by Harris (1982).] The weather during the first 4 days was dominated by a strong low pressure system moving out to sea off the Middle Atlantic States. The strong counterclockwise circulation around the center of the low resulted in northwest winds of 15 m/s. Curve 1 is the average of six size distributions taken between 2100 GMT on 11 March and 2200 on 12 March. The total aerosol concentration calculated from curve 1 is about 5300 $cm^{-3}$. During this period the skies were overcast with stratocumulus and the radon count dropped from about 50 to 25 pCi $m^{-3}$. One to five hours earlier the air had been over land.

Curve 2 is the average of 18 similar size distributions taken between 1100 and 2100 on 14 March. The total number of particles was 1740 $cm^{-3}$. The wind was from the west and had decreased to about 3 to 5 m/s. Skies were nearly clear with about 10% cumulus cloud cover. The back trajectory for this period (Fig. 2) shows the air had been over the water for about 18 hours. The radon concentration had dropped to 7 pCi $m^{-3}$. Satellite photos (together with the air trajectories) indicate that the air came from a very clear area over Tennessee and behind the low pressure system. Balloon soundings showed a strong inversion at about 1400 m with a relative humidity maximum reaching only 90% at the inversion. The surface relative humidities were between 50% and 60%, which are very low for the marine surface layer. All indications are that the size distributions represented by curve 2 were taken in a subsident air mass originating over the U.S.A., and advecting to the location of the ship along a trajectory free of clouds.

Curve 3 is the average of 14 similar size distributions taken on 15 March between 0930 and 1800 GMT. The total concentration is 1150 $cm^{-3}$. The local wind was from the SW at about 2 m/s. The skies during this period were overcast with 80% to 100% strato-cumulus cloud coverage. The soundings showed a saturated region between 1200 and 1700 m, capped by an inversion that was rising and broke up later in the day. Air trajectories indicate that the air passed over the Georgia and northern Florida coasts 2.5 days earlier. The satellite photos show little cloud coverage along the trajectory until 15 March when significant cloud coverage was observed locally. The radon concentrations remained at about the 7- to 8-pCi $m^{-3}$ level during this period. We believe that the hump at 0.1 $\mu$m, which is evident in curve 3, is associated with the effect of non-precipitating clouds on the size distribution. This hump did not occur during the time period represented by curve 2 where the air had traversed a cloud-free region between land and the measurement location. The possible effects of nonprecipitating clouds on the size distribution will be discussed later.

During the night of 15 March and morning of 16 March a front passed over the ship bringing rain and heavy overcast, which lasted most of the 16th. After the passage of the disturbance the local winds were from the southeast for several days with speeds generally below 7 m/s, and the trajectory analysis (Fig. 2) showed that the air had been over water for many days. The radon concentration dropped to 2 pCi $m^{-3}$, which is normal for the Atlantic marine background level. Figure 3 shows the composite of 51 size distributions taken during the 35-hour period from 18 March 0800 till 19 March 1900. The composite is shown to emphasize that all of the size distributions exhibited the double-peaked characteristic during this time period. The average of the composite is also shown as curve 4 on Fig. 1 to compare the background marine size distribution with the series showing the transition from continental to marine. The average total aerosol concentration was about 200 particles $cm^{-3}$. The cloud cover during this period varied. Whereas throughout the day on 18 March there was about 85% coverage consisting of cirrus, cumulus, and altostratus, on 19 March there was only about 20% cumulus coverage.

The size distributions shown in Fig. 3 are quite typical of size distributions encountered throughout the cruise when all vestiges of land influence were absent. On several occasions following heavy

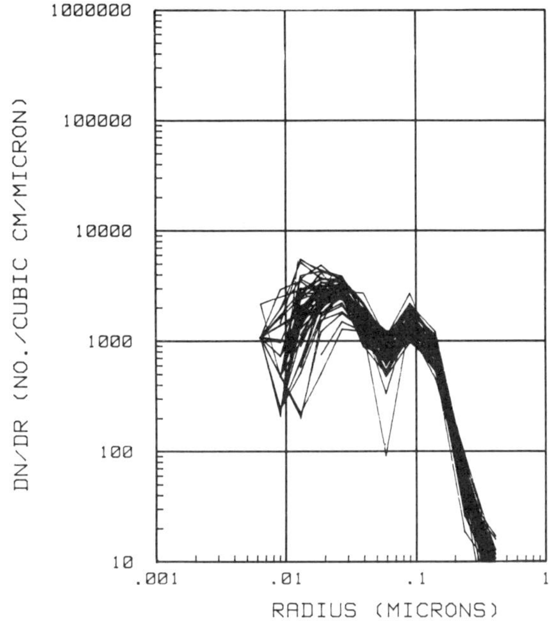


FIGURE 3. Composite of 51 size distributions exhibiting the double-peaked feature.

precipitation or passage of a meteorological disturbance, there was a temporary change but the distribution always recovered to the characteristic shape shown in Fig. 3. Similar double-peaked distributions were also found in the tropical Pacific, from Hawaii to Tahiti, during the 1984 cruise aboard the NOAA ship DISCOVERER. (The total number of particles found in the Pacific was somewhat less.) The data indicate that there is some mechanism in the remote (tropical) marine Atlantic that sustains the type size distribution shown in Fig. 3.

## 3. MECHANISMS

### 3.1. NONPRECIPITATING CLOUD HYPOTHESIS

Four mechanisms that might produce the peak observed in the 0.08-$\mu$m to 0.1-$\mu$m radius range were proposed by Hoppel et al. (1985). The nonprecipitating cloud mechanism best explains the observations and can be described as follows: All cloud droplets are initially formed on a subset of aerosol particles that are called "cloud condensation nuclei." During the cloud phase, the droplets absorb trace gases and convert some of the gaseous material to particulate matter. In addition to chemical conversion of trace gases, the smaller interstitial aerosol diffuse to the cloud droplets, transferring mass from the smaller particles to the larger particles. When the cloud droplets later evaporate the resulting residue is larger than the original cloud condensation nucleus on which the droplets formed. Averaged globally, the nonprecipitating cloud cycles repeat themselves 10 to 25 times before a precipitating cloud system develops (Pruppacher, 1986). Thus the aerosol is cycled through more than 10 nonprecipitating clouds (on the average) before the particles can be removed by precipitation scavenging.

The evidence supporting the nonprecipitating cloud hypothesis has been detailed by Hoppel et al. (1986) and includes:

1) volatility measurements that show the particles under both peaks are much too volatile to be sea salt,

2) case studies that show the double-peaked characteristic is associated with air that had passed through regions containing low-level clouds,

3) the minimum between the two peaks occurs in the radius range 0.04 to 0.07 μm, which corresponds to a critical supersaturation in the 0.1%- to 0.4% - range (Fitzgerald and Hoppel, 1984), in good agreement with expected cloud supersaturations,

4) the number of particles under the 0.09 -μm peak is in the 50- to 130 $cm^{-3}$ range, which is a reasonable number for cloud droplet concentrations, and

5) the source of convertible gases emanating from the sea surface is adequate to sustain the observed aerosol size distribution against removal mechanisms.

We therefore propose that the minimum observed in the 0.05-μm to 0.07-μm range and the peak in the 0.08-μm to 0.1-μm range result from the cycling of the aerosol through nonprecipitating clouds many times prior to the removal of the particles by precipitation scavenging.

## 3.2 GAS-TO-PARTICLE CONVERSION

When the product of a gas phase reaction has a low volatility, the environment will become supersaturated with respect to that species at very low concentrations. Condensation of that species will first occur on existing particles. If the concentration continues to increase homogeneous nucleation of new embryos will occur at some critical concentration. If the nucleation process involves more than a single molecular species, as is usually the case in the atmosphere, then the nucleation is heteromolecular and the resulting particle is a solution or compound of unique proportions determined by the environmental vapor pressures of the condensing species. The simplest type of hetermolecular nucleation is binary nucleation of solution droplets. Of particular interest in the marine atmosphere is the possible conversion of dimethyl sulfide (DMS) to methane sulfonic acid (MSA) and sulfuric acid. Both of these acids have very low volatilities and strong affinities for water vapor.

The yearly global injection of DMS from the oceans is estimated by Andreae and Raemdonck (1983) and Andreae et al. (1985) to be 40±20 Tg S $yr^{-1}$, whereas Bates et al. (1987) estimate the global flux to be about 17 Tg S $yr^{-1}$. In either case DMS is clearly a significant portion of the global natural sulfur flux of 60-90 Tg S $yr^{-1}$ (Andreae et al., 1985). The natural and anthropogenic sources of sulfur are about equal but most anthropogenic sulfur is emitted as $SO_2$ over continental regions of the Northern Hemisphere. The importance of DMS as a source of sulfur in remote oceanic regions must be relatively large and could account for most of the non-sea salt sulfate in marine aerosols. Hatakeyama et al. (1985) carried out experiments that indicated that the photo-oxidation of DMS gives about 21% $SO_2$ with the remaining sulfur being converted to MSA. Experiments by the same authors show that the photo-oxidation of DMS in humid air (r.h.= 35%) containing 12 ppb DMS and 6 ppb NO produced large concentrations of particles in which both MSA and sulfuric acid were detected.

The water-sulfuric acid system has been explored by a number of investigators (for example: Reiss, 1950; Mirable and Katz, 1974; Hoppel, 1975). At 80% r.h., a sulfuric acid vapor pressure of only $10^{-16}$ atm is required to saturate the environment and $10^{-10}$ atm to supersaturate the environment sufficiently for homogeneous nucleation to occur (Hoppel, 1975). At intermediate supersaturations, sulfuric acid solution will condense on existing particles. The MSA-water system has been investigated recently by Hoppel (1987) and shows that an MSA vapor pressure less than $10^{-12}$ atm saturates the atmosphere at 80% r.h., while about $10^{-8}$ atm will cause homogeneous nucleation. It is clear that sulfuric acid has a greater nucleation potential than does MSA. Furthermore, it has been experimentally observed that the nucleation threshold for sulfuric acid is even less when ammonia is present. Once new embryos are formed, they grow slowly to sizes observable as cloud condensation nuclei, at a rate determined by the very low concentrations of the condensing species. It is not our purpose here to champion any particular reaction, but to illustrate that substances that have a high potential to nucleate new particles do exist in the marine environment. To make a quantitative estimate of particle formation and growth, we will henceforth assume that oxidation of DMS is the only source of condensable material and that the sulfur is converted to MSA or sulfuric acid. For most of our arguments (except for the threshold concentration required for nucleation), it will make little difference how the sulfur partitions between the two acids.

If the flux of DMS from the ocean surface is 140 $\mu g$ S $m^{-2}$ $day^{-1}$ (Bates et al., 1987) and if DMS is converted uniformly to a strong acid within a 1 km thick marine boundary layer, then the volume production rate would be about 0.14 $\mu g$ S $m^{-3}$ $day^{-1}$. If all the acid condenses onto aerosols, this would replenish 1.4 $\mu$g of S (4.2 $\mu$g of acid) $m^{-3}$ in 10 days. Our data taken in the remote tropical regions indicate that the mass loading of particles smaller than 0.3 $\mu$m is typically in the 1-to 4- $\mu$g $m^{-3}$ range.

### 3.2.1. Ambient Concentration of Vapors

If a steady state exists between production of acid vapor and condensation on aerosols, then the gaseous concentrations can be estimated by

$$\rho(\infty) - \rho(a) = \frac{Q}{4\pi\, D'(a)\, a\, Z} \tag{1}$$

where $\rho(\infty)$ is the vapor density of the acid in the atmosphere, $\rho(a)$ is the surface vapor density of the acid in the solution droplet, Q is the acid production rate (0.42 $\mu$g $m^{-3}$ $day^{-1}$); and, a and Z are the mean size and number density of existing aerosol particles. D' is the effective diffusion coefficient given by

$$D'(a) = \frac{D}{\frac{a}{a+\lambda} + \frac{4D}{av}} \tag{2}$$

where D is the molecular diffusion coefficient for the acid, and $\lambda$ and v are the molecular mean free path and mean velocity. We have taken the mean radius of particles less than a half micron to be 0.08 $\mu$m and the concentration of particles to be 300 $cm^{-3}$. The vapor concentration calculated from Eq. (1), neglecting the surface vapor pressure, is about 3 ng $m^{-3}$ or a mixing ratio of $6\times10^{-13}$ atm. For sulfuric acid such an environment would be highly supersaturated and neglecting the surface vapor pressure in Eq. (1) is justified. For MSA the atmosphere would be slightly supersaturated at 90% r.h. and undersaturated at 80% r.h. Since the only removal process is assumed to be condensation on existing particles, subsaturated conditions cannot be maintained. The subsaturation results from neglecting the surface vapor pressure of MSA in calculating the environmental acid vapor pressure from Eq. (1).

### 3.2.2 Growth Rate of Particles

The growth time of aerosols due to acid condensation can also be estimated. The growth time for a newly formed particle growing through a background population of absorbing aerosols of average size a and number density Z is given by

$$t = 4\pi\rho_p\, C\left(\frac{aZ}{Q}\right)\frac{D'(a)}{D}\left[1/2(r-\lambda)^2 - 1/2(r_c-\lambda)^2 + \lambda^2\, \mathrm{Ln}\left(\frac{r+\lambda}{r_c+\lambda}\right) + \frac{4D}{V}\,(r-r_c)\right] \tag{3}$$

where $\rho_p$ is the density of the particle, C the weight concentration of acid in the droplet, and $r_c$ is the starting radius (assumed to be 0.005 μm in the following calculations). Using the values already cited for Q, Z, and a, the time to grow to 0.02 μm is about 0.7 days; whereas the time to grow to 0.04 μm and 0.08 μm are 1.7 and 3.8 days, respectively.

### 3.2.3 Nucleation Rate

The ambient vapor density of the monomer was calculated to be of the order of $10^{-12}$ atm. Since the concentration required for homogeneous nucleation in the water-sulfuric acid system is $10^{-10}$ atm, this would indicate that homogeneous nucleation of acid particles in the marine atmosphere is unlikely. However, there may be other constituents in the atmosphere that encourage nucleation. Also the ambient calculations are based on a steady-state balance between production and condensation of the monomers. Since the disintegration of DMS is photochemically induced, large temporal and spatial inhomogeneities are likely to exist and local concentrations may exceed the nucleation threshold. Assuming that nucleation does occur, we now seek to estimate the average nucleation rate required to sustain the observed small particle concentrations shown in Fig. 3.

The small size portion ($r<.03$ μm) of the distribution shown in Fig. 3 exhibits a great deal of variability. Since growth to 0.02 μm was estimated to be less than one day, it is reasonable to assume that this portion of the size distribution is controlled by: 1) the rate of production of embryos, 2) condensational growth, and 3) coagulation. The loss of small particles by coagulation with each other is much less than the rate of coagulation of small particles with the particles under the peak at 0.09 μm. The equation for the change in the size distribution n(r) under these assumptions is:

$$\frac{\partial n(r)}{\partial t} = - \frac{\partial}{\partial r} [n(r)f(r) - N\, n(r)K(r,R)] \tag{4}$$

where K(r,R) is the coagulation coefficient between small particles of radius r and larger particles of mean radius R and concentration N, where f(r) is the growth law given by

$$f(r) = \frac{dr}{dt} = \frac{D'(r)}{r} \frac{\rho(\infty)}{\rho_p} \tag{5}$$

where $\rho(\infty)$ is the environmental vapor density of the monomer. Equation (4) has the steady state solution

$$\frac{n(r)}{n(r_o)} = \frac{f(r_o)}{f(r)} \mathrm{EXP} \left[ -N \int_{r_o}^{r} \frac{K(r,R)}{f(r)} dr \right] \tag{6}$$

where $n(r_o)$ is the size distribution at some small starting radius $r_o$. The first factor accounts for the effect of growth on the size distribution and the second for coagulation. For the remote oceanic regions the effect of coagulation is small and condensational growth dominates in the small size range. The nucleation rate F is related to $n(r_o)$ by $F=f(r_o)*n(r_o)$.

The measurements presented in Fig. 3 indicate that $n(r_o)$ is about $10^7$ $cm^{-4}$ at r=0.01 μm, which gives a nucleation rate of only 40 cm $^3$ $day^{-1}$. The threshold concentrations for nucleation cited earlier were for a rate of 1 $cm^{-3}$ $sec^{-1}$. Had the nucleation threshold been based on the rates estimated here they would have been smaller by a factor of two or three; thus increasing the probability of new particle formation.

## 4. CONCLUSIONS

Within the marine boundary layer over the remote tropical oceans there appears to be some mechanism that sustains the submicron size distribution with a shape similar to that shown in Fig. 3. We have argued that the minimum between the two peaks and the peak at 0.09 μm result from cycling marine boundary layer aerosol through non-precipitating clouds before eventual removal of particles by a precipitating cloud. It is our opinion that particles below 0.02 μm are formed by nucleation at a rate on the order of 50 $cm^{-3}$ $day^{-1}$ and subsequently grow through this region in less than a day. By the time they grow to 0.04 μm (which would require about two days) they are large enough to be cloud condensation nuclei and become cloud droplets during the cloud cycles. As cloud droplets they scavenge additional trace material so that when they evaporate the residue forms a particle with an average radius of about 0.09 μm.

## ACKNOWLEDGMENT

Partial support for this work was provided by Code 12 of the Office of Naval Research.

## REFERENCES


Andreae, M. O., and H. Raemdonck, 1983: Dimethyl Sulfide in the Surface Ocean and the Marine Atmosphere: A Global View, Science 221, 744-747.

Andreae, M.O., R. Ferek, F. Bermond, K. P. Byrd, R.T. Engstrom, S. Hardin, P. D. Houmere, F. LeMarrec and H. Raemdonck, 1985: Dimethyl Sulfide in the Marine Atmosphere, J. Geophys. Res., 90, 12891-12900.

Bates, T. S., J. D. Cline, R. H. Gammon and S. R. Kelley-Hanson, 1987: Regional and Seasonal Variations in the Flux of Oceanic Dimethylsulfide to the Atmosphere, J. Geophys. Res., 92, 2930-2938.

Fitzgerald, J. W., and W. A. Hoppel, 1984: Equilibrium size of atmospheric particles as a function of relative humidity: calculations based on measured aerosol properties. In L. Ruhnke and A. Deepak (Eds.), Hygroscopic Aerosols, A. Deepak Publishing, Hampton, VA., 21-34.

Harris, J. M., 1982: NOAA Tech. Memo, ERL-ARL-116, The GMCC Atmospheric Trajectory Program, NOAA Environmental Research Laboratories, Boulder, CO.

Hatakeyama, S., K. Izumi and H. Akimoto, 1985: Yield of $SO_2$ and Formation of Aerosol in the Photo-oxidation of DMS Under Atmospheric Conditions, Atmos. Environment, 19, 135-141.

Hoppel, W. A., 1975: Growth of Condensation Nuclei by Heteromolecular Condensation, J. Rech. Atmos., 9, 41-54.

Hoppel, W. A., 1978: Determination of the Aerosol Size Distribution From the Mobility Distribution of the Charged Fraction of Aerosols, J. Aerosol Sci., 9, 41-54.

Hoppel, W. A., 1987: Nucleation in the MSA-water Vapor System. Atmos. Environment, 21, 2703-2709.

Hoppel, W.A., J.W. Fitzgerald and R.E. Larson, 1983: Measurements of Atmospheric Aerosols: Experimental Methods and Results of Measurements Off the East Coast of the United States, NRL Report 8703, NTIS-ADA 130998, 66 pp.

Hoppel, W. A., J. W. Fitzgerald and R. E. Larson, 1985: Aerosol Size Distributions in Air Masses Advecting Off the East Coast of the United States, Geophys. Res., 90, 2365-2379.

Hoppel, W. A., G. M. Frick and R. E. Larson, 1986: Effect of Nonprecipitating Clouds on the Aerosol Size Distribution in the Marine Boundary Layer, Geophys. Res. Letts., 13, 125-128.

Mirabel, P., and J. L. Katz, 1974: Binary Homogeneous Nucleation as a Mechanism for the Formation of Aerosols, J. Chem. Phys., 60, 1138-1144.

Pruppacher, H. R., 1986: The Role of Cloud Physics in Atmospheric Multiphase Systems: Ten Basic Statements. In W. Jaeschke (Ed.), Chemistry of Multiphase Atmospheric Systems, NATO ASI Series G: Ecological Sciences, Vol. 6, Springer Verlag, 133-190.

Reiss, H., 1950: The Kinetics of Phase of Transitions in Binary Systems, J. Chem. Phys., 18, 040-848.

# DEAD SEA EVAPORATION RATES IN RELATION TO AEROSOL NUMBER DENSITY VARIATIONS

*A. Cohen, Y. Mahrer, M.A. Weiss and M. Litvak-Neumann*[1]
The Hebrew University of Jerusalem
Jerusalem, Israel

ABSTRACT

Evaporation rates of the Dead Sea are calculated for different summer meteorological conditions, using a radiation transfer model. Several values of Dead Sea salinity and surface temperatures are examined in order to study the effect of adding Mediterranean Sea water to the Dead Sea. The effect of changing surface salinity on the aerosol content above the Dead Sea and consequently on the solar radiation contribution to the radiation balance equation is also presented and discussed.

## 1. INTRODUCTION

The Dead Sea is a highly saline body of water located about 70 km east of the Mediterranean. It is about 90 km long and 10-15 km wide and its surface is about 400 m below sea level. The lake consists of two basins: the northern one has a maximum depth of 330 m, whereas the southern basin is only a few meters deep.

The main sources that feed the Dead Sea are the Jordan River and its tributaries, springs, floods, and rainfall. Since there is no outflow of water from the lake, evaporation is the most critical physical process that balances the inflow of water and governs the level of the lake. In the last twenty years, the level of the lake has been decreasing at a rate of about half a meter per year. This has happened mainly because since 1964 the Jordan River has been utilized intensively for irrigation.

The possibility of replacing the fresh water by Mediterranean water has been investigated in recent years. The basic idea is the use of the elevation difference between the two bodies of water (about 400 m) for the generation of electric power (Weiner, 1980). As a consequence of a permanent flow of the Mediterranean water, the salinity of the upper layer of the Dead Sea is predicted to decrease. This, in turn, will affect the number density of highly active condensation nuclei in the atmosphere of the Dead Sea basin. Under normal prevailing atmospheric conditions a relatively constant thick aerosol layer is present and it is expected that the distance, being dependent on the salinity of the upper Dead Sea water layer, would change. As a result, the direct and diffused solar light would also change.

---

[1] Present address is: Technion, Haifa, Israel.



ISBN 0-937194-11-5

A decrease in the salinity would cause a decrease in the concentration and a change in the size distribution of the aerosol. A chemical analysis of the aerosol showed that it is composed of minerals found in the Dead Sea. Samples of aerosols were taken also from a spot 450 m above Dead Sea level. These samples were also chemically analysed and it was shown that they are a mixture of Dead Sea aerosol and a typical desert aerosol.

According to Segal et al. (1983), the induced mesoscale flows in the Dead Sea area, with the interaction of the synoptic Mediterranean sea breeze, results in a typical daily cycle of the surface variables during the summer months. Figure 1 illustrates a three-dimensional simulation of the wind field. In the late afternoon hours the windspeeds rise to values of 8-10 m/s, followed by a generation of a boundary layer of several hundred meters. The aerosol generated on the sea surface is lifted up and is found even in Mazokei-Dragot (450 m above Dead Sea level).

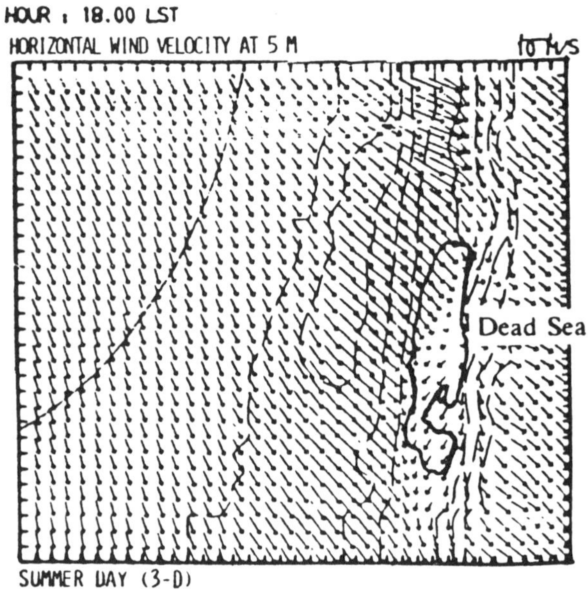


FIGURE 1. Three-dimensional simulation of the wind field at 18.00 (LSY). (Taken from Segal et al., 1983.)

## 2. MODEL DESCRIPTION

The radiative model was built according to the following schemes:

1) Calculation of the extinction coefficients of the aerosol particles in the atmosphere of the Dead Sea basin, on the basis of the Mie theory.
2) Calculation of the visual range, based on the extinction coefficients.
3) Representation of the radiation flux at the Dead Sea surface, by using a simple radiation transfer equation.
4) Calculation (estimation) of sea-surface temperature.
5) Computation of the evaporation rate by solving the heat balances equation.

### 2.1 RADIATION SCATTERING IN A CLOUD OF PARTICLES

In order to simplify the model as much as possible we made the following assumptions:

1) All the particles in the medium scatter the direct solar radiation. The particles concentration is low enough to neglect higher order of scattering.
2) Each particle is totally independent in scattering the radiation. Therefore, it is possible to integrate the individual contributions of the single particles.
3) All the particles scatter the radiation in its whole spectrum range.
4) All the particles are homogeneous spheres; each particle is characterized by its radius (r) and one diffraction coefficient (m).

It is essential to take into account the size distribution of the scattering particles because of the high sensitivity of the scattering efficiency to the size of the scattering particle. According to the Mie theory, the scattering cross-section per volume unit at wavelength $\lambda$ is represented by:

$$\sigma_\lambda = \int_{r_0}^{r_1} \pi r^2 Q(\gamma)[dn/dr]dr \ , \qquad (1)$$

where $Q(\gamma)$ is the Mie scattering efficiency, n the number of particles per volume unit, and r the radius of the particle.

### 2.2 VISUAL RANGE

The sensitivity of the human eye at wavelength $\lambda(\Psi_\lambda)$ is defined by the ratio of intensity at the wavelength of 0.55 μm (where the eye has maximum sensitivity), to intensity at wavelength $\lambda$ required to yield equal impression of brightness.

Monochromatic brightness is defined by:

$$B_\lambda = \Psi_\lambda I_\lambda \ , \tag{2}$$

where $I_\lambda$ is a monochromatic intensity that is emitted or reflected by the object.

Brightness is represented by:

$$B = \int_0^\lambda B_\lambda d\lambda \ . \tag{3}$$

An object is seen when the contrast $C$ is larger than 0.02.

$$C = [B-B_0]/B_0 \ , \tag{4}$$

where $B_0$ is the brightness of the background.

The brightness of the object is represented by:

$$B_\lambda = B_{s\lambda}(1-e^{-\sigma_\lambda x}) \ , \tag{5}$$

where $B_{s\lambda}$ is the monochromatic brightness of the sky and $x$ is the visual range, and the total brightness of the object is represented by:

$$B = \int_0^\infty B_{s\lambda}(1-e^{-\sigma_\lambda x})d\lambda \ . \tag{6}$$

Contrast is represented by:

$$C = - [\int_0^\infty B_{s\lambda}(1-e^{-\sigma_\lambda x})/B_s \ . \tag{7}$$

If it is assumed that $\sigma_\lambda$ is independent of $\lambda$, $x$ may be written by:

$$x = \ell nC/\sigma \ . \tag{8}$$

Taking $C = 0.02$, the visual range will be:

$$x_v = 0.912/\sigma \ . \tag{9}$$

## 2.3 THE EXTINCTION COEFFICIENT AND THE RADIATION TRANSFER

Since all the particles are scattering the light on the whole spectrum, it is possible to calculate the Mie cross-sections for all the combinations of $\lambda$, $m$, and $r$, and then to integrate them in the proper proportions, in order to get a total extinction coefficient $\sigma$. At visible wavelengths it is adequate to use a simple equation for the radiation transfer:

$$I = I_0 e^{-\int_0^{x_1} \sigma dx} , \qquad (10)$$

where $I_0$ is the radiation flux at the top of the aerosol layer, and $I$ is the flux at the point represented by $x_1$. If $\sigma$ is independent of $x$:

$$I = I_0 e^{-\sigma x_1} , \qquad (11)$$

where $x_1$ is the path of the radiation inside the aerosol layer, which is independent of the thickness of the layer and the angle of the Sun.

$$I = I_0 e^{-\sigma h \cdot \sec \alpha} , \qquad (12)$$

where $h$ is the thickness of the layer and $\alpha$ is the zenith angle of the Sun.

## 2.4 SEA-SURFACE TEMPERATURE

The heat balance at the surface is represented by:

$$F_{SR} - F_R + F_{AR} - F_{WR} - F_{LH} - F_{SH} - F_S = 0 , \qquad (13)$$

where $F_{SR}$ is the incoming shortwave radiation, $F_R$ is the reflected shortwave radiation, $F_{AR}$ the atmospheric longwave radiation, $F_{WR}$ the water emitted longwave radiation, $F_{LH}$ the latent heat of evaporation, $F_{SH}$ the sensible heat, and $F_S$ is the heat storage in the water body.

$F_{WR}$ is given by:

$$F_{WR} = \varepsilon \sigma T_0^4 , \qquad (14)$$

where $\varepsilon$ is the water emissivity coefficient, $\sigma$ the Stephen-Bolzman constant, and $T_0$ the sea-surface temperature.

The latent heat of evaporation is described by:

$$F_{LH} = 0.62 \rho L C_E U (e_0 - e_a)/p , \qquad (15)$$

where $\rho$ represents the air density, $L$ the latent heat of evaporation, $C_E$ the eddy diffusivity of the vapor mass, $U$ the wind velocity, $p$ the atmospheric pressure, $e_0$ the saturated vapor pressure and $e_a$ is the vapor pressure in the air.

The sensible heat may be described by:

$$F_{SH} = \rho_a C_H C_p U(T_0 - T_a) , \tag{16}$$

where $C_H$ represents thermal eddy diffusivity, and $C_p$ the specific heat at constant pressure of air. The eddy diffusivities near the surface of the Earth increase with increasing windspeed (Haltiner and Martin, 1957).

The salinity in the upper layer of the water reduces the vapor pressure and therefore $e_0$ has to be multiplied by a "salinity coefficient."

## 2.5 EXTINCTION COEFFICIENTS AND VISUAL RANGE

The Dead Sea aerosol has a size range from 0.1-15 μm. We calculated the range of the size parameter $\gamma$ $(2\pi r/\lambda)$ . We assume that the real refraction index is between 1.33 (the refraction index of yellow light in pure water and 1.55 (for water with dissolved minerals). For a very concentrated aerosol layer of 1.700 particles per $cm^3$ and size distribution that varies as $r^{-2}$ , we obtain $\sigma \approx 1.0\ km^{-1}$ .

In this case, the visual range is $x_v = 3.912 \approx 4$ km. This value represents the situation in the Dead Sea at certain hours of the day during summertime.

## 2.6 HEAT FLUX CHANGES FOR DIFFERENT VISUAL RANGES

Assuming that the penetration depth is 500 m, which means that $h \cdot \sec \alpha = 0.5$ km, we calculated the ratio of the radiation at the sea surface and the radiation at the top of the aerosol layer:

$$I_1/I_0 = e^{-\sigma h \cdot \sec \alpha} \approx 0.6 \tag{17}$$

When the concentration of particles is reduced and the visual range is 5 km, $\sigma$ will be $3.912/5 = 0.78$, so the ratio between the radiation fluxes will be $I_2/I_0 = 0.68$.

One may immediately get $I_2/I_1 = 1.13$; the aerosol's concentration is then reduced and the radiation flux rises.

Table 1 represents the changes in visual range and the corresponding changes in the radiation flux for penetration depths of 0.5 and 0.6 km.

TABLE 1. CHANGES IN VISUAL RANGE

| Radiation Flux Changes (%) | Visual Range | |
|---|---|---|
| | Penetration Depth of 0.6 km | Penetration Depth of 0.5 km |
| -- | 4 | 4 |
| 10 | 5 | 5 |
| 20 | 6 | 6 |
| 30 | 7 | 8 |
| 40 | 10 | 12 |
| 50 | 13 | 22 |

## 3. SEA-SURFACE TEMPERATURE AND THE RATE OF EVAPORATION IN DIFFERENT SITUATIONS OF RADIATION

To calculate the evaporation rate, the following values for the constants of the heat balance equation were taken:

1) Stephen Bolzman coefficient: $\sigma = e.910^{-3}$ J/m² day degree,
2) emissivity coefficient of the water surface, $\varepsilon = 0.97$,
3) air density, $\rho = 1.19$ kg/m³ ,
4) latent heat of evaporation for water (depending on the water temperature), $L = 2.511 - 2.3\cdot(T_0-273.16)\cdot10^3$ J/kg,
5) heat and mass eddy diffusivities of the vapor, $C_H \approx C_E \approx 1.06\cdot10^{-3}$ ,
6) the average atmospheric pressure at the Dead Sea level, $p = 1.0628\cdot10^5$ Pascal,
7) specific heat of air at constant pressure, $C_p \approx 1000$J/kg,
8) the activity of salinity coefficient.

We took into account two values that represent the salinity at the present time and a lower salinity at a probable higher sea level situation. At present, with a density of 1.23 gr/cm, the salinity coefficient is assumed to be 0.68, and for a lower salinity situation the activity will be 0.81 (Krumgalz and Millero, 1982).

The results of the changes in the evaporation rates due to changes in the aerosol number density, which cause changes in the visual range and hence changes in the net incoming radiation at the Dead Sea level, are presented in Tables 2 and 3 for the months of June and September, respectively. We assume that the average air temperature is 31° C and the wind velocity is 3.3 m/s, the net solar radiation flux is 25.7 MJ/(m² day), the longwave radiation flux from the atmosphere to the Sun is 32,5 MJ/(m² day), and the heat storage in the sea is 4.6 MJ/(m² day) (Stanhill, 1982).

TABLE 2. CHANGES IN EVAPORATION RATES (JUNE)

| Activity | Visual Range (km) | Radiation Flux (%) | Outcoming Heat Fluxes (MJ/m$^2$d) Sensible | Evaporation | Sea-Surface Temperature ( C) | Monthly Evaporation (%) | Changes in Evaporation (%) |
|---|---|---|---|---|---|---|---|
| 0.68 | 4 | -- | 0.7 | 11.4 | 33.0 | 14.1 | -- |
| 0.81 | 4 | -- | 0.0 | 12.6 | 31.0 | 15.5 | 10 |
| 0.81 | 5 | 10 | 0.5 | 14.2 | 32.5 | 17.5 | 24 |
| 0.81 | 6 | 20 | 0.9 | 15.4 | 33.5 | 19.0 | 35 |
| 0.81 | 7 | 30 | 1.4 | 17.2 | 35.0 | 21.2 | 50 |
| 0.81 | 10 | 40 | 1.8 | 18.5 | 36.0 | 22.9 | 62 |
| 0.81 | 13 | 50 | 2.2 | 19.9 | 37.0 | 24.6 | 74 |

TABLE 3. CHANGES IN EVAPORATION RATES (SEPTEMBER)

| Activity | Visual Range (km) | Radiation Flux (%) | Outcoming Heat Fluxes (MJ/m$^2$d) Sensible | Evaporation | Sea-Surface Temperature ( C) | Monthly Evaporation (%) | Changes in Evaporation (%) |
|---|---|---|---|---|---|---|---|
| 0.68 | 4 | -- | 0.9 | 12.2 | 34.0 | 15.1 | -- |
| 0.81 | 4 | -- | 1.3 | 12.2 | 35.0 | 15.1 | -- |
| 0.81 | 5 | 10 | 1.4 | 12.8 | 35.5 | 15.9 | 5 |
| 0.81 | 6 | 20 | 1.8 | 14.2 | 36.5 | 17.6 | 11 |
| 0.81 | 7 | 30 | 2.2 | 15.6 | 37.5 | 19.3 | 28 |
| 0.81 | 10 | 40 | 2.5 | 17.1 | 38.5 | 21.2 | 40 |
| 0.81 | 13 | 50 | 2.7 | 17.8 | 39.0 | 22.1 | 46 |

## 4. CONCLUSIONS

The different rates of evaporation due to a change in activity represent the influence of the salinity of the surface layer on vapor pressure and on evaportion.

A change in the aerosol number density that would most probably happen when the salinity of the sea-surface layer decreases would increase immediately the solar radiation flux. This, in turn, increases in all of the physical terms that represent the different phenomena that release heat from the Dead Sea: latent heat flux, sensible heat flux, sea-surface temperature, and as a result, the longwave radiation. An increase in the relative humidity moderates the increase in evaporation while it decreases the changes in the other physical terms. In Table 4 we present the annual evaporation rates and the changes in evaporation for an average increase of 20% in the relative humidity.

A 20% increase in relative humidity influences the evaporation more than the change in the activity factor. As a result of the increase in vapor pressure, the evaporation decreases at a rate of 5%. When the incoming solar radiation increases there is not only a full compensation for the decrease of the Dead Sea evaporation but also an addition of 5% in the evaporation rate. At higher values of incoming radiation, the evaporation rates increase dramatically in spite of the 20% rise in the relative humidity.

TABLE 4. DEAD SEA RELATIVE HUMIDITY AND EVAPORATION RATES

| Activity | Visual Range (km) | Radiation Flux Increase (%) | Annual Evaporation (cm) | Evaporation Change (%) |
|---|---|---|---|---|
| 0.68 | 4 | -- | 134.3 | -- |
| 0.81 | 4 | -- | 127.6 | -5 |
| 0.81 | 5 | 10 | 141.7 | 5 |
| 0.81 | 6 | 20 | 158.6 | 18 |
| 0.81 | 7 | 30 | 174.0 | 30 |
| 0.81 | 10 | 40 | 189.2 | 41 |
| 0.81 | 13 | 50 | 205.2 | 53 |

## REFERENCES

Haltiner G.J., and F.L. Martin, 1957: Dynamical and Physical Meteorology, McGraw-Hill Book Company, New York, Toronto and London.

Krumgalz, B.S., and F.J. Millero, 1982: Marine Chemistry, 11 (3), 209-222.

Segal, M., Y. Mahrer and R.L. Pielke, 1983: A Study of Meteorological Patterns Associated with a Lake Confined by Mountains -- The Dead Sea Case, Quart. J. Met. Soc., 109, 549-564.

Stanhill, G., 1982: Radiation at the Dead Sea, Report submitted to the Ministry of Energy, Jerusalem, Israel.

Weiner, D., 1980: The Mediterranean Sea Project: A Mathematical Model and Dynamic Optimization of Solar Hydroelectric Power Plant, J. of Solar Energy, 102, 281-286.

# THE PRIMARY BIOGENIC FRACTION OF THE ATMOSPHERIC AEROSOL

*R. Jaenicke and S. Matthias*
Institute of Meteorology
D6500 Mainz, Federal Republic of Germany

## ABSTRACT

In the past, aerosol particles of primary biogenic origin have been studied only for selected species such as pollen and spores. Now a method has been developed for distinguishing the primary biogenic particles from other aerosol particles and the size distribution for primary biogenic particles larger than 2 µm is presented. Up to 30% of the aerosol might be of biogenic origin. This amount seems to be sufficient for placing the biogenic particles as top candidates for acting as cloud condensation nuclei.

## 1. INTRODUCTION

The term "biogenic particles" might cause some misunderstanding in atmospheric aerosol physics and chemistry. Therefore we define "primary biogenic particles" as having their origin in the disintegration or dispersal of biological bulk material. Consequently, "secondary biogenic aerosol particles" might be produced by other processes such as gas-to-particle-conversion of gases exhaled from the biosphere.

As hay fever proves, primary biogenic particles are part of the atmospheric aerosol. The aerosol portion responsible for allergies has been investigated rather comprehensively by biologists and physicians. Aerosol allergens mostly are pollen and spores. Other particles such as bacteria, yeast, algae, plant debris, fragments of insects, and hair belong to primary biogenic particles, too.

In addition to their medical relevance, primary biogenic particles might be of importance in cloud physics. Dingle (1966) experimented with ambrosia artemisiifolia and found it to be hygroscopic in the range of 54% to 75% relative humidity, thus having the potential of acting as cloud condensation nuclei. Schnell and Vali (1975) showed that decaying leaves can act as ice nuclei at temperatures of -4 °C.

Once discharged from the surface-bound biosphere into the atmosphere, the primary biogenic particles might be transported into the elevation of clouds. This transport is even possible for rather large or even giant particles because primary biogenic particles exhibit a remarkable low bulk density. An example that might serve is pinus sylvestris with a density of 0.39 g/cm³. The density of decaying leaves is even smaller because of their porous structure.



ISBN 0-937194-11-5

Giant particles of 10 μm size and larger can have the potential of starting the rain-forming process in warm clouds.

Besides the specific topics mentioned above, the primary biogenic aerosol has never been investigated in its entirety for concentration and size distribution. First exploratory measurements were carried out by Matthias (1987) and will be presented in this paper.

## 2. COLLECTION METHOD AND EVALUATION

For the investigation of primary biogenic particle size distribution, particles must be collected from the air and be tagged to distinguish them from other aerosol particles. Due to the selected tagging procedure (see below) the collected particles have to be investigated with an optical microscope and thus the particle size is limited to those larger than 2 μm. In order to collect with full efficiency all particles larger than 2 μm on glass slides for microscopic inspection, two different types of impactors were selected:

1) a rectangular slit impactor, and
2) a rotating wing impactor.

Impactors have the advantage of concentrating the aerosol particles on small surfaces, thus resulting in an appropriate area density for those particles we expected to form a minor fraction of the aerosol.

According to previous impactor investigations (Jaenicke and Blifford, 1973; Jaenicke and Junge 1966) the parameters of the impactors have been calculated and determined for a particle density of 0.95 g/cm$^3$. This density has been calculated as an average for the respective primary biogenic particles (Gregory, 1973). The calculated parameters of the impactors are:

| | cut-off radius, μm | volume flow, m$^3$/h |
|---|---|---|
| slit impactor | 1.9 | 3.0 |
| rotating wing impactor | 7.0 | 42.4 |

The rotating wing impactor was selected because of its unique ability to collect particles with radii greater than some 10 μm with full efficiency and without problems of non-isokinetic sampling. Both devices were run simultaneously. The slit impactor needed a shorter time for optimal particle collection than the rotating wing impactor. To compensate for the different time lengths, the run of the slit impactor was interrupted several times during each collection period.

The particles were collected on microscope glass slides. In order to avoid particle bounce-off during collection, the glass slides were covered with a colourless and transparent adhesive. This adhesive was carefully selected to avoid any interference with the staining substance used for tagging. The adhesive glycerine gelatin was heated to 60 °C and applied at 30 °C.

Tagging of biological particles is usually done by staining or cultivation on a culture medium, if appropriate. Cultivation does not permit determination of the original particle size, thus staining seems to be the only appropriate method. In staining and selecting the particles by colour the investigation was limited to particle sizes larger than 2 µm. The results show that an unexpected high portion of the primary biogenic aerosol is smaller than 2 µm in radius and thus it was unintentionally excluded from the investigation.

Alexander (1980) stated the staining of biological material was pH-dependent. Following Gerlach (1977) it seems as if a combination of basic and acidic stains tags most biological material.

A comparison of two different staining substances showed COOMASSIE-blue (Knox, 1979) to stain 30% more particles than GIEMSA. Separate experiments showed the loss of particles from the slides through the application of the staining liquid to be negligible. At this point, however, we have no proof if all primary biogenic particles are stained properly and further experiments will be necessary in the future.

The procedure for obtaining the staining liquid is:

> 0.125 g COOMASSIE-blue (MERCK #15444) is added to 40 ml glacier acidic acid, 10 ml methyl alcohol, and 50 ml distilled water. This solution has to be filtered and mixed at a ratio of 5:1 with 10% saccharose.

After staining the collected particles they were sized and counted with an optical microscope. Two different counts were made: the primary biogenic (stained) particles and the nonstained particles. Thus it became possible to calculate a primary biogenic and a total particle size distribution.

Due to the exploratory nature of the experiments the collection of aerosol particles was carried out at the city border of Mainz, Germany. Mainz is located on the west border of the rather densely populated area around Frankfurt, Germany. In the vicinity of the collection site agriculture prevails. The measurements in the time period discussed in this paper were carried out each day at noon. The sampling site was located on the roof of the Institute of Meteorology at an elevation of 37 m above ground.

In addition to the measurements in Mainz some measurements are available for particles larger than 10 µm from an area of the Atlantic Ocean between 17°W-23°W and 47°N-50°N.

## 3. RESULTS

Figure 1 shows the concentration of primary biogenic particles and the percentage in terms of the total particle concentration for the size range larger than 2 µm in radius from 9 June to 25 June,

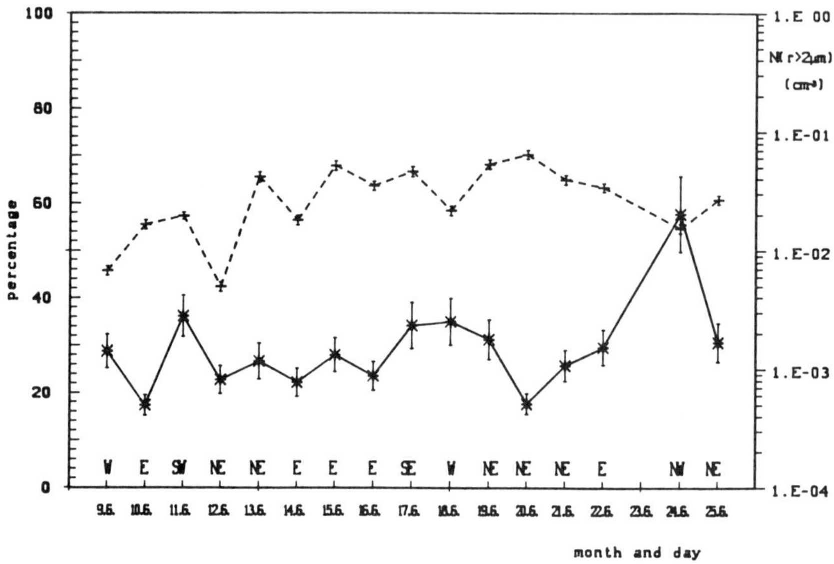


FIGURE 1. The concentration of biogenic particles larger than 2 μm in radius (dashed line) and its percentage of the atmospheric aerosol (solid line) during June 1986 in Mainz, Germany. The wind directions are indicated on the abscissa.

1986. In Germany during this period several phenological dates generally occur, so the ejection of pollen from willow, birch, and poplar. During the period in 1986 high atmospheric pressure prevailed, however, air mass changes occurred several times, e.g., on 18 June, 23 June, and 25 June. Most remarkable was the advent of Saharan dust after 20 June with a steep increase in the total particle concentration.

While the primary biogenic particle concentration (0.004 $cm^{-3}$ to 0.07 $cm^{-3}$ meaning a change of a factor of 18) and the total particle concentration varied widely depending on the meteorological and biological conditions, the percentage of the primary biogenic particle concentration varied only within a factor of 3 from 17% to 56%, averaging 28%, if we include earlier (April) and later (October) measurements. This rather uniform result is surprising. However, one has to keep in mind the opposite biological behavior of some plants with regard to meteorological conditions. While precipitation certainly has a depressing effect on the release of some pollen, fungi-spores are often expelled after rain.

Selected size distributions are shown in Figs. 2 and 3. The distributions have several features in common. More or less parallel to the total size distribution the primary biogenic particle size distribution shows a concentration decrease for particles larger than

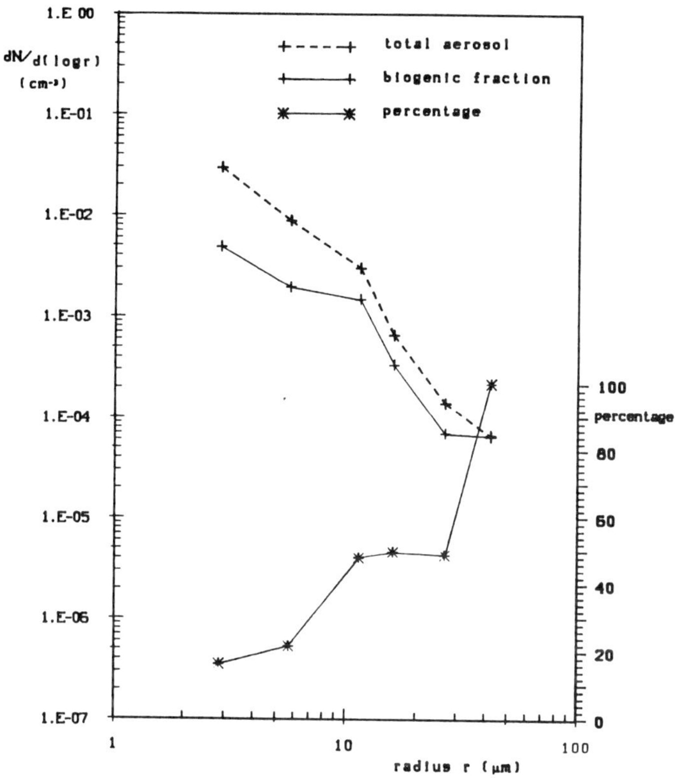


FIGURE 2. The number size distribution of the total and the primary biogenic aerosol (and its percentage) on 6 May 1986 in Mainz, Germany. The left-hand ordinate is the differential size distribution normalised to the logarithmic unit of the abscissa while the right-hand ordinate indicates the percentage of the primary biogenic aerosol.

2 µm. The maximum of the primary biogenic size distribution could not be observed and must be expected for particles smaller than 2 µm. This is a rather important result and must be verified in the future. Additional measurements are needed for the size range below 2 µm.

Substantial changes in the percentage of the primary biogenic material in the different particle size ranges can be observed. Figure 2 shows a measurement in May 1986. The particles with 40 µm in radius consist exclusively of biological particles. In June 1986 (Fig. 3) the percentage has decreased to 40% and thus the biological fraction is more or less constant with particle size.

Without presenting data, it can be mentioned that over the Atlantic Ocean, 800 km downwind of Europe, the biological particle concentration was only about one-tenth that in Mainz. However, the

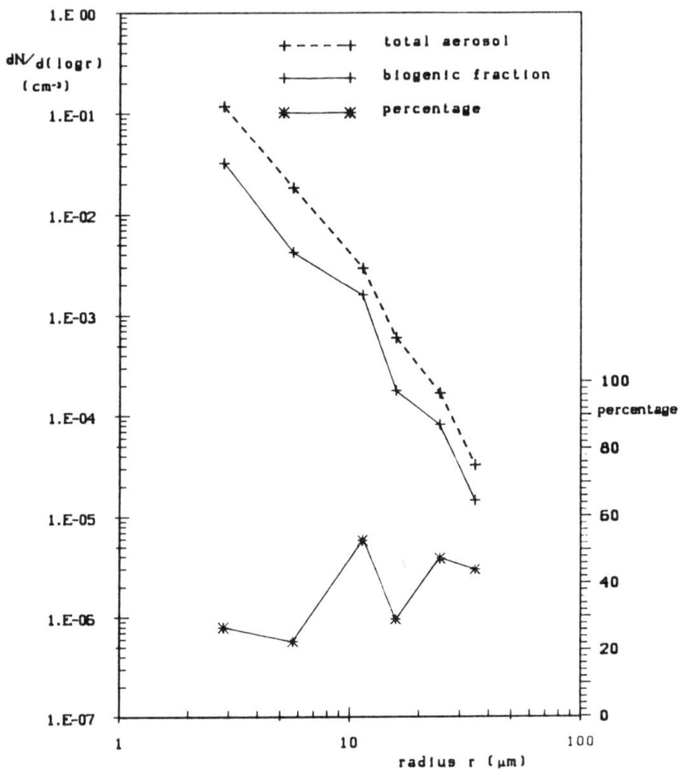


FIGURE 3. The number size distribution of the total and the primary biogenic aerosol (and its percentage) on 13 June in Mainz, Germany. Ordinate and abscissa as in Fig. 2.

primary biogenic percentage decreased only to 7% of the total aerosol concentration in the observed size range larger than 12.5 µm in radius.

To determine if the biogenic aerosol changes with altitude, simultaneous collection of aerosol particles was carried out at 2 and 37 m above ground. Figure 4 shows the result of a typical day. The ordinate values of the size distribution of the total aerosol are roughly 2 times and those of the primary biogenic aerosol 5 times higher in 2 m elevation than in 37 m elevation indicating a concentration gradient. The percentage of the primary biogenic aerosol decreased with altitude by a factor of two. However, it is remarkable that the percentage of the 25 µm particles remains unchanged and that of the 40 µm particles increases with altitude by a factor of two. Consequently, the ordinate value of those particles at 37 m was even higher than that at 2 m elevation. It remains to be confirmed in further experiments whether this observation can be seen

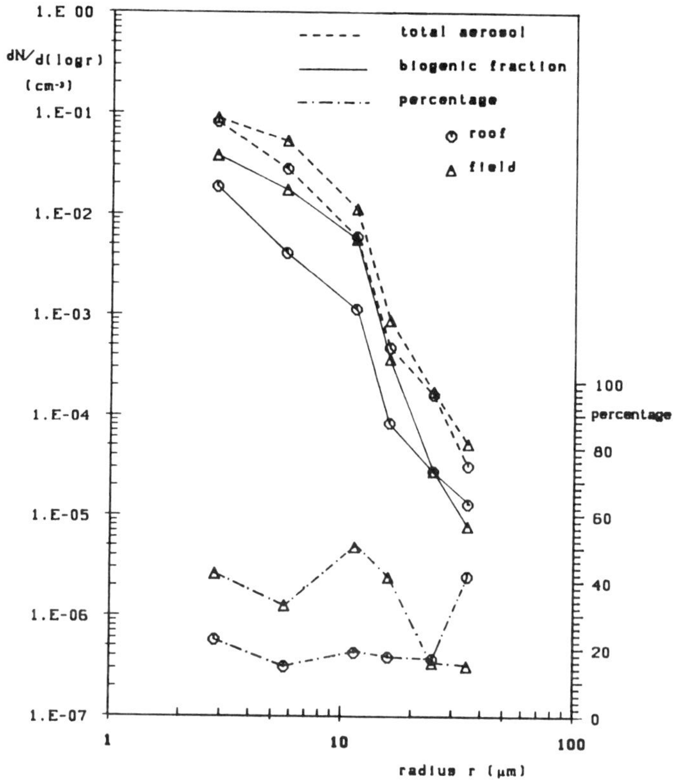


FIGURE 4: Simultaneous measurements in 2 m (field) and 37 m (roof) elevation on 15 July, 1986 in Mainz, Germany. Ordinate and abscissa as in Fig. 2.

as a confirmation of the above statement for giant primary biogenic particles being easily carried aloft.

## 4. DISCUSSION AND CONCLUSION

The primary biogenic aerosol forms a remarkable constant portion of the atmospheric aerosol. While the observations were limited to particles larger than 2 µm in radius, the maximum of the size distribution must be expected for smaller particles. The size distribution of the primary biogenic particles shows a concentration decrease similar to the total aerosol with increasing particle size. The constant presence of primary biogenic particles even over oceans is certainly of importance in cloud physics, most notedly is the indication of a comparatively large concentration of giant biogenic particles at higher elevations.

## REFERENCES

Alexander, M.P., 1980: A Versatile Stain for Pollen, Fungi, Yeast, and Bacteria, Stain Technology, 55, 13-18.

Dingle, A.N., 1966: Pollen as Condensation Nuclei, J. de Recherches Atmosphérique, 2, 231-237.

Gerlach, D., 1977: Botanische Mikrotechnik, Einführung, Thieme Verlag, Stuttgart.

Gregory, P.H., 1973: The Microbiology of the Atmosphere, Leonard Hill, Aylesbury.

Jaenicke, R., and I.H. Blifford, 1973: The Influence of Aerosol Characteristics on the Calibration of Impactors, J. of Aer. Sci., 5, 457-464.

Jaenicke, R., and C. Junge, 1966: Studien zur oberen Grenzgröße des natürlichen Aerosols, Beiträge zur Physik der Atmosphäre, 40, 129-143.

Knox, R.B., 1979: Pollen and Allergy, Camelot Press, Southampton.

Matthias, S., 1987: Ein Beitrag zur biogenen Komponente des atmosphärischen Aerosols in Mainz, Master Thesis, University of Mainz/Germany.

Schnell, R.C., and G. Vali, 1975: Biogenic Ice Nuclei: Part I, Terrestrial and Maritime Sources, J. Atmos. Sci., 33, 1554-1564.

# REMOTE SENSING OF ATMOSPHERIC AEROSOLS OVER THE OCEANS USING MULTISPECTRAL RADIANCES MEASURED WITH THE ADVANCED VERY HIGH RESOLUTION RADIOMETER ONBOARD THE NOAA METEOROLOGICAL SATELLITES

*K. Masuda and T. Takashima*
Meteorological Research Institute
Ibaraki, 305 Japan

and

*C.R. Nagaraja Rao*
National Oceanic and Atmospheric Administration
Washington, DC 20233, USA

ABSTRACT

We have examined various aspects of the feasibility of remote sensing of aerosols over the oceans from the NOAA environmental satellites. The technique is based on matching the radiances measured in channels 1 (0.58–0.68 μm) and 2 (0.725–1.1 μm) of the Advanced Very High Resolution Radiometer (AVHRR) with computations in a realistic model of a cloud-free, plane parallel turbid atmosphere. An extensive model library, comprising broadband and monochromatic radiances in the two channels for different combinations of aerosol size distribution functions, aerosol refractive indices, and total amounts and vertical distributions of atmospheric ozone and water vapour, has been assembled. It is observed that both the broadband and monochromatic radiances in these two channels increase linearly with atmospheric turbidity. The radiance in channel 1, which encompasses the Chappuis bands (0.45–0.75 μm) of absorption by atmospheric ozone, decreases by approximately 1.6% as the amount of ozone increases from 0.324atm-cm to 0.403atm-cm. Similarly, the radiance in channel 2, which encompasses atmospheric water vapour absorption bands, decreases by amounts ranging from ~5.5% to 8.5% as the amount of precipitable water is increased from 0.85 cm to 2.92 cm in the model atmosphere. It is also observed in the model computations that the radiances in the two channels are highly correlated. The linear relationship between the two is not sensitive to changes in the refractive index of the aerosol material; however, changes in the aerosol size distribution function lead to different linear regression relationships between the two radiances. These computational results have been used to examine briefly some of the effects of the proposed modifications of the spectral characteristics of AVHRR channels 1 and 2 for the NOAA K, L, and M satellites on remote sensing of aerosols.




ISBN 0-937194-11-5

## 1. INTRODUCTION

The impact of changes in atmospheric turbidity on weather and climate has been well documented since the late eighteenth century when Benjamin Franklin tried to explain the unusually severe winter of 1783/84 as a result of the dense haze resulting from the eruption of the volcano Laki in Iceland in 1783 (Mitchell, 1982). It has been proposed that the climatic and environmental impacts of the anomalous increases in atmospheric turbidity due to bolide impacts could have been responsible for the transition from the age of dinosaurs to the age of mammals (Alvarez, 1987). It has now been recognized (e.g., Coakley and Cess, 1985) that atmospheric aerosols, along with clouds, are the major modifiers of the shortwave or solar( 0.3-5 μm) radiation regime of the earth-atmosphere system, and that, depending on their composition, they could also affect the longwave radiation regime. It is therefore necessary to study the spatial and temporal variations of atmospheric aerosols on regional and global scales, and monitor their physical and radiative properties such as total columnar abundance, size distribution, and absorption and scattering of radiation. The advantages of earth-orbiting satellites, with their potential for global coverage, as a platform to monitor atmospheric aerosols are obvious in this context. They are especially suitable for such monitoring of aerosols over the oceans where conventional techniques of aerosol measurements are not practicable. More importantly, because of the low albedo of the oceans, practically all of the upwelling radiation originates from within the atmosphere, and is thus very sensitive to changes in atmospheric turbidity.

A brief review of the suitability of a satellite-borne instrument, such as the Advanced Very High Resolution Radiometer (AVHRR), for remote sensing of atmospheric aerosols over the oceans, and of the development of aerosol retrieval algorithms based on look-up tables is found in Rao et al. (1988). We shall discuss here some aspects of this method of aerosol retrieval in the light of model computations.

## 2. THE ATMOSPHERIC MODEL

We have computed the upwelling (diffusely reflected) radiation at the top of a realistic model of a cloud-free, plane parallel, vertically inhomogeneous turbid atmosphere. The model atmosphere is made up of ten homogeneous layers among which the gaseous and particulate (aerosol) constituents of the atmosphere have been distributed in accordance with their vertical distributions in different versions of a model atmosphere proposed by McClatchey et al. (1970). We have assumed the ocean at the bottom of the model atmosphere to be a Lambertian reflector with an albedo of 0.015.

Atmospheric aerosols are considered to be spherical Mie scatterers with either the Junge power law size distribution of the form $n(r) = Cr^{-4.5}$, or with the modified gamma size distribution (Deirmendjian, 1969) of the form $n(r) = Ar\exp(-8.9443r^{0.5})$, where n(r) is the number of aerosols per unit volume per unit increment in radius r and A and C are normalization factors. Wavelength-independent values of 1.50 and 1.50–i0.01 have been used for the refractive index of the aerosols; in addition, some computations have also been made with the wavelength-dependent value of the refractive index of water substance (~1.33).

We have used the "doubling" (Hansen, 1971) and the "adding" (Takashima, 1975) methods to compute the upwelling radiation at the top of the model atmosphere, which is illuminated by parallel radiation from the Sun with spectral irradiance as given in Neckel and Labs (1981).

## 3. RESULTS AND DISCUSSION

### 3.1 GENERAL

Radiative transfer computations for various combinations of aerosol size distributions, refractive indices, total amounts of ozone and water vapour, and spectral intervals within the passbands of the two channels have resulted in 71 look-up tables. We shall present here only the salient features of the dependence of the upwelling radiation on atmospheric turbidity and on gaseous absorption deduced from these look-up tables. Towards this end, for purposes of comparison and reference, we have shown in Fig.1 the two aerosol size distributions used in our computations, and the corresponding wavelength dependence of the normal attenuation optical thickness of the aerosols, with the aerosol optical thickness normalized to 0.10 at the wavelength of 0.55 µm. Broadband radiances in the two channels have been computed using the standard technique of expressing them as the sum of radiances over a finite number of spectral intervals, properly weighted by the normalized spectral response functions of the two channels. We have used 9 unequal spectral intervals in channel 1 and 17 in channel 2. The broadband radiances thus computed, when divided by the equivalent width of the channel(Rao, 1987), yield spectral radiances in units of $Wm^{-2} \mu m^{-1} sr^{-1}$. It has, however, been the usual practice (e.g., Quenzel and Koepke, 1984) to simulate the radiance over the passband of a satellite sensor in terms of the monochromatic radiance at an effective wavelength. We have therefore computed monochromatic radiances at the wavelengths of 0.635 µm and 0.8675 µm for purposes of comparison. We have used the normal aerosol attenuation optical thickness at the wavelength of 0.55 µm to label the state of atmospheric turbidity; it has been given values of 0.05 (very clear), 0.10, 0.175, 0.20, 0.350, and 0.698 (very turbid).

In practice, in order to avoid the adverse effects of Sun glint, only radiance measurements made in the antisolar half of the AVHRR cross-track scan are used to retrieve aerosol information (Rao et al., 1988). We shall therefore present and discuss here only the computational results for the relative azimuths from $140^{\circ}$ to $180^{\circ}$(antisolar azimuth).

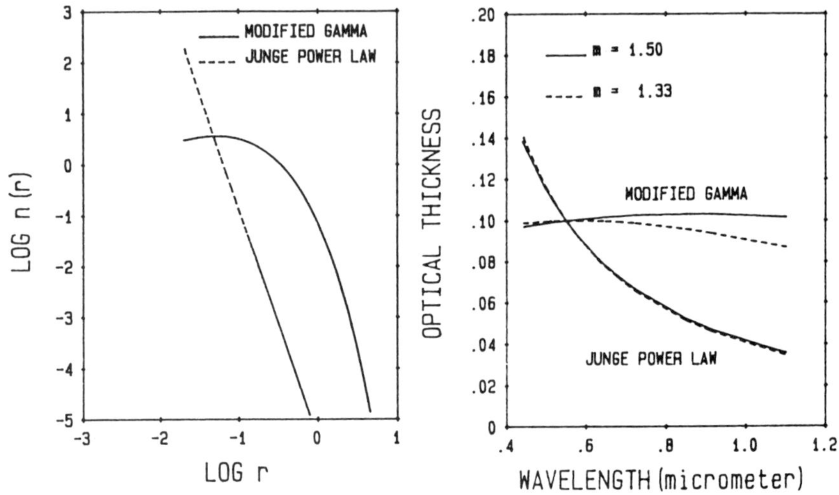


FIGURE 1. Aerosol size distributions (left) and wavelength dependence of aerosol scattering optical thickness (right). The radius r is expressed in units of micrometer; and n(r) is the number of aerosols per cubic centimeter per micrometer. The aerosol optical thickness has been normalized to 0.10 at 0.55 μm. The two curves for the modified gamma size distribution correspond to refractive index of 1.50 (upper) and 1.33 (lower) at 0.55 μm. The corresponding curves for the Junge power law size distribution are practically indistinguishable. Absorption by aerosols does not affect this general trend of the wavelength dependence of the aerosol scattering optical thickness.

## 3.2 RADIANCES IN CHANNELS 1 AND 2

We have shown in Fig.2 the variation of the upwelling radiance in channels 1 and 2 with atmospheric turbidity for the Junge and modified gamma (Model M) size distributions for different values of the refractive index of the aerosol material in the azimuthal plane of $140^{\circ}$, with the solar and satellite zenith angles, $\theta_0$ and $\theta$, being respectively $48^{\circ}$ and $42^{\circ}$. The linear dependence of tne upwelling radiance in the two channels on atmospheric turbidity is very apparent for the two aerosol size distributions and refractive indices. Absorption by aerosols (m = 1.50–i0.01), while lowering the radiances, does not affect the linear dependence. Simple regression of the upwelling radiance on the normal aerosol optical

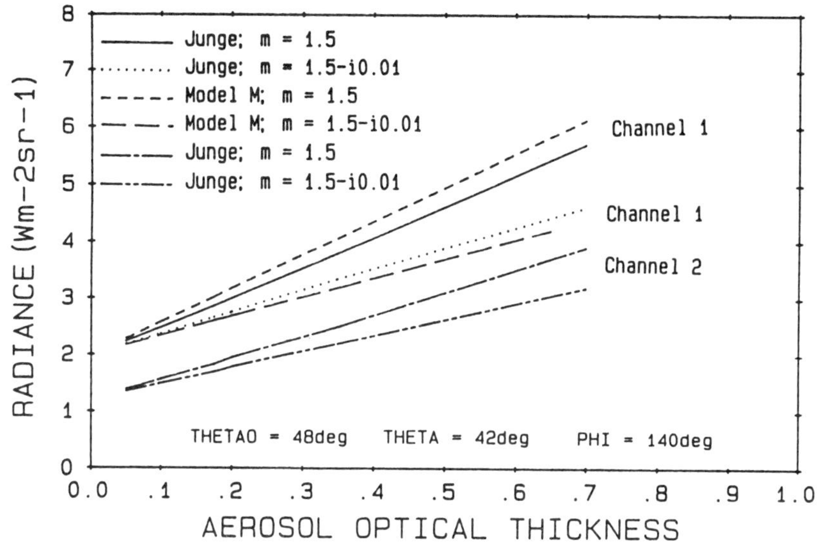


FIGURE 2. Dependence of the diffusely reflected radiation at the top of the atmosphere on atmospheric turbidity.

thickness at 0.55 μm yields values for the coefficient of determination in excess of 0.99. It is observed that such linear regression relationships, with different intercepts and slopes, can be established between the upwelling radiance and aerosol extinction optical thickness at 0.55 μm over the entire range of solar zenith [42°-84°], satellite zenith [0°-60°], and relative azimuth [140°-180°] angles used in the model computations.

The displayed data also point to the inherent ambiguity in the interpretation of remotely sensed radiance measurements, namely, that different combinations of the total number, size distribution, and refractive index of aerosol particles can yield the same upwelling radiance. We shall discuss the implications of this ambiguity for remote sensing of aerosols later.

The spectral passband of channel 1(0.58-0.68 μm) falls within the Chappuis bands (~0.45-0.75 μm) of absorption by atmospheric ozone. Thus, in order to study the effects of variations in atmospheric ozone on channel 1 radiance, we have computed the diffusely reflected radiation in the model atmosphere with 0.324atm-cm and 0.403atm-cm of ozone. It is found that this increase of ~25% (of 0.324atm-cm) in ozone causes a decrease of about 1.6% in the upwelling radiance when the aerosols have the Junge power law size distribution; slightly lower decreases are observed with the modified gamma size distribution for the aerosols. We have also examined the effects of variations of atmospheric water vapour (precipitable water) on the upwelling radiance in channel 2 whose spectral passband encompasses water vapour absorption bands centered around 0.740, 0.796, 0.825, 0.846, 0.889, 0.910 and 0.950 μm. Our computations indicate that as the amount of precipitable water is increased from 0.85 cm to 2.92 cm, the upwelling radiance in channel 2 decreases by amounts ranging from 5.5% to 8.5% of the value corresponding to 0.85 cm of precipitable water for the Junge size distribution. The larger decreases are observed at higher values of atmospheric turbidity. The decrease in radiance is slightly larger when the aerosols have the modified gamma size distribution. Comparison of the monochromatic radiance at the wavelength of 0.8675 μm, which lies outside the water vapour absorption bands, with the broadband radiance in the model atmosphere with 2.92 cm of precipitable water indicates that the ratio of the spectral radiance to the monochromatic radiance ranges from 0.88 to 0.98, with the lower value being observed at larger atmospheric turbidity and at large solar and satellite zenith angles.

We have shown in Fig.3 the proposed modifications in the spectral characteristics of channels 1 and 2 for the NOAA K, L, and M satellites, along with typical features of atmospheric transmission. It is apparent that the proposed modification in channel 2, with the centre wavelength shifted to 0.865 μm and the passband considerably narrowed (~0.035 μm), would help minimize the adverse effects of the variations in atmospheric water vapour in the interpretation of the upwelling radiance. However, our computations also indicate that the energy detected by the AVHRR in the modified channel would decrease by a factor of about 7 compared to the present values-- nearly an order of magnitude. The proposed modification in channel 1 mainly renders the spectral response symmetrical about the centre wavelength of 0.635 μm.

### 3.3 INTERRELATIONSHIP BETWEEN THE RADIANCES IN THE TWO CHANNELS

We see from our model computations that the radiances in the two channels are highly correlated; simple regression of channel 2 radiance on channel 1 radiance yields values for the coefficient of determination in excess of 0.99. This high degree of correlation is clearly seen in the data displayed in Fig. 4 and is traceable to the fact that the spectral radiative effects of the aerosols are prescribed by the size distribution function and the refractive index of the aerosols. The most noticeable feature is that absorption by aerosols, while reducing the radiances, hardly affects the regression relationship. Conversely, changes in the aerosol size distribution function, while maintaining the high degree of correlation between the two radiances, alter both the slope and intercept of the regression relationship. It is thus likely that the slope of this regression relationship could be utilized as a criterion to distinguish between different aerosol size distributions in any dual-channel aerosol retrieval algorithm.

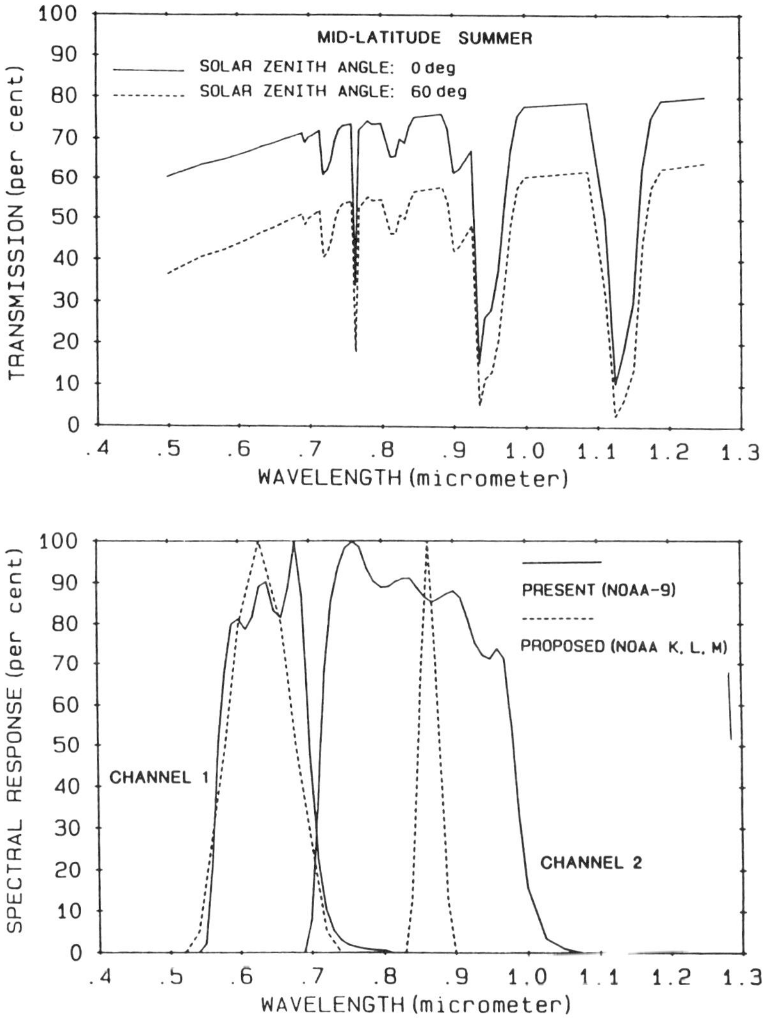


FIGURE 3. Typical atmospheric transmission (top) and the present and proposed spectral response functions of channels 1 and 2.

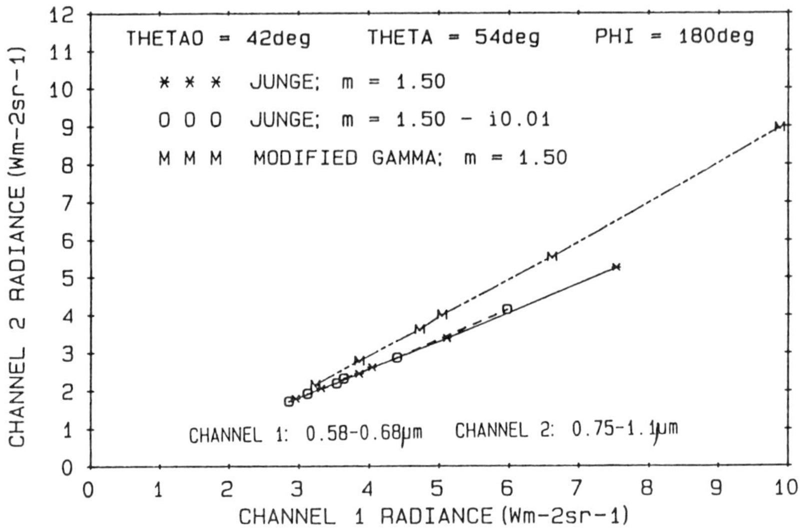


FIGURE 4. Interrelationship between radiances in channels 1 and 2.

3.4 MODEL UNCERTAINTY ANALYSIS

The major source of uncertainty or ambiguity in the retrieved aerosol information obtained by matching measured and modeled radiances is that the model atmosphere used in the computations may not be truly representative of the actual conditions under which the upwelling radiances are measured. Thus, the total number of aerosols in the atmospheric column, their size distribution, and refractive index could be different from the model-specified values. Similarly, there could be differences in the total amounts and vertical distributions of atmospheric ozone and water vapour. And, the reflectional properties of the ocean are non-Lambertian. In the light of these uncertainties, and of the inherent ambiguity in the interpretation of measured radiances referred to earlier, it has been recognized (e.g., Rao et al., 1973) that attention should be directed towards the construction of a model that would be radiatively equivalent to the actual atmosphere.

The effects of the departure of the reflectional properties of the ocean surface from those of a Lambertian reflector used in the model will now be briefly examined. We have shown in Fig. 5 the ratio of the upwelling radiance at 0.63 μm from an atmosphere resting upon a Lambertian ocean to the upwelling radiance when the wind-driven ocean surface is represented as being made up of multifaceted waves, reflecting radiation according to the Fresnel

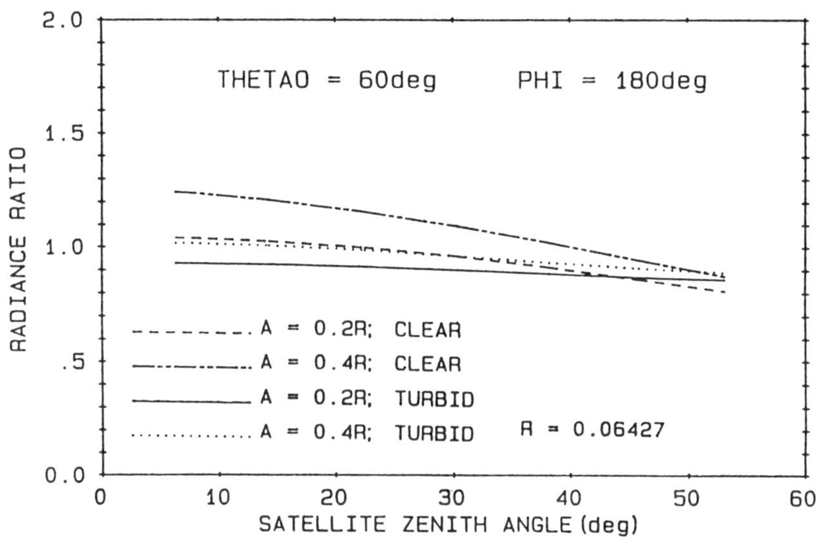


FIGURE 5. Comparison of upwelling radiances at the top of a model atmosphere resting upon Lambertian and non-Lambertian ocean surfaces.

laws. The Lambertian ocean has been assigned albedos of 0.2R and 0.4R, where R is the average reflectance of the simulated, wind-driven ocean surface when the surface windspeed is 5m/s. R can have values of up to 0.07 (Takashima and Takayama, 1981) over the greater part of the range (up to 70$^{o}$) of solar zenith angles used in aerosol remote sensing from AVHRR data. Radiances have been computed for two different values of atmospheric turbidity, with the aerosol extinction optical thickness at 0.55 μm having values of 0.255 (clear) and 0.891 (turbid). It is clearly seen that, when the Lambertian ocean albedo is about 0.013 for the "clear" case and about 0.026 for the "turbid" case, the radiance ratio is within ±10% of unity for satellite zenith angles up to 50$^{o}$. It can therefore be concluded that our computations with a surface albedo of 0.015 may lead to uncertainties of the same order of magnitude in the retrieved aerosol optical thickness so long as measurements are made away from the region of Sun glint.

It is also likely that the presence of white caps may affect the upwelling radiance and lead to erroneous estimates of atmospheric aerosol. However, whitecap coverage at winds of up to 5m/s is very small and insignificant (Monahan and O'Muircheartaigh,

1980). At greater windspeeds, when white cap coverage can be significant, clouds are usually present, and measurements made under cloudy skies are excluded from analysis.

Errors in the determination of the solar zenith, satellite zenith, and relative azimuth angles could also affect the retrieved aerosol information. It is seen from our computational results that, generally, the variation of the upwelling radiance with azimuth angle is less pronounced than its variation with the solar and satellite zenith angles; however, it is expected that correct values of these angles will be available. Our computations also indicate that the relative uncertainty in the retrieved aerosol extinction optical thickness will be of the same order of magnitude as the relative uncertainty in the measured radiances.

## 4. CONCLUDING REMARKS

The computational results we have presented here point to the need for and the relevance of exhaustive model sensitivity analyses in the development of aerosol retrieval algorithms using multi-spectral radiances. Such model sensitivity analyses would play an important role in minimizing uncertainties in aerosol retrieval from remotely sensed visible and near-infrared radiances and in the optimization of the design of future radiometers.

Examples of aerosol optical thickness fields over the global oceans, generated from weekly, composited aerosol optical thickness values retrieved from AVHRR channel 1 measurements using a look-up table algorithm, are found in the companion paper by Rao et al., (1988). It is expected that such a product, when validated on a continuing basis by ground truth measurements of aerosol optical thickness at strategically located ocean sites, and supplemented by information on aerosol size distribution and refractive index, will find application in several areas. Some examples are: climate modeling, studies of horizontal and slant path visibilities over the oceans, contrast transmission properties of the atmosphere, and the development of corrections for aerosol-induced atmospheric effects in the determinations of sea surface temperature and atmospheric composition and structure.

## ACKNOWLEDGMENT

This work was completed while one of the authors (C.R.N.R) held a NOAA/NRC resident research associateship at the National Environmental Satellite, Data, and Information Service, National Oceanic and Atmospheric Administration, Washington,D.C. 20233.

## REFERENCES

Alvarez, L.W., 1987: Mass Extinctions Caused by Large Bolide Impacts, Physics Today, 40, 24-33.

Coakley, J.A., and R.D.Cess, 1985: Response of the NCAR Community Climate Model to the Radiative Forcing of

Naturally Occurring Tropospheric Aerosol, J. Atmos. Sci., 42, 1677-1692.

Deirmendjian, D., 1969: Electromagnetic Scattering on Spherical Polydispersions, American Elsevier Publishing Company, New York.

Hansen, J.E., 1971: Multiple Scattering of Polarized Light in Planetary Atmospheres. Part 1: The Doubling Method, J. Atmos. Sci., 28, 120-125.

McClatchey, R.W., R.W.Fenn, J.E.A.Selby, J.S.Garing and F.E.Volz, 1970: Optical Properties of the Atmosphere, Environmental Research Papers, No.331(AFCRL-70-0527), Air Force Cambridge Research Laboratories, L.G.Hanscom Field, Bedford, Massachusetts, 98pp.

Mitchell, J.M., Jr., 1982: El Chichon, Weather-Maker of the Century, Weatherwise, 35, 252-259.

Monahon, E.C., and I.G.O'Muircheartaigh, 1980: Optimal Power Law Description of White Cap Coverage Dependence on Wind Speed, J. Phys. Oceanogr., 10, 2094-2099.

Neckel, H., and D.Labs, 1981: Improved Data for Solar Spectral Irradiance from 0.33 μm to 1.235 μm, Solar Phys, 74, 231-237.

Quenzel, H., and P.Koepke, 1984: Tropospheric Aerosol Optical Depth Inverted From Upwelling Radiances. In H.E.Gerber and A.Deepak(Eds.), Aerosols and Their Climatic Effects, A.Deepak Publishing, Hampton, Virginia, 227-240.

Rao, C.R.N., L.L.Stowe, E.P.McClain and J.Sapper, 1988: Development and Application of Aerosol Remote Sensing With AVHRR Data From the NOAA Satellites. In P.V.Hobbs and M.P.McCormick (Eds.), Aerosols and Climate, A.Deepak Publishing, Hampton, Virginia.

Rao, C.R.N., 1987: Prelaunch Calibration of Channels 1 and 2 of the Advanced Very High Resolution Radiometer, NOAA/NESDIS Technical Report No.36, U.S.Department of Commerce, Washington, D.C., 62pp.

Rao, C.R.N., T.Takashima and R.B.Toolin, 1973: Measurements and Interpretation of the Polarization of Radiation Emerging from the Atmosphere at an Altitude of 28 km over Southwestern New Mexico (USA), Quart. J. Roy. Meteor. Soc., 99, 294-302.

Takashima, T., and Y.Takayama, 1981: Emissivity and Reflectance of Model Sea Surface for the Use of AVHRR Data of NOAA Satellites, Papers in Meteorology and Geophysics, 32, 267-274.

Takashima, T., 1975: A New Approach of the Adding Method for Computations of Emergent Radiation of an Inhomogeneous Plane Parallel Atmosphere, Astrophys. Spa. Sci., 36, 319-328.

# SATELLITE MAPPING OF AEROSOL LOADING OVER VEGETATED AREAS

*Yoram J. Kaufman*
Department of Physics, Haifa, Israel 32000, and
NASA Goddard Space Flight Center, Greenbelt, Maryland 20771, USA

and

*Claudia Sendra*
Besares 3240 (1429) Cap. Fed., Argentina

## ABSTRACT

An algorithm is developed for automatic mapping of the aerosol loading from a single satellite imagery of the Earth's surface over vegetated areas. The algorithm utilizes the satellite imagery and climatology of the area. It is applicable to both low resolution (1 km field of view) and high resolution (10-80 m field of view) imagery of land areas for the solar spectrum. The algorithm requires that some pixels in the image should have dense dark vegetation as the surface cover. Once the presence of such pixels is established, the algorithm automatically chooses these pixels and derives the atmospheric optical thickness (a measure of the aerosol loading). The algorithm is sensitive to the assumed reflectance of the dense dark vegetation in the red and blue bands (if present in the imagery), to the assumed aerosol characteristics (scattering phase function and single scattering albedo), and to the accuracy of satellite calibration. It is not very sensitive to the knowledge of the exact fraction of the image covered by the dense dark vegetation, as long as some fraction of the pixels are fully covered by the vegetation.

The algorithm was applied to "clear" and "hazy" images of the same area in the Washington D.C., and the Chesapeake Bay region. Landsat multispectral scanning radiometer (MSS) imagery with reduced resolution (resolution of ~ 700 m) was used. The aerosol optical thickness ($\tau_a$) derived from the imagery shows good agreement with simultaneous sunphotometer measurements from the ground within $\Delta\tau_a=\pm0.20$ in band 1 (0.5-0.6 μm) and $\Delta\tau_a=\pm0.05$ in band 2 (0.6-0.7 μm). These optical thicknesses were used successfully to predict the optical thickness in the near IR band (0.7-0.8 μm) with an error of $\Delta\tau_a=\pm0.02$. In the near-IR the surface reflectance is large and $\tau_a$ cannot be determined otherwise from satellite imagery. Application to the Advanced Very High Resolution Radiometer (AVHRR) type of sensor (one broad channel in the visible) would need some modifications.


A. DEEPAK Publishing
ISBN 0-937194-11-5

## 1. INTRODUCTION

Satellite imagery has been used to determine the spatial distribution of the aerosol optical thickness (a measure of the aerosol loading). Using a single image for its determination the method required a dark surface and, therefore, was applied only over water (Griggs, 1975; Mekler et al., 1977; Carlson, 1979; Koepke and Quenzel, 1979). The water reflectance was estimated from theoretical computations. Over land the high and variable land reflectivity required a different technique. In this case two images of the same area were used (Fraser et al., 1984; Kaufman et al., 1985a; Kaufman, 1987), one representing the "hazy day" for which the optical thickness should be derived and a second one from a "clear day" that served as the reference to derive the surface reflectance. As a result the difference in the optical thickness between the hazy and clear days was derived by this method. The use of two images of the same area to derive $\tau_a$ restricts the application of the method only to cases where the appropriate images are manually chosen. It would be useful to find the appropriate conditions where a single image of the land surface can be used to derive $\tau_a$, similar to the methods applied over the water. We suggest images of vegetated areas be used for this purpose, due to the low reflectivity of dense vegetation in the blue and red parts of the spectrum. An automatic algorithm is suggested in which pixels that correspond to dense dark vegetation are chosen and the optical thickness of the atmosphere above them is determined.

Dense dark vegetation has a low reflectance in the visible part of the spectrum. In Table 1 measured reflectances of pasture land are summarized. The reflectances are in the range of 2.5%-4%. This reflectance can be decreased by the effect of shadows present in tall vegetation, such as trees and bushes, to 1%-3%. The reflectance of forests is in the range of 1%-2% in the blue (0.4-0.5 μm) and red (0.6-0.7 μm) parts of the spectrum and 2%-3% in the green (0.5-0.6 μm). For atmospheric corrections of satellite imagery, it is suggested to use the upper limit of these reflectances due to the possible presence of bare soil. These low reflectances are unique for land cover types. Forested areas can be found in many locations of the world and their optical characteristics are rather stable (except for deciduous trees in the winter). Therefore, dense dark vegetation is suggested as the controlled dark surface used to derive the aerosol optical thickness.

## 2. DERIVATION OF THE AEROSOL OPTICAL THICKNESS

The basic philosophy in the derivation of the aerosol optical thickness from radiances recorded in a single image acquired over the land is to determine pixels in the image for which the surface reflectance is low and its value can be a priori determined within some small uncertainty. The detected radiance is then affected mainly by $L_0$ and the aerosol optical thickness is derived based on auxiliary information on the atmospheric optical characteristics (gaseous absorption, aerosol scattering phase function and single scattering albedo). The auxiliary information can be obtained from three different sources:

**TABLE 1: VEGETATION REFLECTANCE (NADIR VIEW)**
Reflectances (solar zenith angle 40$^{o}$)

| Vegetation \ λ(μm) | 0.4-0.5 | 0.5-0.6 | 0.6-0.7 | Source |
|---|---|---|---|---|
| Pasture | 0.026 | 0.030 | 0.039 | Kriebel (1977)* |
| Pasture | 0.027 | 0.039 | 0.032 | Kaufman (1986)** |
| Coniferous forest | - | 0.011 | 0.017 | Kriebel (1977)* |
| Decideous forest | 0.010 | 0.022 | 0.012 | Kaufman (1986)** |
| Hardwood forest | - | - | 0.025 | Kimes et al. (1986)*** |
| Pine forest | - | - | 0.028 | Kimes et al. (1986)*** |

The exact wavelengths used are: (*) - 0.43 μm, 0.52 μm, 0.61 μm; (**) - 0.46 μm, 0.56 μm, 0.62 μm; (***) - 0.58-0.68 μm.

## 2.1 CLIMATOLOGY

Documented information on the atmospheric characteristics and their variation can be used to estimate the expected atmospheric effect for a specific part of the world and a specific season. Such documentation can be obtained from the analysis of measurements taken from the ground, and partially from the analysis of satellite data (Fraser et al., 1984).

## 2.2 MEASUREMENTS FROM THE GROUND

The aerosol optical thickness can be obtained from sunphotometer measurements (King et al., 1978; Kaufman and Fraser, 1983). The phase function can be determined from inversion of solar almucantar measurements (Kaufman et al., 1985b) and the single scattering albedo from the collection of particles on filters, preferably by aircraft sampling of the entire atmospheric boundary layer. The application of such measurements for atmospheric corrections is useful for intense field measurements or for establishing the climatology of a given area.

## 2.3 DETERMINATION FROM SATELLITE IMAGERY

Determination of the aerosol single scattering albedo and particle size, which can be used to compute the scattering phase function, was suggested by Kaufman et al. (1985a) and Kaufman (1987). This method is useful to determine the aerosol characteristics (from imagery that includes water/land interfaces) in areas that suffer from substantial aerosol outbreaks.

# 3. ALGORITHM

The following algorithm is developed assuming that the satellite image contains some patches of dense dark vegetation (e.g., trees, shrubs) larger than the pixel size. These patches are used as a controlled dark surface for the estimation of the aerosol optical thickness. Although this requirement limits the applicability of the algorithm, it results in an automatic procedure in which the atmospheric haziness is estimated from a single satellite image. The algorithm is based on the following assumptions:

1) There is at least a fraction $f_1$ of the pixels in the image that are covered fully by dense dark vegetation. The algorithm will determine in the image a fraction $f_1$ of the pixels that best fit the criterion of dense dark vegetation.
2) The surface reflectance of these patches of dense dark vegetation in the spectral bands used in the algorithm is known. For example, for the blue and the red parts of the spectrum we can assume $\rho=0.02\pm0.01$ and for the green part of the spectrum $\rho=0.03\pm0.01$ (see Table 1).
3) The aerosol scattering phase function and single scattering albedo are known.
4) The surface can be considered as Lambertian for the purpose of the correction (for a discussion of this assumption see Lee and Kaufman, 1986). This assumption is valid for most irradiation and viewing directions. Large errors can be expected for high solar zenith angles, for viewing directions that are close to specular reflection and backscattering.

The sensitivity of the algorithm to these assumptions will be tested in the latter part of this paper. The following are the steps of the algorithm:

1) Estimate the gaseous absorption based on climatology of the area by the LOWTRAN 6 code.
2) Remove cloudy areas by masking out pixels with high wavelength independent reflectance or low IR radiative temperature. The remaining subpixel clouds that cannot be removed by this method are considered as part of the aerosol layer, and will be treated as such.
3) Estimate the aerosol single scattering phase function (P), single scattering albedo ($\omega_o$) and the height of the aerosol above the surface from climatology of the area, or any other information. Use ground measurements if they are available.
4) (Optional) If the area has a substantial aerosol loading (e.g., location where Saharan dust or forest fire smoke is present, anthropogenic air pollution as in the summer in the Eastern USA), update (3) by the derivation of aerosol characteristics from the difference in the brightness between satellite images during clear and hazy conditions (Kaufman et al., 1985a).
5) Generate a lookup table of the upward radiance L as a function of the aerosol optical thickness $\Delta\tau_a$ and surface reflectance $\rho$: $L_\lambda(\tau,\rho)$, for the solar direction and the observation direction, and for the local aerosol characteristics (phase function P and single scattering albedo $\omega_o$). The radiances are computed here by the Dave and Gazdag (1970) radiative transfer code.
6) Identify pixels covered by dense dark vegetation. This determination is performed by choosing from the image a fraction $f_1$ (e.g., $f_1$=0.2) of the pixels with the highest vegetation index, and from this group choosing a fraction $f_2$ (e.g., $f_2$=0.5) of pixels with the lowest radiance in the near IR (thus choosing pixels covered with dark vegetation from the vegetated pixels). The fractions $f_1$ and $f_2$ are determined a priori based on an estimate of the occurrence of such vegetation.
7) Compute the aerosol optical thickness ($\tau_a$) for each pixel identified as covered by

dense dark vegetation. The optical thickness is found by interpolation of $\tau_a$ from the library $L_\lambda(\tau,\rho)$ for the radiance $L_m$ measured for each pixel, for the assumed reflectance in each band. The determination of the optical thickness is performed for the red and if available also for the blue parts of the spectrum where vegetation is dark.

8) Generate a map of $\tau_a(\lambda,x,y)$ by interpolating and smoothing the derived values.

9) For each pixel compute the slope $\nu$ defined by:

$$\nu=-\ln(\tau_{a1}/\tau_{a2})/\ln(\lambda_1/\lambda_3) \qquad (1)$$

The average value of $\nu$ on the entire subarea ($\nu'$) can be used to test and possibly change the assumed phase function in step (5) (Kaufman et al., 1985a).

## 4. APPLICATION AND VALIDATION

The suggested algorithm was applied to LANDSAT MSS images taken over the Washington D.C., and the Chesapeake Bay area. Cloud-free images during relatively clear (20 August 1982 - referred to here as a clear day) and hazy (2 August 1982 - referred to here as a hazy day) conditions were used. For each day two subsets of the image were used, each containing 50x100 pixels. Each pixel in the subset is an average of 10x10 pixels in the original Landsat image and thus each subset represents an area of 30 km X 70 km, with a resolution of ~700 m. This resolution is close to the 1 km resolution of the AVHRR. The two subsets are aligned in a north-south direction. The northern subset is between latitudinal lines $38^o30'$ W - $39^o15'$ W and longitudinal lines $76^o35'$ N - $76^o50'$ N. The southern subset is between latitudinal lines $37^o50'$ W - $38^o30'$ W and about the same longitudinal lines. The images in the two days were navigated to overlap each other.

In Fig. 1 an example is shown of the relation between the brightness in band 2 (600-700 nm) and the vegetation index (VI - is the ratio of radiances in band 3 - $L_3$: 700-800 nm to radiances in band 2 - $L_2$) as well as the normalized difference NDVI:

$$NDVI = (L_3-L_2)/(L_3+L_2) \qquad (2)$$

for the northern subset. The brightness is given by the normalized radiance $\pi L/F_o\mu_o$ ($F_o$ is the extraterrestrial solar flux and $\mu_o$ is cosine of the solar zenith angle). This normalization allows us to interchange the terms "normalized radiance" and "apparent reflectance" since the normalized radiance is equal to the surface reflectance for no atmosphere. Therefore, any difference between the normalized radiance and the corresponding surface reflectance is due to the atmospheric effect. This relation is plotted for the clear day and for the hazy day. For 20% of the pixels ($f_1$=0.2) with the highest VI (NDVI=0.46 for the clear day and 0.2 for the hazy day) the radiance $L_2$ varies in each day within $\pi\Delta L/F_o\mu_o=\pm0.01$, which corresponds to reflectance ($\rho$) of $\Delta\rho=\pm0.01$. Note that the value of VI is different in the hazy day from that in the clear day. The radiance is more dispersed in band 1 (500-600 nm) for pixels with high vegetation index (Fig. 2), and its value is also higher. This is

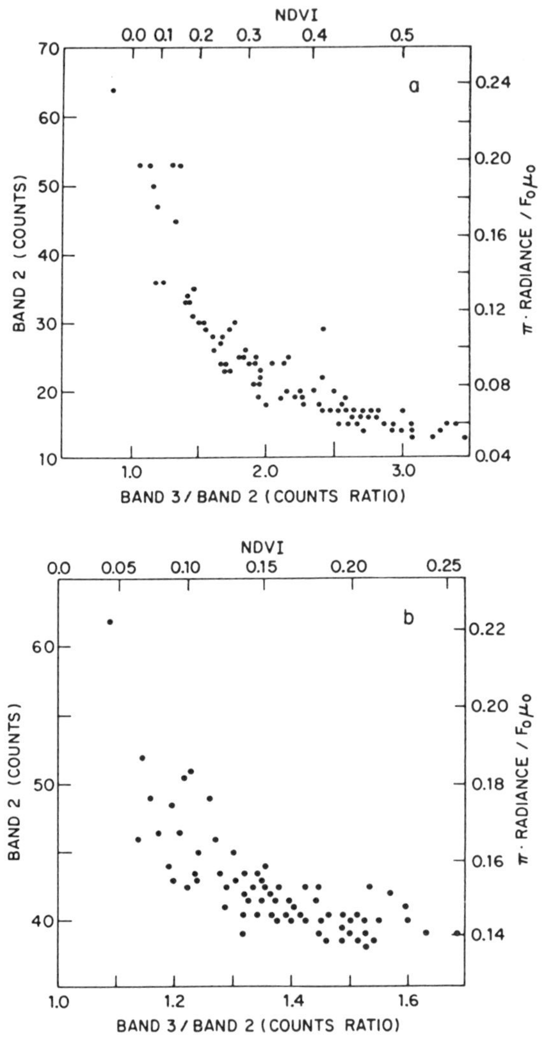


*FIGURE 1. A scatter diagram of the radiances (in counts) from Landsat MSS band 2 (600–700 nm) and the vegetation index (ratio between the counts in band 2 and band 3). Each point represent an average of 10 × 10 pixels. (a) – 20 August 1982 – clear day, (b) – 2 August 1982 – hazy day.*

due to the higher reflectivity of vegetation in the green part of the spectrum and stronger atmospheric scattering. Unfortunately no blue band is present in the MSS and therefore band 1 (in addition to band 2) is used to derive the aerosol optical thickness from the Landsat MSS image. In Fig. 3 a comparison is shown between the

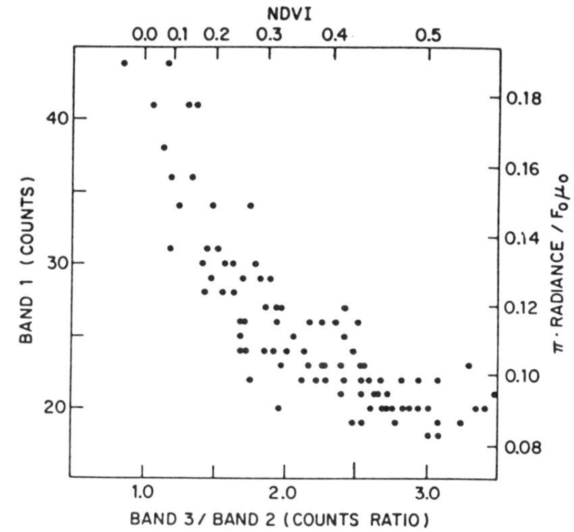


*FIGURE 2. Same as in Fig. 1a but for Landsat MSS band 1 (500-600 nm).*

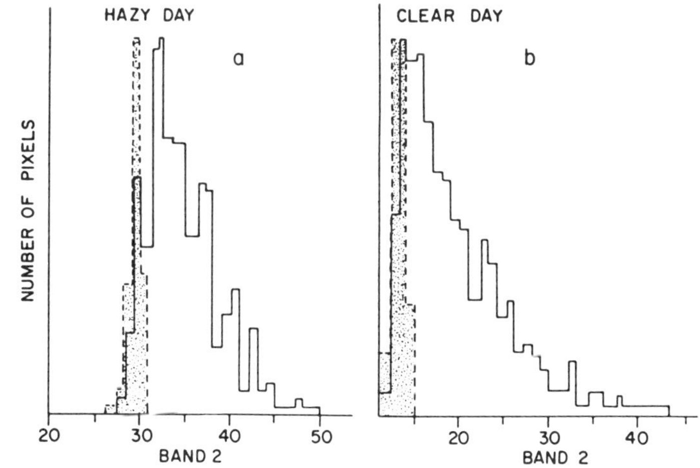


*FIGURE 3. Histogram of the radiance (in counts) from Landsat MSS band 2 (600-700 nm) for clear day (20 August 1982) and hazy day (2 August 1982). (———) - original histogram and (- - - -) - shaded histogram after the selection algorithm was applied (see text).*

histograms of the radiances in band 2 of the original northern subset (solid lines) and histograms of these dark pixels selected by the procedure (step 6). The application of the procedure results in a much narrower histogram, shown in the figure (shaded area). These histograms of the selected pixels correspond to the lowest radiances in the image.

The histograms of the aerosol optical thickness computed from the chosen pixels in bands 1 and 2, for the northern subset (step 7) are shown in Fig. 4. The average and standard deviations of the histograms are given in Table 2b. Comparison between the satellite derived optical thickness and the optical thickness simultaneously measured from the ground (arrows in Fig. 4), show good agreement between them within $\Delta\tau=\pm0.20$ in band 1 and $\Delta\tau=\pm0.05$ in band 2 (see Table 2b). In this case it was assumed that the surface reflectance is $\rho_1=0.03$ and $\rho_2=0.02$ in the two MSS bands (see Introduction). We can see the superiority of band 2 (red) over band 1 (green).

The clear day (20 August 1982) represents here a background aerosol that is present in the Eastern USA during the summer (Kaufman and Fraser, 1983; Peterson et al., 1981). The hazy day (2 August 1982) represents the effect of forest fire smoke

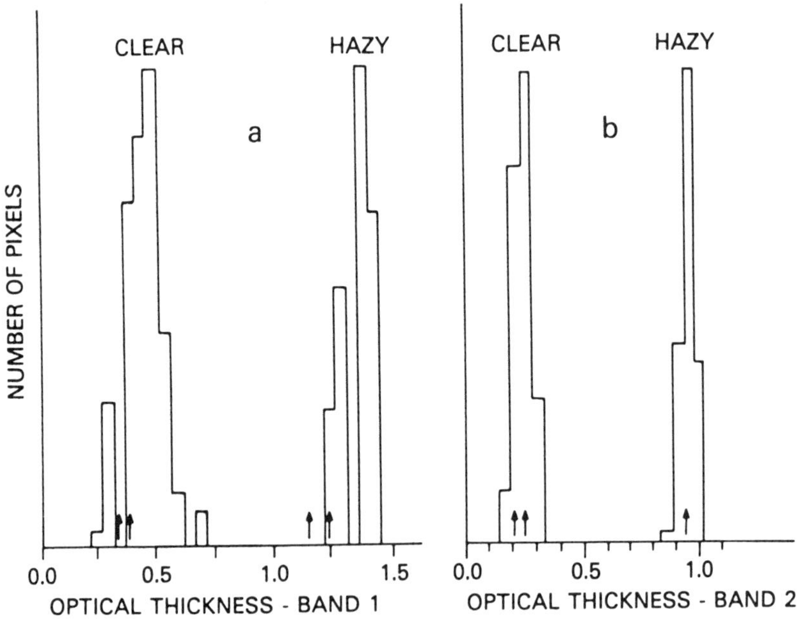


*FIGURE 4. Histograms of the optical thickness derived from Landsat MSS: (a) band 1 (500-600 nm) and (b) band 2 (600-700 nm) for clear day (20 August 1982) and hazy day (2 August 1982). The arrows show corresponding measurements from sunphotometers (see also Table 2).*

## TABLE 2: RESULTS OF THE ALGORITHM

**a.** Aerosol Optical Thickness Measured by Sunphotometer and the Slope($\nu$):

| Subarea | Date | Aerosol Optical Thickness band 1 ($\tau_1$) | band 2 ($\tau_1$) | band 3 ($\tau_1$) | Slope ($\nu$) |
|---|---|---|---|---|---|
| northern | 8/20/82 | 0.34 ± 0.08 | 0.23 ± 0.02 | 0.17 ± 0.01 | 2.9 ± 0.3 |
| northern | 8/02/82 | 1.20 ± 0.05 | 0.96 ± 0.02 | 0.77 ± 0.01 | 1.7 ± 0.2 |

**b.** Results of the Algorithm

| subarea | date | aerosol optical thickness band 1 ($\tau_1$) | band 2 ($\tau_1$) | band 3 ($\tau_1$) | Slope ($\nu$) |
|---|---|---|---|---|---|
| northern | 8/20/82 | 0.49 ± 0.08 | 0.279 ± 0.05 | 0.177 ± 0.03 | 3.41±0.6 |
| northern | 8/02/82 | 1.40 ± 0.06 | 0.977 ± 0.03 | 0.747 ± 0.08 | 2.18 ±0.1 |
| southern | 8/20/82 | 0.44 ± 0.05 | 0.275 ± 0.04 | 0.197 ± 0.03 | 2.72 ±0.3 |
| southern | 8/02/82 | 1.04 ± 0.04 | 0.716 ± 0.03 | 0.573 ± 0.08 | 2.17 ±0.1 |

**c.** Effect of Assumed Surface Reflectance (Southern Subarea)

| $\rho$ band1 | band2 | date | $\tau_1$ | $\tau_2$ | $\tau_3$ | $\nu$ |
|---|---|---|---|---|---|---|
| 0.03 | 0.02 | 8/02/82 | 1.04 | 0.52 | 0.57 | 2.2 |
| 0.03 | 0.03 | 8/02/82 | 1.03 | 0.63 | 0.47 | 2.9 |
| 0.04 | 0.03 | 8/02/82 | 0.95 | 0.63 | 0.49 | 2.4 |
| 0.03 | 0.02 | 8/20/82 | 0.44 | 0.27 | 0.20 | 2.7 |
| 0.04 | 0.02 | 8/20/82 | 0.32 | 0.27 | 0.25 | 0.8 |
| 0.03 | 0.03 | 8/20/82 | 0.45 | 0.15 | 0.07 | 6.1 |

Table 2. (continued)

d. Effect of Error in Satellite Calibration (Northern Subarea)
(A) - Degradation of 10% in all channels
(B) - Degradation of 10% in channel 1 only

| Case | Date | $\tau_1$ | $\tau_2$ | $\tau_3$ | $\nu$ |
|---|---|---|---|---|---|
| No error | 8/20/82 | 0.496 | 0.279 | 0.177 | 3.41 |
| A | 8/20/82 | 0.356 | 0.200 | 0.127 | 3.41 |
| B | 8/20/82 | 0.356 | 0.279 | 0.231 | 1.50 |
| No error | 8/02/82 | 1.400 | 0.977 | 0.747 | 2.18 |
| A | 8/02/82 | 1.190 | 0.845 | 0.630 | 2.14 |

e. Effect of Size Distribution (Northern Subarea)

| $R_m$ (μm) | Date | $\tau_1$ | $\tau_2$ | $\tau_3$ | $\nu$ |
|---|---|---|---|---|---|
| Power Law | 8/20/82 | 0.496 | 0.279 | 0.177 | 3.41 |
| 0.20 | 8/20/82 | 0.765 | 0.467 | 0.311 | 2.95 |
| 0.60 | 8/20/82 | 0.894 | 0.562 | 0.388 | 2.76 |
| Power Law | 8/02/82 | 1.400 | 0.977 | 0.747 | 2.18 |
| 0.20 | 8/02/82 | 1.671 | 1.205 | 0.950 | 1.95 |
| 0.60 | 8/02/82 | 2.140 | 1.487 | 1.130 | 2.20 |

f. Effect of Single Scattering Albedo (Northern Subarea)

| $\omega_0$ | Date | Derived | | | |
|---|---|---|---|---|---|
| | | $\tau_1$ | $\tau_2$ | $\tau_3$ | $\nu$ |
| 0.95 | 8/20/82 | 0.496 | 0.279 | 0.177 | 3.41 |
| 0.85 | 8/20/82 | 0.724 | 0.390 | 0.238 | 3.70 |
| 0.95 | 8/02/82 | 1.400 | 0.977 | 0.747 | 2.18 |
| 0.85 | 8/02/82 | 2.170 | 1.490 | 1.089 | 2.24 |

originating in Northwest Canada and transported to the Washington D.C., area (Kaufman et al., 1985a). A plot of the spatial variation of the optical thickness derived in the two bands, in the clear and hazy days is shown in Fig. 5. In this plot the values of the optical thickness derived in each east-west line was averaged to yield a single value representing the line. The plots show the variation of the optical thickness along 150 km of the longitudinal line. For the clear day the optical thickness is very stable across the area represented here. For the hazy day, a gradient in the smoke density can be seen in the plot. In the case of such a gradient the procedure tends to select pixels with a lower optical thickness because the vegetation index, used as a criterion in the selection, is reduced with an increase in the optical thickness. This dependence is a limitation of the proposed correction procedure. It can be explained by the help of the theoretical computations shown in Fig. 6.

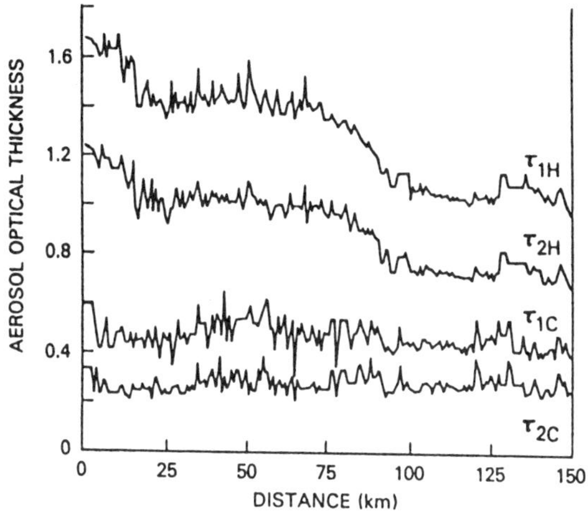


*FIGURE 5. Variation of the aerosol optical thickness derived from Landsat MSS bands 1 and 2 for 2 August, 1982 (upper two plots - $\tau_{1H}$ and $\tau_{2H}$) and for 20 August 1982 (lower two plots - $\tau_{1C}$ and $\tau_{2C}$).*

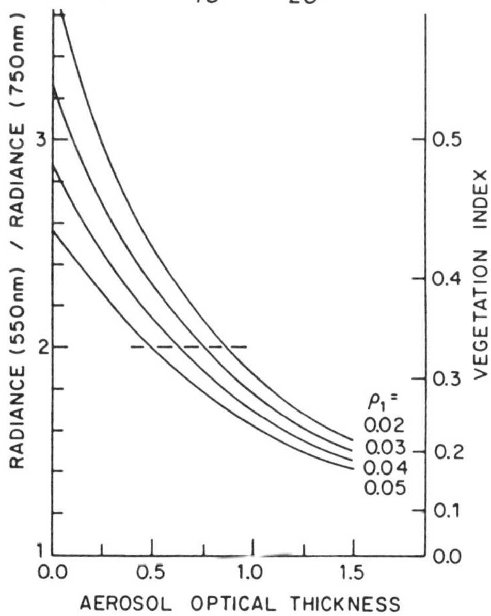


*FIGURE 6. Computations of the radiance ratio between the radiance at 750 nm and 550 nm and the corresponding vegetation index as a function of the aerosol optical thickness. Same optical thickness was used for both bands. The computations are shown for four values of the surface reflectance in 550 nm ($\rho_1$) and for $\rho_2$=0.2 for $\lambda$750 nm and solar zenith angle 34$^o$.*

In Fig. 6 the vegetation index is plotted as the ratio of radiances in band 3 to radiances in band 1 (also translated to NDVI - the difference between band 3 and band 1 divided by their sum) as a function of the optical thickness. The value of the surface reflectance in band 2 ($\rho_2$=0.2) is kept fixed and the value of the reflectance in band 1 is variable. Let us consider an image with a variable optical thickness ($\tau_a$) and a variable surface reflectance in band 1 (with a fixed reflectance in band 2). Specifying a given fraction $f_1$ for an image corresponds to choosing all pixels with a radiance ratio larger than a given threshold ($NDVI_t$). Let us take an example of a threshold of $NDVI_t$=0.33. This threshold may result from surface reflectance in band 1 of $\rho_1$=0.05 for a moderate amount of haze - $\tau_a$=0.5 or from $\rho_1$=0.02 for a heavy haze - $\tau_a$=0.9. Therefore, for this fraction $f_1$, the procedure will choose the appropriate pixels by minimizing the aerosol optical thickness and the surface reflectance simultaneously. As a result application of the procedure to the two subsets of the image, for the hazy day, resulted in areas where the optical thickness was not computed at all, due to the higher values of the optical thickness in these areas. This higher optical thickness reduced the value of the NDVI in these areas under the threshold value. In order to add the missing information, the values of the optical thickness were recomputed for these portions separately. The procedure can compute the values of the optical thickness with a spatial nonhomogeneity of $\Delta\tau_a$ = 0.2. Beyond this range there is a need to apply the procedure to each section of the image that has a semiuniform $\tau_a$. Computations in the different sections of the sub-areas resulted in an optical thickness that varies smoothly across the area (Fig. 5).

The values of the optical thickness are interpolated (and if necessary extrapolated) across the whole scene (step 8) to generate mapping of the aerosol loading. From the derived optical thickness in each band the slope $\nu$ is computed. Values of the average slope $\nu$ for each subarea and day are tabulated in Table 2b. For the northern subarea, the derived slope $\nu$ can be compared with the value of $\nu$ derived from optical thicknesses measured by sunphotometers from the ground. The values of $\nu$ derived from satellite measured radiances are larger by $\Delta\nu$=0.5 from values measured from the ground. This difference is due mainly to the higher optical thickness derived in band 1 from the satellite imagery. This difference is reasonable and allows the extrapolation of the optical thickness to the near IR (band 3 - $\tau_3$). A comparison between the extrapolated value of $\tau_3$ (Table 2b) and the value measured by the sunphotometer from the ground (Table 2a) shows excellent agreement.

The interpolation and the smoothing process did not change significantly the average optical thickness in the image ($\Delta\tau_a \leq 0.01$). Its standard deviation was decreased (by 10-30%), by eliminating the high spatial frequency variations in the data, which are considered as noise.

## 5. EVALUATION

In this section we shall discuss the applicability of the algorithm to different parts of the world and the sensitivity of the method to the accuracy of the satellite calibration and to the three major assumptions in the procedure: the fraction of dense dark vegetation in the image ($f_1$), the assumed value of the surface reflectance of the dense dark vegetation, and the aerosol characteristics.

The major uncertainty in the suggested algorithm results from the uncertainty in the assumed reflectance of the dense dark vegetation. Assuming an error of $\delta\rho=0.01$, the expected error in the optical thickness would be $\delta\tau_a=0.13$ for optical thickness in the range: $\tau_a=0.20\text{-}0.40$ and $\delta\tau_a=0.09$ for $\tau_a=0.8\text{-}1.0$ (see Table 2c).

Results of the sensitivity study of the derived optical thickness to calibration error are tabulated in Table 2d. Two kind of errors are tested: (A) - error in all the bands (degradation of 10%), and (B) - degradation of 10% in band 1 only. Degradation in all the bands (case A) decreased all the optical thicknesses by 29% for the clear day and by 15% for the hazy day, resulting in the same slope $\nu$. Calibration error in one band only (case B - error in band 1) resulted in a change in the slope $\nu$.

The sensitivity of the algorithm to the fraction $f_1$, is shown in Fig. 7. Five different values of the fraction $f_1$ were used. If the fraction $f_1$ is larger than the actual fraction of dense dark vegetation ($f_1$= 0.5-1.0), the resultant optical thickness is too high. For $0.1<f_1<0.5$ the sensitivity to $f_1$ is not substantial, resulting in $\delta\tau\leq0.04$. This sensitivity is not a general result and it depends on the actual fraction of dense dark vegetation in a given image. Therefore, if the fractional cover of dense vegetation is unknown, a small value of $f_1$ will be safe. One method to test whether the chosen value of $f_1$ is correct is to perform the computations for several values of $f_1$ and to choose a value in a range where the resultant optical thickness is not sensitive to the fraction $f_1$ (such as $f_1<0.5$ in Fig. 7). A very small value of $f_1$ ($f_1<0.05$) produced poor statistics and noisy data.

Uncertainty in the aerosol scattering phase function, due to uncertainty in its size distribution or composition, may result in large errors in the derived optical thickness. The sensitivity shown in Table 2e shows the dependence of the derived optical thickness to the uncertainty in the aerosol size distribution. Rather large uncertainty in the size distribution was used. In addition to the power law distribution with exponent of $\nu$=3.7, representative to this climate (Kaufman and Fraser, 1983), a lognormal distribution was used with mass mode radius $R_m$=0.2 and 0.60 μm (standard deviation of dlnR - $\sigma$=0.61). As a result the error in the

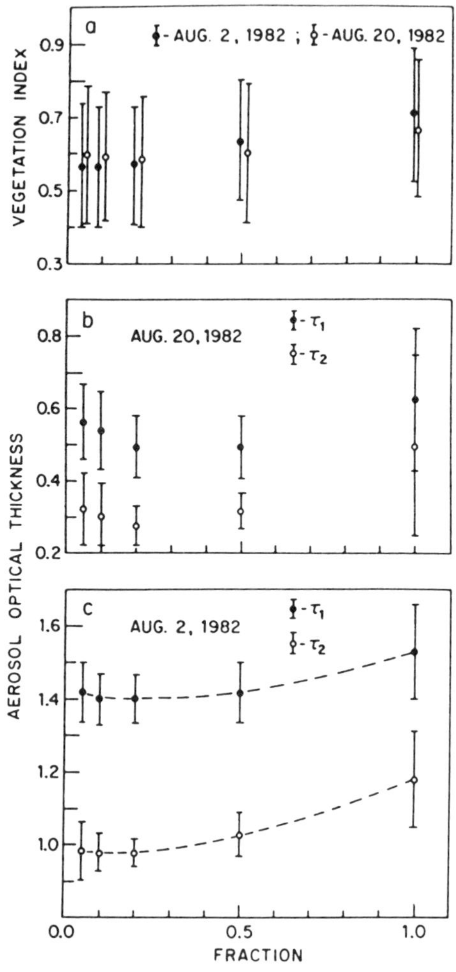


*FIGURE 7. The dependence of the aerosol optical thickness (in band 1 - $\tau_1$ and in band 2 - $\tau_2$) on the chosen fraction $f_1$. Results are given for the clear (20 August 1982) and the hazy (2 August 1982) days.*

derived optical thickness is large (e.g., for the clear day $\tau_1$ increases from $\tau_1$=0.50 for the power law to $\tau_1$=0.76 for $R_m$=0.20 μm and $\tau_1$=0.89 for $R_m$=0.60 μm). Therefore, application of the procedure requires a priori determination of the type of aerosol being sensed (dust, urban aerosol, etc.). Application of the method to several spectral bands results in the power exponent ν that can be used to check whether the right aerosol was chosen.

The sensitivity of the algorithm to the single scattering albedo $\omega_o$ is shown in Table 2f. A relatively large error in $\omega_o$ was simulated $\Delta\omega_o$=0.1. This difference may result from a change from a rural to an urban aerosol. The "true" value in the simulation was $\omega_o$=0.95 and the value used in the correction was $\omega_o$=0.85. This underestimation of $\omega_o$ causes an overestimation in $\tau_a$ by 50%.

Figure 8 shows a preliminary suggestion of the areas around the world where:

1) the algorithm is expected to be successful,
2) there is an uncertainty in the application, and
3) the successful application is very unlikely.

In the first type of area, forests with evergreen and deciduous trees as well as high shrubs are included. This area seems suited for the application of the procedure. The second type of area includes similar vegetation but growing in patches or singly, thus a full cover of a pixel is not certain. The applicability of the algorithm to this type of area depends also on the resolution. In the third type of area the trees or bushes are far apart or completely absent.

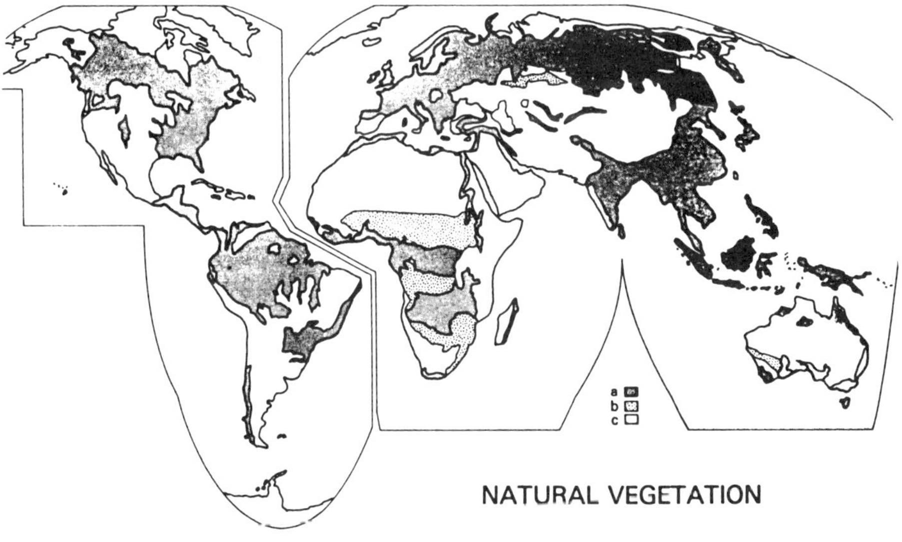


*FIGURE 8. Regions where: (a) the algorithm is expected to be successful (forests with evergreen and deciduous trees as well as high shrubs), (b) there is an uncertainty in the application (same vegetation as in (a) but growing in patches or singly).,(c) the successful application is very unlikely (areas where trees or high shrubs if found grow singly). The figure is based on a map of natural vegetation (Kühler, 1979).*

## 6. CONCLUSIONS

An algorithm is suggested for satellite mapping of aerosol loading over vegetated land. By utilizing the low reflectance of dense vegetation in several locations in the image (in several spectral bands in the visible part of the spectrum), the aerosol optical thickness can be derived from the same image. As a result an automatic processing of imagery can be performed to derive routinely the aerosol loading, for geographic locations indicated in Fig. 8. The algorithm has to assume a given aerosol model. It was applied to Landsat MSS imagery of the Washington D.C., and Chesapeake Bay area, with a reduced resolution (~700 m). The derived aerosol optical thickness ($\tau_a$) shows good agreement with simultaneous sunphotometer measurements from the ground within $\Delta\tau_a=\pm0.20$ in band 1 (0.5-0.6 μm) and $\Delta\tau_a=\pm0.05$ in band 2 (0.6- 0.7 μm). Introduction of future satellite systems with several narrow spectral bands may reduce the errors resulting from this assumption. Mapping of the aerosol loading may help in the understanding of atmospheric processes related to the aerosol and the effect of air pollution on the environment.

## ACKNOWLEDGMENTS

I would like to thank Jim Irons from NASA/GSFC and Ruth Kenard for their help in obtaining the Landsat data used in this work. It is my pleasure to thank Robert Fraser from NASA/GSFC and Yuri Mekler from Tel-Aviv University for helpful discussions regarding this work.

## REFERENCES

Carlson T.N., 1979: Atmospheric Turbidities in Saharan Dust Outbreaks as Determined by Analysis of Satellite Brightness Data, Mon. Wea. Rev., *107*, 322-335.

Dave, J. V., and J. Gazdag, 1970: A Modified Fourier Transform Method for Multiple Scattering Calculations in a Plane Parallel Mie Atmosphere, Appl. Opt., 9, 1457-1466.

Fraser, R.S., Y.J. Kaufman and R.L. Mahoney, 1984: Satellite Measurements of Aerosol Mass and Transport, Atmos. Environ., 18, 2577-2584.

Griggs, M., 1975: Measurements of Atmospheric Aerosol Optical Thickness Over Water Using ERTS-1 Data, J. Air Pollut. Control Ass., 25, 622- 626.

**Kaufman, Y.J., 1986: Atmospheric Effect on Spectral Signature – Measurements, COSPAR XXVI plenary meeting, Toulouse, France, and accepted for publication in IEEE Trans. Geosci. and Rem. Sens.**

Kaufman, Y.J., 1987: Satellite Sensing of Aerosol Absorption, J. Geoph. Res., 92, 4307-4317.

Kaufman, Y.J., and R.S. Fraser, 1983: Light Extinction by Aerosols During Summer Air Pollution, J. Appl. Meteor., 22, 1694-1706.

Kaufman Y.J., R.S. Fraser and R.A. Ferrare, 1985a: Remote Sensing of Forest Fire Smoke Characteristics, IAMAP/IAPSO Assembly, Honolulu, Hawaii.

Kaufman Y.J., R.S. Fraser and R.A. Ferrare, 1985b: Derivation of Aerosol Scattering Phase Function and Absorption From Almucanter Measurements, Joint

IAMAP/IAPSO Assembly, Honolulu, Hawaii.

Kimes, D.S., W.W. Newcomb, R.F. Nelson and J.B. Schutt, 1986: Directional Reflectance Distributions of a Hardwood and Pine Forest Canopy, IEEE Trans. on Geosci. and Rem. Sens., GE-24, 281-297.

King M.D., D.M. Byrne, B.M. Herman and J.A. Reagan, 1978: Aerosol Size Distribution Obtained by Inversion of Optical Depth Measurements, J. Atmos. Sci., 35, 2153-2167.

Koepke, P. and H. Quenzel, 1979: Turbidity of the Atmosphere Determined From Satellite Calculation of Optimum Viewing Geometry, J. Geoph. Res., 84, 7847-7855.

Kriebel, K.T., 1977: Reflection Properties of Vegetated Surfaces: Tables of Measured Spectral Biconical Reflectance Factors, Munchener Universitats-Schriften, Meteorologisches Ins., Wissenschaftl. Mitteilung #29.

Kühler, A.W., 1979: Map of Natural Vegetation. In E.B. Espenshade and J.L. Morrison (Eds.), World Book Atlas, World Book Encyc. Inc., Chicago.

Lee, T., and Y.J. Kaufman, 1986: The Effect of Surface Nonlambertianity on Remote Sensing, IEEE Trans. Geosci. and Rem. Sens. , GE-24 , 699–708.

Mekler Y., H. Quenzel, G. Ohring, and I. Marcus, 1977: Relative Atmospheric Aerosol Content from ERTS Observations, J. Geoph. Res.,82, 967-972.

Peterson, J. T., E. C. Flowers, G. J. Berri, C. L. Reynolds, and J. H. Rudisil, 1981: Atmospheric Turbidity Over Central North Carolina, J. Appl. Meteor. ,20, 229-241.

# DEVELOPMENT AND APPLICATION OF AEROSOL REMOTE SENSING WITH AVHRR DATA FROM THE NOAA SATELLITES

*C.R. Nagaraja Rao, L.L. Stowe and E.P. McClain*
National Oceanic and Atmospheric Administration
Washington, DC 20233, USA

and

*J. Sapper*
Science Systems and Applications, Inc.
Seabrook, Maryland 20706, USA

ABSTRACT

We have used the visible radiance measurements in channel 1 (0.58-0.68 μm) of the Advanced Very High Resolution Radiometer (AVHRR) onboard the NOAA polar-orbiting environmental satellites for the remote sensing of aerosols over the oceans. The optical thickness of the aerosol particles in the total atmospheric column--the aerosol content--is derived by matching the measured radiances with computations in a realistic model of a cloud-free, plane-parallel turbid atmosphere. Quality control measures to detect the presence of clouds within the field of view of the radiometer and to reject cloud-contaminated data from global analyses have been developed and implemented in the aerosol retrieval algorithm. These tests are based on the sensitivity of channel 3 (3.55-3.93 μm) to reflected solar radiation. Objective analysis of the aerosol content thus determined has been employed to generate a 100-km-grid-analyzed aerosol field over the global oceans on a weekly basis. The utility of these aerosol fields to the improvement of atmospheric corrections to the sea surface temperature retrieval algorithms and to climate modeling studies will be mentioned briefly.

## 1. INTRODUCTION

Atmospheric aerosols, along with clouds, are the major modifiers of the shortwave or solar (~0.3-5 μm) radiation regime at the Earth's surface and within the atmosphere and thus play a very important role in atmospheric energy balance (Spankuch, 1983). It has been observed, for example, that the insertion of the effects of tropospheric aerosols on solar radiation into climate models would cause the models to point towards desertification over parts of Africa (e.g., Coakley and Cess, 1985). Depending upon their composition and location within the atmosphere and upon the surface albedo, atmospheric aerosols may either augment or partly offset the greenhouse effect (e.g., Rasool and Schneider, 1971). Although such climatic effects of aerosols have been recognized for a long time, their incorporation in a physically meaningful manner in climate and weather studies on regional and global scales has been hampered by


A. DEEPAK Publishing
ISBN 0-937194-11-5

the sparsity of information. There is a noticeable lack of data, especially over the oceans, on the total columnar abundance and vertical distribution of aerosols in the atmosphere; on their physical and optical properties such as size distribution, and the real and imaginary parts of the refractive index; and on their spatial and temporal variations (Toon and Pollack, 1980). In this context, it is apparent that an earth-orbiting meteorological satellite, with its potential for global coverage, offers itself as an eminently suitable platform for monitoring atmospheric aerosols. It is against this background that we wish to present and discuss here the development and application of remote sensing of aerosols with the Advanced Very High Resolution Radiometer (AVHRR) data from the NOAA polar-orbiting environmental satellites.

## 2. THE NOAA AEROSOL REMOTE SENSING ACTIVITY

### 2.1 BACKGROUND

It has been demonstrated by Rao et al. (1973), among others, that effective aerosol and surface properties can be inferred from an interpretation of measurements of the diffusely reflected radiation at the top of the atmosphere. More recently, Norton et al. (1980), Griggs (1983), Griggs and Stowe (1984), Quenzel and Koepke (1984), Lee et al. (1985), and Durkee (1985) have investigated the feasibility of retrieving the normal aerosol attenuation optical thickness and, subsequently, some of the aerosol properties mentioned earlier, from an analysis of the visible and near-infrared radiances measured with satellite-borne radiometers. It is thus natural that, as part of the NOAA global atmospheric remote sounding and sea surface temperature programs, an experimental aerosol retrieval algorithm has been developed at the National Environmental Satellite, Data, and Information Service (NESDIS). This algorithm presently utilizes the radiances measured in channel 1 (0.58-0.68 $\mu$m) of the AVHRR and has the potential to utilize radiances measured in the present and proposed near-infrared channels. We shall describe in what follows the basic features of this algorithm, with some emphasis on cloud-screening quality control measures that have been implemented, and illustrate the use of the algorithm with examples of detection and mapping of recent anomalous aerosol episodes over the oceans.

### 2.2 THE NOAA AEROSOL RETRIEVAL ALGORITHM

The physical basis for the NOAA aerosol retrieval algorithm is the linear dependence of the diffusely reflected radiation at the top of the atmosphere on the aerosol attenuation optical thickness over surfaces of low albedo, such as the oceans. Over such surfaces practically all of the diffusely reflected radiation originates from within the atmosphere and is thus very sensitive to changes in atmospheric turbidity. This dependence has been observed both in the analysis of satellite radiance measurements (e.g., Griggs, 1983) and in the computations of radiative transfer in realistic models of the Earth's atmosphere (e.g., Masuda et al., 1988). Thus, the algorithm is based on establishing correspondence between the measured radiance in channel 1 and computed radiances available in the form of a look-up table to determine the corresponding aerosol attenuation optical thickness in the model atmosphere for given values of the solar zenith, satellite zenith, and azimuth angles of observation. The associated aerosol properties such as size distribution and refractive index are specified by the atmospheric model.

The look-up table consists of radiance computations at the wavelength of 0.65 μm in a cloud-free, plane-parallel, vertically inhomogeneous turbid atmosphere for aerosol attenuation optical thicknesses (at 0.65 μm) ranging from 0 to 1.18. We have assumed that the ocean can be simulated by a Lambertian surface of albedo 0.015. The aerosol particles are considered to be spherical Mie scatters with a modified power law size distribution (e.g., McClatchey et al., 1970) of the form

$$n(r) = C \qquad \text{for } r_{min} \leq r \leq r_m$$

$$n(r) = C(r/r_m)^{-4.5} \qquad \text{for } r_m \leq r \leq r_{max} \tag{1}$$

$$n(r) = 0 \qquad \text{for } r < 0.02\ \mu m \text{ and } r > 10\ \mu m$$

where n(r) is the number of aerosols of radius r per unit volume per unit increment in radius r. We have used values of 0.02 μm, 0.1 μm and 10 μm for $r_{min}$, $r_m$, and $r_{max}$ respectively. We have taken the refractive index of the aerosols to be 1.5. This choice of model parameters was governed by earlier work directed towards the interpretation of radiances measured over coastal and island sites in channel 5 (~0.60-0.70 μm) of the Multispectral Scanner (MSS) onboard LANDSAT. (Personal communication, December 1986, M. Griggs, Science Applications International Corporation, San Diego, California.) Model error analysis has shown that the relative error in the aerosol optical thicknesses thus determined is proportional to and is of the same order of magnitude as the relative error in the measured radiances.

As the look-up table is based on computations in a plane-parallel atmosphere, we have excluded from our analyses measurements made at solar zenith angles larger than 70 deg to avoid the effects of the sphericity of the atmosphere. To avoid the adverse effects of sun glitter, we confine our attention to measurements made in the antisolar half of each cross-track scan. Thus, we composite measurements made over the 8-day repeat cycle of the satellite global coverage.

Implementation of the algorithm requires that only radiances measured under cloud-free conditions be used. In addition to cloud screening criteria, which are based on visible and near-infrared reflectance thresholds and uniformity of radiances and brightness temperatures over adjacent pixels derived from the operational

Multi-Channel Sea Surface Temperature (MCSST) procedure (McClain et al. 1985), cloud screening tests based on the effects of the cloud-enhanced reflectance in channel 3 have been developed. An analysis of representative global brightness and MCSST-derived sea surface temperature has revealed that, when clouds are present within the field of view of the radiometer, the difference between the triple- and split-window sea surface temperatures is in excess of 5 deg C. However, this difference is observed to be well within 5 deg C over areas of high turbidity. We have, therefore, used a threshold of 5 deg C for this difference to distinguish between enhanced reflectances from high turbidity and from clouds. The resulting aerosol analysis indicates that this threshold may vary with the composition of the aerosols. To the extent practicable, we have used high resolution (~1 km) local area coverage (LAC) AVHRR imagery and available meteorological data to ascertain the presence or absence of clouds within the field of view of the radiometer in the development of this cloud screening test.

It has been the usual practice to express the state of atmospheric turbidity in terms of the aerosol attenuation optical thickness at the wavelength of 0.5 $\mu$m. Therefore, we infer this model-prescribed value from the aerosol attenuation optical thickness at the wavelength of 0.65 $\mu$m deduced from satellite radiance measurements. The same value is used in the generation of the weekly, composite 100-km aerosol fields, using computer-based objective analysis techniques.

## 3. RESULTS

We shall now illustrate the application of the NOAA aerosol retrieval algorithm with examples of detection and mapping of aerosol episodes over the Atlantic and Pacific Oceans and the Arabian Sea. We have shown in Fig. 1 a section of the analyzed aerosol optical thickness field over the Pacific Ocean for the period 21-28 April 1987; the contour labels are 100 x optical thickness at 0.50 $\mu$m. The most noticeable feature is an area of high turbidity to the west of the Isthmus of Panama attributed to smoke resulting from slash burning in the Central American States before spring planting. Information on winds and on surface observations of dust, haze, and smoke, corroborates the existence of the smoke cloud during this period. The smoke cloud is also clearly seen in the GOES-EAST visible image obtained on 26 April 1987 (Fig. 2). Aerosol optical thickness values as high as 0.8 were obtained with the algorithm in the middle of the smoke cloud.

We have shown in Figs. 3 and 4 portions of the analyzed aerosol field over the tropical Atlantic and the Arabian Sea, respectively, for the period 18-25 June 1987. The presence of the Saharan dust cloud is very obvious. The latitudinal extent of the cloud and the general morphology are in agreement with earlier observations (e.g., Norton et al., 1980). Also, the flow of monsoon-borne dust off Arabia and Somalia towards India is clearly seen. An analysis of prevalent meteorological situations supports such monsoon-borne dust

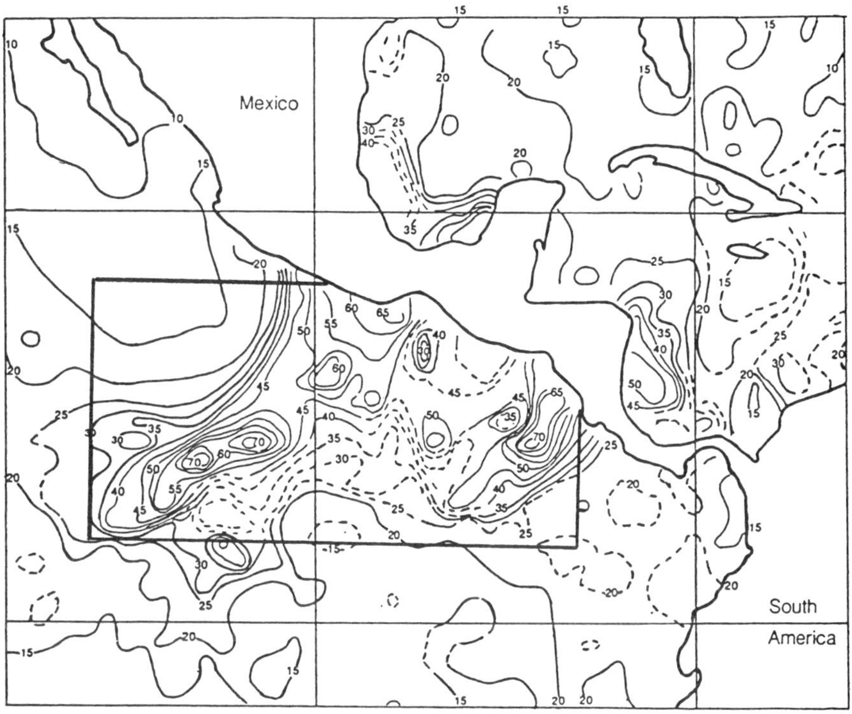


FIGURE 1. Analyzed aerosol field for the period 21-28 April 1987. The contour labels are 100 x aerosol optical thickness at 0.50 μm. The smoke cloud (enclosed area) resulting from slash burning in the Central American States is clearly seen to the west of the Isthmus of Panama.

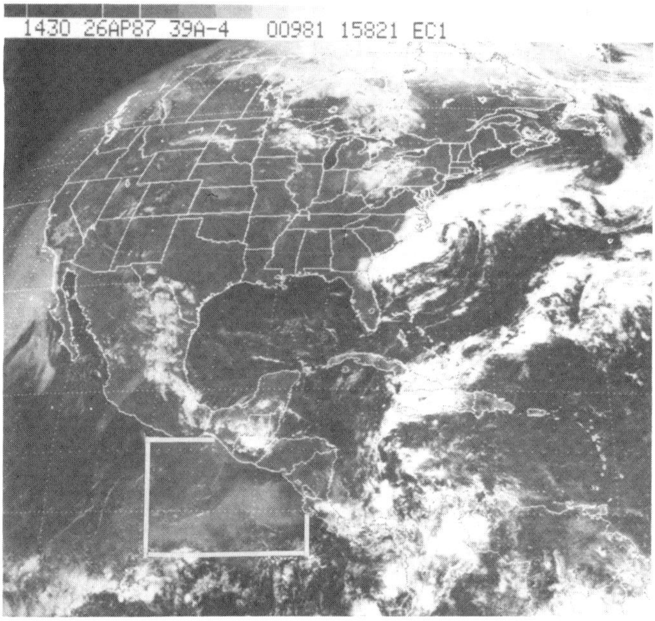


FIGURE 2. GOES-EAST image, obtained on 26 April 1987, showing the smoke cloud (enclosed area) to the west of the Isthmus of Panama.

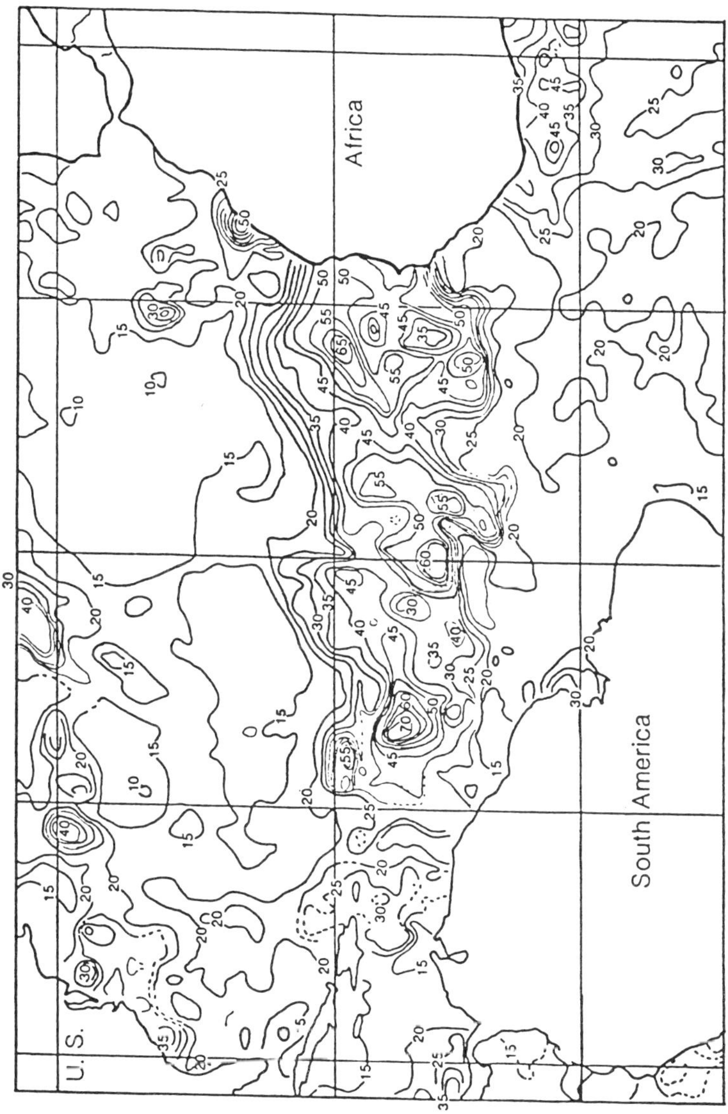


FIGURE 3. Analyzed aerosol field for the period 18-25 June 1987. The Saharan dust cloud is seen over the Atlantic.

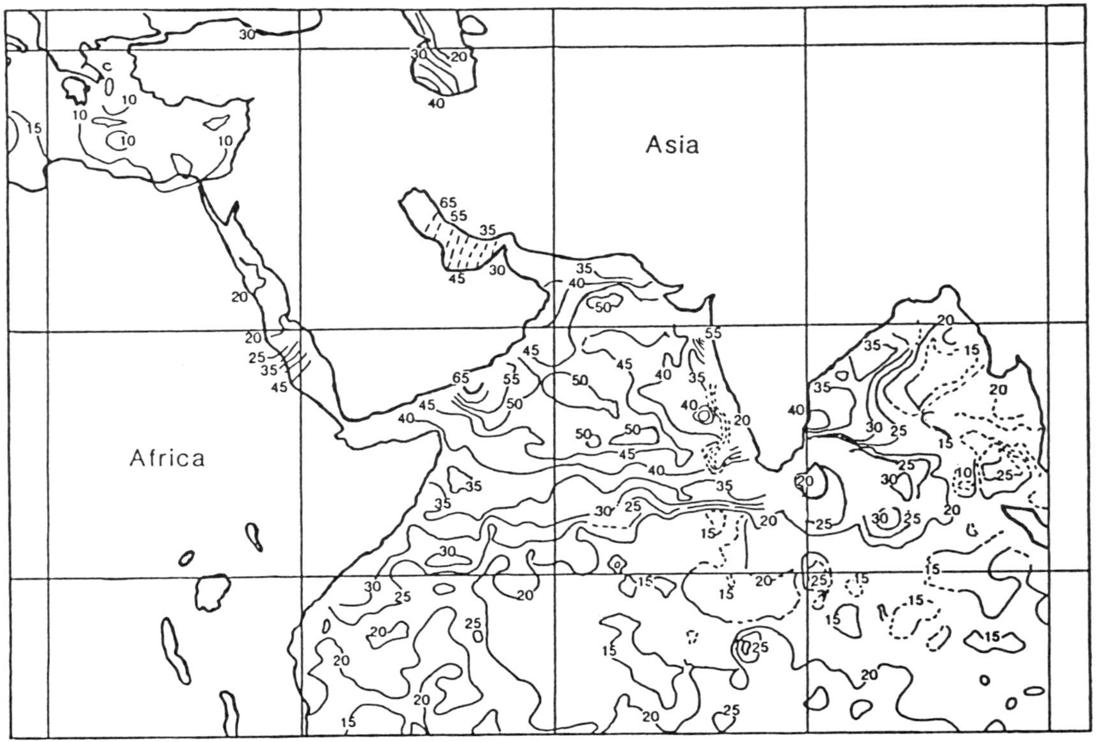


FIGURE 4. Analyzed aerosol field for the period 18-25 June 1987. The monsoon-borne dust cloud is seen over the Arabian Sea.

flows; good agreement has been observed between satellite-derived dust front propagation and modeled winds over the Arabian Sea (Chen, 1986).

## 4. CONCLUDING REMARKS

We have demonstrated the basic features and application of a viable aerosol retrieval and field generation algorithm based on radiances measured in AVHRR channel 1. It is felt that multichannel algorithms utilizing radiances measured in channels 1, 2 (0.84-0.89 μm), and 3A (1.58-1.64 μm), proposed for the NOAA K, L, and M satellites, may enable us to retrieve more detailed aerosol information with minimal ambiguity.

We believe that the information on aerosol properties inferred from remotely sensed radiance data can be useful, in the immediate context, in developing corrections for aerosol thermal effects on the present operational sea surface temperature retrieval procedure. Such thermal effects have been documented by Lee et al. (1985) with reference to the Saharan dust, which, because of its absorption properties in the visible and infrared regions of the spectrum and of its location within the atmosphere, has been observed to cause brightness temperature depressions of as much as 6 deg C in the window channels. These thermal effects were also observed by Griggs (1985), who has studied the anomalies caused by the El Chichon volcanic cloud in the retrieval of sea surface temperature using the MCSST algorithm. We are presently investigating the radiative effects of tropospheric aerosols in the thermal infrared, for representative air mass conditions over the oceans, with the objective of establishing effective correction procedures to the operational MCSST algorithm. It is also our belief that the aerosol climatology that can be constructed from a long-term series of such satellite-based global aerosol charts will find application in climate models and, subsequently, in the perceived and proposed activities associated with global change studies. However, in order to establish the viability of the aerosol retrieval algorithm on a long-term basis so that the derived aerosol information can be used with confidence, it is essential that such information be continually validated and associated aerosol properties made at strategically located island and coastal sites. The complexities of this long-term algorithm validation experiment are such that it will be possible only as a collaborative effort of the international scientific community.

## ACKNOWLEDGMENTS

The authors are thankful to Messrs. William Pichel, Charles Walton and Robert Warner, NOAA/NESDIS, for their contributions to the aerosol field generation program. Special appreciation is expressed to Messrs. Paul Pellegrino and Robert Ryan, NOAA/NESDIS for their help in data analyses. This work was completed while one of the authors (C.R.N.R.) held a NOAA/NRC resident research associateship at the National Environmental Satellite, Data, and Information Service, Washington, D.C.

## REFERENCES

Chen, F.H., 1986: Analysis of a Major Dust Outbreak Over the Arabian Sea During MONEX., M.S. Thesis, Department of Meteorology, University of Wisconsin, Madison, 94 pp.

Coakley, J.A., and R.D. Cess, 1985: Response of the NCAR Community Climate Model to the Radiative Forcing of Naturally Occurring Tropospheric Aerosol, J. Atmos. Sci., 42, 1677-1692.

Durkee, P.A., 1985: Aerosol Characterization With Dual Wavelength Radiance Measurements. In Preprints: Second Conference on Satellite Meteorology, Remote Sensing and Applications, American Meteorological Society, Boston, 298-302.

Griggs, M., 1985: A Method to Correct Satellite Measurements of Sea Surface Temperature for the Effects of Aerosols, J. Geophys. Res., 90, No. D7, 12951-12959.

Griggs, M., and L.L. Stowe, 1984: Measurements of Aerosol Optical Parameters From Satellites. In G. Fiocco (Ed.), IRS' 84: Current Problems in Atmospheric Radiation, A. Deepak Publishing, Hampton, Virginia, 42-45.

Griggs, M., 1983: Satellite Measurements of Tropospheric Aerosols, Adv. Spa. Res., 2, 109-118.

Lee, T.Y., R.S. Fraser and Y. Kaufman, 1985: Satellite Measurements of Physical Properties of Saharan dust. In Preprints: Second Conference on Satellite Meteorology, Remote Sensing and Applications, American Meteorological Society, Boston, 294-297.

Masuda, K., T. Takashima and C.R.N. Rao, 1988: Remote Sensing of Atmospheric Aerosols Over the Oceans Using Multispectral Radiances Measured With the Advanced Very High Resolution Radiometer (AVHRR) Onboard the NOAA Meteorological Satellites. In P.V. Hobbs and M.P. McCormick (Eds.), Aerosols and Climate, A. Deepak Publishing, Hampton, Virginia.

McClain, E.P., W.G. Pichel and C.G.Walton, 1985: Comparative Performance of AVHRR-Based Multichannel Sea Surface Temperatures, J. Geophys. Res., 90, No. C6, 11587-11601.

McClatchey, R.A., R.W. Fenn, J.E.A. Selby, J.S. Garing and F.E. Volz, 1970: Optical Properties of the Atmosphere, Environmental Research Papers, No. 331 (AFCRL-70-0527), Air Force Cambridge Research Laboratories, L.G. Hanscom Field, Bedford, Massachusetts, 98 pp.

Norton, C.C., F.R. Mosher, B. Hinton, D.W. Martin, D. Santek and W. Kuhlow, 1980: A Model for Calculating Desert Aerosol Turbidity Over the Oceans From Geostationary Satellite Data, J. Appl. Meteor., 19, 633-644.

Quenzel, H and P. Koepke, 1984: Tropospheric Aerosol Optical Depth Inverted From Upwelling Radiances. In H.E. Gerber and A. Deepak (Eds.), Aerosols and Their Climatic Effects, A. Deepak Publishing, Hampton, Virginia, 227-240.

Rao, C.R.N., T. Takashima and R.B. Toolin, 1973: Measurements and Interpretation of the Polarization of Radiation Emerging From the Atmosphere at an Altitude of 28 km Over Southwestern New Mexico (USA), Quart. J. Roy. Meteor. Soc., 99, 294-302.

Rasool, S.I., and S.H. Schneider, 1971: Atmospheric Carbon Dioxide and Aerosols, Science, 173, 138-141.

Spankuch, D., 1983: Aerosols Within the World Climate Research Program, Adv. Spa. Res., 2, 1-3.

Toon, O.B., and J.B. Pollack, 1980: Atmospheric Aerosols and Climate, American Scientist, 268-278.

# A SPACEBORNE BACKSCATTER LIDAR PROPOSED FOR A GERMAN SPACE-SHUTTLE MISSION

*Heinrich Quenzel, Matthias Wiegner and Martina Kästner*
Meteorologisches Institut der Universität München
D-8000 München 2, Federal Republic of Germany

*Wolfgang Renger*
Institut für Physik der Atmosphäre, DFVLR
D-8031 Oberpfaffenhofen, Federal Republic of Germany

and

*Martin Endemann*
Battelle-Institut e.V.
D-6000 Frankfurt 90, Federal Republic of Germany

ABSTRACT

An atmospheric lidar experiment in space (ALEXIS) is proposed to be launched with a Space Shuttle in the early 1990s. The purpose of this experiment is to explore the measurement capabilities of a spaceborne lidar and to assess the performance potential of future operational instruments. A first flight opportunity may arise with the German D2-mission.

## 1. INTRODUCTION

Remote sensing of the Earth's atmosphere by passive radiometry is now widely applied: It allows the monitoring of temperature distribution, energy budget components, cloud cover, etc., with almost global coverage and with high temporal resolution if polar-orbiting and geostationary instruments are combined (e.g., Wiegner and Raschke, 1987). The main shortcoming of these methods is the rather poor vertical resolution. For this purpose, active methods (lidar) have been developed for atmospheric studies, at first with ground-based instruments (e.g., Collis, 1969), and for more than a decade with airborne systems (Grams et al., 1974; Melfi et al., 1974).

Encouraged by our experience with airborne lidar systems (Mörl et al., 1981) and the results of theoretical studies (Quenzel et al., 1984), we have proposed a backscatter lidar experiment ALEXIS for the second German Spacelab mission (D2-mission). ALEXIS should be understood as an intermediate step between an airborne lidar and an operational spaceborne lidar on future Earth observation satellites. This step is necessary to gain more insight into the technological design and to develop and test inversion algorithms for the evaluation of the lidar signals. Hence, ALEXIS is not only a demonstration of technological potentials, but also an opportunity to acquire unique measurements of atmospheric backscatter from space. Its data should be used to:

1) evaluate applications of spaceborne lidar measurements,
2) demonstrate the expected benefits of active remote sensing techniques alone and in combination with passive sensors,
3) validate lidar data by independent measurements, and
4) verify cloud heights, gained from passive techniques.




ISBN 0-937194-11-5

The design goal is to build an instrument that can be added to the Spacelab facility. A decision was made to use readily available components wherever possible, and to use available space-qualified pressurized containers for the different critical subcomponents (e.g., laser, power supplies) to minimize the required space-qualification testing.

In the subsequent sections the scientific objectives of the mission are outlined, the inversion of lidar measurements, the verification experiment, and the instrument design.

## 2. SCIENTIFIC OBJECTIVES AND MEASUREMENT PROGRAM

ALEXIS will allow a number of different measurements to obtain performance data for the design of future operational spaceborne lidar instruments. The data sets can be used to determine the following parameters with very high vertical resolution:

1) cloud top height,
2) vertical structure of thin and medium clouds (e.g., cirrus clouds),
3) height of the planetary boundary layer (PBL),
4) spatial distribution of tropospheric aerosols,
5) spatial distribution of stratospheric aerosols, and
6) tropopause height.

The determination of these parameters with an accuracy of up to 100 m in a vertical direction will constitute important progress for many different meteorological applications, for weather forecasts as well as for research activities. It will help to improve, in combination with passive radiometric measurements and the already existing retrieval algorithms, our knowledge about temperature profile (surface temperature), humidity profile, wind field, cloud statistics, radiation budget, and physics of the PBL. Hence, the expected scientific return of a spaceborne lidar is very promising.

The measurement program is designed to cover these different meteorological aspects of lidar applications as far as possible during an experiment of less than 2 hours' duration (the main purpose of the 8-day Shuttle flight is zero-g experiments), to demonstrate that the measurements can be performed by means of the proposed instrument concept, and to fulfill the scientific tasks. Consequently, the plan is to run the experiment in two distinct phases: the first concentrates on investigations concerning aerosol particles, the second on cloud parameter studies.

The measurement of aerosol parameters, in particular the vertical structure of aerosol extinction, requires complex numerical inversion algorithms. Hence ALEXIS requires a ground-truth experiment to validate the inverted $\sigma$-profiles ($\sigma$: extinction coefficient). It must include independent (ground-based and airborne) measurements of those meteorological quantities that are derived by the space lidar. Both experiments should measure as far as possible simultaneously the same atmospheric volume. For these and economical reasons a fixed location for the ground-truth experiment should be selected, where high aerosol amounts and cloudfree conditions coincide.

Spatial homogeneity of the aerosol distribution is required for technical reasons. It must be regarded that the signal-to-noise ratio of ALEXIS and its pulse repetition rate are very important parameters that determine whether the expected scientific results can be acquired or not. The signal-to-noise ratio determines the number of laser shots that have to be integrated to receive a significant signal. This number corresponds to a certain averaging area on the Earth's surface according to the velocity of the spacecraft and the pulse repetition frequency. The size of the area is expected to be in the order of 10 - 25 km, and, hence, for this spatial extent, the homogeneity of the atmosphere must be proved.

At the present status of planning, this site will be in the Sahara, probably in the central desert (Agadez) or near the western coast (Dakar). In view of the orbital characteristics of the flight, the lidar crosses the ground-truth area within a few minutes (one minute corresponds to a subsatellite track of 420 km). For an area at 15°N, 21 crossings of the satellite above the test site occur, which seems to be sufficient to select 4 or 5 periods under favourable weather conditions to make simultaneous measurements from the space lidar, airborne, and ground-based instruments. These measurements will be approximately at 1050 and 2140 local time for the ascending and descending nodes, respectively, if the Shuttle is launched at 1730 GMT at Cape Canaveral. The ground truth experiment is briefly discussed in Section 4.

The second objective of ALEXIS is dedicated to measurements of cloud parameters. The main goals are to study cloud topography (cloud statistics) and to verify cloud top heights derived from passive sensors. Backscatter data are collected along a flight track from northern latitudes to the equator, to cover a wide range of climatic conditions. The orbit for these measurements will be selected according to the availability of simultaneous data from passive (multispectral) radiometers. For cloud studies one or two continuous observation periods of about 20 minutes each (corresponding to 8400 km) are envisaged.

To fulfill the scientific objectives, theoretical and numerical studies are required in addition to the technical realization. First, the inversion algorithms (including multiple scattering and particle shape effects) to derive $\sigma$ as a function of height must be improved and tested (e.g., Spinhirne, 1982); the present state of our theoretical studies is discussed in the next section. Second, sampling studies are necessary to assess the accuracy of large-scale averages determined from lidar measurements, which represent (only) areas of about 150 m in diameter and 700 m distance between them. They are important for comparisons between lidar and passive measurements.

## 3. INVERSION OF LIDAR MEASUREMENTS

The determination of the altitude of sharp steps in particle concentration (e.g., cloud tops, strong aerosol layers) by means of a lidar is rather simple. But, if vertical profiles of the extinction coefficient are desired--as in case of ALEXIS--more complex inversion algorithms are necessary. The most common algorithms (Klett, 1985; cf. Kästner, 1986) and a new iterative scheme were compared to assess the advantages and restraints of the different approaches. These studies utilized simulated data and the "single scattering approximation".

Problems in the inversion arise because there are more unknowns than equations. Therefore, additional information is required and the accuracy of the inverted extinction profiles depends on the accuracy of this information. In case of Klett's solution, a boundary value of the aerosol extinction coefficient is required; in general at the top of the atmosphere in case of clear conditions ("near-end solution") or at the bottom of the atmosphere in case of turbid conditions ("far-end solution"). In contrast, the iterative scheme requires calibrated lidar data.

The sensitivity of the retrieved $\sigma(z)$-profiles to errors in the above mentioned parameters (boundary value or calibration constant) has been investigated for various atmospheric situations. The main conclusions are:

1) For clear atmospheres (optical depth less than about 0.35 for $\lambda$ = 550 nm), the near-end solution requires an accuracy of 20% in the boundary value to get reasonable results. In the lower atmosphere the retrieved $\sigma(z)$ profile is better than in the stratosphere (10% compared to 50%).
2) For turbid atmospheres the far-end solution is appropriate and the uncertainty of the boundary value (near the surface) should not exceed 5%. This seems difficult to fulfill for a spaceborne lidar.
3) The iterative solution can be used for both clear and turbid conditions. The calibration constant should be known better than $\pm$10%. This problem requires further consideration.
4) Errors in the lidar-ratio of the atmospheric aerosols (which in general are unknown without additional information) cause fewer problems, if the iterative scheme is applied. In particular this improvement appears for turbid conditions.

So, the iterative scheme seems to be quite promising, if calibrated signals are available. Furthermore, it is possible to extend this algorithm to multiple scattering effects.

## 4. VERIFICATION OF LIDAR MEASUREMENTS

We want to emphasize that the goals of ALEXIS are twofold: On the one hand the Shuttle lidar experiment is a technological demonstration of the feasibility of active remote sensing from space; on the other hand the scientific benefits to improve our understanding of the atmosphere should be investigated. For the latter reason an international field experiment is proposed to assess the scientific progress of the mission.

The first priority is to verify the space lidar measurements related to aerosol parameters by (direct) in-situ measurements. The inverse idea, the verification of (passive) remotely sensed data (e.g., cloud top height determination from multispectral radiance measurements) by means of the ALEXIS-observations should be addressed by additional studies. With regard to the latter goal, combinations of the already existing passive spaceborne systems, the Space lidar, and the field experiment offer the opportunity of many intercomparisons between different methods; so these "by-products" should not be overlooked during the campaign.

The technical characteristics (signal-to-noise ratio, pulse repetition frequency and laser intensity) will probably not cause problems because of the rather homogeneous structure of the aerosol layers in the Sahara. Nevertheless, one of the tasks of the ground-truth experiment is to check this assumption by monitoring the aerosol distribution as a function of time and space.

To verify a $\sigma(z)$ profile derived from the Shuttle lidar, different approaches are possible.

The first method utilizes in-situ measurements of the microphysical characteristics of the aerosol layer. Either, the quantity, which has to be validated (extinction coefficient as a function of height), is measured with nephelometers and aethalometers (Hansen et al., 1982), and the parameter, which is relevant for the inversion of the lidar signal, (lidar ratio $\sigma/\beta$; $\beta$: backscatter coefficient) is derived from them, or, measurements of the most fundamental aerosol parameters--size and index of refraction--are combined with model calculations to determine $\beta$ and $\sigma$. Particle spectrometers onboard aircrafts are employed to measure the size distribution of the aerosol particles at different heights. The index of refraction is determined from the particles that have been collected. With these input data the optical parameters $\beta$ and $\sigma$ can be calculated very precisely and hence their $\sigma/\beta$ ratio. Also, the independent determination of the vertical profile $\sigma(z)$ for validation of the profile, resulting from the ALEXIS measurements, is accomplished.

A second approach to check the $\sigma(z)$ profiles derived from the spaceborne lidar is the comparison with other active systems. It is intended to have lidars onboard aircrafts as well as at various locations in the ground-truth area. These lidars provide a better signal-to-noise ratio than the space lidar because of the smaller distance from the scattering atmospheric constituents. The $\sigma/\beta$ ratios (which are needed for the inversion) are taken from the measurements described above. Furthermore it is possible to monitor the changes in the vertical and horizontal distribution of the aerosol layers and hence their homogeneity. So it might be possible to decide during the campaign, whether an orbit is adequate for the proposed measurements or not.

A third approach takes the optical depth for validation. The optical depth is derived from integration of the $\sigma(z)$ profile, so this approach is rather a consistency check than a strict verification. The optical depth can be measured by ground-based instruments such as sunphotometers (d'Almeida, 1985) or the aureolephotometer, developed by the Meteorological Institute of Munich. Furthermore, radiance data from passive satellite sensors can be used to complete these comparisons.

The estimation of the optical depth has been studied theoretically (e.g., Koepke and Quenzel, 1981) as well as using Landsat data (e.g., Fraser, 1976) or Advanced Very High Resolution Radiometer data (e.g., Griggs, 1983). Corresponding algorithms for ATSR (Along Track Scanning Radiometer) will be available at the time of the ground-truth experiment.

Validation of the retrieved aerosol data can be accomplished by various independent methods. In view of the great interest of the scientific community in aerosol parameters, participation of international groups in this experiment can be anticipated, as well as the coverage of all different approaches, and hence, the validation of the lidar measurements can be achieved with an high degree of accuracy.

## 5. INSTRUMENT DESIGN

In this section we will give a brief description of some technical details of the instrument.

Figure 1 shows the Spacelab lidar as it is located in the Space Shuttle. It comprises two Getaway-Special (GAS-) containers that are mounted to the Unique Support Structure (USS) of the Spacelab facility. The GAS-containers have been selected as the primary structure for the lidar since they provide an appropriate environment for housing nonspace-qualified parts in orbit. Thus the time- and cost-consuming qualification procedures for each lidar component can be greatly reduced.

The first GAS-container holds the complete transmitter assembly with laser and beam expanding telescope, as well as the power supply. Also included is a passive infrared radiometer. The transmitter laser is a flashlamp-pumped Nd:YAG laser with frequency doubler. The transmitted energy is 300 mJ at the fundamental wavelength at 1.064 μm and about 200 mJ at the harmonic at 0.532 μm; the pulse rate is 10 Hz. The beam divergence behind the telescope can be as low as 0.2 mrad at 1.064 μm. For eye-safety reasons, it can be switched to 2 mrad at night. Power consumption of the laser is 600 W on average.

The receiving telescope with signal and alignment detectors is mounted in the second GAS-container with a lid that will be opened during the measurement. This container has an extension on the rear side for the detectors and receiver electronics, as well as the video camera that is used for data navigation. A compact Cassegrainian telescope, measuring 0.48 m in diameter and 0.65 m in overall length, is used as a photon collector. The radiation at 1.064 μm is detected by a Si-Avalanche photodiode, the 0.532 μm radiation by a photomultiplier with S20 photocathode. Both signal channels use narrow band filters with 1 nm bandwidth to reduce the background signal. The extra alignment channel is necessary since transmitter and receiver are mounted in separate containers. This is done by means of a quadrant detector using the strong return signal from the ground or optically dense clouds. The signals from both spectral channels are digitized with a 12-bit, 3MHz ADC, while the alignment channel can operate with reduced accuracy and bandwidth (8-bit, 0.5 MHz). The digitized signals are transmitted by a serial bus to the recorder module in the Spacelab. Measurement data are

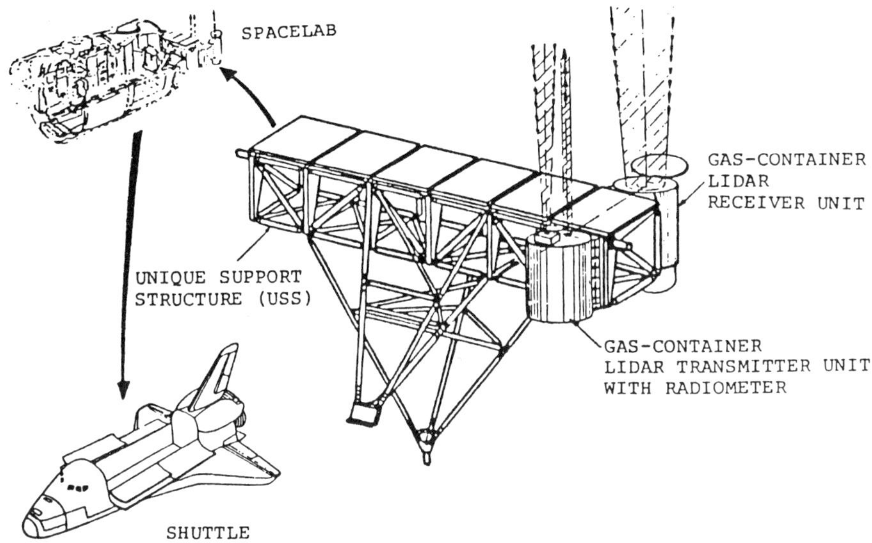


FIGURE 1. Construction of the lidar transmitter and receiver unit at USS in the rear of the Spacelab. The units are mounted in separate GAS-containers.

stored on a Winchester disk that is housed in the control unit. The data from the video camera are stored in the Spacelab on a video recorder.

The Spacelab-lidar will operate as an autonomous instrument, but can also be operated manually from Spacelab. The control unit is rack mounted and comprises a signal scope, an indicator for the alignment detector signal and a device for manual realignment.

## 6. CONCLUSION

A German backscatter lidar experiment from space is proposed. It should be demonstrated that the technical concept is feasible, that the data can be inverted by numerical codes, and that the results prove to be correct by means of a verification campaign. If funding is provided, we hope ALEXIS can be realized in the early 1990s.

REFERENCES

Collis, R.T.H., 1969: Lidar, Advances in Geophysics, 13, Academic Press, 113-139.

d'Almeida, G.A., 1985: Report on the Study of Transport of Sahelian Particulate Matter Using Sunphotometer Observations, WMO/TR No. 45, Geneva.

Fraser, R.S., 1976: Satellite Measurement of Mass of Sahara Dust in the Atmosphere, Appl. Opt., 15, 2471-2479.

Grams, G.W., E.M. Patterson and C.M. Wyman, 1974: Airborne Laser Radar for Mapping Two-Dimensional Contours of Aerosol Concentration, LAS 6, paper 1-3.

Griggs, M., 1983: Satellite Measurements of Atmospheric Aerosols, Annual Report, ONR Contract No. N00014-77-C-0489.

Hansen, A.D.A., H. Rosen and T. Novakov, 1982: Real-Time Measurements of the Absorption Coefficient of Aerosol Particles, Appl. Opt., 21, 3060-3062.

Kästner, M., 1986: Lidar Inversion with Variable Backscatter/Extinction Ratios: Comment, Appl. Opt., 25, 833-835.

Klett, J.D., 1985: Lidar Inversion With Variable Backscatter/Extinction Ratios, Appl. Opt., 24, 1638-1643.

Koepke, P., and H. Quenzel, 1981: Turbidity of the Atmosphere Determined from Satellite: Calculation of Optimum Wavelength, J. Geophy. Res., 86, C10, 9801-9805.

Melfi, S.H., D. Bundy, J. Eckert, J. Guagliardo and J.L. McElroy, 1974: Boundary Layer Investigations Using a Down Looking Airborne Lidar System, LAS 6, paper 6-1.

Mörl, P., M.E. Reinhardt, W. Renger and R. Schellhase, 1981: The Use of the Airborne Lidar System 'ALEX F1' for Aerosol Tracing in the Lower Troposphere, Beitr. Phys. Atmosph., 54, 403-410.

Quenzel, H., E. Thomalla and K. Nodop, 1984: Performance Simulations of the Backscatter Lidar. In M. Endemann (Ed.), Orbiting Lidars for Atmospheric Sounding, Vol. 1, ESTEC/Contract No 5186/82/NL/HP, Battelle e.V., Frankfurt, FRG.

Spinhirne, J.D., 1982: Lidar Clear Atmosphere Multiple Scattering Dependence on Receiver Range, Appl. Opt., 21, 2467-2468.

Wiegner, M., and E. Raschke, 1987: Planetary Radiation Budget over North Africa from Satellite Data, Theor. Appl. Climatol., 38, 24-36.

# OPTICAL REMOTE SENSING OF ATMOSPHERIC AEROSOL, WATER VAPOR, AND TRACE GAS $NO_2$

*Zhao Bolin, Mao Jietai, Zhang Chengxiang and Wang Qiang*
Department of Geophysics, Peking University
Beijing, People's Republic of China

ABSTRACT

Using radiometers in the visible and infrared bands, a number of ground-based observations of the intensity of the solar spectrum and atmospheric optical thickness have been made and methods for the remote sensing of atmospheric aerosol, water vapor, and $NO_2$ have been studied. Measurements of atmospheric aerosol and $NO_2$ over Beijing are reported.

## 1. INTRODUCTION

By analysing the attenuation of solar radiation in the atmosphere, the amount of aerosol, water vapor, and trace gases in the atmosphere can be evaluated. In order to obtain the amounts of aerosol, water vapor, and trace gases, some suitable channels have been selected that avoid other gas absorptions and reveal the effects of aerosol, water vapor, and $NO_2$.

The solar radiation of earth surface $I_\lambda$ may be expressed as follows:

$$I_\lambda = I_{\lambda o} e^{-\tau_\lambda m} \tag{1}$$

where $I_{\lambda o}$ is the solar radiation at the edge of the outer atmosphere, m the atmospheric optical mass number and the atmospheric optical thickness, which may be expressed as:

$$\tau_\lambda = \tau_R + \tau_{O_3} + \tau_a + \tau_w \left(+ \tau_{NO_2}\right) \tag{2}$$

and where $\tau_R$, $\tau_{O_3}$, $\tau_w$, $\tau_a$, and $\tau_{NO_2}$ are the optical thicknesses of molecules, ozone, water vapor, aerosol, and nitrogen dioxide in the atmosphere, respectively. $\tau_R$ and $\tau_{O_3}$ are relatively small in the band selected here, and can be computed from climatological data. Hence, Eq. (2) can be simplified to a relationship between $\tau_\lambda$, $\tau_a$, and $\tau_\omega$ (or $\tau_{NO_2}$). In the visible band (0.445 $\mu$m), absorption of nitrogen dioxide is important, while absorption by water vapor may be neglected.

AEROSOLS AND CLIMATE
Peter V. Hobbs and M. Patrick McCormick (Eds.)




ISBN 0-937194-11-5

## 2. ATMOSPHERIC AEROSOL AND WATER VAPOR CONTENT

To obtain the atmospheric optical thickness in the 0.5-1.06 $\mu m$ wavelength band, a multiple wavelength radiometer was developed. It is composed of an object lens, a detector, a set of interference filters, amplifier, and recorder. The aperture diameter of the telescope is 64 mm and the focal length of the object lens is 840 mm. This provides a small view angle of about 4′ and avoids scattering in the measurement. All the filters have a width of 0.075-0.225 $\mu m$; the central wavelengths are 0.5000, 0.5940, 0.7465, 0.8420, 0.9195 and 1.0600 $\mu m$. The total error arising from the instrument and the observational method is < 10%. The observation station (Peking University) is located in the northwest suburb of Beijing, at an elevation of 50 m. Observations were carried out on 115 days, from July 1980 to July 1981, under clear, anticyclone weather conditions.

### 2.1 ATMOSPHERIC AEROSOL OPTICAL THICKNESS

The morning data were chosen and analysed. The annual variation of atmospheric optical thickness is shown in Fig. 1. In the 0.5 $\mu m$ band, $\tau_a$ is 0.452 and $\tau_R$ is 0.151, therefore the atmospheric attenuation was due mainly to particles.

In order to investigate the particle characteristics, four channels were chosen: 0.5000, 0.5940, 0.7465, and 0.8420 $\mu m$. The optical thickness of the particles $\tau_a$ in these bands obeys the formula:

$$\tau_a = \beta e^{-\alpha\lambda} \tag{3}$$

with a correlation coefficient of $\gamma \geq 0.95$ (Fig. 2), where $\alpha$ and $\beta$ are parameters.

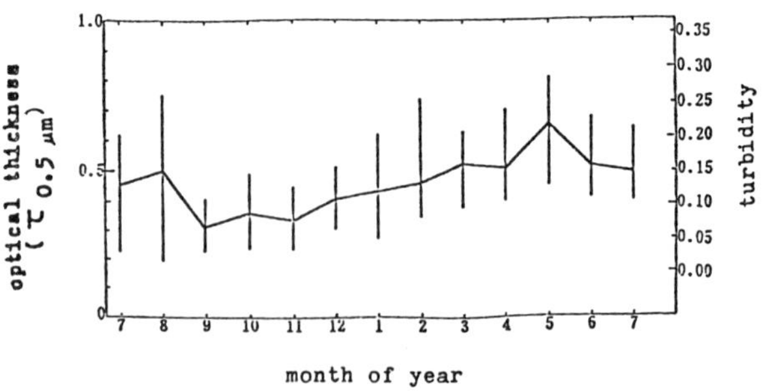


FIGURE 1. Annual variation of atmospheric optical thickness and turbidity from July 1980 to July 1981, Beijing.

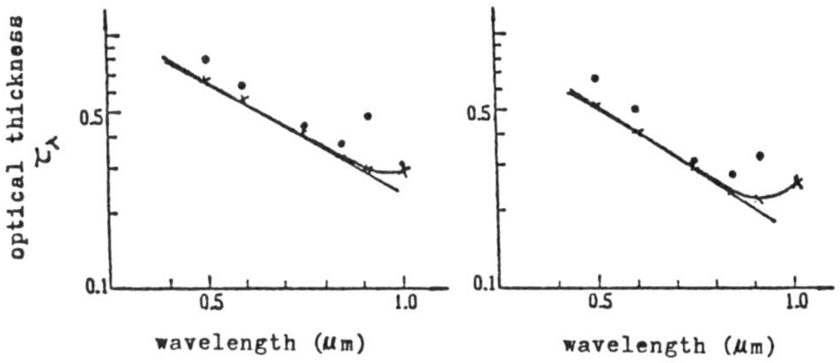


(a) 8:55 a.m. 4 June 1981, Beijing. (b) 7:45 a.m. 18 May 1981, Beijing.

FIGURE 2. $\tau_\lambda$ and empirical formula $\tau_a = \beta e^{-\alpha\lambda}$
$\cdot\tau_a$ (observed value)
$*\tau_a$
straight line: $\tau_a = \beta e^{-\alpha\lambda}$

$\tau_a$ also obeys the formula:

$$\tau_a = \beta'\lambda^{-\alpha'}$$

The average values are $\beta'$=0.116 and $\alpha'$=1.448. The relationship between $\alpha'$ and $\beta'$ has been found from

$$\beta' = -0.0899 + 0.41145\alpha'^{-1} - 0.0895\alpha'^{-2}$$

correlation coefficient $\gamma \sim$ 0.66 (Fig. 3).

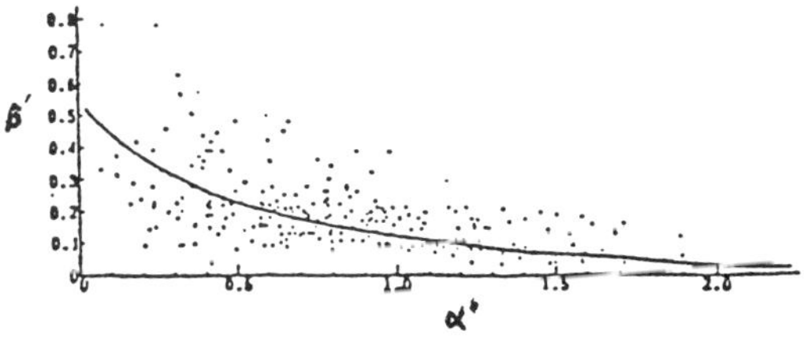


FIGURE 3. The relationship between $\alpha'$ and $\beta'$ in the formula $\tau_a = \beta'\lambda^{-\alpha'}$. For July 1980 to July 1981, Beijing.

## 2.2 AEROSOL SPECTRUM

The optical thickness of atmospheric particles may be expressed as follows:

$$\tau_a = \int_0^{\infty} \eta(r)\pi r^2 Q_{sc}(r,\lambda,m)dr \tag{4}$$

where n(r) is the columnar size distribution, r the radius of the particle, $\lambda$ the wavelength, m the complex refractive index of the particle (m = 1.53-i0.015), and $Q_{sc}(r,\lambda,m)$ is the efficiency factor of extinction from the Mie theory. In order to obtain the particle spectrum, two inverse methods were used.

The particle size distribution was assumed to follow:

$$\eta(r) = \frac{1}{2}ab^3r^2e^{-br} \tag{5}$$

where a and b are parameters. Substituting (5) into (4), we obtain:

$$\tau_a = \frac{1}{2}ab^3 \int_0^{\infty} \pi r^4 e^{-br} Q_{sc}(r,\lambda,m)dr \simeq \beta e^{-\alpha\lambda} \tag{6}$$

where parameters $\alpha$ , $\beta$ depend on a, b only.

$\tau_a$ could be obtained from observation. To investigate the particle characteristics, four channels were chosen. They were 0.5000, 0.5940, 0.7465, and 0.8420 $\mu$m. Then the parameters a and b in Eq. (5) were obtained from Eq. (6). The aerosol can be characterized by the following parameters: N(=a) is the total number of particles in the air column, $r_m$(=2/b) is the maximum population radius, and $N_m$ the spectrum density of the population at $r_m$. The influence of meteorological factors on the particle spectrum distribution has been shown and the diurnal variation of particle characteristics has been analysed (Fig. 4).

The randomized minimization search technique (Heintzenberg et al., 1981) was used to invert the particle spectrum distribution for the purpose of expressing the precise structure of atmospheric particle size distribution. According to Eq. (4), the atmospheric particle optical thickness of j channel can be calculated from:

$$\tau_{ja} = \sum_{i=1}^{n} n(r_i)\pi r_i^2 Q_{sc}(r_i,\lambda_j,m)\Delta r = \sum_{i=1}^{n} n_i G_{ij} \tag{7}$$

where $n_i$=n($r_i$), $G_{ij}=r_i^2\ Q_{sc}(r_i,\lambda_j,m)\Delta r$. By taking six particle radii $r_i$, i=1, 2,..n, and channel wavelength $\lambda_j$, j=1, 2,..m, choosing parameters randomly, and iterating Eq. (7) continuously until convergence, the maximum likelihood criterion is:

$$\sum_{j=1}^{m}\left[\tau_{ja} - \sum_{i=1}^{n} n_i G_{ij}\right]^2 \Big/ \sum_{j=1}^{m} \tau_{ja}^2 < \delta \tag{8}$$

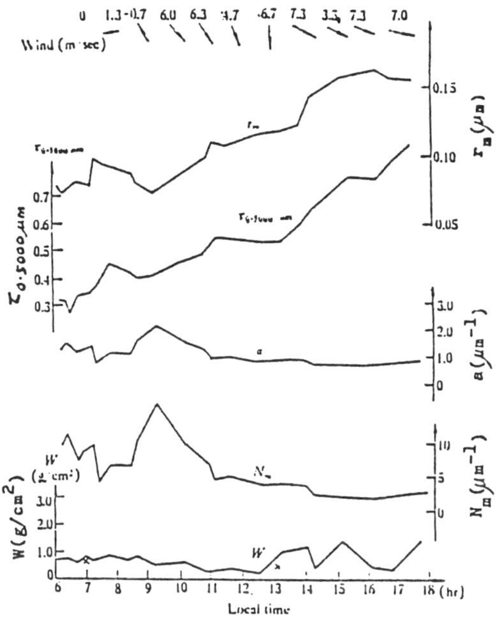


FIGURE 4. Diurnal variation of atmospheric optical thickness and aerosol. (P=1013.3mb, T=18°C, f=18%). For 24 April 1981, Beijing.
W derived from Eq. (9).
x derived from radiosonde data.

where $\delta$ is a given error, say $\delta$ = 0.1. Six channels--0.5000, 0.5940, 0.7465, 0.8420, 0.9195, and 1.0600 $\mu$m were used. $\tau_{ja}$ was obtained from the observation. The diurnal and seasonal variations of the atmospheric particles spectrum are shown in Figs. 5 and 6, respectively.

The results obtained from the distribution model Eq. (5) from the four channels are in good agreement with those obtained from the six channels by the randomized minimization search technique (Fig. 7).

## 2.3 ATMOSPHERIC WATER VAPOR CONTENT

Two channels are 0.7150 $\mu$m and 0.7465 $\mu$m. The former is the water vapor absorption region and the latter is outside the above region.

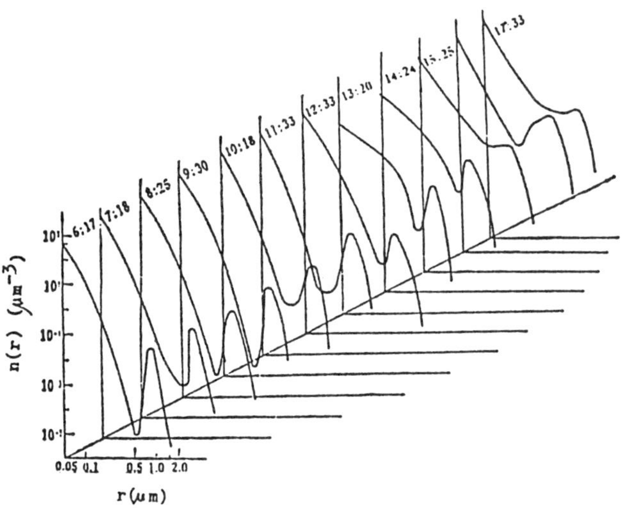

FIGURE 5. Diurnal variation of atmospheric aerosol spectrum for 24 April 1981 in Beijing.

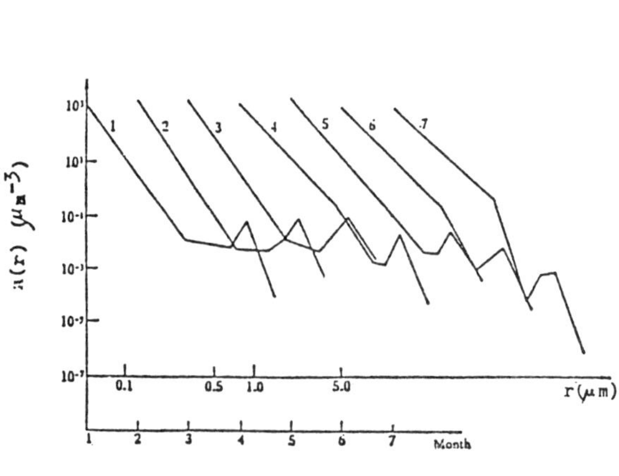

FIGURE 6. Seasonal variation of atmospheric aerosol spectrum in Beijing.

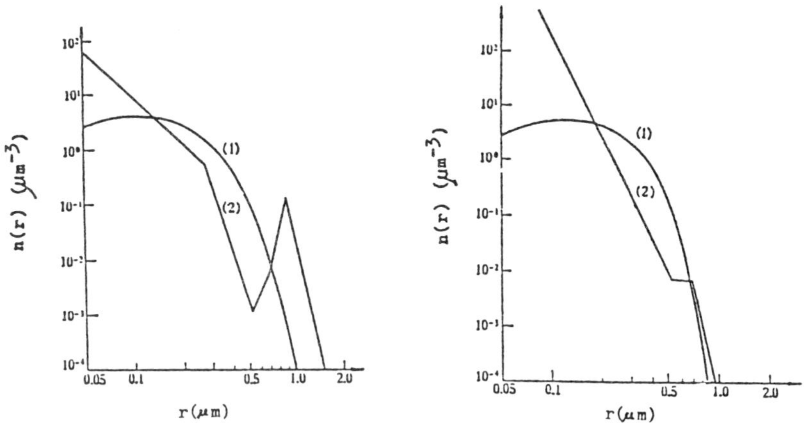

FIGURE 7. Comparison of the result obtained from the 4-channel model size distribution inverse method (curve 1) with that obtained from the 6-channel randomized minimization search technique (curve 2).

Let $\Delta\tau_\lambda = \tau_{0.7150\ \mu m} - \tau_{0.7465\ \mu m}$, where $\Delta\tau_\lambda$ was dependent on the atmospheric water vapor content. The particle effect was corrected using the formula $\tau_a=\beta e^{-\alpha\lambda}$, which was obtained from the other four channels mentioned in the above section. After the aerosol correction was made for $\Delta\tau_\lambda$, we arrived at $\Delta\tau_\omega$. The relationship between atmospheric water vapor content w and $\Delta\tau_\omega$ is approximated by a 2-order function, which may be expressed as follows (correlation coefficient $\gamma$ = 0.94):

$$\omega = 0.18 + 12.4\ \Delta\tau_\omega + 149.2\ \Delta\tau_\omega^2 \tag{9}$$

which is shown in Fig. 8.

## 3. ATMOSPHERIC TRACE GAS $NO_2$

An automatic solar spectrophotometer controlled by an Apple II computer was used for the direct solar radiation measurement. The spectrum scanning measurement in 0.445-0.450 μm was carried out. The columnar concentration of $NO_2$ can be obtained from the measured data. The structure of this automatic solar spectrophotometer is shown in Fig. 9. The sunlight is received and transmitted by a heliostat. It is then reflected by a plane mirror and a spherical mirror and finally enters into a grid spectrometer. The parameters of the grid spectrometer are f=302.5 mm, line density of raster=1200 line/mm. The detector is a multiplier tube. The control part consists of an Apple II computer, interface, digital voltmeter, and scanning control circuit. The sample output from the multiplier tube is taken by the

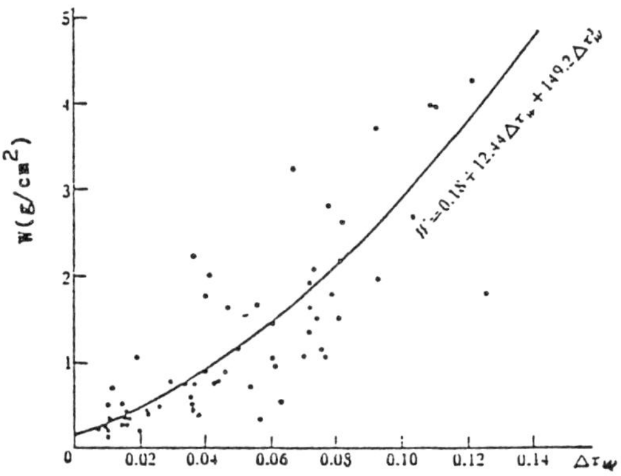

FIGURE 8. Atmospheric water vapor content w against water vapor optical thickness $\Delta\tau_\omega$ in Beijing.

digital voltmeter; the scanning measurement is realized by a step-by-step motor, which is controlled by the scanning control circuit. The samples are taken and dealt with at the same time, then the $NO_2$ columnar concentration can be obtained immediately. The $NO_2$ absorption spectrum is measured with an $NO_2$ cell, which was put in the light path at the entrance of the spectrometer (Fig. 10).

The absorption of $O_3$ is 1 or 2 orders smaller than that of $NO_2$, so that the $O_3$ absorption may be neglected. In the $\lambda_k$ wave band, the surface solar radiation $I_k$ is:

$$I_k = K \int_{\lambda_k-\Delta\lambda}^{\lambda_k+\Delta\lambda} \sigma(\lambda_k-\lambda')I_o(\lambda')e^{-\tau_r m-\tau_a m-a_{NO_2}xm}\, \alpha\lambda \tag{10}$$

where K is the instrument constant, $\Delta\lambda$ the half wavewidth of the slit of the spectrometer, $\sigma(\lambda_k-\lambda')$ a function of the spectrometer slit (Stewart, 1970), $I_o(\lambda')$ the solar radiation, m the optical mass number of the atmosphere, x the $NO_2$ columnar amount, and $a_{NO_2}$ the absorption of $NO_2$.

Using the maximum likelihood estimator method, the $NO_2$ columnar concentration may be obtained from Eq. (10) (Noxon et al., 1979). The examples of $NO_2$ observation are shown in Fig. 11. The diurnal variation is shown in Fig. 12. The daily variation of $NO_2$ concentration is shown in Fig. 13. The average columnar concentration is $1.8x10^{16}$ molecular $cm^{-2}$. The results were compared with those obtained by other scientists (Noxon, et al., 1979) and found to be consistent.

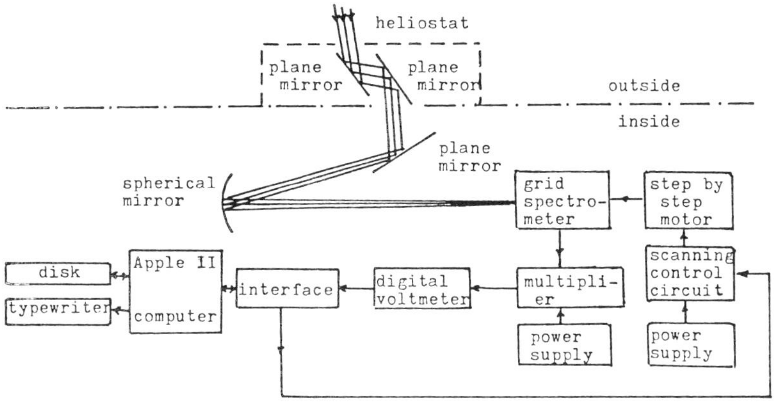


FIGURE 9. Components of the automatic solar spectrophotometer.

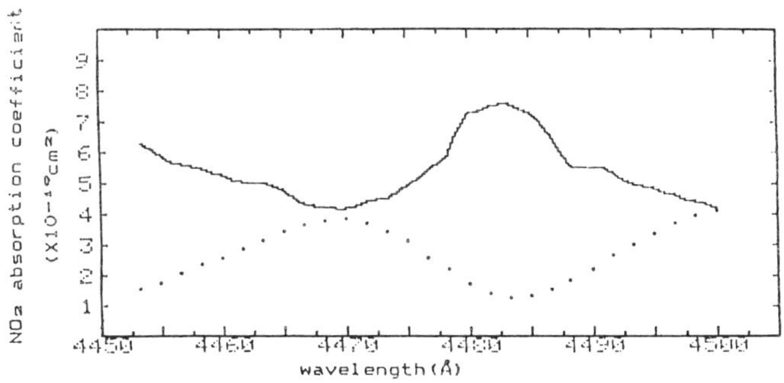


FIGURE 10. $NO_2$ absorption coefficient from Shibasaki (1979) (solid line) and observed $NO_2$ absorption spectrum measured in the laboratory with an automatic solar spectrophotometer system, transmission value (dotted line).

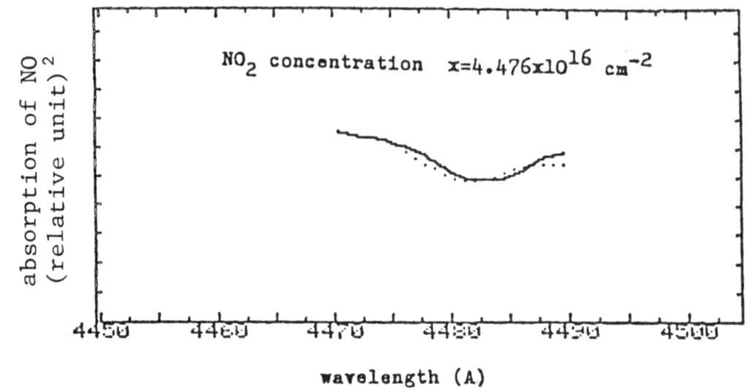


FIGURE 11. Examples of columnar concentration of atmospheric trace gas $NO_2$ measured at 11:15 a.m. on 31 March 1986 in Beijing.
Theoretical values (line), observation value (dots).

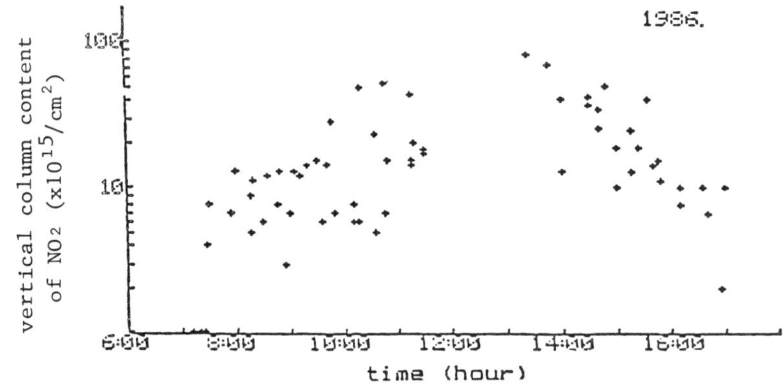


FIGURE 12. The diurnal variation of columnar concentration of atmospheric trace gas $NO_2$ in Beijing.

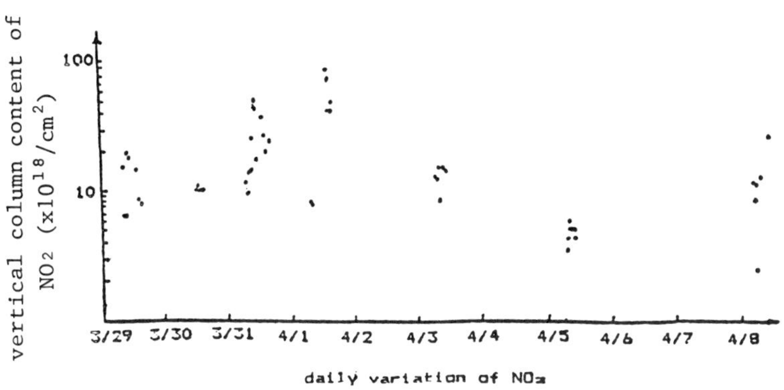


FIGURE 13. The day-by day variation of columnar concentration of atmospheric trace gas $NO_2$ in Beijing.

## 4. CONCLUSIONS

It has been shown that a multichannel radiometer operating at visible and infrared wavelengths can be used to remotely sense atmospheric particles and water vapor content, and that an automatic solar spectrophotometer can be used to remotely sense atmospheric $NO_2$. These instruments may be used for the monitoring of atmospheric pollution. Some observational results on aerosol and $NO_2$ obtained in a suburb of Beijing, which have been described in this paper, can be used to assess the atmospheric condition at that location.

## REFERENCES

Heintzenberg J., H. Müller, H. Quenzel and E. Thomalla, 1981: Information of Optical Data With Respect to Aerosol Properties, Appl. Optics, 20, 1308-1315.

Noxon J.F., E.C. Whipple Jr., and R.S. Hyde, 1979: Stratospheric $NO_2$, J. Geophys. Res., 84, 5047-5065.

Shibasaki W., S. Suzuki and T. Ogawa, 1979: $NO_2$ Distrubution With Height in Stratosphere Measured by Balloon Sounding, Rep. Inst. Space & Astronaut. Sci., 15, 243-254.

Stewart J., 1970: Infrared Spectroscopy, Dekker, New York, 639 pp.

# INFERENCE OF ATMOSPHERIC AEROSOL SIZE DISTRIBUTION FROM GROUND MEASUREMENTS OF SOLAR RADIATION

*U. Amato*
Istituto per Applicazioni della Matematica–C.N.R.
Napoli, Italy

*V. Cuomo*
Instituto di Fisica, Fac. di Ingengeria, Univ. della Basilicata
Potenza, Italy

and

*F. Fontana, M. Macchiato and C. Serio*
Dip. Fisica Nucleare, Struttura della Materia e Fisica Applicata
Napoli, Italy

ABSTRACT

This paper deals with the problem of retrieving aerosol size distribution from optical measurements of solar radiation at the ground. At present no inversion method seems completely successful. We believe that more attention ought to be paid to the hypotheses that are made in the inversion techniques; moreover a unified test procedure should be developed in order to evaluate the performance of the methods. For this reason we analyze the influence of some hypotheses that are commonly assumed; furthermore we propose some test problems to evaluate the validity of the different inversion techniques, according to the indications of the International Meeting on Atmospheric Transparency for Satellite Data, held at Capri in September 1986. Finally we propose a new inversion method, based on the Maximum Entropy Principle, which overcomes some problems of conventional methods.

## 1. INTRODUCTION

The knowledge of aerosol properties in the atmosphere is assuming a growing importance in the field of atmospheric physics. It is well known that the aerosol distribution in the atmosphere is restored from ground measurements of solar spectral irradiance, solving an inverse problem. This technique is quite reliable and cheap from an experimental point of view, but the inverse problem is generally difficult to solve and several physical hypotheses about the interaction radiation-matter are made that are not always met. Present techniques for restoring do not seem completely satisfactory. They are essentially based on a non-parametric regularization technique (e.g., King, 1982 for a partial

AEROSOLS AND CLIMATE
Peter V. Hobbs and M. Patrick McCormick (Eds.)


ISBN 0-937194-11-5

review). Generally inversion is performed for aerosol size ranging within an interval $r_{MIN}$ to $r_{MAX}$. The main problems of these techniques are related to the presence of unphysical solutions, to the dependence of the solution on the choice of the refractive index and the limit radii $r_{MIN}$ and $r_{MAX}$, and to the error due to the use of a quadrature formula. Moreover, experimental data are of course affected by instrumental errors, so that it is quite important to assess the dependence of retrieved solutions on experimental error. Consequently it is not easy to compare the various retrieving algorithms and assess their power for many reasons; the main ones are:

1) the influence of instrumental errors on the retrieved solution depends on the entity of the error itself;
2) the retrieved solution depends of course on the number of measuring channels, and
3) the retrieved solution depends of course on the refractive index of the aerosol.

The test of the inverse methods proposed by the various authors is generally made comparing the restored aerosol size distribution with some experimental distributions obtained by impactors. However, these distributions are generally affected by an error, which is very difficult to determine. For this reason it is desirable to define some precise test problems to be solved under controlled steps by the various methods in simulation. In this case it is necessary:

1) to generate synthetic distributions that reflect typical aerosol distributions in the atmosphere and can be used as reference distributions, and
2) to calculate input data to inverse problems, starting from reference distributions.

In this way it is possible to assess the dependence of inversion techniques on the number of measuring channels and on the instrumental error. In this approach comparison with experimental distributions, obtained by impactors, represents a second level of intercomparison procedure.

Section 2 of this paper contains a proposal of test problems, as obtained by suggestions made in the International Meeting on Atmospheric Transparency for Satellite Data held at Capri in September 1986. Sections 3 and 4 contain a new approach to the problem, based on the Maximum Entropy Principle. It uses a parametric regularization technique, which avoids unphysical solutions. We present some preliminary results obtained in simulation using as reference distributions the ones proposed by the World Meteorological Organization (1986).

## 2. DEFINITION OF THE TEST PROBLEMS

### 2.1 AEROSOL DISTRIBUTIONS

The most complete and updated data on aerosol distributions in the atmosphere are reported in WMO (1986). Five typical aerosol components are defined, whose distributions follow a lognormal

$$\frac{dN(r)}{d\log r} = \frac{N}{\sqrt{2\pi}\,\log\sigma^*} \exp\left(-\frac{(\log r - \log r^*)^2}{2(\log\sigma^*)^2}\right) \tag{1}$$

or a modified gamma

$$\frac{dN(r)}{d\log r} = Nr^2b^2\exp(-br) \tag{2}$$

law:

1) water soluble: lognormal, $r^* = 0.005$ µm, $\sigma^* = 2.99$ µm
2) soot : lognormal, $r^* = 0.0118$ µm, $\sigma^* = 2.00$ µm
3) dust-like : lognormal, $r^* = 0.5$ µm, $\sigma^* = 2.99$ µm
4) 75%$H_2SO_4$ : modified gamma, $b = 18$ µm$^{-1}$
5) oceanic : lognormal, $r^* = 0.3$ µm, $\sigma^* = 2.51$ µm

In all cases N is the total number of particles and is estimated such that the extinction coefficient is 1 km$^{-1}$ at 0.55-µm wavelength. Three basic aerosol models are defined by mixing the five components with the corresponding percentage by aerosol particle volume; they are shown in Table 1.

Our knowledge of the aerosol composition of the atmospheric column allows us to obtain the columnar aerosol distributions for three typical cases, with optical depths at 0.55-µm wavelength as shown in Table 2. In Figs. 1, 2, and 3 we show the particle distribution as obtained for the three cases.

### 2.2 THE PHYSICAL PROBLEM

There are two different methods for retrieving aerosol size distributions from ground measurements of solar radiation: optical extinction and angular scattering. Both of them involve a First Kind Fredholm Integral Equation:

$$\tau(\lambda) = -\ln\frac{I(\lambda)}{I_o(\lambda)} = \int_0^\infty \pi r^2 Q_{ext}(\lambda,r)\; n(r)\; dr \tag{3}$$

for optical extinction and

TABLE 1. EXTINCTION COEFFICIENT FOR THE THREE BASIC AEROSOL MODELS

| Basic Aerosol Model | Aerosol Component | % by Aerosol Particle Volume | Extinction Coefficient ($km^{-1}$) |
|---|---|---|---|
| Continental Aerosol | Water Soluble | 29% | 0.8688 |
| | Soot | 1% | 0.0588 |
| | Dust-like | 70% | 0.0724 |
| Urban/Industrial Aerosol | Water Soluble | 61% | 0.5820 |
| | Soot | 22% | 0.4123 |
| | Dust-like | 17% | 0.0056 |
| Maritime Aerosol | Water Soluble | 5% | 0.2560 |
| | Oceanic | 95% | 0.7440 |

TABLE 2. OPTICAL DEPTH FOR THE THREE AEROSOL PROFILES

| Aerosol Component | Urban Profile | Continental Profile | Maritime Profile |
|---|---|---|---|
| Water Soluble | 0.6037 | 0.1955 | 0.0345 |
| Soot | 0.4138 | 0.0132 | 0.0015 |
| Dust-like | 0.0074 | 0.0163 | 0.0018 |
| 75%$H_2SO_4$ | 0.0048 | 0.0048 | 0.0048 |
| Oceanic | 0.0 | 0.0 | 0.0372 |
| TOTAL | 1.0297 | 0.2298 | 0.0798 |

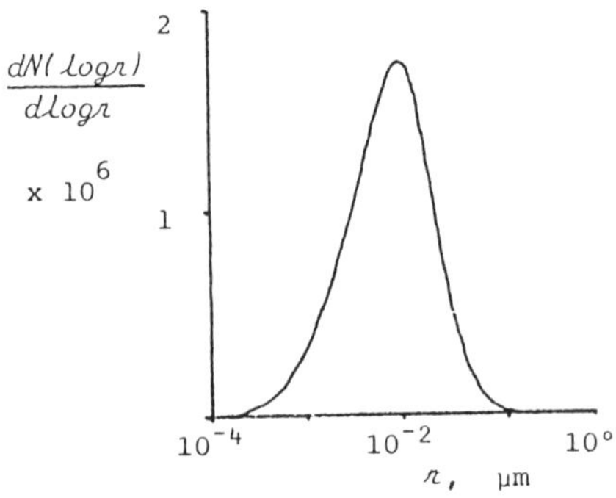


FIGURE 1. Aerosol size distribution for the urban profile.

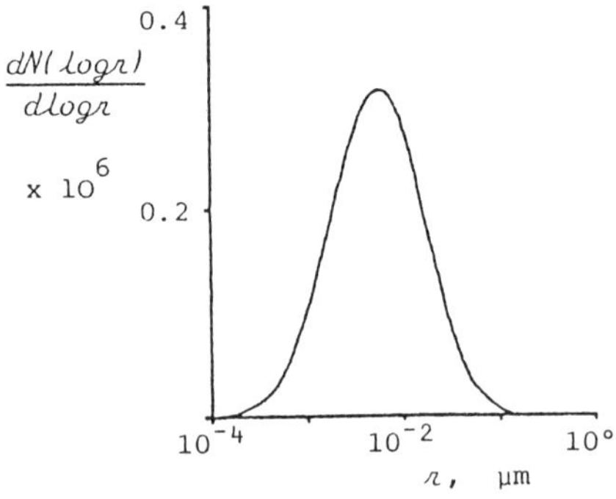


FIGURE 2. Aerosol size distribution for the continental profile.

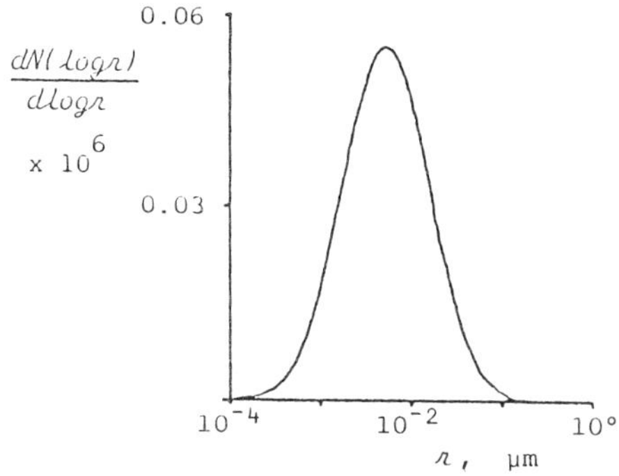


FIGURE 3. Aerosol size distribution for the maritime profile.

$$s_\lambda(\vartheta) = \frac{I(\lambda,\vartheta)}{I_o(\lambda)} = \int_0^\infty \frac{r^2}{x^2} \frac{i_1(\lambda,r,\vartheta) + i_2(\lambda,r,\vartheta)}{2} n(r)\, dr \tag{4}$$

for angular scattering

with $x = 2\pi r/\lambda$ and $I_o$ and $I$ being extra-atmospheric and ground spectral irradiance.

The analytical expression of the kernels is well known in the literature (e.g., Van De Hulst, 1957). In the optical extinction problem it is assumed that the attenuation of solar radiation is ruled by the Theory of Mie and that the particles can be considered spherical. Moreover, the angular scattering problem also assumes that no multiple scattering occurs in the atmosphere; the hypothesis is valid only for small scattering angles. Actually, the evaluation of the optical depth and the angular scattering is more complicated than Eqs. (3) and (4), since the transmission and scattering functions depend not only on the number but also on the kind of particle, by means of the refractive index. For this reason, Eqs. (3) and (4) ought to be rewritten more correctly as:

$$\tau(\lambda) = \sum_j \int_0^\infty \pi r^2 Q^j_{ext}(\lambda,r)\, n^j(r)\, dr \tag{5}$$

$$s_\lambda(\vartheta) = \sum_j \int_0^\infty \frac{r^2}{x^2} \frac{i^j_1(\lambda,r,\vartheta) + i^j_2(\lambda,r,\vartheta)}{2} n^j(r)\, dr \tag{6}$$

with the summation being extended to all the different kinds of particles that are present in the atmosphere. In Figs. 4 and 5 we show the plots $Q_{ext}(\lambda,r)$ vs. r for a 0.55-μm reference wavelength and two aerosol components (water soluble and soot); the other components have plots similar to the water soluble one. In Figs. 6 and 7 we show the

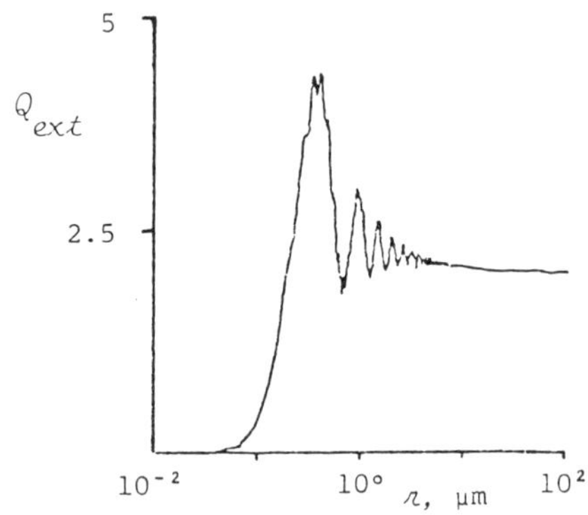


FIGURE 4. Optical extinction kernel for the water soluble aerosol.

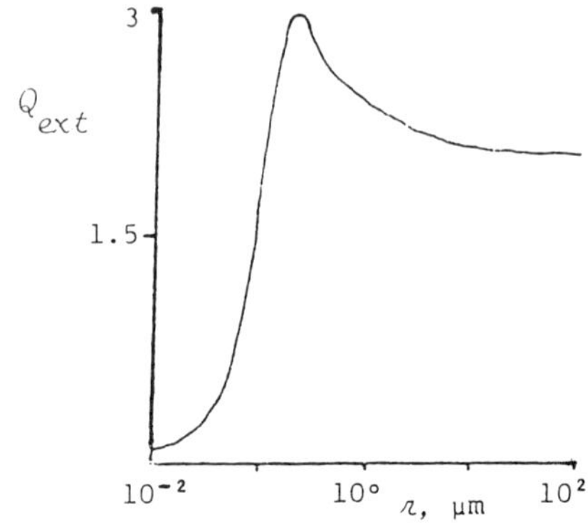


FIGURE 5. Optical extinction kernel for the soot aerosol.

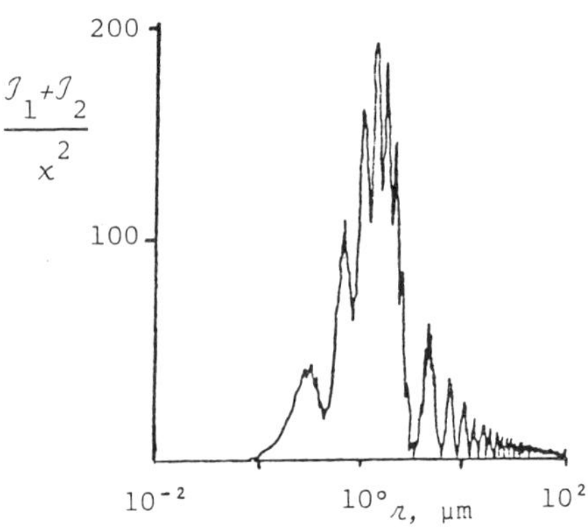


FIGURE 6. Angular scattering kernel for the water soluble aerosol.

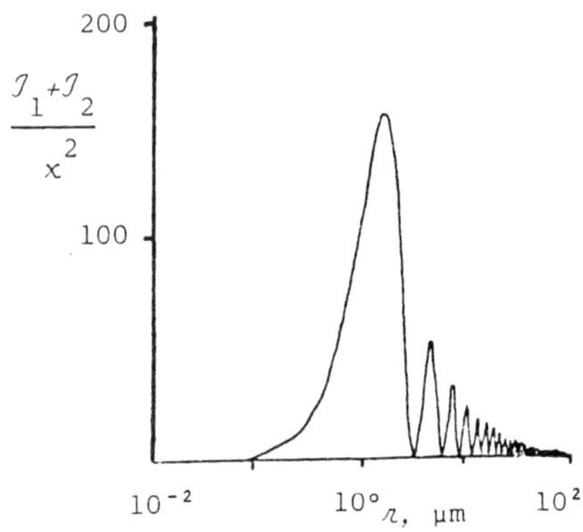


FIGURE 7. Angular scattering kernel for the soot aerosol.

plots of the kernel function (divided by $\pi r^2$) for the angular scattering problem and the same aerosol components at 0.4-μm wavelength and 4° scattering angle. When the values $\tau(\lambda)$ or $s_\lambda(\vartheta)$ are measured by suitable experimental equipment, the aerosol size distribution in the atmosphere can be retrieved solving the corresponding integral equation, provided that the hypotheses made in the physical model are met.

## 2.3 THE TEST PROBLEM

As mentioned in the Introduction, we think that a serious analysis of inversion methods can only be performed on synthetic data that simulate actual measurements. We suppose that the experimental equipment is able to get up to eight measurements (either of $\tau$ at different wavelengths or of s at different angles for a fixed wavelength). For the optical extinction problem the wavelengths are chosen according to Amato et al. (1986a, 1987a, 1987b): a range 0.4 to 2.0 μm is first chosen since it has been shown that it is essential to have a first wavelength as small as possible, but it is not greatly advantageous to get measurements in the infrared region. The wavelengths are chosen in the 0.4 to 2.0 μm range following a $1/\lambda$ dependence; this corresponds to the fact that the kernel function $Q_{ext}$ only depends on the size parameter x and the refractive index of the particle. The calculated values of the wavelength are then adjusted in order to represent spectral windows with low absorption of the other constituents of the atmosphere, so that the eight wavelengths are 0.4, 0.45, 0.52, 0.73, 1.0, 1.25, 1.60 and 2.0 μm. As far as the angular scattering problem is concerned, the wavelength for measurements must be as small as possible (Amato et al., 1986a; 1987a and 1987b), so that it has been chosen at 0.4 μm. The range of measurement for the angle is chosen below 8°, in order to keep minimal the effect of multiple scattering in the atmosphere. Thus the resulting eight angles are 1°, 2°, 3°, 4°, 5°, 6°, 7° and 8°.

Based on the three atmospheric aerosol distributions, the synthetic optical depths and angular scattering functions can be calculated by Eqs. (5) and (6). The optical depths are given by:

| Profile λ: | 0.40 | 0.45 | 0.52 | 0.73 | 1.00 | 1.25 | 1.60 | 2.00 |
|---|---|---|---|---|---|---|---|---|
| Urban | 1.518 | 1.321 | 1.110 | 0.620 | 0.458 | 0.332 | 0.232 | 0.150 |
| Continental | 0.322 | 0.286 | 0.245 | 0.160 | 0.110 | .0817 | .0601 | .0386 |
| Maritime | .0959 | .0896 | .0824 | .0672 | .0580 | .0525 | .0472 | .0471 |

and the angular scattering function by:

| Profile angle: | 1° | 2° | 3° | 4° | 5° | 6° | 7° | 8° |
|---|---|---|---|---|---|---|---|---|
| Urban | 1.789 | 1.298 | 1.153 | 1.065 | 0.995 | 0.935 | 0.881 | 0.831 |
| Continental | 1.605 | 0.645 | 0.450 | 0.376 | 0.334 | 0.306 | 0.284 | 0.265 |
| Maritime | 1.842 | 0.791 | 0.435 | 0.278 | 0.199 | 0.153 | 0.125 | 0.106 |

In order to simulate actual data, the synthetic unperturbed data are to be corrupted by noise, which takes account of the experimental error and of the error in removing contributions of the other constituents of the atmosphere (molecules, ozone, water vapor, etc.). The simple hypothesis is made that the error is distributed as a Gaussian distribution around the unperturbed data and that the percentage error is constant for all wavelengths or angles. Summarizing, corrupted data are calculated for the optical extinction problem as

$$\tau^p(\lambda) = \tau^\circ(\lambda)\ (1 + \varepsilon_p(\lambda)) \tag{7}$$

where $\tau^p(\lambda)$ = simulated optical depth, assuming that an error p percent affects the unperturbed data;

$\tau^\circ(\lambda)$ = unperturbed optical depth;

$\varepsilon_p(\lambda)$ = random number extracted from a Gaussian distribution having zero mean and variance $\sigma_p^2$ given by

$$\sqrt{\sigma_p^2(\lambda)} = 0.01\ p\tau^\circ(\lambda) \tag{8}$$

Analogously, the corrupted data for the angular scattering problem are calculated as

$$s_\lambda^p(\vartheta) = s^\circ(\vartheta)\ (1 + \varepsilon_p(\vartheta)) \tag{9}$$

with obvious notation.

A test problem consists in generating a sequence $\tau^p(\lambda_i)$ or $s_\lambda^p(\vartheta_i)$ for a fixed percentage error and performing the inversion by some method. The validity of the restored solution can be evaluated by the following two indexes

$$I_1 = \sqrt{\frac{\int_0^\infty (n(r) - n^\circ(r))^2\, dr}{\int_0^\infty (n^\circ(r))^2\, dr}} \tag{10}$$

with $n^\circ(r)$ being the true solution and $n(r)$ the retrieved solution

and

$$I_2 = \frac{\int_0^\infty n(r)\, dr}{\int_0^\infty n^\circ(r)\, dr} \tag{11}$$

which give the percentage error in retrieving the particle size distribution and the total number of particles, respectively. It is suggested that several sequences (about 50-100) be generated for the same percentage error, in order to evaluate the validity of the retrieval in a significant way from the statistical point of view.

Moreover, performing the analysis with different values of p (1%, 2%, 3%, etc.,) gives the capability of the inversion method with respect to the error on input data. Finally it is advisable to perform again the same whole analysis using only four data, in order to check how the number of data affects the retrieval.

## 3. TECHNIQUE OF INVERSION

Equations (1) and (2) are known in the mathematical framework as First Kind Fredholm Integral Equations. It is known that they are ill-posed problems, that is small variations in some data can produce large variations in the solution, which makes the problem very difficult to solve. Present techniques of solution are essentially based on this method: First, the First Kind Fredholm Integral Equation is transformed into a Second Kind by parameter $\Lambda$ and smoothing operator $O$ (typically zero-th or second-order derivative) (regularization):

$$\int_0^\infty K(x,r)\ n(r)\ dr + \Lambda \int_0^\infty O(n(r))\ dr = g(x) \tag{12}$$

where $x$ stands for $\lambda$ or $\vartheta$, with regard to the optical extinction or angular scattering problem, and $K(x,r)$ and $g(x)$ are the corresponding kernel functions and right-hand sides. Second, the solution is retrived in a finite set of radii $n_1 \equiv n(r_1) \ldots n_m \equiv n(r_m)$; when the integrals are substituted by suitable quadrature formulas and the derivatives by finite differences, the linear system is obtained

$$\sum_{i=1}^{m} w_i\ K_{ij}\ n_i \ + \ \Lambda \sum_{i=1}^{m} w'_i\ n_i \ = \ g_j, \qquad j = 1 \ldots n \tag{13}$$

where $w_i$, $w'_i$ are suitable weights,
$K_{ji} = K(x_j, r_i)$,
$g_j = g(x_j)$,
$n$ = the number of available data.

Finally the system is solved and the solution of the problem found in correspondence with a value of $\Lambda$ chosen by some empirical method. Actually, the methods of inversion are more complicated, since a previous factorization of the solution is performed and several starting solutions are considered; the procedure is generally iterated in order to obtain better solutions. This procedure presents some approximations, whose influence on the solutions is to be carefully examined, due to the ill-conditioning of the problem. First of all, the solution can be meaningfully retrieved only in a small number of points (10 to 15); this implies that the error of the quadrature formula and the finite difference can be not negligible and could be greatly amplified, due to the ill-conditioning of the problem. It also follows that the choice of lower and upper radii for the range of particle size becomes essential. Moreover, the method implicitly assumes statistical hypotheses that do not seem to fit the experimental equipment completely. Namely, the solution of Eq. (13) is obtained in the sense of least squares, which works if the error on the data has a Gaussian distribution with a variance independent of the data. Unphysical solutions are often obtained during the inversion, since $n_i$'s can assume negative values when inverting matrices. Finally, the influence of the mean refractive index assumed for the calculation of the kernel functions should be tested.

## 3.1 INFLUENCE OF QUADRATURE ERROR

As an example of the care to be taken in the assumption of the previous simplifications, we propose an analysis of the error induced by the quadrature formula in the calculation of the optical depth. We again stress that the ill-conditioning of the problem runs the risk of amplifying a small error dramatically, so that it is essential to limit approximations and wrong hypotheses as much as possible.
In this section we assume that I data $n(r_i)$ for each radius decade are to be retrieved, so that the total number of data n(r) will be $M=D\cdot I$ (D is the number of decades over which the integration is performed); the global range of particle radius is decomposed in M equispaced subintervals (on a logarithmic scale), each one centered at $r_i$,

$$r_i = 10^{\mathrm{Log}(r_{MIN}) + (i-0.5)\,\Delta\mathrm{Logr}}$$

and of length $\Delta\mathrm{Log}\ r = 1/I$. Since n(r) is assumed constant within its own subinterval, the integral (Eq. (3)) is decomposed as the sum of M integrals, each one over the proper subinterval. Note that these M integrals can be evaluated with any precision, since they do not contain the unknown function. Thus the integral is evaluated as

$$\pi\sum_{i=1}^{M} n(r_i) \int_{\Delta r_i} r^2 Q_{ext}(\lambda,r)\ r\ d\mathrm{Logr} + \eta_1 \tag{14}$$

with $\eta_1$ being the error due to the quadrature formula. Actually, we could perform other factorizations of the function to be integrated, which lead to the following evaluations:

$$\pi \sum_{i=1}^{M} n(r_i)\, r_i \int_{\Delta r_i} r^2 Q_{ext}(\lambda,r)\, d\mathrm{Log}r + \eta_2 \tag{15}$$

$$\pi \sum_{i=1}^{M} r_i^2\, n(r_i)\, r_i \int_{\Delta r_i} Q_{ext}(\lambda,r)\, d\mathrm{Log}r + \eta_3 \tag{16}$$

We have calculated the percentage error $\eta/\tau$ in calculating $\tau$ at 0.55-μm wavelength as a function of the number of data per decade for the five aerosol components. The results is that the choice of the factorization affects the quadrature error. Table 3 shows the number of points per decade needed in order to make the percentage error in the calculation of $\tau$ less than 1% for the five aerosol components and the three factorizations. The plot percentage error vs. I is shown in Fig. 8 for the water soluble component.

TABLE 3. QUADRATURE ERROR FOR THE DIFFERENT FORMULAS CONSIDERED

| Formula | Water Soluble | Soot | Dust-like | 75% $H_2SO_4$ | Maritime |
|---|---|---|---|---|---|
| Eq. (14) | 19 | 21 | 15 | 19 | 14 |
| Eq. (15) | 15 | 16 | 10 | 14 | 10 |
| Eq. (16) | 5 | 6 | 1 | 4 | 3 |

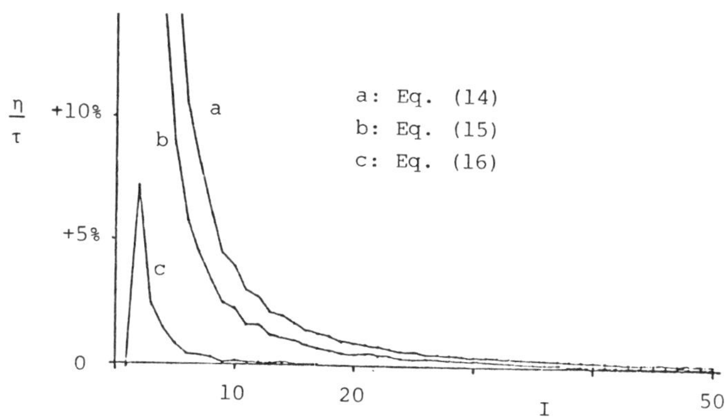


FIGURE 8. Plot percentage error vs. I for different quadrature formulas.

## 4. MAXIMUM ENTROPY METHOD

In this section we propose a new approach to the inversion of aerosol size distribution. It is based on the Maximum Entropy Principle, which is well known in some particular fields (see Jaynes, 1983 for its introduction and other topics), but whose use in solving integral equations by continuous functions is new (Amato et al., 1986b; 1987b). In this section we briefly recall the Maximum Entropy Principle; then we apply it to the integral equations, in order to estimate the parameters of a known distribution. In the present case we look for the best parameters N and b of a modified gamma function (Eq. (2)) that satisfy the synthetic optical depths obtained with the integral equation. We want to stress that the modified gamma function does match many unimodal distributions; however we are improving the method, in order to consider also bimodal distributions. Moreover, the use of a parametric technique, rather than a nonparametric one, is certainly limiting; however it avoids the quadrature error, which, as we have seen, is all but negligible.

The entropy of a continuous probability function p(x) is given by Jaynes (1983):

$$H = - \int_a^b p(x) \log \frac{p(x)}{h(x)} dx \tag{17}$$

where h(x) is a function representing infomation that could be known about the function p(x), and p(x) subject to the normalization constraint

$$\int_a^b p(x) dx = 1 \tag{18}$$

If other constraints are known about the function p(x),

$$F_j(p(x)) = 0 \tag{19}$$

the Jaynes principle of maximum entropy (Jaynes, 1983) affirms that: "The prior probability assignment that describes the available information but is maximally noncommittal with regard to the unavailable information is the one with maximum entropy."
In order to apply the principle to the solution of the integral equation, let us decompose the whole problem with 8 data $(\tau_1,\ldots,\tau_8)$ into $28 = \binom{8}{2}$ subproblems, each one obtained with a different couple $(\tau_i, \tau_j)$, and solve each subproblem independently. Each subproblem $\ell$,

formed by data $\tau_i$ and $\tau_j$, is solved using the Maximum Entropy Principle:

$$\max_{b_\ell,\, p_\ell(r)} \left\{ H_\ell = -\int_0^\infty p_\ell(r) \log \frac{p_\ell(r)}{g(b_\ell;r)} \, dr \right\}$$

$$\text{s.t.} \quad \int_0^\infty p_\ell(r)\, dr = 1 \tag{20}$$

$$\frac{\int_0^\infty \pi r^2 Q_{ext}(\lambda_i,r) \not{N} p_\ell(r)\, dr}{\int_0^\infty \pi r^2 Q_{ext}(\lambda_j,r) \not{N} p_\ell(r)\, dr} = \frac{\tau_i}{\tau_j}$$

where $g(b_\ell;r) = b_\ell^2 r \exp(-b_\ell r)$

$N$ = total number of particles.

Note that the problem only depends on the shape of the particle distribution and that the normalization condition on $p_\ell(r)$ holds, so that the problem is correctly posed in the framework of the Maximum Entropy Principle. It is easy to verify that the maximim value of entropy ($H_\ell$=0) is obtained when $p_\ell(r) = g(b_\ell;r)$, so that the solution of the problem is the value of the parameter $b_\ell$, which satisfies the nonlinear equation

$$\int_0^\infty r^2 (\tau_i\, Q_{ext}(\lambda_j,r) - \tau_j\, Q_{ext}(\lambda_i,r))\, g(b_\ell;r)\, dr = 0 \tag{21}$$

By solving each subproblem, we obtain 28 estimates of $b_\ell$ and consequently of $p_\ell(r)$. The final estimate of N and b will be obtained by minimizing the functional

$$\sum_{i=1}^{8} \left( \int_0^\infty \pi r^2 Q_{ext}(\lambda_i,r)\, N\, g(b;r)\, dr - \tau_i \right)^2 + \\ + \Lambda \sum_{\ell=1}^{28} \int_0^\infty p_\ell(r) \log \frac{p_\ell(r)}{g(b;r)}\, dr \tag{22}$$

with respect to N and b.

This minimum problem corresponds to finding the solution of the integral equation by regularization and choosing the entropy of the system as smoothing operator. $\Lambda \geqq 0$ is the regularization parameter, to be evaluated someway. It represents a trade-off between the fidelity of the restored optical depths to the input ones ($\Lambda \rightarrow 0$) and the complete smoothness of the solution, regardless of the input data ($\Lambda \rightarrow \infty$). In order to test the capability of the method, we have inverted the optical depths relative to the 75%$H_2SO_4$ aerosol component (N=9088; b=18), corrupted by Gaussian error 5%, 10%, 15% and 20%. The validity of the retrieved solution is evaluated by the index $I_1$ (Eq. (10)). We have performed the analysis on 100 samples. For each one we have estimated N, b, and calculated the index $I_1$. The values obviously depend on $\Lambda$. In Fig. 9 we show the plot $I_1$ vs. $\Lambda$ for the different errors on input data. It can be seen that the best choice of $\Lambda$ is about 0.1 to 0.2 and is independent of the error affecting the data. In Table 4 we show the mean value of N, b and $I_1$ over the 100 samples, and the root mean square error for $\Lambda$=0.2 and $\Lambda \rightarrow 0$, which corresponds to a classical least squares method. It is seen that the improvement obtained by the Maximum Entropy Principle is significant. As an example we show in Figs. 10 and 11 the true and retrieved distributions for $\Lambda$=0.2 relative to the values of the parameters shown in the table, for the cases with 10% and 20% error in input data, respectively.

Finally we want to stress that the method we are proposing is quite young, so that the very promising results we present still need a great effort in order to apply the method to more general cases.

## 5. CONCLUSIONS

This paper deals with the problem of retrieving aerosol size distribution from ground measurements of solar radiation. We think that too many uncertainties still affect the approach to the inverse problem. This paper is devoted to the definition of some test problems, to be solved under controlled conditions, so that the actual capacity of the inversion methods can be evaluated. In addition, the influence of some approximations involved in some methods is analyzed. This kind of analysis is more important, since the ill-conditioning of the problems greatly amplifies any source of error present in the inversion methods. Finally, a new inversion method based on the Maximum Entropy Principle is proposed, which overcomes some problems of the conventional methods. It has been applied to a particular case; the results are very promising and suggest its extension to general cases.

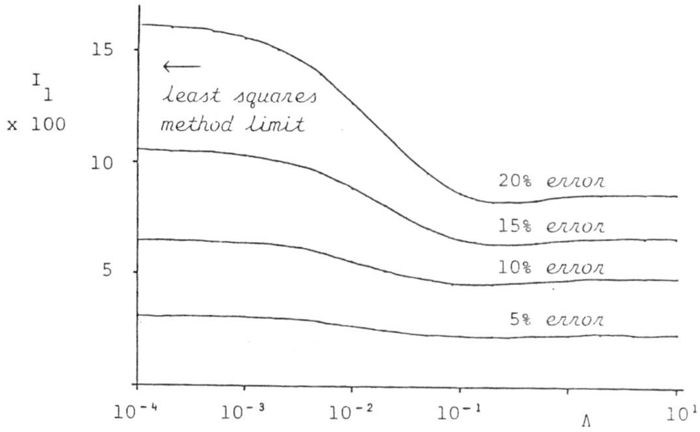


FIGURE 9. Plot error in retrieval vs. Λ for different errors.

TABLE 4. ERROR IN RETRIEVING THE SOLUTION

| Method | error | N | (RMSE) | b | (RMSE) | $I_1 \times 100$ | (RMSE) |
|---|---|---|---|---|---|---|---|
| Least | 5% | 9200 | (1200) | 18.0 | (0.7) | 3.1 | (2.4) |
| Squares | 10% | 9600 | (2600) | 18.1 | (1.4) | 6.5 | (5.6) |
| Method | 15% | 10300 | (4500) | 18.3 | (2.2) | 10.5 | (11.1) |
| | 20% | 11500 | (8100) | 18.5 | (3.1) | 16.2 | (22.4) |
| Maximum | 5% | 8900 | (750) | 17.8 | (0.4) | 2.2 | (1.5) |
| Entropy | 10% | 8800 | (1400) | 17.7 | (0.7) | 3.7 | (2.7) |
| Method | 15% | 8600 | (2000) | 17.5 | (1.0) | 5.4 | (4.0) |
| (Λ=0.2) | 20% | 8500 | (2700) | 17.3 | (1.3) | 7.2 | (5.5) |

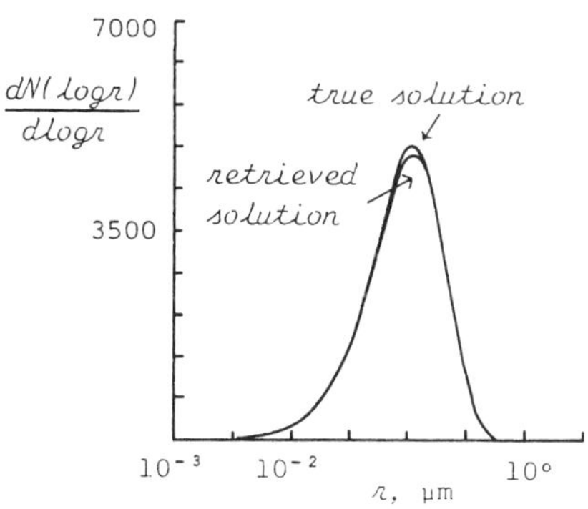


FIGURE 10. True and retrieved solutions (Λ=0.2) for 10% error.

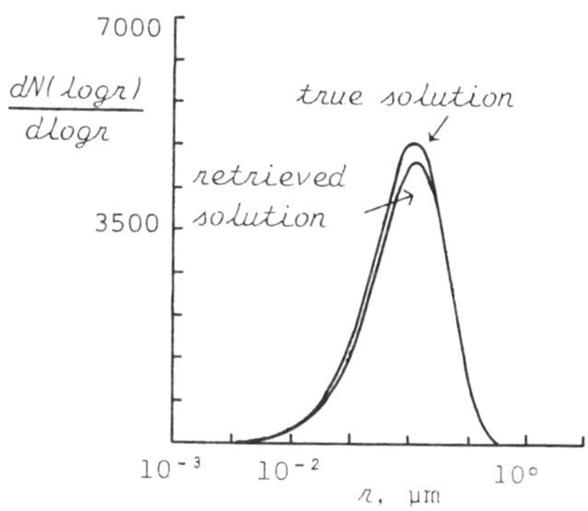


FIGURE 11. True and retrieved solutions (Λ=0.2) for 20% error.

## REFERENCES

Amato, U., V. Cuomo, F. Fontana and C. Serio, 1986a: Some Mathematical Aspects of Retrieving Aerosol Size Distribution. In CUEN (Ed.), _Proceedings of the International Meeting on Atmospheric Transparency for Satellite Data_, Napoli.

Amato, U., V. Cuomo, F. Fontana and C. Serio, 1986b: A Maximum Entropy Techniques to retrieve Aerosol Size Distribution. In CUEN (Ed.), _Proceedings of the International Meeting on Atmospheric Transparency for Satellite Data_, Napoli.

Amato, U., V. Cuomo, F. Fontana, M. Macchiato and C. Serio, 1987a: A Maximum Entropy Method in Inverse Problems to Retrieve Aerosol Size Distributions, XII General Assembly of the European Geophysical Society, Strasbourg.

Amato, U., V. Cuomo and C. Serio, 1987b: _A General Overview to Inverse Problems: The Case of Aerosol Size Distribution_, IAM Technical Report 29/87, Istituto per Applicazioni della Matematica, V. P. Castellino 111, 80126 Napoli, Italy.

Jaynes, E.T., 1983: _Papers on Probability, Statistics and Statistical Physics_, Riedel, Holland.

King, M.D., 1982: Sensitivity of Constrained Linear Inversions to the Selection of the Lagrange Multipliers, _J. Atm. Sci._, _39_, 1356-1369.

Van De Hulst, H.C., 1957: _Light Scattering by Small Particles_, J. Wiley & Sons, New York.

World Meteorological Organization, 1986: _A Preliminary Cloudless Standard Atmosphere for Radiation Computation_, WCP-112 WMO/TD - No. 24.

# ROCKET AND BALLOON MEASUREMENTS OF THE VERTICAL DISTRIBUTION OF AEROSOLS IN THE INDIAN TROPICAL REGION

*A. Jayaraman and B.H. Subbaraya*
Physical Research Laboratory
Ahmedabad 380009, India

ABSTRACT

During the past few years an experimental programme has been initiated at the Physical Research Laboratory, Ahmedabad, India to study the vertical distribution of aerosols in the tropical middle atmosphere. Several rocket and balloon measurements have been made of the angular distribution of scattering of the solar radiation in different wavelength bands from the Thumba Equatorial Rocket Launching Station (8.5° N) and the TIFR Balloon Facility at Hyderabad (17.5° N), respectively. The data have been inverted to estimate the aerosol characteristics in the altitude region of 5 km to about 35 km. It is found that the aerosol size distribution is altitude dependent. Smaller values of the size parameter $\nu$ indicate that a relatively larger portion of bigger size particles are found in the altitude region 20 to 25 km than at lower and higher altitudes. The aerosol attenuation coefficients generally show the existence of structures in the height profile and also a certain amount of variability from one measurement to another. A set of measurements made during February 1980 gave large aerosol attenuation of about $7.6 \times 10^{-4}$ ($km^{-1}$) at 20 km altitude, presumably due to the aerosol loading by the Sierra Negra volcanic eruption. Measurements made during 1984 and 1985 show that the 1984 values are somewhat larger than the 1985 values in the altitude region 20 to 25 km. This difference is attributed to the El Chichon eruption in 1982 whose effect had not yet completely subsided during 1984.

## 1. INTRODUCTION

Aerosols affect the climate by modifying the earth's radiative balance. Several theoretical studies (e.g., Luther, 1976; Harshvardhan, 1979) show that the climatic effects of aerosol are due to the enhanced reflectivity of solar radiation (cooling), with relatively less increase in the infrared opacity (warming). The amount of reflected solar radiation and the infrared opacity of aerosol strongly depend on the refractive index/types of aerosols (Herman and Browning, 1975). Uncertainties in the various physical and chemical processes (Turco et al., 1982) that govern the vertical distribution of aerosol concentration and its size distribution function, and the large spatial and temporal variation of aerosol abundances due to volcanic eruptions, make atmospheric aerosol study complex. Attempts are being made to characterise the atmospheric aerosols

AEROSOLS AND CLIMATE
Peter V. Hobbs and M. Patrick McCormick (Eds.)


ISBN 0-937194-11-5

using satellite (e.g., McCormick et al., 1979), in situ (e.g., Rosen and Hofmann, 1977), and ground-based lidar (e.g., Fiocco and Grams, 1964) observations.

A programme of in situ measurements of atmospheric aerosol parameters in the tropical middle atmosphere using rocket-borne photometers was initiated at the Physical Research Laboratory, Ahmedabad, India, in 1980. Later, under the Indian Middle Atmosphere Programme, several measurements were made on the vertical distribution of aerosol concentration and the size distribution function using rocket as well as balloon-borne photometric techniques over Thumba (8.5° N) and Hyderabad (17.5° N), respectively. The instrument details, the data analysis procedure, and some preliminary results were reported earlier (Subbaraya and Jayaraman, 1982; Jayaraman et al., 1987). In this paper the results obtained so far, both over Thumba and Hyderabad for the period 1980 to 1986, are presented and discussed.

## 2. MEASUREMENTS

Our first aerosol measurements using rockets were made from the Thumba Equatorial Rocket Launching Station (TERLS), Thumba, India. The altitude variation of the direct solar radiation intensity and the angular distribution of scattered radiation intensity in the 310 nm wavelength region was measured on three rocket flights conducted on 16 February 1980 at 1454 hr, 1522 hr, and at 1522 hr on 17 February 1980. Scattered radiation could be measured only up to an altitude of about 23 km. Derived aerosol concentrations and the slope of the power-law size distribution curve corresponding to the three measurements were reported earlier (Subbaraya and Jayaraman, 1982). For comparison with other data, the mean of the three measurements is shown in the present paper.

A balloon-borne suntracking multichannel photometer was developed (Acharya et al., 1985), which tracks the sun in elevation and scans the sky in the azimuthal direction for direct and scattered radiation measurements. The instrument was first flown on a balloon platform from the Balloon Facility at Hyderabad on 20 October 1983. However, due to the premature bursting of the balloon at the tropopause level and noise in the telemetered signals, the experiment was not successful. A repeat launch was successfully made on 18 April 1984. The balloon reached a ceiling altitude of 32 km and measurements were made in four wavelength regions, namely, 290 nm, 310 nm, 450 nm and 800 nm. Measurements made in the UV regions (290 nm and 310 nm) are inverted to obtain the ozone concentration profile. While the data corresponding to 450 nm was found to be noisy, the 800 nm data was inverted to obtain aerosol concentration and size distribution function in the 10 to 32 km altitude region. However, there was an intermittent data gap in the altitude region 17 to 22 km due to high undulations of the balloon platform.

An improved version of the instrument with a six-channel photometer was successfully launched on 22 October 1985. Aerosol parameters were obtained in the altitude region of 10 to 35 km using data from 450 and 600 nm channels. Also a vertical distribution of ozone concentration was obtained in the altitude region 10 to 35 km (Subbaraya, 1987).

Recently in an attempt to compare the data from the rocket experiment and the ground-based lidar measurement, made by the Space Physics Laboratory, Trivandrum (Parameswaran et al., 1984), an aerosol intercomparison experiment was conducted at Thumba during the first week of October 1986. Data on aerosols could be obtained from the rocket-borne photometer in the altitude region 5 to 35 km from the radiation measurements in the 750 nm wavelength region. Data on ozone concentrations could be obtained in the altitude region 18 to 67 km using the attenuation data at 250 and 290 nm. A detailed analysis of the data from the rocket and the lidar experiments is still in progress and will be reported at a later stage. In the following section we present the results from the rocket experiments and compare them with the earlier rocket measurements in February 1980 from the same site, as well as the balloon measurements of 1984 and 1985 from the Hyderabad site.

## 3. RESULTS

The aerosol concentration values are estimated from the ratio of the aerosol scattering function to the Rayleigh scattering function derived from the direct as well as scattered radiation measurements at different scattering angles and at different altitude levels The power-law size distribution defined as $dn(r) = kr^{-(\nu+1)}$ is assumed for the aerosol size distribution. $\nu$ , the slope of the power law curve is determined by comparing the experimentally observed angular variation of the aerosol scattering function with that of the theoretical estimation for different values. A detailed description of the data analysis procedure can be found in Subbaraya and Jayaraman 1982, Jayaraman 1985.

The derived power law size distribution parameters and k were found altitude dependent. The obtained $\nu(z)$ values along with the associated uncertainties are presented in Table 1. The uncertainties in $\nu$ are due to the field of view of the photometer (typical value is 9°), assumptions of an altitude independent refractive index value of 1.5, and the lower and upper size limit of the particles, taken as 0.04 µm and 10.0 µm respectively. The parameter k, is a function of aerosol concentration and could be estimated from the derived aerosol concentrations and $\nu$ values at each altitude level.

The $\nu$ values corresponding to the February 1980 period over Thumba are found to be lower than over Thumba on 3 October 1986 and over Hyderabad, both on 18 April 1984 and 22 October 1985, showing the presence of a relatively large number of bigger

TABLE 1. THE SLOPE $\nu(z)$ OF THE POWER LAW SIZE DISTRIBUTION CURVE OBTAINED FROM THE ROCKET AND BALLOON EXPERIMENTS

| Alt. km. | Thumba Mean, Feb.'80 $\nu(z)$ | Hyderabad 18 Apr.'84 $\nu(z)$ | Hyderabad 22 Oct.'85 $\nu(z)$ | Thumba 3 Oct.'86 $\nu(z)$ |
|---|---|---|---|---|
| 6 | 2.9 ± 0.2 | - | 3.0 ± 0.2 | 2.9 ± 0.1 |
| 8 | 2.8 ± 0.2 | - | 3.0 ± 0.2 | 2.9 ± 0.1 |
| 10 | - | 3.0 ± 0.2 | 3.1 ± 0.1 | 3.0 ± 0.1 |
| 12 | 2.5 ± 0.2 | 3.0 ± 0.2 | 3.1 ± 0.1 | 3.0 ± 0.1 |
| 14 | 2.4 ± 0.2 | 3.0 ± 0.2 | 3.1 ± 0.1 | 3.0 ± 0.1 |
| 16 | 2.2 ± 0.2 | 2.9 ± 0.2 | 3.1 ± 0.1 | 3.0 ± 0.1 |
| 18 | 2.2 ± 0.2 | - | 3.1 ± 0.1 | 2.9 ± 0.1 |
| 20 | 2.2 ± 0.2 | - | 3.1 ± 0.1 | 2.8 ± 0.1 |
| 22 | 2.2 ± 0.2 | 2.8 ± 0.2 | 2.9 ± 0.1 | 2.8 ± 0.1 |
| 24 | - | 2.9 ± 0.2 | 2.9 ± 0.1 | 2.7 ± 0.1 |
| 26 | - | 3.0 ± 0.2 | 2.9 ± 0.1 | 2.7 ± 0.1 |
| 28 | - | 3.3 ± 0.2 | 3.2 ± 0.2 | 2.8 ± 0.2 |
| 30 | - | 3.5 ± 0.2 | 3.4 ± 0.2 | 3.0 ± 0.2 |
| 32 | - | 3.6 ± 0.2 | - | 3.3 ± 0.2 |
| 34 | - | - | - | 3.6 ± 0.2 |

particles at all altitudes. However, in general, all four measurements show a decrease in $\nu$ value in the altitude region 20 to 25 km, showing the existence of a layer of aerosol characterised by a relatively larger number of bigger particles. The $\nu$ values increase continuously above about 26 km, showing a relative increase in the number of smaller particles. $\nu$ values corresponding to 3.5 and more represent mostly Aitken type or condensation nuclei (CN) type of particles.

Figure 1 shows altitude profiles of the aerosol concentrations obtained over Thumba and Hyderabad. It should be noted here that particles of radii smaller than about 0.04 µm behave more like air molecules and will not be sensitive to Mie scattering observations. Also, for deriving the aerosol concentration values from the measured radiation intensities, independently determined $\nu$ values at various altitude levels are necessary. Hence, any uncertainty in the determined $\nu$ value will introduce an error in the derived aerosol concentrations. It is found that the maximum error in the computed aerosol concentrations is ±40%, corresponding to the maximum error ±0.2 in the $\nu$ value.

Large variations were found in the aerosol concentration values both at tropospheric and stratospheric levels. Also the concentration values obtained over Thumba both during February 1980 and on 3 October 1986 were higher than the Hyderabad values of 1984 and 1985. There is a possibility that both the results obtained over Thumba were influenced by volcanic eruptions that

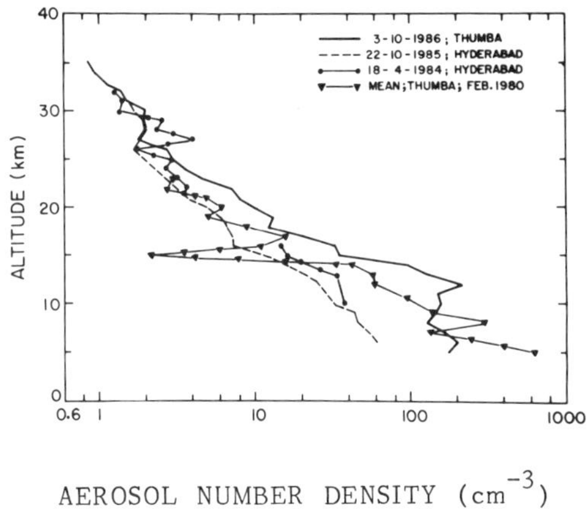


FIGURE 1. Altitude profiles of the aerosol concentrations obtained over Thumba and Hyderabad.

occurred in the equatorial region a few months prior to the measurements. Also, all of the measurements show an increase in the aerosol concentration values in the altitude region 26 to 30 km. This increase in aerosol concentration associated with an increase in the $\nu$ values shows the increased abundance of CN type particles in the 26 to 30 km region over the equatorial region. As shown by Turco et al. (1982), these high concentrations of CN type particles observed over the equatorial region are transported polewards and become the source of bigger particles at lower altitudes over the high latitude regions.

Aerosol scattering coefficients were computed from the aerosol concentrations and $\nu$ values using the procedure described by Subbaraya et al. (1984). Figure 2 shows the altitude profiles of the computed aerosol scattering coefficients for 1 µm for the four sets of measurements made over Thumba and Hyderabad. It is seen that the scattering coefficients corresponding to the February 1980 measurements show larger values at all altitudes. It has been shown (Jayaraman et al., 1987) that these large aerosol scattering coefficient values represent the influence of the Sierra Negra volcano eruption in November 1979, about three months prior to the measurements.

Comparison of the two aerosol scattering coefficient profiles obtained over Hyderabad on 18 April 1984 and on 22 October 1985 show that the former is higher at all altitudes below about 25 km. As it has been shown by several workers (e.g., Hofmann and Rosen, 1987), the effect of the El Chichon eruption in April

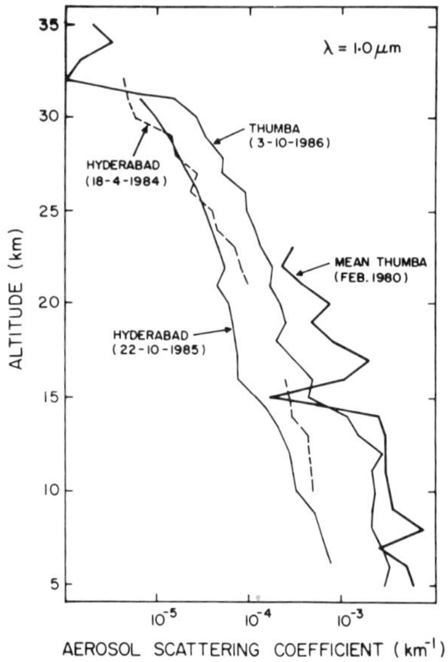


FIGURE 2. Altitude profiles of the computed aerosol scattering coefficients for 1 μm.

1982 had not completely subsided, even two years after the eruption. Also, there have been no other volcanic eruptions during the 1983-84 period. The larger values of aerosol scattering obtained on 18 April 1984 compared to those of 22 October 1985 demonstrate the decay of the effects of the El Chichon eruption.

The aerosol scattering coefficients computed for the 3 October 1986 measurement also show larger values compared to the Hyderabad measurements. Since there was another volcanic eruption, the eruption of the Nevado del Ruiz (5° N) volcano situated in Columbia, on 13 November 1985, it is suspected that the recent Thumba measurement, made about ten months after the eruption, is also representative of a volcanically disturbed condition. It is worth noting here that, in the case of El Chichon, the perturbation caused in the optical properties of the atmosphere reached the maximum about ten months after the eruption, as shown by Jagar and Carnuth, 1987.

## 4. CONCLUSIONS

The following features can be summarised from the Thumba and Hyderabad measurements on the atmospheric aerosols over tropical India :

1) Large variations well beyond the estimated maximum possible error of ±40% are found in the aerosol characteristics obtained over Thumba and Hyderabad at all altitude levels.

2) All measurements show the existence of a layer of aerosol particles characterised by a relatively larger number of bigger particles in the altitude region 20 to 25 km.

3) Above about 26 km, there is an increase in the concentration of the condensation nuclei type particle with altitude, in the altitude region 26 to 30 km.

A few more balloon and rocket measurements, planned for the 1988-89 period, along with a few new in situ measurements on the optical properties of aerosol, such as polarizability and wavelength dependence of aerosol scattering, are expected to reveal additional information on the types of aerosols and their variations over the tropical regions.

## REFERENCES

Acharya, Y.B., A. Jayaraman and B.H. Subbaraya, 1985: A Balloon-Borne Suntracking Multichannel Photometer for Atmospheric Aerosol Measurement, Adv. Space Res., 5, 65-68.

Fiocco, G., and G.Grams, 1964: Observation of Aerosol Layer of 20 km by Optical Radar, J. Atmos. Sci., 21, 323-324.

Harshvardhan, 1979: Perturbation of the Zonal Radiation Balance by a Stratospheric Aerosol Layer, J. Atmos. Sci., 36, 1274-1285.

Herman, B.M., and S.R. Browning, 1975: The Effects of Aerosol on the Earth-Atmosphere Albedo, J. Atmos. Sci., 32, 1430-1445.

Hofmann, D.J., and J.M. Rosen, 1987: On the Prolonged Lifetime of the El Chichon Sulfuric Acid Aerosol Cloud, J. Geophys. Res., in press.

Jager, H., and W. Carnuth, 1987: The Decay of the El Chichon Stratospheric Perturbation Observed by LIDAR at Northern Midlatitudes, Geophys. Res. Letts., 14, 696-699.

Jayaraman, A., B.H. Subbaraya and Y.B. Acharya, 1987: The Vertical Distribution of Aerosol Concentration and Size Distribution Function over the Tropics and Their Role in Radiative Transfer, Physica Scripta, 36, 358-361.

Jayaraman, A., 1985: Measurement of Atmospheric Minor Constituents Using Rocket-Borne Techniques, Ph.D. Thesis, Gujarat University, Ahmedabad, India.

Luther, F.M., 1976: Relative Influence of Stratospheric Aerosols on Solar and Longwave Radiation Fluxes for a Tropical Atmosphere, J. Appl. Meteor., 15, 951-955.

McCormick, M.P., P. Hamill, T.J. Pepin, W.P. Chu, T.J. Swissler and L.R. McMaster, 1979: Satellite Studies of Stratospheric Aerosols, Bull. Amer. Meteor. Soc., 60, 1038-1046.

Parameswaran, K., K.O. Rose and B.V. Krishnamurthy, 1984: Aerosol Characteristics from Bistatic LIDAR Observations, J. Geophys. Res., 89, 2541-2552.

Rosen, J.M., and D.J. Hofmann, 1977: Baloon-Borne Measurements of Condensation Nuclei, J. Appl. Meterol., 16, 52-56.

Subbaraya, B.H., and A. Jayaraman, 1982: Aerosol Concentrations and Size Distribution in the Troposphere and Lower Stratosphere over Thumba, Pure and Appl. Geophys., 120, 407-421.

Subbaraya, B.H., A. Jayaraman and Y.B. Acharya, 1984: Aerosol Size Distribution and Number Densities at the Tropospheric and Lower Stratospheric Altitudes Over the Tropics and Their Role in Radiative Transfer. In G. Fiocco (Ed.), IRS '84: Current Problems in Atmospheric Radiation, A. Deepak Publishing, Hampton, Virginia, 89-92.

Subbaraya, B.H., 1987: Vertical Distribution of Minor Constituents in the Tropical Middle Atmosphere, Indian J. Radio and Space Phys., 16, 25-35.

Turco, R.P., R.C. Whitten and O.B. Toon, 1982: Stratospheric Aerosols: Observation and Theory, Rev. Geophys., 20, 233-279.

# AN APPROACH TO A GLOBAL OPTICAL AEROSOL CLIMATOLOGY

*G.A. d'Almeida and P. Koepke*
Meteorologisches Institut der Univeristät München
D-8000 München 2, Federal Republic of Germany

ABSTRACT

Ten tropospheric aerosol types have been considered and characterized by size distributions fitted by independent lognormal components. The influence of the ambient relative humidity, between 50% and 99%, has been assessed using the so-called mixture rule for the chemically heterogeneous particles. As for the monosource aerosol types the particle growth has been computed separately for each mode of the lognormal distribution with a simultaneous consideration of the change in particle refractive index and a subsequent addition of the modal contributions. The pertaining radiative characteristics of both dry and wet particles are computed using the Mie theory for 60 wavelengths in the spectral range between 0.3 µm and 40 µm, and for a spatial and temporal resolution of 5 deg latitude, 5 deg longitude, and 1 month respectively. Space and time variations of the physical and radiative properties follow the seasonal migration of the climatic belts that act as an effective barrier to meridional mixing. In the light of these prerequisites global scattering coefficients and absorption coefficients are computed and discussed for the wavelengths 0.500 µm and 10 µm, for January, and for dry aerosol particles as examples of the resulting large data set. Maritime and mineral aerosols turn out to be the most efficient scatters in the visible region and the main absorbers in the thermal infrared window.

## 1. INTRODUCTION

While there exists among scientists a wide consensus on the potential importance of atmospheric particles on solar and terrestrial radiations, and on cloud formation processes, the magnitude of these effects has not been well quantified. The degree to which atmospheric aerosols will warm or cool the atmosphere depends on their absorptivity and backscattering properties as well as the reflection characteristics of the underlying ground. Due to different sources and production mechanisms, the atmospheric aerosol represents a physically and chemically heterogeneous "system" of components with highly variable optical behaviour. A more or less realistic computation of aerosol-induced effect on global climate, as well as of the influence on atmospheric dynamics, is contingent upon the extent to which the properties of the aerosol components in the system are taken into account. A recent sensitivity study conducted by Tanré et al. (1984), using the European Centre for Medium Range Weather Forecasts (ECMWF)



ISBN 0-937194-11-5

low resolution model, indicates that the "effect of aerosol on climate is significant for temperature but is just at the limit of the noise level for the dynamic." The model by Tanré et al. (1984) considers, however, only three tropospheric aerosol types. For many reasons it is, indeed, a complex task to properly introduce aerosol properties in global models:

1) There is a considerable gap in the spatial distribution as the data are not available.

2) There are some deficiencies in the temporal representativeness and lack of coherence in the observed data. The latter is generally due to the different methods used to measure quantities such as the particle size distribution.

3) Very little is known about the high spatial and temporal variability of the parameters upon which tropospheric aerosol depends.

Probably, in the awareness of these deficiencies, Shettle and Fenn (1979) have presented average properties for aerosol models by splitting tropospheric aerosols into their components. Following the work by Shettle and Fenn (1979) the aerosol models of the Standard Radiation Atmosphere (SRA model), reported by Deepak and Gerber (1983), were set up, consisting of three main tropospheric aerosol types and four aerosol components. The SRA-aerosol models designed for initial sensitivity tests in climate models need some improvements to reflect the current knowledge on aerosols. The present work discusses the results of a comprehensive literature study of tropospheric aerosol models, their subsequent use for the computation of the radiative characteristics, and their distribution in a global net considering seasonal variations. Stratospheric aerosols are not considered here, since they have been widely investigated (Hofmann and Rosen, 1980; McCormick et al., 1981; Russell et al., 1981a, 1981b).

## 2. AEROSOL MODEL

Atmospheric aerosol components are characterized by their production sources that determine their chemical composition and make up their refractive index and by their production mechanisms that determine size spectrum and shape of the particles. The transport processes induced by large-scale processes (general circulation pattern, convective processes, etc.) and microscale processes (Brownian diffusion, coagulation, etc.) enable the mixture of the airborne aerosol components to form a given aerosol type. Relying on the above-mentioned work by Shettle and Fenn (1979) and a recommendation of the International Radiation Commission (Deepak and Gerber, 1983) that gives the size distribution of an aerosol type in a bimodal or trimodal lognormal distribution, each component is characterized by a single mode with an assigned refractive index; 10 aerosol components and 10 aerosol types are defined. Table 1 shows an overview of the different aerosol types considered in the present study, their components, and examples of locations in which they occur. Besides continental and urban aerosols, defined by Shettle and Fenn (1979), and reported in SRA models (Deepak and Gerber, 1983), clean-polar aerosol (Arctic in summer and Antarctic throughout the year), polar-polluted (Arctic in winter), desert aerosol, clean-maritime, maritime-mineral aerosol, maritime-polluted aerosol, clean-continental, and clean-forest aerosols have been introduced.

Polar-polluted aerosol refers to that type of heavily polluted particulate matter encountered in wintertime in the arctic region and consisting of dust-like substances, sea salt particles, and aged continental or anthropogenic particulate matter from midlatitudes (Borys and Rahn, 1981; Heintzenberg, 1981; Rahn, 1981; Rosen et al., 1981). Clean-polar aerosol identifies both the aerosol types in the Antarctic throughout the year and the Arctic in summertime. Antarctic aerosol consists mostly of sea salt particles, desert originated substances, meteorite, sulfuric acid droplets, and sulfate particles generated in the stratosphere (Hogan, 1975; Parungo et al., 1979; Shaw, 1979). The general circulation pattern prevents the Eurasian anthropogenic sources from reaching the arctic region in summertime and facilitates a poleward transport in wintertime. Observations in different remote oceanic areas made by Junge and Jaenicke (1971) and Meszaros and Vissy (1974) and compiled by Jaenicke (1984) have been chosen to describe the aerosol properties in a clean-maritime environment. That aerosol model differs from the SRA-maritime aerosol model (Deepak and Gerber, 1983). Desert aerosol consists of purely monosource aerosol with mineral substances generated in arid and semiarid regions. Depending on the prevailing weather conditions, desert aerosol exhibits the highest variability of all aerosol types with respect to its microphysical parameters (d'Almeida, 1987). The ability of desert aerosol to travel over long distances and, among others, over the world oceans, leads to a mixing with sea spray aerosol (Prospero et al., 1979; Schütz, 1980; Péwé, 1981; Tsunogai and Kondo, 1982; Uematsu et al., 1983). We call that type of mixture "maritime-mineral" aerosol. In the vicinity of heavily polluted regions and near industrialized harbour areas, soot particles and other anthropogenic substances are transported towards maritime environments and interact with the background aerosol (Gerber, 1979). We refer to that type of particulate matter as "maritime-polluted" aerosol. Clean-continental or rural aerosol and clean-forest aerosol are mixtures of particles encountered in remote continental environments. They consist of naturally produced components. Carbonaceous aerosol, converted sulfate, and similar species containing aerosol particles of anthropogenic origins are referred to as soot in Table 1. Similarly, natural sulfate, sulfuric acid, and other sulfuric aerosol particles produced by nucleation of natural trace sulfur-bearing gases are just termed sulfate.

TABLE 1: TROPOSPHERIC AEROSOL TYPES, COMPONENTS, AND LOCATIONS

| Aerosol types | Components | Locations |
|---|---|---|
| Clean-continental or Remote-continental | * water-soluble<br>* dust-like | Rural environment |
| Clean-forest | * water-soluble<br>* biogenic<br>* dust-like | Amazon, African, and Asian thick forests |

| | | |
|---|---|---|
| Continental | * water-soluble<br>* soot<br>* dust-like | Continental environment slightly influenced by man-made pollution |
| Desert | * mineral | Arid and semiarid regions |
| Urban or Industrial | * water-soluble<br>* soot<br>* dust-like | highly polluted environment |
| Clean-maritime | * sea-salt<br>* NSS-sulfate | Southeast Pacific, South equatorial Atlantic, Indian Ocean |
| Maritime-mineral | * sea-salt<br>* mineral | maritime environment influenced by desert aerosol |
| Maritime-polluted | * water-soluble<br>* soot<br>* sea-salt | Mediterranean, North Atlantic |
| Polar-polluted | * soot<br>* mineral<br>* sea-salt | Arctic in wintertime |
| Clean-polar | * mineral<br>* sea-salt | Arctic in summertime |
| | or | |
| | * sea-salt<br>* mineral<br>* sulfate<br>* (meteorite) | Antarctic |

The available literature has been critically reviewed to derive the needed raw quantities such as size distribution or size distribution-related quantities, elemental or chemical composition, refractive index, and profile. Additionally, the data generated in the framework of the WMO-BAPMoN-Programme (WMO, 1974-1984) that cover a considerable part of the Earth's land surface have been used as a valuable source to augment the available data. The bulk refractive index of the considered components were tabulated by Volz (1972, 1973), Palmer and Williams (1974), and Shettle and Fenn (1979). More details on the aerosol models will be published elsewhere.

Assuming that the particle shape is spherical, the Mie theory is applied to compute the radiative characteristics of the different aerosol types. These quantities are the extinction, absorption, and scattering coefficients as well as the optical depth, the asymmetry factor, the single scattering albedo, and the phase function. Refractive index and the aforementioned radiative characteristics are functions of the wavelength of the radiation. The computation is performed for discrete values in the spectral range between 0.3 $\mu$m and 40 $\mu$m. Aerosol types and radiative characteristics are horizontally distributed in a global grid of 5 deg latitude and 5 deg longitude. A temporal variation is considered in terms of aerosol composition, mixing ratio, and number density on a monthly basis. Space and time variations are included in compliance with the seasonal migration of the different climatic belts that determine the latitudinal extent of pollutants with sources located in the lowest atmospheric layers.

The high spectral, spatial, and temporal resolution gives rise to a considerable data set structured in modules to allow, if required, the consideration of current knowledge. The data set consists of:

1) raw (primary) data including geographical coordinates, aerosol type, profile type, relative humidity class, particle number density, number of components of a given aerosol type, type of component, and component-mixing ratio for each month;

2) computed monochromatic radiative characteristics: extinction coefficient, scattering coefficient, absorption coefficient, single scattering albedo, asymmetry factor, phase function for a selected scattering angle, and backscattering ratio as functions of wavelength, relative humidity, location, and time (all these quantities are normalized to a number density 1 $cm^{-3}$);

3) absolute radiative characteristics: same quantities as 2) multiplied by the considered particle number density, and

4) computed optical depth as a function of wavelength, relative humidity, location, and time.

A specific vertical distribution has been set up for each aerosol type as the computed optical characteristics refer to some meters above ground. The water uptake by atmospheric particles is considered following essentially the model by Hänel (1976), Hänel and Lehmann (1981), and the approach by Shettle and Fenn (1979). The transition point at which a solid particle becomes saturated or supersatured depends on the particle size and its chemical composition. For 0% (dry particles), 50%, 70%, 80%, 90%, 95%, 98%, and 99% relative humidity values, the growth of the aerosol particles is computed separately for the different components by modifying the mode radius of the partial lognormal distribution by humidity changes and assuming negligible possible variations in the standard deviation. The mean refractive index is obtained by means of the so-called mixture rule. A detailed discussion on relative humidity dependence of the aerosol optical properties will be published elsewhere.

## 3. GLOBAL DISTRIBUTION OF AEROSOL OPTICAL PROPERTIES

Half-tone pictures have been chosen to present the spatial and temporal variations of the physical and optical quantities. The radiative characteristics, as far as they depend upon the highly variable particle number density, exhibit a large gradient in magnitude. It may then happen that two adjacent grid points depict appreciable discrepancies. A considerable number of contour lines or half-tone pictures are then drawn between the two points that lead to the artifact around the continental outlines.

Figures 1a and 1b show the computed global scattering coefficients for January, for dry particles and for the wavelengths 0.5 µm and 10.0 µm. The first wavelength is thought to represent the solar spectrum while the second one is located in the middle of the infrared atmospheric window. For 0.5 µm, high scattering features appear in the

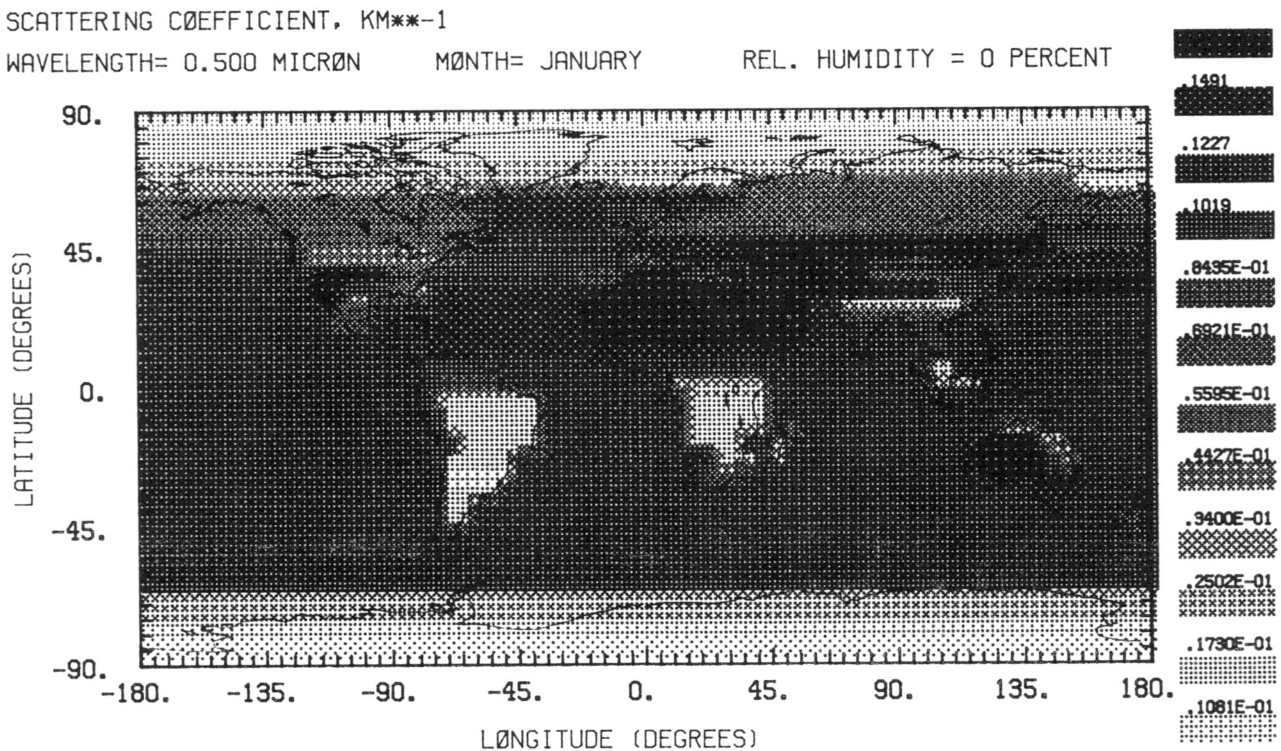


FIGURE 1a: Global scattering coefficients for January, for dry aerosol and for the wavelength 0.5 µm, presented in half-tone picture along with the continental outlines. All values lower than $0.1838 \times 10^{-2}$ are plotted with the lowest half-tone picture intensity. The intensity levels are the same for all figures.

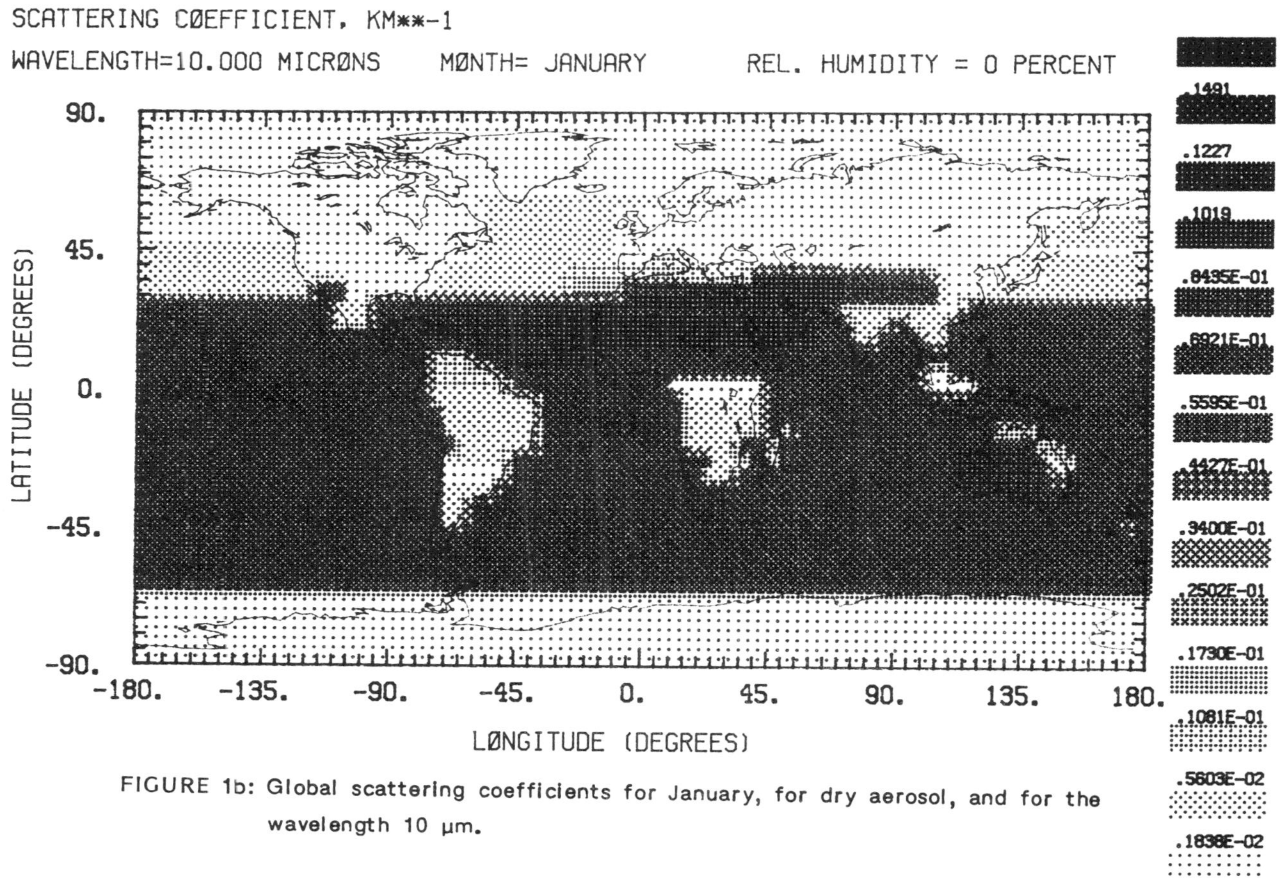


FIGURE 1b: Global scattering coefficients for January, for dry aerosol, and for the wavelength 10 μm.

entire maritime environment. That characteristic is strengthened by the presence of mineral particles in the trade wind regions of the Atlantic and Pacific Oceans. The arid regions depict higher scattering coefficient values primarily due to the large and effective scattering particle diameter. Rural environment as well as polar regions present the lowest scattering properties.

Similar to the smaller wavelength, the global scattering coefficients in the thermal infrared atmospheric window at 10 μm (Fig. 1b) show a relatively high scattering coefficient in the clean-maritime environment and in arid regions. In contrast to clean-maritime environment, polar regions and urban environments display low scattering features with values below $1*10^{-5}$ $km^{-1}$ and $1*10^{-3}$ $km^{-1}$, respectively. Furthermore, it becomes apparent that even for large wavelengths the monosource natural aerosols, i.e., clean-maritime and desert aerosols, show relatively high scattering coefficients with values between 0.05 $km^{-1}$ and 0.15 $km^{-1}$.

Figure 2a shows the global absorption coefficients for 0.5 μm wavelength. The figure points out a relatively low absorption feature in the Arctic, although the tremendous amount of graphitic carbon from the Eurasian sources (Heintzenberg, 1980; Heidam, 1981, 1984; Rosen et al., 1981 and others) has been included in the present model. For January and for the 0.5 μm wavelength, an absorption coefficient value of $0.428*10^{-3}$ $km^{-1}$ and a single scattering albedo of 0.967 have been computed respectively. A lower single scattering albedo value is expected in February and a higher value in the summer months in compliance with the different concentrations of graphitic carbon. Apparently, the absorption coefficient of the clean-maritime environment ($9.48*10^{-6}$ $km^{-1}$) is the lowest of all aerosol types considered in the present study and exhibits the highest single scattering albedo of 1, a characteristic of the whole visible spectrum. The fact that maritime-mineral aerosol (about $0.9*10^{-2}$ $km^{-1}$) and maritime-polluted aerosol ($1.38*10^{-2}$ $km^{-1}$) indicate stronger absorption features than the clean-maritime aerosol supports the idea of distinguishing between different aerosol types in the maritime environment. The same holds for the differentiation between clean-continental and continental aerosols.

Figure 2b represents the computed global absorption coefficient for 10.0 μm with predominant absorption features for the two natural aerosol types: clean-maritime ($2.41*10^{-2}$ $km^{-1}$) and mineral (about $8.36*10^{-2}$ $km^{-1}$) and negligible in the polar regions: Arctic ($2.30*10^{-4}$ $km^{-1}$) and Antarctic ($2.56*10^{-4}$ $km^{-1}$). Since the spatial and temporal variability in particle number density is more pronounced in arid regions than in the maritime environment and the absorption coefficient is a function of particle number density, desert aerosol turns out to be the strongest absorber in the atmospheric window. This finding is corroborated by previous investigations by Levin et al., (1980).

## 4. CONCLUDING REMARKS

The present study is an attempt to supply researchers with

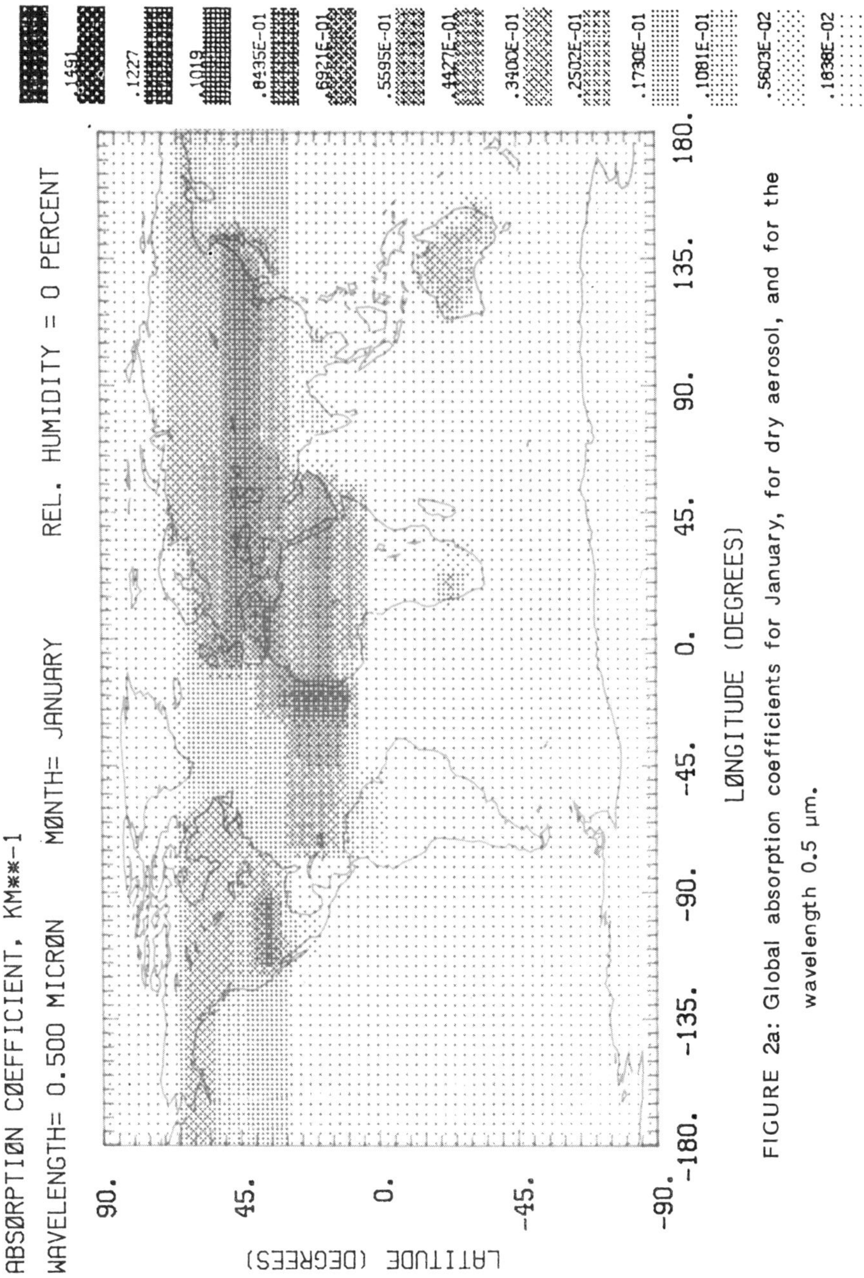


FIGURE 2a: Global absorption coefficients for January, for dry aerosol, and for the wavelength 0.5 µm.

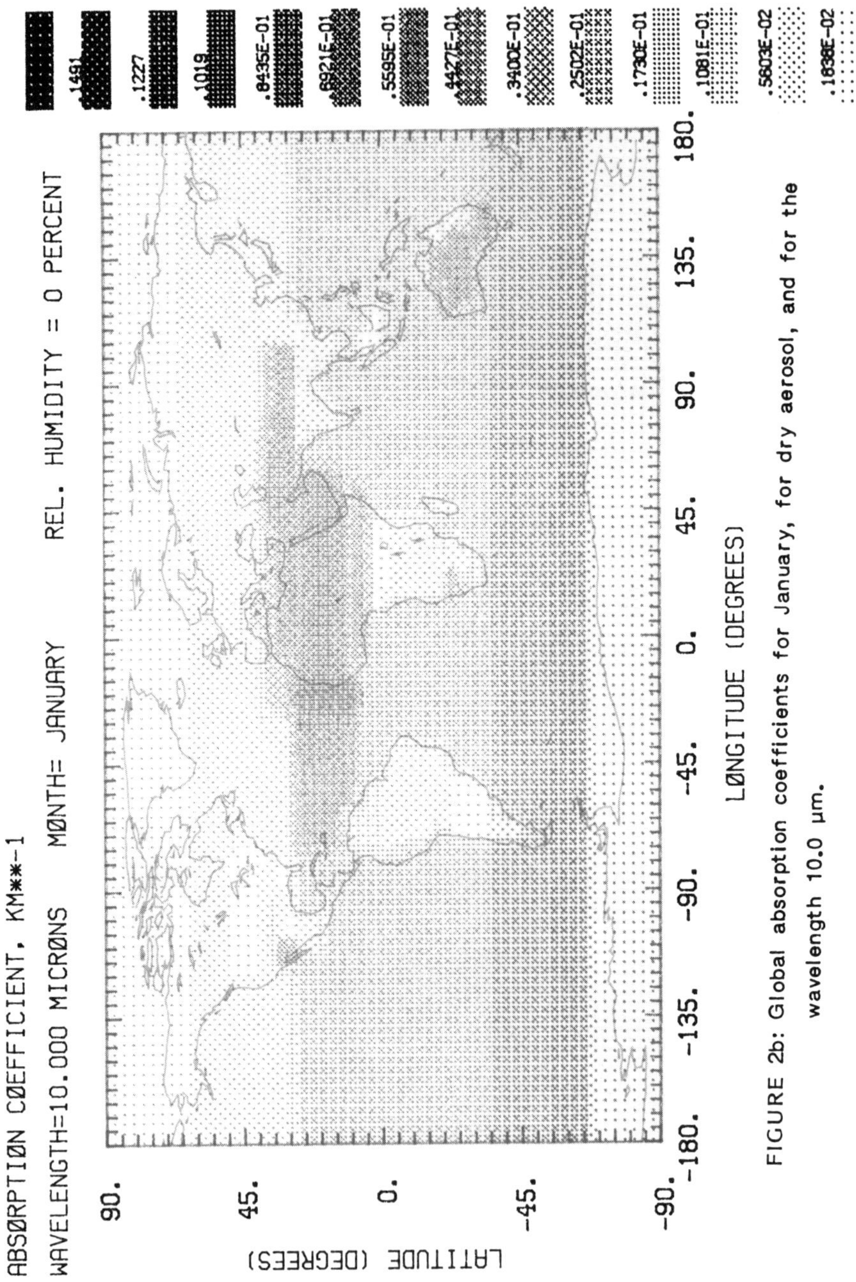


FIGURE 2b: Global absorption coefficients for January, for dry aerosol, and for the wavelength 10.0 μm.

aerosol data useful for climate modeling purposes and, to a certain extent, for remote sensing applications. The results shown in the present paper are only a tiny choice of the large data set generated. We believe that the study is comprehensive enough to fulfill the assigned requirement. Although these data, probably, describe the present state of knowledge of atmospheric aerosol optical properties, they hold at the same time some shortcomings due to simplifications, assumptions, and the lack of information needed. The last is a real drawback for an unbiased aerosol climatology. For instance, very little is known about the aerosol properties on the Russian subcontinent, the Australian deserts, the space and time variability of aerosols in the maritime environment, as well as the productivity, the characteristics, and the space and time variability of the biogenic compound. The scientific community is required for its appreciable endorsement.

## REFERENCES


d'Almeida, G.A., 1987: On the Variability of Desert Aerosol Radiative Characteristics, J. Geophys. Res., 92, 3017-3026.

Borys, R.D., and K.A. Rahn, 1981: Long-range Atmospheric Transport of Cloud-active Aerosol to Iceland, Atmos. Environ., 15, 1491-1501.

Deepak, A., and H.E. Gerber (Eds.), 1983: Report of the Experts Meeting on Aerosols and Their Climatic Effects (WCP-55), World Meteorological Organization, Geneva, Switzerland.

Gerber, H.E., 1979: Absorption of 632.8 nm Radiation by Maritime Aerosols Near Europe, J. Atmos. Sci., 36, 2502-2575.

Hänel, G., 1976: Influences of Physical and Chemical Properties of Atmospheric Particles on the Activation Process in Fog and Clouds, Contrib. Atmos. Phys., 54, 159-170.

Hänel, G., and M. Lehmann, 1981: Equilibrium Size of Aerosol Particles and Relative Humidity: New Experimental Data from Various Aerosol Types and Their Treatment for Cloud Physics Application, Contrib. Atmos. Phys., 54, 57-71.

Heidam, N.Z., 1981: On the Origin of the Arctic Aerosol: A Statistical Approach, Atmos. Environ., 15, 1421-1427.

Heidam, N.Z., 1984: The Components of the Arctic Aerosol, Atmos. Environ., 18, 329-343.

Heintzenberg, J., 1980: Particle Size Distribution and Optical Properties of Arctic Haze, Tellus, 32, 251-260.

Heintzenberg, J., 1981: The Chemical Composition of Arctic Haze at Ny-Alesund, Spitzbergen, Tellus, 33, 162-171.

Hofmann, D.R., and J.M. Rosen, 1980: Sulfuric Acid Layer; Evidence for Anthropogenic Components, Science, 208, 1368-1370.

Hogan, A.W., 1975: Antarctic Aerosols, J. Appl. Meteor., 14, 550-559.

Jaenicke, R., 1984: Aerosol Physics and Chemistry, Meteorology Volume in Landholt-Bornstein, to be published.

Junge, C.E., and R. Jaenicke, 1971: New Results in Background Aerosols Studies from the Atlantic Expedition of the R.V. Meteor, Spring 1969, J. Aerosol Sci., 2, 305-314.

Levin, Z.L., J.H. Joseph and Y. Mekler, 1980: Properties of Sharav (Khamsin) Dust - Comparison of Optical and Direct Sampling Data, J. Atmos. Sci., 37, 882-891.

McCormick, M.P., W.P. Chu, G.W. Grams, P. Hamill, B.M. Herman, L. R. McMaster, T.J. Pepin, P.B. Russell, H.M. Steele and

T.J. Swissler, 1981: High Latitude Stratospheric Aerosols Measured by the SAM II Satellite System in 1978 and 1979, Science, 214, 328-331.

Meszaros, A., and K. Vissy, 1974: Concentration, Size Distribution and Chemical Nature of Atmospheric Aerosol Particles in Remote Oceanic Areas, J. Aerosol Sci., 5, 101-109.

Palmer, K.F., and D. Williams, 1974: Optical Properties of Water in the Near Infrared, J. Geophys. Res., 80, 1107-1110.

Parungo, F., E. Ackerman, W. Caldwell and H.K. Weickmann, 1979: Individual Particle Analysis of Antarctic Aerosols, Tellus, 31, 521-529.

Péwé, T.L., 1981: Desert Dust: Origin, Characteristics, and Effect on Man, Special paper 186, Geophysical Society of America, Boulder, Co., 303 pp.

Prospero, J.M., D.L. Savoie, T.N. Carlson and R.T. Nees, 1979: Monitoring Saharan Aerosol Transport by Means of Atmospheric Turbidity Measurements. In C. Morales (Ed.), Saharan Dust, Scope 14, John Wiley & Sons, Chichester, New York, Brisbane, Toronto, 171-193.

Rahn, K., 1981: Relative Importance of North America and Eurasia as Sources of Arctic Aerosol, Atmos. Environ., 15, 1447-1457.

Rosen, H., T. Novakov and B.A. Bodhaine, 1981: Soot in the Arctic, Atmos. Environ., 15, 1371-1374.

Russell, P.B., T.J. Swissler, M.P. McCormick, W.P. Chu, K.M. Livingston and T.J. Pepin, 1981a: Satellite and Correlative Measurements of the Stratospheric Aerosol: I. An Optical Model for Data Conversions, J. Atmos. Sci., 38, 1279-1294

Russell, P.B., M.P. McCormick, T.J. Swissler, W.P. Chu, J.M. Livingston, W.H. Fuller, J.M. Rosen, D.J. Hofmann, L.R. McMaster, D.C. Woode and T.J. Pepin, 1981b: Satellite and Correlative Measurements of the Stratospheric Aerosol: II. Comparison of Measurements Made by SAM II, Dustsondes, and Airborne Lidar, J. Atmos. Sci., 38, 1295-1302.

Schütz, L., 1980: Long-range Transport of Desert Dust with Special Emphasis on the Sahara, Ann. N.Y. Acad. Sci., 338, 515-532.

Shaw, G.E., 1979: Considerations on the Origin and Properties of the Antarctic Aerosol, Rev. Geophys. Space Phys., 8, 1983-1998.

Shettle, E.P., and R.W. Fenn, 1979: Models for the Aerosols of the Lower Atmosphere and the Effects of Humidity Variations on Their Optical Properties, AFGL-TR-79-0214, Environ. Res. Paper, No. 675, 94 pp.

Tanré, D., J.F. Geleyn and J. Slingo, 1984: First Results of the Introduction of an Advanced Aerosol-Radiation Interaction in the ECMWF Low Resolution Global Model. In: H.E. Gerber and A. Deepak (Eds.), Aerosols and Their Climatic Effects, A. Deepak Publishing, Hampton, Virginia, 133-177.

Tsunogai, S., and T. Kondo, 1982: Sporadic Transport and Deposition of Continental Aerosols to the Pacific Ocean, J. Geophys. Res., 87, 8870-8874.

Uematsu, M., R.A. Duce, J.M. Prospero, L. Chen, J.T. Merrill and R.L. McDonald, 1983: Transport of Mineral Aerosol From Asia Over the North Pacific Ocean, J. Geophys. Res., 88, 5343-5352.

Volz, F.E., 1972: Infrared Refractive Index of Atmospheric Aerosol Substance, Appl. Opt., 11, 755-759.

Volz, F.E., 1973: Infrared Optical Constants of Ammonium Sulfate, Saharan Dust, Volcanic Pumice, and Flyash, Appl. Opt., 12, 564-568.

World Meteorological Organization, 1974-1984: Global Atmospheric Background Monitoring for Selected Environmental Parameters, Background Air Pollution Monitoring Network (BAPMoN)-data, WMO-Environment Division, Geneva, Switzerland.

# LONG-TERM SERIES OF ATMOSPHERIC TURBIDITY, ESTIMATED FROM RECORDS OF SUNSHINE DURATION AND CLOUD COVER

*L. Helmes and R. Jaenicke*
Institute of Meteorology
6500 Mainz, Federal Republic of Germany

ABSTRACT

Atmospheric turbidity as a measure of the load of the atmosphere with substances other than clean air -- especially aerosols -- is an important element in the climatic system. As long-term series of atmospheric turbidity are rare, a method is presented estimating turbidity from daily records of sunshine duration and cloud cover. Data of these meteorological parameters have been available for a period up to 100 years. Using this method we obtained series of turbidity for several stations of the globe. These climatological series can be examined for trends or abrupt changes as may be caused, for example, by volcanic eruptions. While the effects of major volcanic eruptions can be clearly seen, the presence of a general long-term trend in turbidity cannot be established.

## 1. INTRODUCTION

Atmospheric turbidity, as a measure of the atmospheric composition, is an important element in the climatic system. Due to its influence on solar radiation, changes in turbidity may involve variations of the global energy budget and climate.

Changes in atmospheric turbidity are mainly caused by the varying aerosol content of the atmosphere. Two large aerosol bodies, the stratospheric and tropospheric aerosol, are able to influence solar radiation through scattering and absorption. The stratospheric aerosol in volcanic quiet periods produces very minor optical and thermal effects. However, after an explosive volcanic eruption the amount of aerosol in the stratosphere is greatly enhanced and stratospheric turbidity can then be as large as that of the tropospheric aerosol. Such a large explosive eruption injects fine ash and gases into the stratosphere. Ash particles settle out and simultaneously the gases react to form sulfate aerosol particles. Thus the effects of volcanic eruptions last for several years (see e.g., Turco et al., 1982). During this time the particles are being transported by stratospheric winds and the volcanic dust veil slowly diminishes. The dust is spread into a latitudinal band that slowly diffuses away from the latitude of the eruption (Robock, 1981). Unusual optical phenomena like Bishop's rings or colourful twilights are an indication of the presence of the enhanced stratospheric dust layer and its impact on solar radiation (Deirmendjian, 1973). For example, Mitchell (1982) described the effects of the eruption of the



ISBN 0-937194-11-5

Mexican volcano El Chichon in March/April 1982: "Fiery sunsets were being reported from an ever-increasing number of places around the world." The influence of the El Chichon eruption cloud on solar radiation was monitored by several authors, e.g., DeLuisi et al. (1983), Jaenicke and d'Almeida (1983), Hoecker et al. (1985). However, there are only a few stations around the world where measurements of direct beam radiation (and thus turbidity) have been carried out. Even in the United States and Canada there are only a few stations that have reported the effect of the El Chichon volcanic cloud (Wendler and Kodama, 1986). As "historical records of atmospheric turbidity are of particular interest in the detection of long-term trends in atmospheric transmission," but "such records are sparse (particularly in the Southern Hemisphere) and are inevitably limited in duration," Delmas and Boutron (1980) use the sulfate deposition in central Antarctic snow as an indicator of the past fluctuations of the stratospheric sulfate aerosol burden.

Jaenicke and Kasten (1978) proposed a method for estimating atmospheric turbidity from records of the Campbell-Stokes sunshine recorder. This instrument has been used at many meteorological stations since its introduction into service at the end of the last century (Stokes, 1880). Due to the wealth of sunshine duration data several methods exist to estimate other parameters, e.g., global radiation, from these data. We agree with Revfeim (1981) who says: "The potential of Campbell-Stokes records of sunshine hours has not been exploited." Application of the method proposed here to those sunshine duration data results in more information about the climate-relevant parameter turbidity.

## 2. ESTIMATING TURBIDITY FROM DATA OF SUNSHINE DURATION AND CLOUD COVER

One of the most severe volcano-related climate effects in historical times was caused by the 1783 eruption of the Laki crater-row in Iceland (Sigurdsson, 1982). Benjamin Franklin observed and documented the effects while serving as ambassador in France (Mitchell, 1982). Among others, he wrote: "They (the sun rays) were indeed rendered so faint in passing through it (the fog) that when collected in the focus of a burning glass, they would scarce kindle brown paper." (For the complete text see e.g., Mitchell, 1982.)

So Franklin had already recognized the principle our method is based on: in sunshine recorders of the Campbell-Stokes type the sun rays are focused and burn a trace on a registration card. The trace is burned, if the sun is not covered by clouds and the direct beam radiation exceeds a certain threshold. In the case of cloudless sunrises and sunsets atmospheric turbidity weakens the direct beam and the threshold value might not be reached so that no sunshine is recorded even when the sun is clearly visible. Sunshine registration starts when the sun is at a certain elevation angle above the horizon. The larger the atmospheric turbidity, the larger the elevation angle and the shorter the registered sunshine duration. With known threshold value it is possible to calculate turbidity from the elevation angle, which can be derived from the time at the beginning (respectively end) of the burned trace (Jaenicke and Kasten, 1978; Helmes and Jaenicke, 1984).

As mentioned above the method has been developed to obtain information about the past course of turbidity. However, historic records of sunshine duration do not contain any information about the state of cloudiness during the low-sun hours. In order to reevaluate these sunshine data with regard to turbidity, cloudiness must be considered. As during the morning and evening hours sunshine duration may be shortened due to enhanced turbidity and/or clouds in the path of the direct beam, we had to approach this problem by considering the probability of turbidity influence on sunshine registration; only in the case of cloudless days is the probability at unity for turbidity influencing sunshine registration. With increasing cloud cover the probability increases for cloud-covered sunrises and sunsets to occur.

Lund and Shanklin (1973) examined a similar problem: they estimated probabilities of so-called "cloud-free lines-of-sight" through the atmosphere, as a function of cloud cover and solar elevation angle. For solar elevations less than around 30°--this is the range relevant for our method-- their results agree well with those probabilities we estimated for turbidity influence.

Examination of the sum of daily data of cloud cover, sunshine duration, and turbidity (determined independently with photometers), as a function of cloud cover, resulted in the following relationship:

$$B + RS + T = 103 + 0.57\ B \qquad (1)$$

where B = mean daily cloud cover (0-100%); RS = relative sunshine duration (registered sunshine duration in percent of possible duration, ranging from 0 to 100%); T = mean daily turbidity, expressed in percent equivalent sunshine duration loss per possible sunshine duration. Eq. (1) was obtained for $B \leq 60\%$ with the correlation coefficient r = 0.980 (Helmes and Jaenicke, 1985). With known daily values of relative sunshine duration and cloud cover, turbidity T can be reevaluated:

$$T = 103 - RS - 0.43\ B \qquad (2)$$

Comparisons of turbidity values estimated from B and RS with the help of Eq. (2), with turbidities determined from usual direct measurements, showed the feasibility of the proposed method (Helms and Jaenicke, 1986; Helmes, 1987).

## 3. APPLICATION

As the shown method has been developed to obtain information about the course of atmospheric turbidity in the past, we tried to obtain daily data of sunshine duration and cloud cover for the past 80 to 100 years. We knew that this would be no easy task because it is very difficult to obtain daily data, especially for the period prior to 1950. Unfortunately, we could not use data of U.S. stations. The reason for this is not the use of another type of sunshine recorder there (the method could also be applied to other instruments). Rather that observers of the U.S. Weather Service have been instructed to include sunshine duration observed, but unrecorded due to atmospheric turbidity, associated with cloudless sunrises and sunsets in the daily totals for sunshine duration (Doehring and Karl, 1981). As the original registrations of the instruments are not available (personal communication, 30 August, 1983, L. Roy Hoxit, National Climatic Data Center, Asheville, North Carolina), we cannot use U.S. sunshine data.

Figure 1 shows the locations with suitable series of daily sunshine duration and cloud cover information that were made available to us. In Europe and Japan not all stations could be marked (there are 11 stations in Japan with records for the period 1893-1984). In most cases data prior to 1950 were not stored on magnetic tape; these data first had to be transferred from yearbooks or copy prints of original data sheets.

For example, the reevaluated time series of two Japanese stations are shown (Fig. 2). Plotted are the annual mean turbidity values (Linke turbidity factor) along with the 30-year low-pass filtered series (Gaussian method) and the significance levels. Also marked are major volcanic eruptions. Remarkable are the high values around the turn of the century that coincide with the enhanced volcanic activity during that time. More remarkable are the unusually

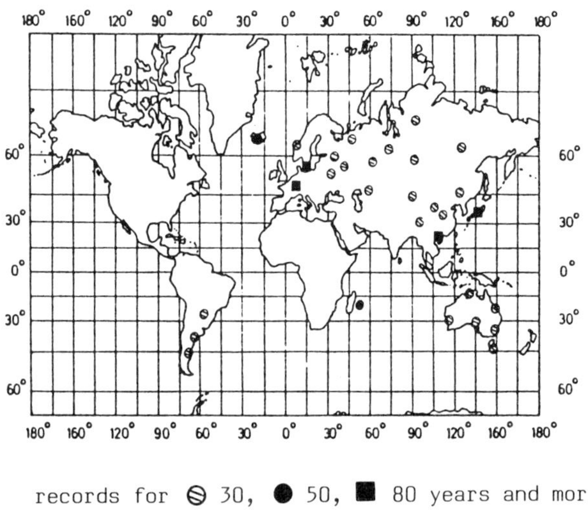


FIGURE 1. Stations with records of daily sunshine duration and cloud cover available at present. Most of them could be reevaluated for turbidity.

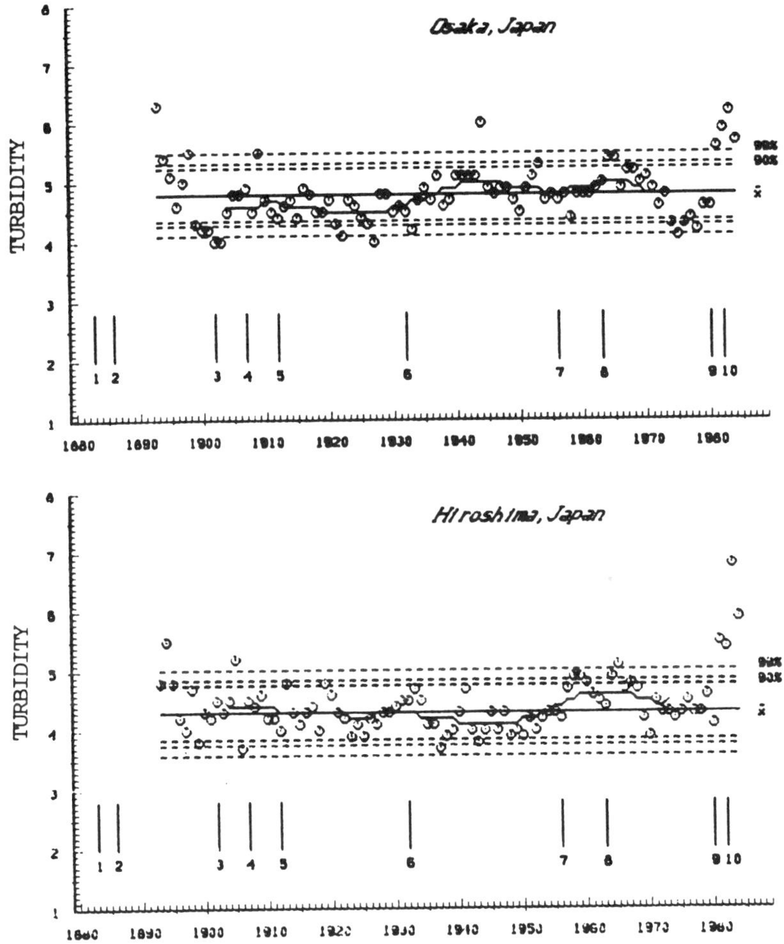


Plotted are the single annual mean values along with the 30-year-low-pass filtered series (Gaussian method) and the significance levels for two Japanese stations, Osaka (34.7°N) and Hiroshima (34.4°N). Also marked are major volcanic eruptions: 1 = Krakatoa (6°S), 2 = Tarawera (38°S), 3 = Santa Maria, Mt.Pelée (15°N), 4 = Ksudach (52°N), 5 = Katmai (58°N), 6 = Quizapu (36°S), 7 = Bezymianny (56°N), 8 = Agung (8°S), 9 = Mount St.Helens (46°N), 10 = El Chichon (17°N).

FIGURE 2. Time series of mean annual turbidity.

high values in the 1980s that were caused by the eruption of El Chichon. It is striking that there were also enhanced turbidities in

1981. This phenomenon can be seen also at the other Japanese stations. At present we have no explanation for this observation. Perhaps there is a relation to "the mystery volcano," which enhanced the stratospheric aerosol layer before the eruption of El Chichon and whose effects were observed at Mauna Loa Observatory, Hawaii (DeLuisi et al., 1983). There cannot be seen a clear long-term trend atmospheric turbidity; the smooth curves vary around the long-term mean and do not leave the significance range. This behaviour is characteristic for most of the cities examined. Only Tokyo shows a positive trend, beginning around the year 1900.

Most of the "comparable" series are correlated. Comparable means: series of such stations are screened that are exposed to similar conditions, like cities, mountain stations, etc. The correlation-coefficients are significant, in spite of the reduced degree of freedoms, due to autocorrelation of the series. The mean turbidity series for Japanese cities show no trend, in contrast to the mean series resulting from the three mountain stations in Europe, Zugspitze (47.4°N), Saentis (47.3°N) and Sonnblick (47.0°N), all around 3000m high (Fig. 3). Around the turn of the century, these data show a downward trend (Helmes, 1987).

## 4. CONCLUSIONS

A proposed method of estimating atmospheric turbidity from records of daily sunshine duration and cloud cover has been applied to a number of stations with suitable records of these two meteorological elements. At the present state a clear long-term trend cannot be seen in atmospheric turbidity. Common to all series available for the past 80 to 100 years are the high values during the beginning of this century that are related to the enhanced volcanic activity during that time. Short time fluctuations often coincide with single volcanic eruptions but there are also variations that cannot be

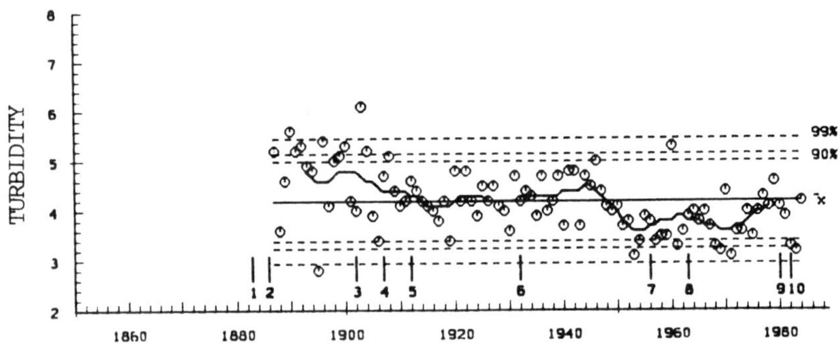


FIGURE 3. Time series of mean annual turbidity, averaged data of mountain stations Zugspitze (47.4°N), Saentis(47.3°N), Sonnblick (47.0°N). (Further details in Fig. 2.)

related to changes in volcanic activity. It seems that during the volcanic quiescent periods changes in turbidity are mainly caused by changes of the tropospheric aerosol. As these tropospheric variations are restricted to specific areas the course of mean hemispheric turbidity is dependent on the selection of the examined stations. Only large volcanic eruptions seem to be able to surpass tropospheric effects and lead to a uniform course of turbidity. Surely, we have not reevaluated all suitable records existing around the globe. However, it is uncertain if the addition of further stations may change the general picture of the course of atmospheric turbidity.

## ACKNOWLEDGMENT

This research was supported by the "Bundesminister für Forschung und Technologie", KF 2010, Germany.

## REFERENCES

Deirmendjian, D., 1973: On Volcanic and other Particulate Turbidity Anomalies, Adv. Geophys., 16, 267-296.

Delmas, R., and C. Boutron, 1980: Are the Past Variations of the Stratospheric Sulfate Burden Recorded in Central Antarctic Snow and Ice Layers?, J. Geophys. Res., 85, 5645-5649.

DeLuisi, J.J., E.G. Dutton, K.L. Coulson, T.E. Defoor and B.G. Mendonca, 1983: On Some Radiative Features of the El Chichon Volcanic Stratospheric Dust Cloud and a Cloud of Unknown Origin Observed at Mauna Loa, J. Geophys. Res., 88, 6769-6772.

Doehring, F., and T. Karl, 1981: A History of Sunshine Data in the United States 1891-1980, National Climatic Center, Asheville, North Carolina, 40pp.

Helmes, L., and R. Jaenicke, 1984: Experimental Verification of the Determination of Atmospheric Turbidity from Sunshine Recorders, J. Climat. Appl. Meteorol., 23, 1350-1353.

Helmes, L., and R. Jaenicke, 1985: Hidden Information Within Series of Measurements - Four Examples from Atmospheric Science, J. Atm. Chem., 3, 171-185.

Helmes, L., and R. Jaenicke, 1986: Atmospheric Turbidity Determined From Sunshine Records, J. Aerosol Sci., 17, 261-263.

Helmes, L., 1987: Klimatologie der Atmosphärischen Trübung - Basierend auf Daten der Sonnenscheindauer und des Bedeckungsgrads, Ph.D. Thesis, University of Mainz, Germany.

Hoecker, W.H., G.F. Cotton and E.C. Flowers, 1985: Variation of Direct Beam Solar Radiation in the United States Due to the El Chichon Debris Cloud, Bull. Amer. Meteor. Soc., 66, 14-19.

Jaenicke, R., and F. Kasten, 1978: Estimation of Atmospheric Turbidity From the Burned Traces of the Campbell-Stokes Sunshine Recorder, Appl. Opt., 17, 2617-2621.

Jaenicke, R., and G.A. d'Almeida, 1983: Atmospheric Turbidity Over Africa - Disturbed by the El Chichon Eruption?, Geophys. Res. Letts., 10, 233-235.

Lund, I.A., and M.D. Shanklin, 1973: Universal Methods for Estimating Probabilities of Cloud-free Lines-of-sight Through the Atmosphere, J. Appl. Meteor., 12, 28-35.

Mitchell, J.M., 1982: El Chichon - Weather-Maker of the Century?, Weatherwise, 35, 252-262.

Revfeim, K.J.A., 1981: Estimating Solar Radiation Income From 'Bright' Sunshine Records, Q.J.R. Meteorol. Soc., 107, 427-435.

Robock, A., 1981: Latitudinally Dependent Volcanic Dust Veil Index, and Its Effect on Climate Simulations, J. Volcanol. Geotherm. Res., 11, 67-80.

Sigurdsson, H., 1982: Volcanic Pollution and Climate: The 1783 Laki Eruption, EOS, 63, 601-602.

Stokes, G.G., 1880: Description of the Card Supporter for Sunshine Recorder Adopted at the Meteorological Office, Q.J.R. Meteorol. Soc., 6, 83-94.

Turco, R.P., R.C. Whitten and O.B. Toon, 1982: Stratospheric Aerosols: Observation and Theory, Rev. Geoph. Space Phys., 20, 233-279.

Wendler, G. and Y. Kodama, 1986: Effect of the El Chichon Volcanic Cloud on the Surface Radiative Regime in Central Alaska, J. Climat. Appl. Meteorol., 25, 1687-1694.

# SIMULATION OF THE DESERT DUST CYCLE USING AN ATMOSPHERIC GENERAL CIRCULATION MODEL

*S. Joussaume and R. Sadourny*
Laboratoire de Météorologie Dynamique
Paris, 75231 Cedex 05, France

ABSTRACT

A modeling of the atmospheric desert dust cycle has been introduced within an atmospheric general circulation model. Source, diffusion, transport, and removal of dust have been included, but, in this first approach, no impact on climate has been considered. Some results of February and August present-day climate are presented. One of the most important features of the simulated results is a strong seasonal dependency of the atmospheric content of dust, with twice as much dust in August as in February.

## 1. INTRODUCTION

Aerosols are important from many viewpoints, e.g., the impact of pollutants on ecosystems and the climatic impact on the hydrologic cycle and on the radiative properties of the atmosphere. Atmospheric general circulation models (AGCMs), which are able to reproduce the main features of climate, can be appropriate tools to study the atmospheric cycles of aerosol particles. They can give an extensive three-dimensional vision of the tracer cycles and can be used for climate sensitivity or forecasting studies.

In this paper, we have modeled the wind-blown dust cycle, which is one important natural source of particles. Moreover, climate modifies the desert dust cycle through changes of source regions and atmospheric circulation. For example, the paleodata in ice cores (Petit et al., 1981) suggest a strong increase of the dust amount for the Last Glacial Maximum (18,000 years before the present time). It is therefore interesting to investigate the sensitivity of the dust cycle with climate using AGCM simulations. We study here the simulation of the present-day climate dust cycle. As a first approach, we have modeled only one size range around 1 $\mu$m, because we are interested in the study of long-range transport. Particles are considered here as passive scalars, which means that their eventual impact on climate, either as condensation nuclei or through their radiative properties, is not included.

## 2. MODELING

Evolution in time of a tracer is governed by source and removal processes, which may be highly dependent on the nature and type of particles, and large-scale transport and diffusion processes, which are the turbulent vertical diffusion within the planetary boundary layer and the convective diffusion within convective clouds. The transport

AEROSOLS AND CLIMATE
Peter V. Hobbs and M. Patrick McCormick (Eds.)


ISBN 0-937194-11-5

and diffusion processes of desert dust have been introduced using modeling schemes, which are similar to those found in our AGCM for the transport and diffusion of water vapor. For a more detailed description of the dust cycle modeling, see Joussaume (1983, 1985) and Joussaume et al. (1984). The AGCM of the Laboratoire de Météorologie Dynamique (LMD) is described in Sadourny and Laval (1984).

## 2.1 SOURCES

Particles are raised by winds in desert areas. The surface flux parameterization over source regions is similar to the flux formulation commonly used for latent and sensible heat. The flux of dust particles depends on a drag coefficient, on the magnitude of wind velocity at the surface, and on the vertical gradient of dust concentration near the surface. At the soil-atmosphere interface, a constant dust concentration $\lambda$ is assumed, which is unknown. However, as all processes in the dust cycle are linear relative to the dust concentration, $\lambda$ is taken to equal 1.0, which defines all dust quantities up to an arbitrary constant.

For source regions, two approaches are possible: either prescribed as known from observations, or generated by the model itself as regions of dry soil. The first approach is ideally the best one for comparisons with observations, but it is somewhat limited because the source regions of fine dust are not well known. The second approach is more interesting for climate sensitivity studies, such as paleoclimatic simulations or when feedback processes are included. We have used the second approach in the simulations presented here. However, it is important to point out that the definition of source areas is certainly the most critical point in the modeling.

## 2.2 REMOVAL

Removal processes for dust particles are highly dependent on the size range of particles. In the 1 $\mu$m range, particles in the atmosphere act as condensation nuclei and are removed by rainout. This process has been parameterized with the simple assumption that dust particles and water vapor are removed in the same proportion by condensation.

Although weak, gravitational settling has been included with a velocity of 30 m/day. Moreover, dry deposition occurs near the ground, for example by interception of dust by obstacles. We have therefore introduced a parameterization of dry deposition that assumes the surface to be a perfect sink for particles, with a flux parameterization similar to the dust source. As a first approach, the same drag coefficient has been used for turbulent fluxes of dust particles and water vapor, neglecting specific properties of particles. This parameterization leads to dry deposition velocities in the range of 0.5-1.0 cm/s over oceans and of 1.0-5.0 cm/s over continents. Such velocities seem to be an overestimation for particles of 1 $\mu$m. However, when a whole size distribution of particles is considered, including bigger particles, such high values are reasonable (Giorgi, 1986). Therefore, our modeling is probably more representative of a size distribution of particles rather than particles of 1 $\mu$m only.

## 3. RESULTS

### 3.1 SIMULATIONS

Simulations of February and August present-day climate (100 days each) have been performed using the LMD AGCM with the standard spatial resolution of 64 points in longitude, 50 points in latitude, and 11 levels, that corresponds to a horizontal mesh size of 400 x 400 $km^2$ at 50° of latitude. In the initial state, dust concentration is set at zero. After 10 days, an equilibrium regime of the atmospheric cycle of dust is achieved. The results presented here are averaged over the last 60 days of the simulations.

### 3.2 DUST TRANSPORT

The extent of source regions (Fig.1) is given by the regions of low simulated soil water content both in February and August, and corresponds fairly well to observed arid regions (Unesco, 1979). However, the Australian desert is underestimated. In the model, the two prevailing source regions are Sahara(≃50%) and Arabia-Asia (≃40%).

Figure 2 displays the vertically integrated mass of dust (arbitrary units) for February and August. In tropical regions, dust plumes are transported at low levels towards the West by trade winds, as is observed for Saharan dust (Prospero et al., 1981). In August, the monsoon regime transports Saharan and Arabian-Asian dust towards the East. In middle latitudes, westerlies dominate, as well as in upper levels. In February, a noticeable northward transport of Saharan dust is simulated over the Atlantic, associated with a persistent transient circulation pattern during the simulated period.

Figure 2 suggests that a strong seasonal dependency exists with, globally, twice as much dust content in August as in February. The intensification of source fluxes (factor 1.5) can only account for part of the large seasonal difference, the other part being certainly due to a difference in the atmospheric circulation.

### 3.3 DUST REMOVAL

The comparison of dry and wet removal charts (Fig.3) generally shows stronger dry removal around source regions and lesser in remote areas. Estimating lifetimes as the ratio of the total amount of dust in the atmosphere to the total removal flux gives a dry deposition lifetime of the order of 3-4 days and a wet removal lifetime of the order of 20 days. Globally, dry deposition dominates wet deposition by a factor of 3.5 in August and 4.5 in February, which might be an overestimation.

### 3.4 VERTICAL STRUCTURE

As dust plumes rise, they spread and the total concentration decreases. The vertical structure exhibits a strong seasonal dependency: in August, the global amount of dust in the free

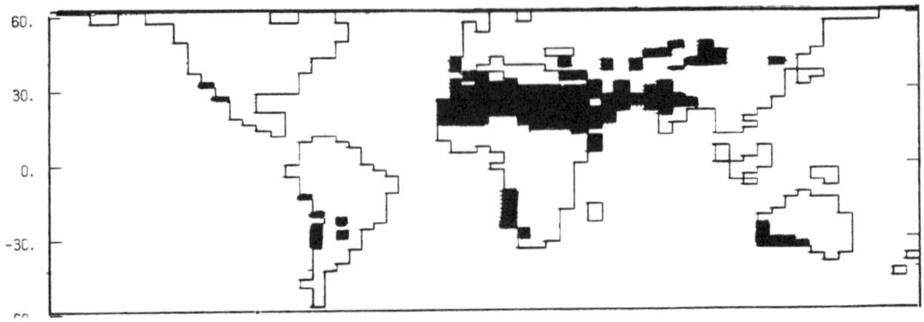


FIGURE 1. Simulated extent of source regions.

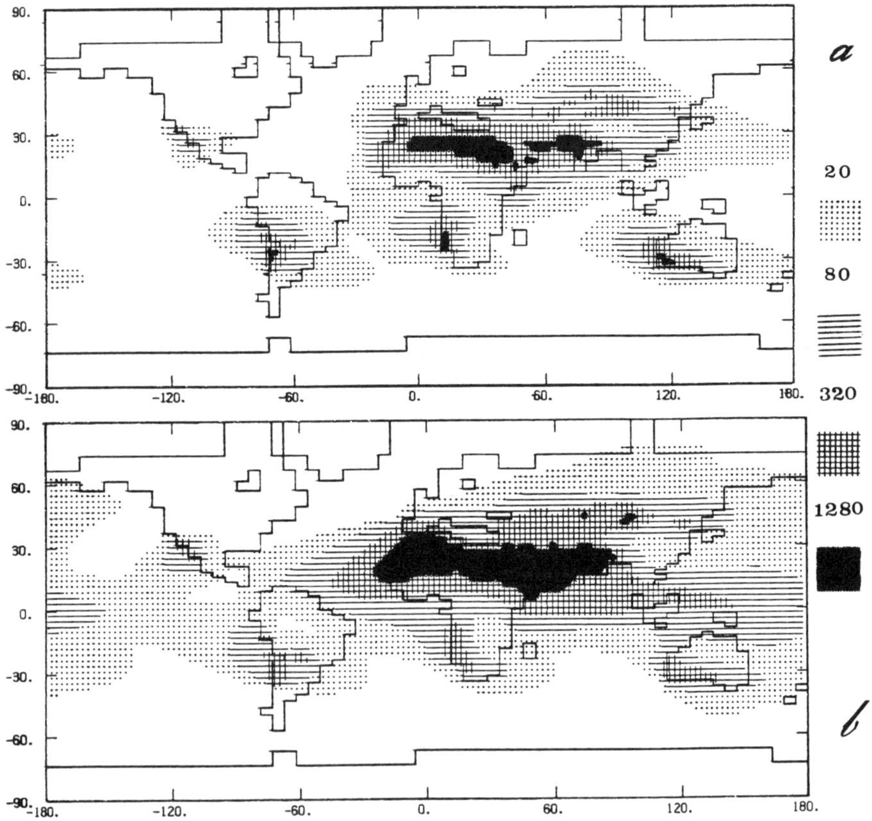


FIGURE 2. Simulated mean vertical mass of dust for present-day climate: a) February and b) August. Unit kg $m^{-2}$, arbitrary constant. Intervals of shading areas: 20- 80- 320- 1280.

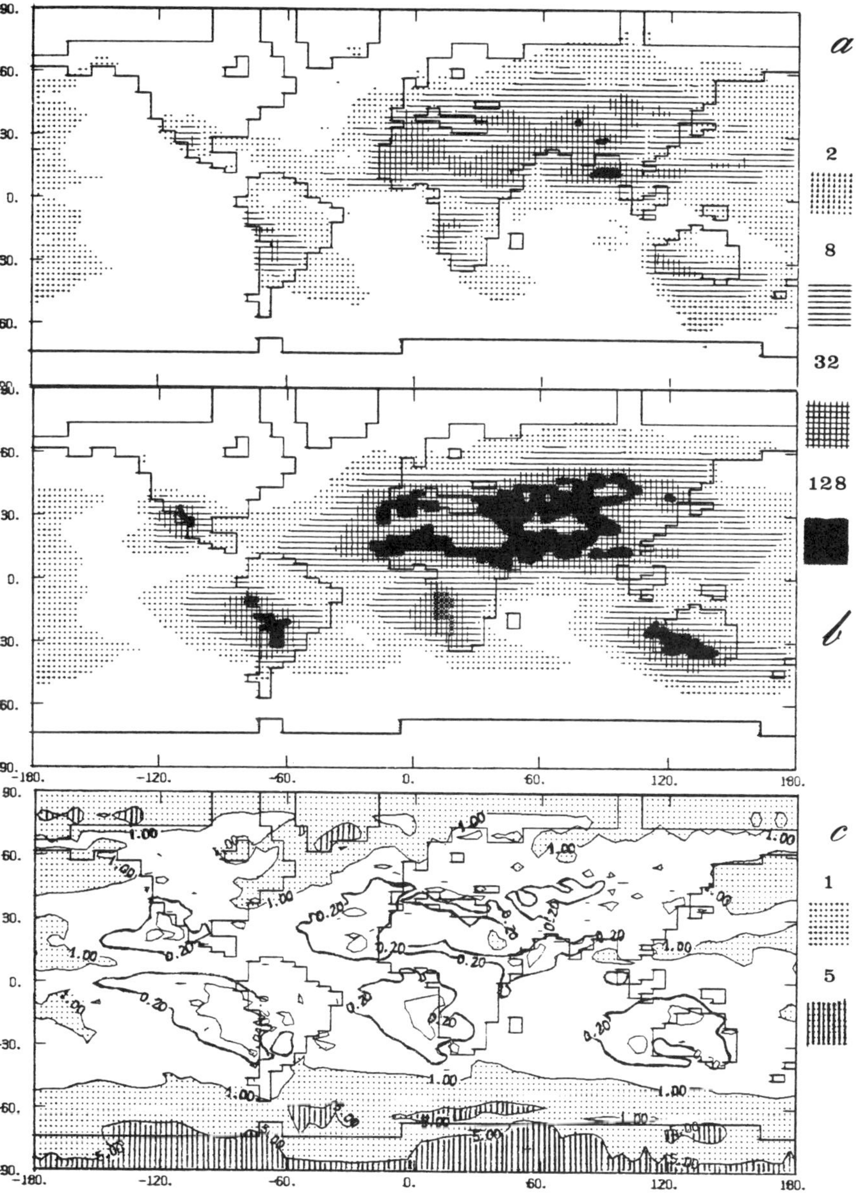


FIGURE 3. Estimated annual average (February + August) of: a) wet and b) dry removal (unit kg $m^{-2}day^{-1}$, arbitrary constant; intervals of shading areas, 2- 8- 32- 128), c) ratio of wet to dry removal (isolines 1/25- 1/5- 1- 5, shading for values of 1).

troposphere and in the boundary layer are of the same order; in February, the free tropospheric content is 4 times less than the boundary layer one (the global budgets essentially reflect the Saharan and Arabian-Asian dust sources). This difference is certainly due to more convective activity in August. The vertical ascent induces an increase in the proportion of wet removal relatively to dry removal, thus enhancing the lifetime of particles, and subsequently the atmospheric dust content.

## 4. CONCLUSION

These results show briefly how general circulation models can help in the knowledge of desert dust cycles: horizontal and vertical dust plumes, removal, and seasonal contrast. Although our results are reasonable, it is, however, hard to tell how realistic they are. Moreover, there still remains a number of critical modeling problems that need further improvement, especially source and removal processes.

## ACKNOWLEDGMENTS

The computer time was contributed by the Centre de Calcul Vectoriel pour la Recherche. We thank G. Rabreau for his contribution in the developments of graphics.

## REFERENCES

Giorgi F., 1986: A Particle Dry-Deposition Parameterization Scheme for Use in Tracer Transport Models, Journal of Geophysical Research, 91, 9794-9806.

Joussaume S., 1983: Modélisation des Cycles des Espèces Isotopiques de l'Eau et des Aérosols d'Origine Désertique dans un Modèle de Circulation Générale de l'Atmosphère. (Thèse de 3ème cycle, Université de Paris VI,1983).

Joussaume S., S.I. Rasool, R. Sadourny, J-R. Petit, 1984: Simulation of Desert Dust Cycles in an Atmospheric General Circulation Model, Annals of Glaciology, 5, 204-207.

Joussaume S., 1985: Simulation of Airborne Impurity Cycles Using Atmospheric General Circulation Models, Annals of Glaciology, 7, 131-137.

Petit J-R., M. Briat and A. Royer, 1981: An Ice Age Aerosol Content from East Antarctic Ice Core Samples in Past Wind Strength, Nature, 289, 391-394.

Prospero J.M., R.A. Glaccum and R.T. Nees, 1981: Atmospheric Transport of Soil Dust from Africa to South America, Nature, 289, 570-572.

Sadourny R., and K. Laval, 1984: January and July Performance of the LMD General Circulation Model. In A. Berger and C. Nicolis (Eds.), New Perpectives in Climate Modelling, Developments in Atmospheric Sciences, 16, Elsevier, 173-198.

Unesco, 1979: Carte de la Répartition Mondiale des Régions Arides, Technical Notes MAB, 7.

# PHYSICAL PROPERTIES OF AEROSOLS IN RURAL AND FORESTED AREAS[1]

*Josef Podzimek*
Graduate Center for Cloud Physics Research
University of Missouri-Rolla
Rolla, Missouri 65401, USA

ABSTRACT

In this work we review some of the measurements of the physical properties of atmospheric aerosols in rural and forested areas in the USA and abroad. Some of the results are compared with measurements in the forested savannah and at Rolla, Missouri in the middle of the forested area, which extends to the Lake of the Ozarks in Missouri. The data on aerosol concentration, size distribution, and nuclei activation show the importance of particulates produced over nonindustrialized regions for the total aerosol budget in the troposphere and for the potential impact of particulates on atmospheric processes. There is an indication of strong variability in space and time of the physical properties of aerosols produced in tropical nonindustrialized regions.

## 1. INTRODUCTION

In past and current investigations of the global aerosol, forests were not considered as a major or important source of aerosol, excluding the biomass burning that is hypothesized to be an important source of many trace species of C, N, Cl and S in the troposphere, e.g., 1.6 to 7.0 x $10^{12}$ g/year particulate organic C, (Duce et al., 1983). Forests were, however, considered a significant source of volatile organic material. The emissions of isoprene and terpenes may total 830 x $10^{12}$ g/year carbon (Zimmerman et al., 1978). Of this total, a mean of 350 x $10^{12}$ g/year was estimated to be isoprene, and 480 x $10^{12}$ g/year terpenes. Of each component 500 x $10^{12}$ g/year was supposed to be emitted in the tropics. It was further hypothesized that terpene emitted from leaf and resin is independent of radiation, unlike isoprene, which is mainly emitted by specific plant families such as oaks, pines, vegetation, etc. Due to chemical and photochemical reactions, considerable parts of these volatile substances are converted to particulates.

Particulates emitted by forests were recently studied with regard to their concentration, size distribution, nucleation activity (to serve as condensation or freezing nuclei), and chemical

---

[1]This study was partially supported by National Science Foundation Grant ATM84-13814.



ISBN 0-937194-11-5

composition. Many interesting studies of forest-derived ultrafine particles (Aitken nuclei with $r < 0.1$ μm) have been done by French investigators in Southern France and in the Congo (e.g., Lopez et al., 1974). Most of the measurements in pine forests yielded Aitken nucleus (AN) concentrations between $1.5 \times 10^4$ to $4.0 \times 10^4/cm^3$ with a distinctly pronounced minimum around 3:00 a.m., and maximum counts between 2:00 and 5:00 p.m. This trend indicates clearly the influence of photochemical reactions resulting ultimately in gas-to-particle formation. The importance of the oxidation of organic vapors by ozone, atomic oxygen, or the radical OH was mentioned by Went (1966) and Lopez et al., (1974), who measured an AN production rate up to $1.2 \times 10^5$ AN/cm$^2$/sec in a pine forest. Additionally, Cros et al., (1981), measured AN production rates ranging from 2 to $7 \times 10^4$ cm$^2$/sec above the canopy of a tropical forest in the Congo. In spite of the fact that the emission of hydrocarbons (some of which have strong surface activity due to oriented dipoles) by vegetation is estimated to be about 440 megatons per year (Rasmussen and Went, 1965), several other mechanisms are mentioned as potential sources of particulates originating in forests. Delmas and Servant (1986) analyzed the gas-to-particle conversion in a tropical forest. Tropical vegetation and soil, especially during the rainy season, is a prolific source of gaseous compounds of sulphur, e.g., $H_2S$ and nitrogen, e.g., NO, $NO_2$, $NH_3$. The oxidation process leads to the simplified reactions

$$H_2S \rightarrow SO_2 \rightarrow H_2SO_4; \; NO \rightarrow NO_2 \rightarrow HNO_3 \; .$$

The acids react further with gases, e.g., $NH_3$, and produce particulates composed mainly of sulfates and nitrates. This scheme will be modified close to large cities, industrial centers, and in littoral regions, where specific atmospheric constituents can intervene; in fact, the above described mechanisms might be responsible for 70% of particulates generated in and above the forest. In addition to the generation of "pure" particles, one should consider mixed nuclei (e.g., Junge and MacLaren, 1971), the impact of heterogeneous reactions, and embryo formation on insoluble (partly organic) carrier particles where capillary condensation can initiate the condensation process (Podzimek and Desalmand, 1985).

A different mechanism of nucleus activation or rate of condensational growth can be detected by the analysis of size distribution curves. In this way the size distribution of sulfate and ammonium ions containing particles in the tropical forest was proved to be parallel, with mean diameters below 0.5 μm (Delmas and Servant, 1986). Nitrates, in contrast, often had a bimodal distribution with two distinct modes in particle size distribution: one below 0.5 μm related to the growth mechanism deduced directly from the gas-to-particle conversion, and the other with particle diameters of around 5 μm. The second peak reveals the possible heterogeneous reactions or deposition of nitrate ion containing nuclei on larger particles (Kadowaki, 1977; Delmas and Servant, 1986). The maximum around 0.08 μm in the particle size distribution curve and the high counts of particle concentration in the size range below 0.08 μm --found by Lopez et al., (1983), in the afternoon hours on sunny days in a pine forest--might be related to the specific conditions in a forest where, due to the large amount of emitted gases and vapors (including terpenes and isoprene), photochemical reactions were intense.

## 2. STUDY OF NUCLEUS SIZE DISTRIBUTION AND ACTIVITY IN FORESTED SAVANNAH AND TROPICAL FOREST

Few studies of the activation of cloud condensation nuclei (CCN) have yet been made in forests (for survey, see Desalmand, 1984). Since 1970 many interesting studies of atmospheric aerosols have been conducted at the University of Abidjan, in the forested savannahs, and in the tropical forests of the Ivory Coast. We participated in the field measurements in the forested savannah and in the measurements of CCN and aerosols at Abidjan during the dry season 1981-1982 and were informed about the evaluation of the measurements from the tropical forest and the measurements during the rainy season. The instrumentation used during our measurements was a Gardner AN counter, a Baush & Lomb counter (measuring by light scattering particles with diameters between 0.3 μm and 10 μm) and a thermal gradient diffusion chamber, which measured the concentration of CCN at supersaturations S = 0.25%, 0.50%, and 1.00% (described by Desalmand, 1984).

The main results of these measurements (Desalmand et al., 1985a, 1985b) can be summarized as follows. In the dry season in the forested savannah (Lamto, in Table I) there are very low counts of AN (around $2000/cm^3$), which correspond roughly to the mean concentration of CCN.

There is no generating mode of nuclei and one assumes that the aerosol is well aged. This is revealed also by a low value of the exponent k in Twomey's relationship between activated nuclei concentration, N, and supersaturation, S, ($N = CS^k$). The one day of particle measurements at Lamto - $\phi$ = 6°10'N, $\lambda$ = 5°00'W (Fig. 1) does not reveal a very strong influence of solar radiation on nucleus generation. The counts were probably dominated by a turbulent exchange and transport of particulates from their prolific sources in the Bantama River's lush tropical vegetation near Lamto.

It was speculated that the majority of the active CCN was composed of potassium oxalate, based on the chemical analysis of the filter samples where larger particles had a high content of K (Crozat, 1978). This was confirmed by measurements in the Amazon Basin tropical forest, where the investigators found a large content of potassium in the aerosol particles (Orsini et al., 1982). The particle size distribution (Fig. 2) reveals that at least a part of

TABLE 1. NUCLEI MEASUREMENTS IN THE DRY SEASON AT LAMTO (APPLICATION OF TWOMEY'S RELATIONSHIP $N=CS^k$)

| CONCENTRATIONS ($cm^{-3}$) | LAMTO MEAN | 1-1-82 Max/Min |
|---|---|---|
| $N_A$ | 2060 | 2.50 |
| $N_{1.0}=C$ | 2980 | 2.08 |
| $N_{0.50}$ | 2090 | 2.16 |
| $N_{0.25}$ | 2060 | 2.61 |
| $N_B$ | 38 | 2.52 |
| k | 0.26 | 2.93 |
| $N_A/N_{0.25}$ | 1 | |
| $N_A/N_B$ | 50 | |

$N_A$ means the concentration of AN. $N_{1.0}$, $N_{0.50}$, $N_{0.25}$ mean concentrations of CCN activated at supersaturations 1.0%, 0.5% and 0.25%, $N_B$, the concentration of all particulates with diameters between 0.3 µm and 10 µm.

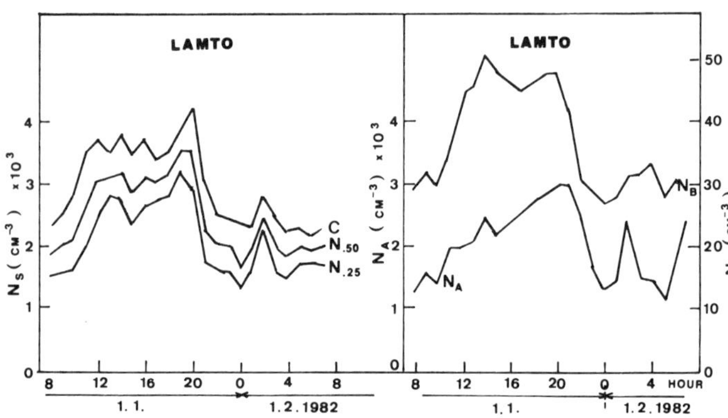


FIGURE 1. Study of nuclei activity in Guinean forested savannah during the dry season. Hourly variations of Aitken nuclei ($N_A$), particles with radii larger than 0.15 µm ($N_B$), and nuclei activated at the supersaturation S = 0.25% ($N_{0.25}$), S = 0.50% ($N_{0.50}$), and S = 1.00% (C) (Desalmand et al., 1985a).

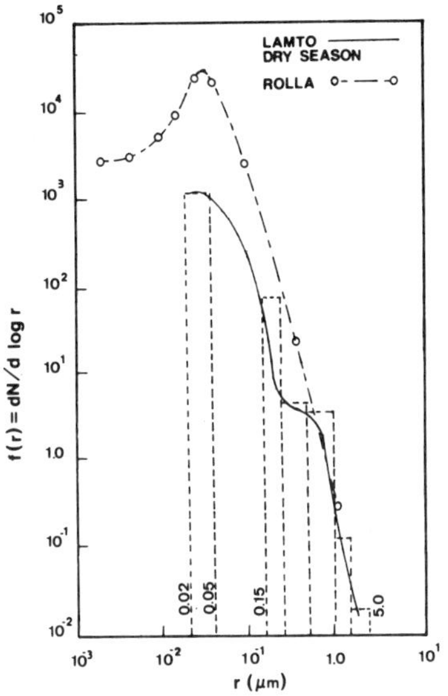


FIGURE 2. Nuclei size distribution: a) from Lamto savannah measurements: Gardner counter (r ⩾ 0.001 μm; thermal gradient chamber (0.02 ⩽ r ⩽ 0.05 μm); Bausch and Lomb counter (r ⩾ 0.15 μm) (Desalmand et al., 1985), b) measurements at Rolla.

the active nuclei might be deposited on carrier larger particles with a tendency to form a bimodal size distribution. The low counts of AN in the forested savannah and the absence of the nuclei generating mode at Lamto is in strong contrast to the counts at Abidjan and also to the AN counts measured in and above the dense tropical forest in the Congo (Cros, 1977). The particle generating mode is strongly expressed at Abidjan during the bush fire season, when the AN concentrations often surpass 2 x $10^5/cm^3$.

3. NUCLEI SIZE DISTRIBUTION AND ACTIVITY MEASUREMENTS AT ROLLA

Aitken nucleus concentrations were measured in our laboratory in Norwood Hall on the University of Missouri-Rolla campus. The large cities (St. Louis and Springfield, Missouri) are located more than 150 km from the sampling site. Rolla is considered a non-industrial town, with a population of 13,500. Forest and agricultural land are prevalent in this area. The nuclei sampling was performed through a metallic duct of 2.54 cm inner diameter protruding 2.0 m outside of the laboratory window on the east side of the building. A permanent flow of 300 liters per minute was maintained during the sampling in the duct. The nuclei concentration was measured by the General Electric and Gardner counters and the size distribution was determined with a GE counter (with a flow stabilizing unit) combined with a Thermo-Systems, Inc.,

(TSI) diffusion battery, and with a Zalabsky electrostatic differential mobility analyzer (Hagen, 1986). The measurements done in the first half of 1987 focused mainly on mean nuclei concentrations and their variability during the day. At the same time, consistency was established between the two techniques used to measure size distribution for particle diameters greater than 0.01 μm.

Most of the 210 measurements performed in the months of March and April 1987, mainly during the daytime, showed Aitken nucleus concentrations ranging from 6623/cm$^3$ to 24667/cm$^3$ as a daily mean value. At the same time 178 readings were made by a Gardner Aitken nucleus counter (which was compared during the measurement with a Nolan-Pollak counter) in order to check the function of the GE counter and to obtain more data points. The mean AN concentration during the main period of measurement was 13252/cm$^3$, with many days with mean concentrations below $10^4$/cm$^3$. One is able to single out from the recording of daily fluctuations the impact of local pollution found at calm and stable meteorological situations with very intense solar radiation (e.g., March 19 and 20, 1987). The peak nucleus concentrations often surpassed 30000/cm$^3$. Intense heating on cold days might be responsible also for AN concentrations surpassing 20000/cm$^3$ in the late evening and night hours (e.g., on March 19 and 25). In spite of these exceptional situations, we observed the following general trend of AN concentrations at Rolla: low counts in the morning hours (often concentrations lower than 9000/cm$^3$), increasing AN concentrations at noon time, maxima in the afternoon hours, and before sunset, often concentrations surpassing 15000/cm$^3$. The maxima have a time period of 10 minutes or more and cannot be related to the sporadic local traffic. There is a strong correlation with the local windspeed but not with wind direction. At night the counts were usually 30% - 50% lower than the daily mean values. This brings to mind the trend of AN concentrations measured at Abidjan in the dry season (Fig. 1), however, with concentrations several times higher at Rolla.

Comparison of the AN size distibutions measured on days with strong solar radiation and those measured under a cloudy sky revealed several interesting features. The first group of data was featured with higher AN counts and a sudden increase in AN concentrations for particles with diameters larger than 0.02 μm (Fig. 3). This increase usually lasted throughout the whole fine particle size spectrum (d > 0.03 μm). The ultrafine particle size distribution measured during the cloudy sky or at night (Fig. 4) is characterized by decreasing nucleus counts with increasing particle size. Usually particles larger than 0.06 μm in diameter had approximately half of the concentration of those with diameters smaller than 0.01 μm.

The measurements of cloud condensation nuclei in the supersaturation range from 0.013 to 0.681% and from 0.0133 to 0.173% (with the use of an alternating gradient chamber operated at the supersaturation of 15.7%) were performed at Rolla almost at the same

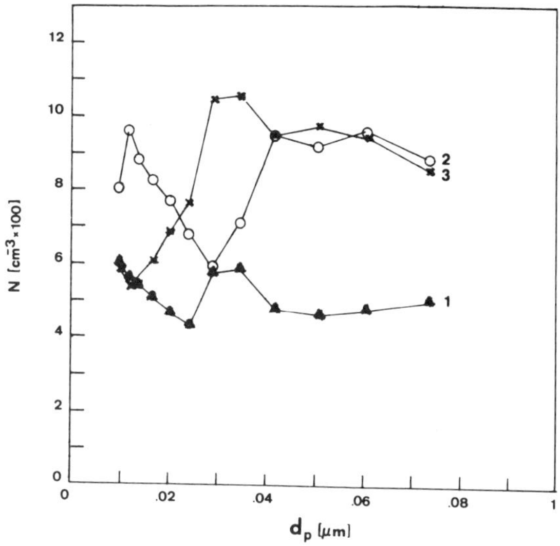


FIGURE 3. AN size distribution measured at Rolla on a sunny day, 30 April 1987 by GE condensation nucleus counter with a Zalabasky electrostatic differential analyzer (1 = morning hours; 2.3 = noon and afternoon data.)

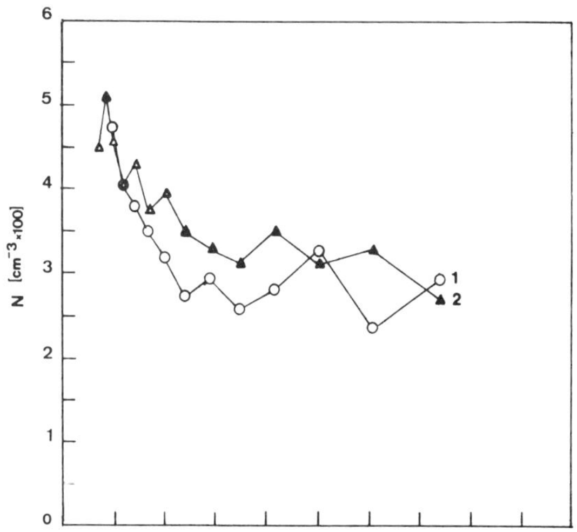


FIGURE 4. AN size distribution measured at Rolla under a cloudy sky in the morning (1) and afternoon (2) hours on 24 April 1987.

location used by Alofs and his fellow workers. Since most of the measurements and the results were described elsewhere (Alofs and Tung-Hsi Lu, 1981; Alofs and Trueblood, 1982), only the main results will be summarized here: Alofs and fellow workers found a two-stage power law well described the average concentration, N, of nuclei activated at supersaturation, S. In mean $N = 3990\ S^{0.643}\ cm^{-3}$ for $0.681\% \geq S \geq 0.0483\%$ and $N = 6.62 \times 10^7\ S^{3.85}\ cm^{-3}$ for $0.013\% \leq S \leq 0.0483\%$. The high value of $C = 3990/cm^3$ at Rolla corresponds reasonably well to our findings at Lamto; however, the authors did not find any significant seasonal or diurnal pattern during the 10-month period in 1978. They concluded that the daily maximum was only several tens of percent higher than the daily minimum. This corresponds to the finding by Vancil (1976) who measured the concentration of cloud condensation nuclei at different supersaturations close to Rolla in forested areas. The other interesting finding was related to the average volume fraction, $\varepsilon$, of water soluble material in the measured nuclei. It appears that it is close to $\varepsilon = 0.6$ and that it is almost independent of the size of nuclei, at least for particles with diameters of 0.2, 0.3, and 0.4 μm. In general, very few nuclei were activated at supersaturations between 0.17% and 15.6%.

## 4. DISCUSSION

Several interesting discussions and hypotheses have been published on the possible nature and role of forest-derived nuclei in atmospheric processes. There seems to be agreement that forest-derived aerosol and aerosol generated over rural areas might play an important role in stimulating water condensation at low supersaturation. Forests (mainly tropical forests) represent a source of larger particles and active nuclei comparable with highly industrialized regions. The large value of the parameter, C, in Twomey's relationship found at Rolla and in the Ivory Coast indicates a possibly similar origin, however, with some distinct differences and large uncertainties. There are small variations of parameters in Twomey's relationship at Rolla and relatively large variations as a function of time of year as well as site of measurement in the Ivory Coast. These facts reflect the difference between particle composition in the dry and rainy seasons. Moreover the measurements in the Ivory Coast (Desalmand, 1985; Desalmand et al., 1985a, 1985b) cover a relatively narrow part of the supersaturation spectrum that refers to particles with radii between 0.02 μm and 0.04 μm. Therefore, the interesting suggestion by Desalmand (1985) to divide the supersaturation spectra in general into two main groups (convex - aged and soluble aerosol; concave - young, less soluble aerosol) awaits corroboration pending further measurements, especially in the tropical forest. A well-coordinated effort in nucleus measurement in the middle latitude forest, savannah, and tropical forest could also contribute to the clarification of the possible function of mixed nuclei active in the tropical forest and savannah at high relative humidities, which is hypothesized from the measurements in onsetting fogs (Podzimek and Desalmand, 1985).

One of the more fascinating subjects in the measurement of forest-derived aerosol is the generation and evolution of ultrafine particles (Aitken nuclei). There is not much difference between the measurements in a pine forest in Europe (AN counts higher than $10^4/cm^3$), in Missouri (AN concentrations around $10^4/cm^3$, in an oak forest with little industry in the environment), and in the Congo (AN concentrations of $10^4/cm^3$ above the dense tropical forest). The absence of a nucleation mode in the dry season in the forested savannah (Desalmand, 1984) is the only seasonal phenomenon that changes during the rainy season.

Further studies will be required to answer questions regarding the AN size evolution during the day, and the impact of solar radiation on AN generation in a forest. In France, Lopez et al. (1983) found a strong correlation between radiation and the aerosol produced in a pine forest. The measurements of AN at Rolla suggest rather peculiar particulate generation in the size range around 0.01 μm, and also in the particle sizes larger than 0.03 μm, under intense solar radiation. This statement certainly requires verification, based on more measurements during the year, possibly at different altitudes over a forest.

This discussion has been primarily concerned with forest-derived aerosol and with the evolution of that component of rural aerosol that is active in condensation processes. It omits certain atmospheric species, such as fungi, pollen, bacteria, algae, viruses, molds, protozoa, phages, and soil-derived large particles consisting mainly of minerals. Although they may play an important role in atmospheric chemistry (see e.g., the survey by Duce et al., 1983) and perhaps optics, their concentrations (highest ~ $10^4/m^3$) are such as to imply a limited role in cloud physics.

## 5. CONCLUSION

Forests represent a source of AN and CCN that might vary with the kind of forest, geographical location, and interaction of the aerosol with the environment that is subjected to seasonal and diurnal variation of meteorological and biological parameters. The AN concentrations might surpass, under exceptional circumstances, $10^5/cm^3$; however, the most probable concentrations in middle latitudes is around $10^4/cm^3$. Somewhat higher values are found above dense tropical forests and values lower than $10^4/cm^3$ are found inside a tropical forest. In forested savannah, the concentration of AN is considerably lower in the dry season. An important factor seems to be the interaction of gaseous and particulate matter with solar radiation and with the closely related air mass exchange in microscale and mesoscale.

Numerous studies have stressed the important function of forest-derived aerosols in the nucleation process, and fog and cloud formation. One study found nucleus concentrations at low supersaturations greater than $10^3/cm^3$. Most of the measurements show that forests produce a large number of active nuclei, but a small number of poor centers of condensation are active at supersaturations larger than 1.0%. This study excludes, however, exceptional seasonal weather variations (e.g., the intertropical convergence zone, ITCZ) and bush fires.

## ACKNOWLEDGMENTS

Most of the data quoted in this work were obtained during a short stay in the Ivory Coast, with the close cooperation of Professor J. Baudet and Dr. F. Desalmand from the University at Abidjan, and Dr. R. Serpolay from the University of Clermont-Ferrand. The author is indebted to Professor D. Alofs and Mr. M. Trueblood, University of Missouri-Rolla (UMR), for the exchange of ideas and data and to his UMR research assistants Messrs. G. Stowell, M. Ianniruberto, and G. Byers for the assistance in AN size distribution measurements and evaluation of data.

## REFERENCES

Alofs, D.J., and Tung-Hsi Liu, 1981: Atmospheric Measurements in the Supersaturation Range 0.013 - 0.681%, J. Atmos. Sci., 38, 2772-2778.

Alofs, D.J., and M.B. Trueblood, 1982: The Correlation Between Size and Critical Supersaturation Exhibited by CCN in Missouri, Proceedings of Conference on Cloud Physics, Nov. 15-18, Chicago, IL, AMS, Boston, pp.1-4.

Cros, B., 1977: Contribution a l'etude des noyaux d'Aitken en Afrique equatoriale, These de Doctorat es-Sciences, Universite de Toulouse (France), No. 789.

Cros, B., A. Lopez and J. Fontan, 1981: Estimation of the Aitken Particle Source Intensity in Rural Equatorial Africa, Atmos. Envir., 15, 83-90.

Crozat, G., 1978: l'Aerosol atmospherique en milieu naturel, etude des differentes sources de potassium en Afrique de l'Ouest (Cote d'Ivoire), These de Doctorat es-Sciences, Universite de Toulouse (France), No. 844.

Delmas, R., and J. Servant, 1986: Echanges Biosphere-atmosphere d'Azote et de Soufre en Zone Intertropicale, Transferts Entre les Ecosystemes Foret et Savanne en Afrique de l'Ouest, accepted for publication in J. Res. Atmos.

Desalmand, F., 1985: An Attempt to Characterize a Relationship Between Supersaturation Spectrum, Size Spectrum and Solubility of CCN Observed Over a Tropical Rain Forest, J. Atmos. Sci., 42, 472-477.

Desalmand, F., 1984: Spectre de sursaturation, spectre dimensionnel et solubilite de l'aerosol atmospherique, Influence du milieu intertropical humide dans le domaine de rayon compris entre 0.02 et 0.04 μm, These de Doctorat es-Sciences, Universite de Clermont-Ferrand (France).

Desalmand, F., J. Podzimek and R. Serpolay, 1985a: A Continental Well-Aged Aerosol in the Guinean Savannah at the Level of a Trough Along the ITCZ, J. Aerosol Sci., 16, 19-28.

Desalmand, F., R. Serpolay and J. Podzimek, 1985b: Some Specific Features of the Aerosol Particle Concentrations During the Dry Season and During a Bushfire Even in West Africa, Atmos. Envir., 9, 1535-1543.

Duce, R.A., V.A. Mohnen, P.R. Zimmerman, D. Grosjean, W. Cantreels, R. Chatfield, R. Jaenicke, J.A. Ogren, E.D. Pellizzari and G.T. Wallace, 1983: Organic Material in the Global Troposphere, Rev. Geoph. and Space Phys., 21, 921-952.

Hagen, D.E., 1986: NASA Final Rep., Contr. No. NAS8-34603, March 11, UMR, 10-13.

Junge, D.E., and E. MacLaren, 1971: Relationship of Cloud Nuclei Spectra to Aerosol Size Distribution and Composition, J. Atmos. Sci., 28, 382-390.

Kadowaki, S., 1977: Size Distribution and Chemical Composition of Particulate Nitrate in the Nagoya Area, Atmos. Environ., 11, 671-675.

Lopez, A., J. Servant and J. Fontan, 1974: Methodes de Mesure de l'Intensite des Sources de Noyaux d'Aitken Dans l'Atmosphere, Atmos. Environ., 8, 733-754.

Lopez, A., G. Lecouteux, S. Prieur and J. Fontan, 1983: Etude de la Formation de Particules a Partir des Hydrocarbures Naturels Degages par la Vegetation, J. Aerosol, Sci., 14, 99-111.

Orsini, C.Q., P.A. Netto and M.H. Tabacniks, 1982: Preliminary Data on Atmospheric Aerosol of the Amazon Basin, Atmos. Envir., 16, 2177-2181.

Podzimek, J., and F. Desalmand, 1985: Properties of Forested Savannah Derived Cloud Condensation Nuclei, J. Rech. Atmos., 19, 203-211.

Rasmussen, R., and F.W. Went, 1965: Volatil Organic Matter of Plant Origin in the Atmosphere, Proc. Nat. Acad. Sci., 53, 215-220.

Vancil, S.P., 1976: Atmospheric Measurements of Cloud Nuclei Active at 0.04% Supersaturation, M.S. Thesis in Mechanical Engineering, University of Missouri-Rolla, 47 pp.

Went, F.W., 1966: On the Nature of Aitken Condensation Nuclei, Tellus, 18, 549-556.

Zimmerman, P.R., R.B. Chatfield, J. Fishman, P.J. Crutzen and P. L. Hanst, 1978: Estimates of the Production of CO and $H_2$ From the Oxidation of Hydrocarbon Emissions From Vegetation, Geophys. Res. Letts., 5, 679-682.

# ON THE SCALING OF ATMOSPHERIC AEROSOL SCATTERING AND EXTINCTION WITH WAVELENGTH

*Eric P. Shettle*
AFGL/OPA, Hanscom AFB, Massachussets 01731, USA

and

*Eli Trakhovsky*
Technion–Israel Institute of Technology
Haifa, 32000, Israel

## ABSTRACT

The possibility of relating the scattering or extinction by atmospheric aerosols at one wavelength to the scattering or extinction properties at another wavelength by scaling laws is examined. These scaling laws were tested using numerical simulations on representative models of the atmospheric aerosols and Mie theory calculations of their expected scattering and extinction properties. The simulations showed that forward scattering at fixed angles in the near-ultraviolet spectral region could provide useful predictions of near-infrared extinction. A good correlation is also presented between calculated forward scattering for a wavelength of 0.25 μm and backscattering at a wavelength of 1.06 μm. In a comparison of the 10.6 μm extinction with backscattering at 10.6 μm and with extinction at a wavelength of 1.06 μm, both showed too much scatter to develop useful scaling laws in those cases.

## 1. INTRODUCTION

This work will examine the possibility of inferring aerosol extinction at one wavelength directly from the angular scattering by the aerosols at another wavelength. In particular we will examine the use of forward scattering data in the ultraviolet (UV), such as that obtained by the instrument described by Trakhovsky & Shettle (1986), to determine the extinction in the infrared (IR). It is possible to invert the forward scattering as a function of angle to obtain the aerosol size distribution (see Trakhovsky & Shettle, 1985), which can be used in Mie scattering calculations to determine the extinction or scattering at any other wavelength. However, if the desired end result is the extinction at a few selected wavelengths, it is more efficient to directly calculate the desired extinction scaling laws, (assuming suitable laws can be derived).

A number of such scaling laws have been derived in the literature. Examples include Angstrom's (1929) classical power law for the wavelength dependence of extinction in the visible and near infrared, relationships between aerosol mass concentration and visible

 A. DEEPAK Publishing
ISBN 0-937194-11-5

(Charlson et al., 1968) or IR extinction (Pinnick et al., 1979), and the relationship between atmospheric backscattering and extinction (Curcio & Knestrick, 1958). Gerber (1985) recently discussed a number of such scaling laws, relating different visible or near-IR scattering measurements to IR extinction. It should be recognized that it is not always possible to develop scaling laws relating two specific aerosol properties such as visible and IR extinction because they may depend on aerosol properties that are only weakly correlated, such as the concentrations of small and large aerosols.

While scaling laws cannot be rigorously derived, an approximation relating the optimal scattering angle for predicting the extinction to given wavelengths for extinction and scattering will be shown. The scaling laws developed below were tested in a series of numerical simulations using various models of the atmospheric aerosols and calculations of their expected scattering and extinction properties.

## 2. THEORY

One of the results of Gerber's (1985) work is that a scaling law can be developed relating two parameters, if the kernels relating the parameters to the aerosol size differ by at most a multiplicative factor over most of the significant size range of the aerosol particles. This suggests that in developing scaling laws, we look for parameters whose kernels have similar dependences on aerosol size, where the scattering or extinction kernels are the scattering or extinction per unit volume as a function of size.

The optical theorem of quantum mechanics or the "extinction" theorem of van de Hulst, 1957, (also see Bohren & Huffman, 1983) implies that extinction is directly related to the scattering amplitude at a $0^\circ$ scattering angle. Therefore the extinction can be measured and accounted for in two ways:

1) extinction = total angular scattering + absorption,
2) extinction is proportional to angular scattering at $0^\circ$.

Unfortunately it is experimentally impossible to distinguish between the incident unscattered and light scattered at $0^\circ$, therefore the optical theorem cannot be directly utilized. However, the forward scattering is a function of the ratio of the scattering angle and the wavelength (see van de Hulst, 1957). Therefore it is possible to scale the scattering kernel for two angles using two appropriate wavelengths. Combining this feature with the optical theorem it might be possible to deduce the extinction at one wavelength using the scattering measurement at another for some specific angle.

Using the above considerations, we will develop a relationship to predict the optimum scattering angle to use for a scaling law relating the extinction at a desired wavelength to the angular scattering at another wavelength. Using the anomalous diffraction approximation for the extinction kernel, $\gamma$, and the Fraunhofer diffraction approximation for the forward-scattering kernel, $\beta$, (van de Hulst, 1957), a relationship can be derived between the wavelengths for the extinction ($\lambda_1$), and the scattering ($\lambda_2$), and the scattering angle (see Trakhovsky and Shettle, 1987 for details):

$$\sin \theta = 0.928 \ (n-1) \ (\lambda_2/\lambda_1) \qquad (1)$$

For $\lambda_2$ = 0.25 μm, $\lambda_1$ = 1.06 μm, and n = 1.33 we obtain θ = 4.14°. Figure 1 shows a comparison between the anomalous diffraction kernel at a wavelength of 1.06 μm and the Fraunhofer diffraction kernel at a wavelength of 0.25 μm and an angle of 4.14°. It is seen that agreement is good for the main lobe. This suggests that the above relationship could be used as the first guess of the optimal scattering angle in order to scale the extinction coefficient.

## 3. SCALING RELATIONSHIPS

Numerical simulations were performed to examine the validity of different scaling laws. Optical properties such as extinction, forward scattering, and backscattering were calculated using the aerosol model data base compiled by Shettle and Fenn (1979). The data base consists of five different aerosol models: tropospheric, rural, urban, maritime, and fog. For all the models but the fog, the calculations were made for relative humidities of RH=0%, 50%, 70%, 80%, 90%, 95%, 98%, and 99%. In addition 4 different types of fog are presented. In doing the numerical simulations, 3 different total number densities were used for each model, for a total of 108 data points. These number densities were the ones used by Shettle & Fenn (1979), and those values increased and decreased by a factor of 2 to represent a range of atmospheric conditions.

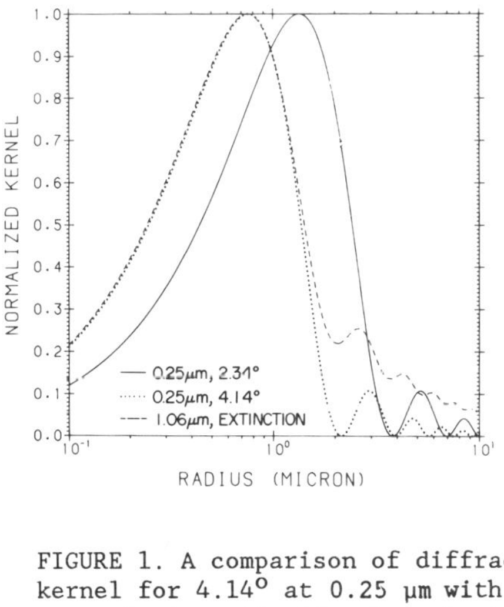


FIGURE 1. A comparison of diffraction kernel for 4.14° at 0.25 μm with the approximate extinction kernel at 1.06 μm. Also shown is the diffraction kernel at 2.34° and 0.25 μm.

In deriving the scaling laws presented below, a quadratic regression equation (in log-log space) of the form:

$$\log \Upsilon(\lambda_1) = a_0(\theta) + a_1(\theta)\log \beta(\lambda_2,\theta) + a_2(\theta)\log^2 \beta(\lambda_2, \theta), \quad (2)$$

was fit to the model calculations of $\Upsilon$ and $\beta$. The "best fit" scattering angle, $\theta$, was the angle for which the rms deviation of $f(\log \beta)$ from $\log \Upsilon$, (over the 108-modeled data points), was a minimum.

### 3.1 FORWARD-SCATTERING/EXTINCTION

Forward-scattering/extinction-scaling relationships were investigated for $\lambda_2$ = 0.25 µm and $\lambda_1$ =1.06 and 3.75 µm. The relationship Eq. (1), was used to determine the approximate location of the best fit angle. Then the angle was varied to determine the one that gives the best fit of the data points to a second degree polynomial in log-log space. The rms deviation of the best fit polynomial from the simulated data points is a weak function of the scattering angle and typically is less than 20% smaller at the best fit angle than at the initial angle given by Eq.(1), even though the angles often differ by a factor of two.

Figure 2 shows the angular scattering at an angle of 2.34° for a wavelength of 0.25 µm versus the extinction at the wavelength of 1.06 µm. This angle is shifted from the first-guess angle of 4.14°, because the exact Mie angular-scattering kernels are different from the approximate diffraction form. This can be seen by comparing the kernels in Fig. 3 with those in Fig. 1. The MIE angular-scattering kernel deviates significantly from the diffraction approximation, especially for the smaller particles. The anomalous diffraction approximation reproduces the exact Mie extinction quite well.

The data points in Fig.2 can be fitted using the following scaling relationship:

$$\log \Upsilon_1 = -1.07 + 0.94 \log \beta_2 + 0.107 \log^2 \beta_2 \quad (3)$$

The rms deviation between the data points and the fitting curve (Eq. 3) is 35% over 5 orders of magnitude.

A similar least squares fit was carried out for forward scattering at 0.25 µm vs. extinction at 3.75 µm. The minimum rms deviation was found for a scattering angle of 0.57°. The data points can be fitted using the following scaling relationships:

$$\log \Upsilon_1 = -2.31 + 1.07 \log \beta_2 - 0.019 \log^2 \beta_2 \quad (4)$$

The rms deviation between the data points and Eq. (4) is 32% over 6 orders of magnitude. As with the 1.06 µm case, the angle for the best fit is shifted from the 1.17° predicted by Eq.(1).

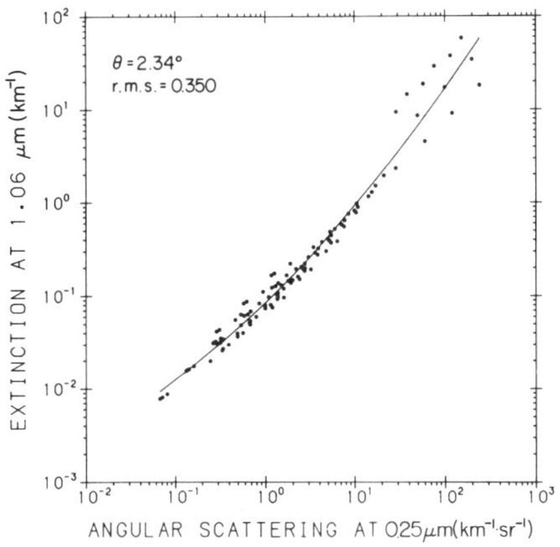


FIGURE 2. A correlation between extinction at 1.06 μm and angular scattering for 2.34° at 0.25 μm.

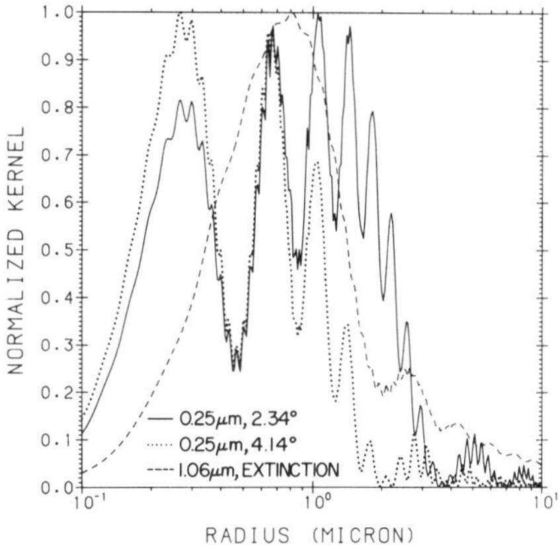


FIGURE 3. A comparison of the Mie angular-scattering kernel for the angles 2.34° and 4.14° at 0.25 μm and the Mie extinction kernel at 1.06 μm.

## 3.2 FORWARD-SCATTERING/BACKSCATTERING

The forward-scattering/backscattering relationship was investigated for the wavelengths of 0.25 and 1.06 μm. A least squares procedure was carried out varying the forward-scattering angle to find the minimum rms deviation from a quadratic function in log-log space as with the extinction and forward scatter. It was found that the best fit is obtained for the angle of 1.02°, which is shown in Fig. 4. The data points can be fitted using the following scaling relationship:

$$\log\beta_1 = -3.05 + 0.70\ \log\beta_2 + 0.100\ \log^2\ \beta_2 \qquad (5)$$

The rms deviation between the data points and the fitting curve is 37% over 5 orders of magnitude.

## 3.3 BACKSCATTERING/EXTINCTION

The correlation between backscattering and extinction at the same wavelength is a critical assumption in the inversion of lidar measurements to obtain extinction data (Klett, 1985). The degree of this correlation for a wavelength of 10.6 μm is examined in Fig. 5. While there is a positive correlation, there is also a large scattering of the results, so the best fit curve through the data points will only predict the extinction from the backscattering within a factor of five. Pinnick et al. (1983), found a similar limited correlation at 10.6 μm, restricting their calculations to water clouds, which correspond to the upper right-hand corner of Fig. 5.

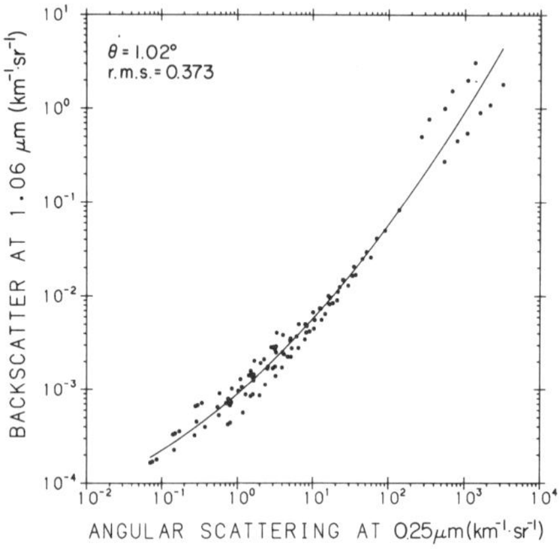


FIGURE 4. A correlation between backscattering at 1.06 μm and angular scattering for an angle of 1.02°.

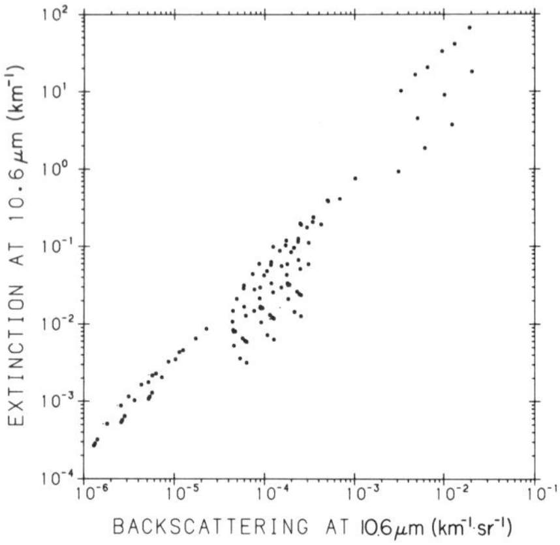


FIGURE 5. A correlation between extinction and 10.6 μm and backscattering at 10.6 μm.

### 3.4 EXTINCTION/EXTINCTION

The possibility of developing a scaling law relating the extinction at two different wavelengths was investigated for the wavelengths of 1.06 and 10.6 μm. The resulting scatter-plot is shown in Fig. 6. It is clear that the correlation is poor. This might be expected since extinction at a given wavelength is most sensitive to particulate size ranges that are similar to the wavelength. Angstrom's power law for the wavelength dependence works well because it is restricted to the visible, and the sensitive size ranges differ by a factor of about two. Also the size ranges fall within the accumulation mode and so can be expected to be reasonably correlated. Whereas, the 10.6-μm extinction is much more sensitive to the coarse mode particles, which are not necessarily well correlated with the accumulation mode aerosols, since in general they represent different sources.

## 4. DISCUSSION

In this paper several possible scaling laws have been examined with mixed results. Good correlations were found between calculated values of forward scattering and calculated values of extinction for a wide range of model parameters, provided suitable angles for the scattering were considered.

A first-guess approximation relating the optimal scattering angle, for predicting the extinction, to given wavelengths for the scattering and extinction was developed. The derivation of this approximation was based on the diffraction approximation to the exact Mie scattering kernel. This approximation does not take into account the wave interference between the transmitted and diffracted light,

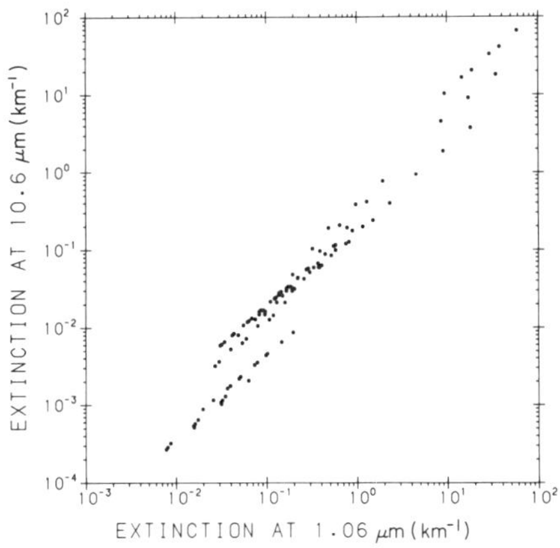


FIGURE 6. A correlation between extinction at 10.6 μm and extinction at 1.06 μm.

which results in the oscillatory behavior of the exact kernel. Therefore, the predicted scaling angle deviates from the optimal angle deduced using the Mie calculations.

A useful scaling relationship between simulated forward scattering at 0.25 μm and backscattering at 1.06 μm was also derived. Attempts to find scaling laws relating simulated 10.6-μm extinction with either backscattering at 10.6 μm or with extinction at 1.06 μm were less successful.

It might be noticed that the scaling laws derived were nearly linear on a log-log plot. This linearity is not surprising and if a wider range of number densities were included in the model calculations the linear behavior might be stronger. This behavior occurs if only the total number of particles is varied while holding the relative number for each size fixed, then the extinction or scattering at a given angle will both vary in direct proportion to the total number of particles. Thus on a plot such as Figure 2 or Figures 4-6, varying the total number density while holding all else fixed, would lead to families of data points along 45-deg straight lines. Concern for artifically producing such linearities limited the range of number densities to that required to produce realistic variations in seeing conditions. The resulting meteorological ranges, varied from 0.07 to nearly 1.0 km for the fogs, and from about 1.0 km for the urban model at 99% relative humidity and maximum number density, to 115 km for the tropospheric model at relative humidity and minimum number density.

In conclusion it should be recognized that the scaling laws discussed in this paper were developed on the basis of calculations using models of the atmospheric aerosols. These scaling laws rigorously apply only to the models used in developing them; they apply to the real world only to the extent that the properties of the models accurately represent the real world, (see Shettle et al., 1984, for a comparison of the models with scattering measurements). Recently Gerber (1987) has presented experimental verification of some scaling laws (relating forward scattering at 0.63 μm to IR extinction) similar to those discussed in the present paper.

ACKNOWLEDGMENTS

The present study was conducted while E. Trakhovsky was at the Air Force Geophysics Laboratory (AFGL) under a National Research Council (NRC) Research Associateship. He is grateful to both the NRC and AFGL for their support.

## REFERENCES

Angstrom, A., 1929: On the Atmospheric Transmission of Sun Radiation and on Dust in the Air ,Geograf. Ann. Deutch., 12, 156-166.

Bohren, C.F. and D.R. Huffman, 1983: Absorption and Scattering of Light by Small Particles, Wiley, New York

Charlson, R.J., N.C. Ahlquist and H. Horvath, 1968: On the Generality of Correlation of Atmospheric Aerosol Mass Concentration and Light Scatter", Atmos. Environ., 2, 445-464.

Curcio, J.A., and G.L. Knestrick, 1958: Correlation of Atmospheric Transmission with Backscattering, J. Opt. Soc. Amer., 48, 686-689.

Gerber, H., 1985: Infrared Aerosol Extinction from Visible and Near-Infrared Light Scattering, Appl. Opt., 24, 4155-4166.

Gerber, H., 1987: Forward-Scatter Meter for Estimating 10.6-μm Aerosol Extinction, Appl. Opt., 26, 539-544.

Klett, J.D., 1985: Lidar Inversion with Variable Backscatter/ Extinction Ratios, Appl. Opt., 24, 1638-1643.

Pinnick, R.G., S.G. Jennings, P. Chýlek and H.J. Auvermann, 1979: Verification of a Linear Relation Between IR Extinction, Absorption, and Liquid Water Content of Fogs, J. Atmos. Sci., 36, 1577-1586.

Pinnick, R.G., S.G. Jennings, P. Chýlek, C. Ham and W.T. Grandy, Jr., 1983: Backscatter and Extinction in Water Clouds, J. Geophys. Res., 88, 6787-6796.

Shettle, E.P., and R.W. Fenn, 1979: Models of the Aerosols of the Lower Atmosphere and the Effects of Humidity Variations on Their Optical Properties, AFGL-TR-79-0214 (ADA 085951).

Shettle, E.P., V. Turner, and L.W. Abreu, 1984: Aerosol Phase Functions: Models and Measurements, Pg 34-37, IRS 84: In G. Fiocco (Ed.) Current Problems in Atmospheric Radiation, A. Deepak Publishing, Hampton, Virginia, 34-37.

Trakhovsky, E., and E.P. Shettle, 1985: Improved Inversion Procedure for the Retrieval of Aerosol Size Distributions Using Aureole Measurements, J. Opt. Soc. Am., A2, 2054-2061.

Trakhovsky, E., and E.P. Shettle 1986: Aerosol Size Distributions From the Inversion of Forward Scattering Measurements, Proceedings of the Sixth Conference on Atmospheric Radiation, American Meteorological Society, Williamsburg, VA, May 1986, 112-115.

Trakhovsky, E., and E.P. Shettle, 1987: Wavelength Scaling of Atmospheric Aerosol Scattering and Extinction, Appl. Opt., 26, 5148-5153.

van de Hulst, H.C., 1957: Light Scattering by Small Particles, Wiley, New York.

# AEROSOL SIZES AND NIGHTTIME COOLING RATES IN THE HARMATTAN[1]

*C.O. Oluwafemi*
University of Lagos
Akoka, Yaba, Lagos, Nigeria

## ABSTRACT

Aerosol size distribution in the harmattan is studied using the Yue-Deepak size-inversion technique. In the episode treated here, about 57% of the extinction of solar wavelengths by dust appears due to particles in the mode $r \geq 1\ \mu m$. Radiative heating, as determined by a model computation for a homogeneous plane-parallel atmosphere, increases approximately proportionately with dust load. Consideration, also, of the observed time-averaged nighttime cooling rate reveals that, overall, the dust plays two opposing roles: By its presence, it increases the cooling rate of the ambient air due to its dryness; it also increases the heating rate of the atmosphere through the radiative process of scattering and absorption. The former role predominates over the latter.

## 1. INTRODUCTION

From November to March, the optical and climatological effects of the Saharan air mass become manifest in the land areas adjacent to the Gulf of Guinea. The associated weather is called the "harmattan." Information on the general meteorology and the transport of the Saharan air mass is quite ample in the literature (see e.g., Hamilton and Archibold, 1945; Carlson and Prospero, 1972; Jaenicke, 1977).

In the first part of this paper, particle size distribution retrieval under harmattan conditions is attempted using biwavelength transmittance data and aerosol size measurements at Lagos (lat. 6.3°N, long. 3.5°E). In the second part, cooling rates during harmattan nights are studied for two locations, Lagos and Kano (lat. 12°N, long. 9°E).

## 2. PARTICLE SIZE DISTRIBUTION

### 2.1 BACKGROUND

In the author's earlier works (Oluwafemi, 1981, 1983), the sizes of the aerosol sampled near ground in Lagos during a moderate harmattan were found to have bimodal lognormal properties and the modal radii ($r_{m1}, r_{m2}$) and the distribution widths ($\ln Sg_1, \ln Sg_2$) were

[1] This work is supported through a grant from the University of Lagos Central Research Fund.



ISBN 0-937194-11-5

obtained. As the particle radii in that data set were essentially within the range $r > 3$ μm (the same size range as used by Yue and Deepak, 1983, in testing their size-inversion procedure), the technique of Yue and Deepak (1983) is adopted here to retrieve total number concentration ($N_{T1}$, $N_{T2}$) in order to obtain absolute size distributions.

The extinction ratio is defined here in terms of the spectral optical depths at two wavelengths of the Volz sunphotometer:

$$R = \frac{\delta_p(\lambda=0,5\ \mu m)}{\delta_p(\lambda=.88\ \mu m)} = \frac{\delta_5}{\delta_8} = \frac{\int_{r_1}^{r_2} n_c(r)Q(\lambda_5,r,m)\pi r^2 dr}{\int_{r_1}^{r_2} n_c(r)Q(\lambda_8,r,m)\pi r^2 dr} \quad (1)$$

where

$\lambda_5$, $\lambda_8$ = 0.5, 0.88 μm respectively,

$Q(\lambda,r,m)$ = spectral efficiency factor for an atmospheric particulate of radius r and refractive index m,

$n_c(r) = H_A \cdot n(r) = \frac{dN_c}{dr}$ is the columnar number concentration in the size-range $r \rightarrow r + dr$ for an atmosphere of height $H_A$.

$$n(r) = \frac{N_T}{(2\pi)^{\frac{1}{2}} \ln Sg} \cdot \frac{1}{r} \exp\left(-\ln^2(r/r_m)/2(\ln Sg)^2\right) \quad (2)$$

where

$N_T$ = total number of particles per unit volume. For the ith mode of a multimodal lognormal columnar size distribution

$$N_{c,T,i} = \frac{\delta_{8,i}}{\int_{r_{1,i}}^{r_{2,i}} n_{c,i}(r)Q(\lambda_8,r,m)\pi r^2 dr} \quad (3)$$

where the numerator on the right-hand side is the contribution of mode i to $\delta_8$ and the denominator is the calculated optical depth at 0.88 μm with $H_A.N_T$ taken as unity.

## 2.2 ANALYSIS

Past results (Oluwafemi, 1981) show that about 85% of the aerosol under consideration have sizes within the first mode. It therefore suffices to compute R versus $\delta_8$ curve for a single mode in order to retrieve $N_{c,T}$ for each of the two modes.

Now, $R_{mea} = \delta_{5,mea}/\delta_{8,mea} = \frac{0.81}{0.53} = 1.53$, where the subscript "mea" refers to measured value. From the theoretical plot of R versus $r_m$, that value of $r_m$ corresponding to $R_{mea}$ is read-off and that value of $r_m$ gives, directly, the expected (theoretical) value of $\delta_8(\delta_{8,cal})$ on the R versus $\delta_8$ plot. Thus,

$$N_{c,T,i} = H_A . N_{T,i} = \delta_{8,mea}/\delta_{8,cal}.$$

## 2.3 RESULTS AND DISCUSSION

The complex index of refraction of desert aerosols, m = n - in' has typical values n, n' in the ranges: $1.1 \leq n \leq 1.7$ and $0.005 \leq n' \leq 0.1$ (Volz, 1973; Pollack et al. 1973; and Fischer, 1976). Figure 1 amply demonstrates the high sensitivity of R to n and n'. However, an earlier work (Oluwafemi, 1983) shows n can be fixed at 1.55, for example, and n' varied; giving rise to a typical result shown in Fig. 2.

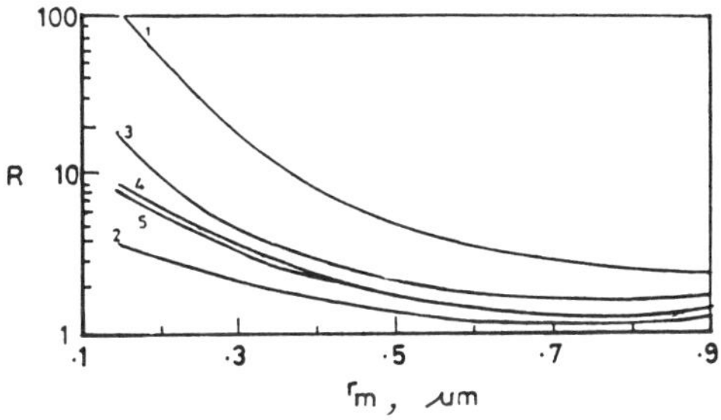


FIGURE 1. Illustrating the variability of the plots of R versus $r_m$ with the complex index of refraction, m: (1) 1.1-0.005i, (2) 1.1-0.1i, (3) 1.35-0.005i, (4) 1.70-0.005i, (5) 1.7-0.1i.

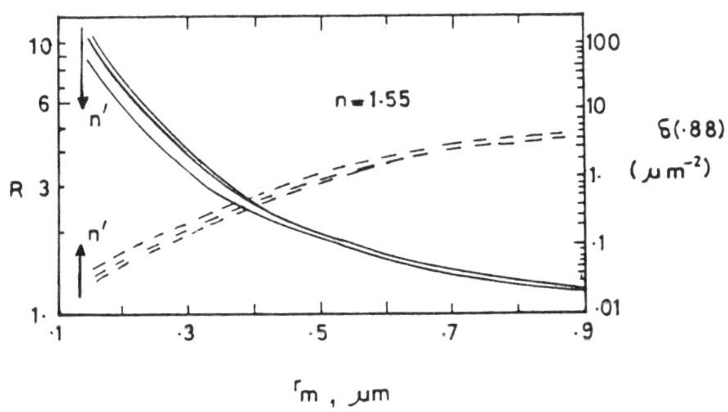


FIGURE 2. Calculated values of R (solid lines) and $\delta_8$ (dashed lines) as functions of $r_m$ for n=1.55 and $0.005 \leq n' \leq 0.1$.

With measurements made with the Volz sunphotometer R is uncertain by about 10%, giving rise to an uncertainty in $N_{c,T}$ of about 67% over the wide range of n'. However, this uncertainty should drop sharply to about 7% should optical depths be accurate to 1%. The mean values of the columnar number concentration retrieved are:

$$N_{c,T} = \begin{cases} 6.8 \times 10^7 cm^{-2}, 0.1 \leq r \leq 1 \ \mu m \\ 1.2 \times 10^7 cm^{-2}, 1 \leq r \leq 3 \ \mu m \end{cases}$$

for $n = 1.55$, $0.005 \leq n' \leq 0.1$. The complete size distribution is shown in Fig. 3.

Assuming an aerosol scale height of between 3.1 km and 4.9 km (Oluwafemi, 1980), the total number $N_T$ ranges between 219 $cm^{-3}$ and 138 $cm^{-3}$ for the class $0.1 \leq r \leq 1$ µm and between 38 $cm^{-3}$ and 24 $cm^{-3}$ for the size-interval $1 \leq r \leq 3$ µm. A simple calculation then shows that some 57% of the extinction of solar wavelengths by dust is due to particles in mode 2 ($1 < r < 3$ µm). In the same subregion, but closer to the Sahara, the results of Cerf et al. (1980) indicate that 75% of solar extinction by dust is due to particles in the mode $r > 1$ µm.

## 3. NIGHTTIME COOLING RATES

### 3.1 BACKGROUND

A plane-parallel homogeneous atmosphere is assumed here. Consequently, neglecting clouds, the external sources of longwave radiation relative to the dust-laden portion of the night atmosphere (assumed existing between the 600 mbar level and ground) are the Earth (flux $F_E$) and the counterradiation (flux, $F_C$) due principally to water vapour and carbon dioxide located above the dust. At any level within the dust-laden portion of the atmosphere, the thermal contributions to the net-flux $F=F\downarrow-F\uparrow$ tend to cancel out; so that the upward-and downward-directed fluxes are sourced effectively, in $F_E$, and $F_C$ respectively. Both $F_E$ and $F_C$ are subject to extinction

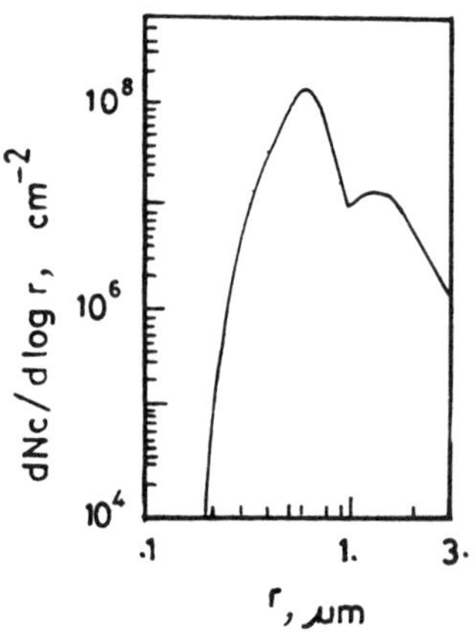


FIGURE 3. Columnar size distribution during a harmattan in Lagos, 5 December 1978; $\delta_5 = 0.81$, $\delta_8 = 0.53$, visibility = 4 km.

(scattering and absorption) by all the atmospheric constituents, namely, the dust particles, water-vapour, and the mixed gases within the dust layer. The fluxes are also part-reflected and part-absorbed by the Earth. These processes are depicted in Fig.4 where the net effect of scattering has been modeled as reflection backward and transmission forward from the centre of the dust layer. As stated earlier, the base of the dust-laden portion of the atmosphere is the ground level.

That fraction of an incident monochromatic flux that is either unscattered as it goes through the layer or scattered forward by the atmospheric constituent type k is given by (Twomey, 1977)

$$t_{\lambda,k} = \tau_{\lambda,k}\,(1 - \tilde{\omega}_{\lambda,k}) + \tilde{\omega}_{\lambda,k}\,g'_{\lambda,k} \tag{4}$$

where $\tilde{\omega}_{\lambda,k}$ = spectral single-scattering albedo of type k constituent,

$\tau_{\lambda,k}$ = transmittance of type k constituent at wavelength $\lambda$,

$g'_{\lambda,k}$ = fraction of the monochromatic radiation scattered forward by type k constituent,

$= \int_0^{\pi/2} P_{\lambda,k}(\cos\theta)\, d\cos\theta \Big/ \int_0^{\pi} P_{\lambda,k}(\cos\theta)\, d\cos\theta$

$= \frac{1}{2}(g_{\lambda,k} + 1)$,

$g_{\lambda,k} = \overline{(\cos\theta)}_{\lambda,k}$ the asymmetry factor of type k constituent at wavelength $\lambda$,

$P_{\lambda,k}$ = the corresponding spectral scattering phase function and

$\theta$ = scattering angle.

A fraction $r_{\lambda,k}$ of any incident flux is reflected backwards; $r_{\lambda,k}$

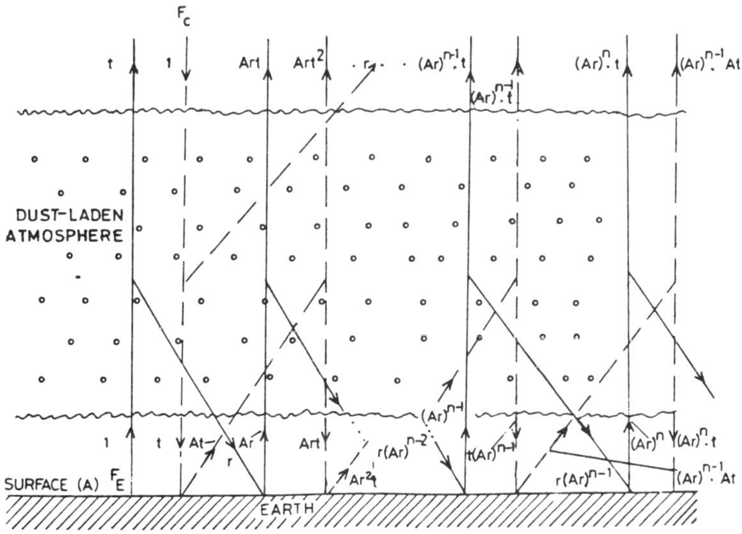


FIGURE 4. A simplified model of the radiative transfer in a homogeneous, cloudless, dust laden atmosphere.

is given by

$$r_{\lambda,k} = (1 - \tau_{\lambda,k})(1-g'_{\lambda,k})\tilde{\omega}_{\lambda,k} \qquad (5)$$

It is straightforward to show that the magnitude of the net flux escaping to space following radiative interaction with type k particles is

$$F_{c,\lambda}(r_{\lambda,k} + At^2_{\lambda,k}/(1 - Ar_{\lambda,k}) + F_{E,\lambda}t_{\lambda,k}/(1 - Ar_{\lambda,k}) \qquad (6)$$

and that absorbed by the Earth is

$$(t_{\lambda,k}\cdot F_{c,\lambda} + r_{\lambda,k}\cdot F_{E,\lambda})/(1 - Ar_{\lambda,k}) \qquad (7)$$

where A is the surface albedo.
Hence the net flux retained in the atmosphere following interaction with type k particles alone has a magnitude

$$E_{\lambda,k} = \alpha_{c,\lambda,k}\cdot F_{c,\lambda} + \alpha_{E,\lambda,k}\cdot F_{E,\lambda} \qquad (8)$$

where $\alpha_{c,\lambda,k} = 1 - r_{\lambda,k} - (t_{\lambda,k} + At^2_{\lambda,k})/(1 - Ar_{\lambda,k})$

$$\alpha_{E,\lambda,k} = 1 - (r_{\lambda,k} + t_{\lambda,k})/(1 - Ar_{\lambda,k}).$$

Following the radiative interaction by all the constituents, the magnitude of the total flux retained within the atmosphere is

$$E_\lambda = F_{c,\lambda}(\alpha_{c,\lambda,p} + t_{\lambda,p}\cdot\alpha_{c,\lambda,w} + t_{\lambda,p}\cdot t_{\lambda,w}\cdot\alpha_{c,\lambda,G})$$
$$+F_{E,\lambda}(\alpha_{E,\lambda,p} + t_{\lambda,p}\cdot\alpha_{E,\lambda,w} + t_{\lambda,p}\cdot t_{\lambda,w}\cdot\alpha_{E,\lambda,G}) \qquad (9)$$

where the subscripts P, W, G refer, respectively, to dust particles, water vapour, and the mixed gases. Thus the total radiative heating rate of the dust-laden portion of the atmosphere, in $Khr^{-1}$ is

$$\frac{\Delta T}{\Delta t}_{rad.} = \frac{3600 g\pi}{C_p \Delta_p} \sum_{i=1}^{N} \frac{1}{\Delta\lambda_i} \left(\int_{\Delta\lambda_i} E_\lambda d\lambda\right) \qquad (10)$$

where, for computational purposes, the quantity inside brackets on the right-hand side has been replaced by the integral of the Planck's function over the band $\Delta\lambda_i$,

N = number of narrow bandwidths, each of width $\Delta\nu = 50\ cm^{-1}$ into which the infrared region (4-28.5 μm) has been divided,
g = acceleration due to gravity,
$C_p$ = specific heat capacity of air at constant pressure and
$\Delta_p$ = the pressure difference across the dust layer.

## 3.2 COMPUTATION OF NIGHTTIME COOLING RATES

In evaluating $F_{c,\lambda}$ in Eq. (9), it is assumed that water vapour and carbon dioxide contribute equally to the counterradiation at the top of the dust layer and that the mean emissivity of carbon dioxide is 0.185 (Haltiner and Martin, 1957). The spectrally-averaged emissivity of the Earth is taken as 0.9 (Griggs, 1968). Using our measured values of $\delta_5$, the normalised spectra of $\delta$, $\tilde{\omega}$, g for Saharan dust over the infrared (Carlson and Benjamin, 1980), and the LOWTRAN 5 data of Kneizys et al. (1980) for the mixed gases and water-vapour absorption coefficients and transmittances, atmospheric heating rates were computed over the infrared region 4-28.5 μm. In applying Eqs. (4) and (5) to the atmospheric constituents, the net transmittances due to water vapour and the mixed gases are taken as

$$\tau_w = \tau'_w \cdot (\tau_{RALI})^{f_w} \qquad (11)$$

and

$$\tau_G = \tau'_G \cdot (\tau_{RALI})^{f_G} \qquad (12)$$

where $\tau_w'$, $\tau_G'$ = transmittances due to absorption by water vapour, mixed gases, respectively,

$\tau_{RALI}$ =transmittance due to Rayleigh scattering by all atmospheric molecules,

$f_w = P_{H_2O}/P_T$ = ratio of water vapour to total atmospheric pressure, and $f_G=(1-f_w)$.

Correction for multiple Rayleigh scattering was effected using Box and Deepak's (1981) result:

$$\delta_{RALI} \rightarrow \delta_{RALI} + 0.02(\delta_{RALI} + \delta_{PS}) + 1.2(\delta_{RALI} + \delta_{PS})^2 (\sec\theta)^{\frac{1}{4}} \qquad (13)$$

where the optical depth on the right-hand side refers to the single scattering case.

In the terminology of LOWTRAN 5 if (C.ω*.DS) is the equivalent optical depth for water vapour, ω*.DS is equated to precipitable water for the purposes of this model; Cω*DS is the effective optical depth and C is the attenuation coefficient.

Hence, considering evaporation and condensation, the total heating rate of the portion of the atmosphere occupied by the dust cloud is

$$\frac{\Delta T}{\Delta t}_{TOT.} = \frac{\Delta T}{\Delta t}_{rad.} + h_c - h_v. \qquad (14)$$

where $-h_v$ = heating rate due to evaporation,

$h_c$ = heating rate due to condensation.

In a typical (dry) harmattan atmosphere $h_c$ is negligible. With the aid of Table 1, typical magnitudes of $h_v$ can be worked out. For molecules, g' is taken as 0.5 and for want of matching molecular data, $\omega_\lambda$ is made the same for both molecules and dust in this model.

TABLE 1. MEAN VALUES OF SOME SURFACE METEOROLOGICAL DATA RELEVANT TO RADIATIVE TRANSFER: FOR SEPTEMBER (A WET MONTH) AND DECEMBER (USUSL HARMATTAN PEAK) OVER A SEVEN-YEAR PERIOD, 1972 TO 1978

| Month | Location | Relative Humidity % | | Night Cooling Rate K/hr | Piche Evaporation ml. |
|---|---|---|---|---|---|
| | | Day | Night | | |
| Sept. | Kano | 61 ± 4 | 82 ± 5 | -(0. 5 ± 0. 1) | 4.3 ± 1.1 |
| | Lagos | 79 ± 3 | 93 ± 2 | -(0. 2 ± 0.03) | 2.2 ± 0.9 |
| Dec. | Kano | 20 ± 1 | 37 ± 3 | -(0.84 ± 0.03) | 12 ± 4 |
| | Lagos | 70 ± 4 | 92 ± 3 | -(0. 3 ± 0.04) | 3 ± 0.7 |

### 3.3 RESULTS AND DISCUSSION

Both gross and net nighttime heating rates for the model in Fig. 4 for the dust-laden atmosphere are shown in Table 2 in variation with dustload, precipitable water within the dust layer, and surface albedo (A=0.03,0.3). Based on our past data, we consider $\delta_5$ = 0.81 as representative of a moderate harmattan and $\delta_5$ = 1.62, a severe one. Temperatures at the surface and at the top of the dust are taken as 296 and 272 K, respectively.

As seen in Table 2 the radiative heating rate decreases with precipitable water for a fixed dustload and a fixed surface albedo, apparently because of the increased "reflectance" or backscattering to space of the counterradiation (see Eqs. (5), (8), and (11)).

For fixed values of precipitable water and surface albedo, atmospheric heating due to scattering and absorption increases with dustload, as expected, because the amount of radiant energy absorbed is proportional to the quantity of absorbers. Of course, there should be an associated "cooling" due to the reemission of the absorbed flux by each dust particle. However, such a reemitted flux is most likely absorbed by a nearby particle; that dynamic situation persists and the originally absorbed radiation appears trapped within the dust cloud. What emerges is effectively the transmitted component of the original incident flux and this decreases with dustload.

For fixed values of dustload and precipitable water, an increase in the reflectivity of the surface can increase the heating of the atmosphere by up to 10% for $0.03 \leq A \leq 0.3$.

Maximum radiative heating rate is about 0.044 $khr^{-1}$ for a severe harmattan and 0.016 $khr^{-1}$ for a moderate one, the same order of magnitude as obtained by Rogers and Walshaw (1966) and Carlson and Benjamin (1980). Subtracting the dust-time actual night cooling rate from that for dust-free nights (Table 1), the observed time-averaged nighttime cooling rate in the dusty atmosphere is 0.34 $khr^{-1}$ in Kano, 0.1 $khr^{-1}$ in Lagos. Using PW = 0.1cm, $\delta_5$=1.62 for Kano and

TABLE 2. MODEL ATMOSPHERIC LONGWAVE RADIATIVE HEATING RATES IN THE HARMATTAN FOR THE WAVELENGTHS 4 TO 28.5 μm

| Precipitable Water (cm) | Surface Albedo, A | Heating Rates ($Khr^{-1}$) Gross | | | Net | |
|---|---|---|---|---|---|---|
| | $\delta_5$: | 0 | .0.81 | 1.62 | 0.81 | 1.62 |
| .1 | .03 | .171 | .187 | .211 | .016 | .040 |
| .1 | .3 | .098 | .114 | .142 | .016 | .044 |
| 1. | .03 | .250 | .256 | .277 | .006 | .027 |
| 1. | .3 | .187 | .194 | .218 | .007 | .031 |

PW = 1.0cm and $\delta_5$ = 0.81 for Lagos as typical cases, the radiative effect of the dust is to reduce the cooling rate (produced by evaporation in the air) by up to 18% and 16% in Kano and Lagos, respectively.

## 4. SUMMARY

Size distribution of the harmattan aerosol near the intertropical convergence zone is retrieved using the inversion technique of Yue and Deepak (1983). The technique seems best suited for particles whose optical and physical properties are invariant in space and time, and this requirement holds for the Saharan dust (Carlson and Benjamin, 1980). Retrieved particle numbers range between 138 and 219 $cm^3$ in the mode $r \leq 1$ µm and between 24 and 38 $cm^{-3}$ in the mode $r > 1$ µm; so that about 57% of solar extinction by dust is due to the mode $r > 1$ µm. The nighttime cooling rate is studied using both a simple radiative transfer model for a plane-parallel, homogeneous atmosphere and also the time-averaged surface moisture and visibility variables over a 7-year period at two extreme locations of the weather regime. Heating by scattering and absorption increases almost proportionately with dustload but slightly with surface albedo. Radiative effects of moisture slightly reduce the heating rate as precipitable water increases, due to increased backscatter to space; but the dry nature of the dust increases atmospheric cooling by evaporation very actively. It appears, therefore, that the dust plays two opposing roles: that of increasing the cooling rate of the ambient air due to its dry nature and that of increasing the heating rate of the atmosphere due to scattering and absorption, with the former role being dominant.

## ACKNOWLEDGMENT

The author wishes to thank the Nigerian Meteorological Department for the back-up surface data.

## REFERENCES

Box, M. A., and A. Deepak, 1981: An Approximation to Multiple Scattering in the Earth's Atmosphere: Almucantar Radiance Formulation, J. Atmos. Sci., 18, 1037-1048.

Carlson, T. N., and S. G. Benjamin, 1980: Radiative Heating Rates for Saharan Dust, J. Atmos. Sci; 37, 193-213.

Carlson, T. N., and J. M. Prospero, 1972: The Large-Scale Movement of Saharan Air Outbreaks over the Northern Equatorial Atlantic, Appl. Meteor., 283-297.

Cerf, A., L. Fouquart and H. Legrand, 1980: Radiation Transfer in the Saharan Dust-haze and Climatological Impact. In Proceedings of International Radiation Symposium, Fort Collins, CO, USA.

Fischer, K., 1976: The Optical Constants of Atmospheric Aerosol Particles in the 7.5-12 µm Spectral Region, Tellus 28, 266-274.

Griggs, M., 1968: Emissivities of Natural Surfaces in the 8-to 14-micron Spectral Region, J. Geophys. Res., 73 (24), 7545-7551.

Haltiner, G.J., and F. L. Martin, 1957: Dynamical and Physical Meteorology, McGraw-Hill Book Company, Inc., New York, Toronto, London, 470 pp.

Hamilton, R.A., and M. A. Archibold, 1945: Meteorology of Nigeria and Adjacent Territory, Quart. J. Roy. Meteor. Soc., 71, 231-265.

Jaenicke, R., 1977: Continental aerosol over the ocean. In H.-J. Bolle (Ed.), Proceedings of the Symposium on Radiation in the Atmosphere, Science Press, Princeton, 29-30.

Kneizys, F. X., E. P. Shettle, W. O. Gallery, J. H. Chetwynd, Jr., L. W. Abreu, J.E.A. Selby and R. A. McClatchey, 1980: Atmospheric Transmittance/Radiance: Computer Code LOWTRAN 5, Air Force Geophysics Laboratory, Hanscom Fields AFB, Mass, 233 pp.

Oluwafemi, C. O., 1980: Some Measurements of Extinction Coefficients of Solar Radiation in Lagos, Pure Appl. Geophys., 118, 775-782.

Oluwafemi, C. O., 1981: On Spectral Extinction of Solar Radiation Within a Tropical Atmospheric Boundary Layer, Pure Appl. Geophys. 119, 831-841.

Oluwafemi, C. O., 1983: Solar Almucantar Radiance and Aerosol Scattering: Comparative Data From One Tropical Station and One Mid-Latitude Station, Pure Appl. Geophys., 121, 1066-1076.

Pollack, J. B., O. B. Toon and B. N. Khare, 1973: Optical Properties of Some Terrestrial Rocks and Glasses, Icarus, 19, 372-389.

Rogers, C. D., and C. D. Walshaw, 1966: The Computation of Infrared Cooling Rate in Planetary Atmospheres, Quart, J. Roy. Meteor. Soc., 92, 67-92.

Twomey, S., 1977: Atmospheric Aerosols, Elsevier Scientific Publishing Company, Amsterdam, 297 pp.

Volz, F. E., 1973: Infrared Optical Constants of Amonium Sulfate, Sahara Dust, Volcanic Pumice, and Flyash, Appl. Opt., 12, 564-568.

Yue, G. K., and A. Deepak, 1983: Retrieval of Stratospheric Aerosol Size-Distribution from Atmospheric Extinction of Solar Radiation at Two Wavelengths, Appl. Opt., 22, 1639-1643.

# IMPACT OF NONSTANDARD CONDITIONS ON VISIBILITY MEASUREMENTS

*D. Weintraub and V.K. Saxena*
North Carolina State University
Raleigh, North Carolina 27695, USA

## ABSTRACT

The presence of nonuniform sky conditions and nonblack targets, which represent nonideal conditions during a visual range observation, can drastically alter the estimate of visibility under an invariant atmospheric pollutant concentration. To investigate these factors, use of the atmospheric extinction coefficient separated into its scattering and absorption components was studied. Instruments deployed to collect the data were: an autotelephotometer, integrating nephelometer, manual dichotomous sampler, and a light transmission device for obtaining the extinction coefficient, scattering coefficient, mass concentration, and absorption coefficient, respectively. Data were collected between 3 July and 5 August 1985 at Research Triangle Park, North Carolina. The scattering component due to fine particulates contributed 74% to the total extinction. A high correlation between telephotometric-derived extinction and the sum of the components indicated good quality assurance. The relationship between light scattering coefficient and fine mass concentration was observed to exhibit a high correlation indicating that particulates of less than 2.5 μm in diameter are mostly responsible for aerosol scattering, which is in accordance with the Mie theory. Sensitivity of the visual range theory was tested by modifying the Koschmieder model, which is used widely in visibility studies. These modifications were performed by replacing the surrogate parameters with actual measured inherent contrast extinction coefficients and a change of the psychometric value of .05 for the liminal contrast. Results showed that the largest difference is caused by the liminal contrast change in conjunction with actual inherent contrast values instead of the customarily assumed value of -1 for the inherent contrast. Measurements of the ideal black box target, made minutes apart, showed the largest deviations in visual range under varying sky conditions. For example, a 51% decrease in the extinction coefficient was observed when a bright cloud passed behind a black box target.

## 1. INTRODUCTION

Since the amendment of the Clean Air Act by the U.S. Congress in 1977, an assessment of atmospheric visibility is required as a part of any environmental impact statement for planning future urban-industrial growth. The eastern U.S.A. experiences relatively low visibilities and strong seasonal variations (Trijonis, 1982). For instance, rural areas of Mississippi and areas lying south of the Great Lakes generally exhibit annual median visual ranges between 15-25 km. Review of the available data indicates that ambient fine particle (about 2.5 μm in diameter or smaller) concentration averages about 29 μg $m^{-3}$ in the rural East, the largest components being water (11 μg $m^{-3}$) and organic matter (4 μg $m^{-3}$). The majority of the water is, indeed, attached to the

AEROSOLS AND CLIMATE
Peter V. Hobbs and M. Patrick McCormick (Eds.)


ISBN 0-937194-11-5

hygroscopic sulfate aerosol. This fact is also supported by the study of Stevens et al., (1980) in the Great Smokey Mountains.

A number of important investigations (for example, Stevens et al., 1984; Dzubay et al., 1982; Lewis, 1981; Horvath, 1971) have discussed the dependence of visibility on environmental parameters. Early work of Koschmieder (1924), which was subsequently improved by Middleton (1952), has traditionally been the basis of visibility theory. Particularly notable is the work of Charlson (1969) and Charlson et al., (1967, 1969) in establishing the relationship between the visual range and atmospheric aerosol parameters such as the light extinction coefficient and mass loading.

From the standpoint of air quality a reduction in visual range may incorrectly be inferred as stemming from an increase in atmospheric pollutant concentration or vice versa, although the cause may be a variation in illumination upon the target itself. It is imperative in visibility monitoring to account for the observational configuration, target characteristics, and sky conditions. Indeed, atmospheric visibility is frequently measured under nonideal conditions (see, for example, Allard and Tombach, 1981).

The objective of this paper is to investigate: 1) the effects of nonstandard visibility conditions upon visual range under varying illumination of the target and changing meteorology, and 2) the difference in reported visual range due to varying target characteristics and sky conditions with respect to invariant aerosol concentration. These investigations are based on the data collected during the period 3 July - 5 August 1985 at Research Triangle Park, North Carolina, using an autotelephotometer, integrating nephelometer, manual dichotomous sampler, and a light transmission device.

## 2. METHODOLOGY

### 2.1 ATMOSPHERIC EXTINCTION COEFFICIENTS

Transmission of monochromatic light through the atmosphere can be described in terms of the extinction coefficient, $b_{ext}$, which may be expressed as

$$b_{ext} = b_{sp} + b_{sg} + b_{ap} + b_{ag}, \tag{1}$$

where $b_{sp}$ and $b_{sg}$ represent scattering coefficients due to aerosol particles and gases, and $b_{ap}$ and $b_{ag}$ denote the absorption coefficients due to aerosol particles and gases, respectively. By far the most significant contributing factor in Eq. (1) is $b_{sp}$. For example, Dzubay and Clubb (1981) showed that in a rural area in North Carolina, $b_{sp}$ accounted for 77% of $b_{ext}$ at relative humidities below 65%. The aerosol scattering depends upon the mass concentration of fine particulate matter; size, shape, and refractive index of aerosol particles; environmental relative humidity; and the wavelength of incident light. The magnitude of $b_{sp}$ is crucially controlled by the mass concentration of particles between 0.1 and 1.0 μm in diameter. If the particle size distribution is represented by $n_d$ ($d_p$,r,t), and the aerosol particles of different sizes are assumed to have the same density, the mass concentration of particulate matter is proportional to the third moment (Friedlander, 1977) of the size distribution.

However, the coefficient $b_{sp}$ is proportional to the sixth moment ($M_6$) when the particles are much smaller than the wavelength of the incident light and may be written as

$$b_{sp} = \frac{2}{3}\frac{\pi^5}{\lambda^4}\left(\frac{m^2-1}{m^2+2}\right) I \int_0^\infty n_d d_p^6 d(d_p), \text{ if } dp \ll \lambda \tag{2}$$

Here m is the refractive index of aerosol particles and λ and I are respectively the wavelength and intensity of the incident light. It has been demonstrated by Waggoner and Charlson (1976) that the spherical aerosol particles with m = 1.50 and at λ=550 nm, the scattering cross-section sharply increases with dp around $d_p$=0.1 μm, goes through a maximum, and becomes smaller as $d_p$ becomes larger than 1.0 μm (Fig. 1).

The most important classes of compounds contributing to $b_{sp}$ are sulfates (sulfuric acid, ammonium sulfate, ammonium bisulfate), nitrates (ammonium nitrate), and organic compounds. The secondary particles, often formed in the atmosphere in situ as

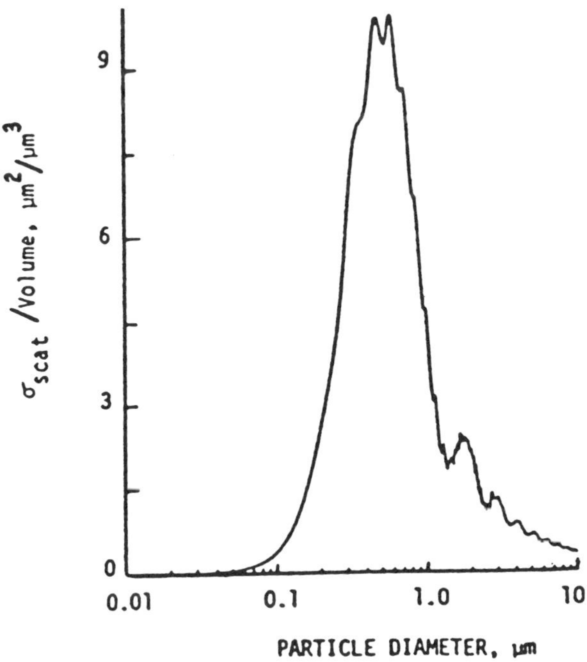


FIGURE 1. Scattering coefficient per particle divided by particle volume plotted as a function of diameter. The particles are assumed spheres of refractive index 1.50 illuminated at 550 nm wavelength (Waggoner and Charlson, 1976).

a result of gas-to-particle conversion (GPC) mechanisms, have a dominating effect on visual range because they conform to the size range that is most efficient per unit mass of the particulate matter for reducing atmospheric visibility.

The scattering coefficient due to gases, $b_{sg}$, in Eq. (1) varies with altitude since it is a function of atmospheric density. At a given altitude, $b_{sg}$ is determined by the oxygen, nitrogen, and trace gases composition of the air. Typically, at sea level $b_{sg} = 1.0 \times 10^{-5}\ m^{-1}$. Absorption of aerosol particles ($b_{ap}$) is of the order of 10% of the $b_{sp}$ or less in clean background air and is dependent upon the chemical composition and size distribution of particles (Bergstrom, 1973). Graphite carbon in the form of soot particles is a very efficient absorber. The soot particles may be emitted from fossil fuel combustion or from biomass. Light absorption of gases ($b_{ag}$) is particularly important in the assessment of atmospheric visibility since its impairment is strongly influenced by anthropogenic production of $NO_2$, which absorbs light at the blue end of the visible spectrum thus producing a reddish-brown haze.

## 2.2 VISUAL RANGE

As is well understood (Middleton, 1952; Dzubay et al., 1982; Horvath, 1981; Allard and Tombach, 1981), the extinction coefficient, $b_{ext}$, is related to the apparent contrast of distant targets. For a homogeneous atmosphere, a target of inherent contrast $C_o$ appears to have a contrast $C_r$ when observed from a distance r:

$$C_r = C_o \left(\frac{{}_sB_o}{{}_sB_r}\right) \exp(-b_{ext}\, r) \tag{3}$$

where ${}_sB_o$ and ${}_sB_r$ respectively represent the brightness of horizon at zero distance and at a distance r from the target. For a black target, $C_o = -1$, and substituting

$$\frac{{}_sB_o}{{}_sB_r} = 1, \text{ and } \varepsilon = \frac{C_r}{C_o} = 0.02,$$

we obtain the Koschmieder relationship, often called the meteorological range

$$V_r = \frac{3.9}{b_{ext}}. \tag{4}$$

Equation (4) is based on the following assumptions: 1) perfectly black objects, 2) homogeneous atmosphere, 3) horizontal viewing, 4) cloudless sky, 5) perfect observer (i.e., $\varepsilon = 0.02$), 6) curvatureless earth, and 7) aerosol particles scatter light individually. Deviations from assumptions 1), 2), and 4) are often crucial to the overall estimation of the visual range. Malm et al., (1981) have shown that ${}_sB_o/{}_sB_r = 1$ is a good approximation for targets located horizontally at less than 15 km from the observer. The numerator of Eq. (4) has been estimated to be 1.8 by Dzubay et al.

(1982) on the basis of their field study in Houston, Texas, 2.9 by U.S. EPA (1980), 3.35 by Horvath and Noll (1969), 1.9 ± 0.4 by Griffing (1980), and 3.0 by Tombach and Allard (1983).

## 2.3 EXPERIMENTAL SITE, INSTRUMENTATION, AND TARGETS

Measurements were made at a rural site about 8 km northwest of the Raleigh-Durham airport in Research Triangle Park, North Carolina. The monitoring site was located in the Research Triangle Institute (35°54'10"N, 78°52'10"W; 102 m MSL). The site was approximately a 91 m clearing off a lightly traveled access road. All the targets were within 50° horizontal view. Two of the targets were black wooden boxes with baffles inside and corrugated cardboard in the back to trap light. These were located at 780 m and 30 m away from the observation point. The nonblack targets were a red brick wall, a white painted masonry wall, a gray concrete wall cobbled with a conglomerate of stones to give a rough texture, and a 15.2 m cylindrical pale-green storage tank with a smooth texture. Two natural targets were an oak tree and a forest, the latter was the farthest available target.

These targets were located so as to be in direct sunlight all morning and part of the afternoon during the month-long study. In this situation, the targets were fully illuminated, thus bright, while the light scattered by the atmosphere was small due to backscatter. This made the targets readily visible. From the midafternoon until sunset, the targets were in their own shadow (i.e., scattering angle being less than 90°), and thus dark. In the afternoon, luminance of the horizon reached a large value since forward scattering is dominant. As Horvath (1981) has pointed out, the decrease in visibility is affected less for all nonblack targets, as long as they stay in their own shadow. In Table 1 are listed the instruments that were used in the study.

TABLE 1. LIST OF INSTRUMENTS, PARAMETERS, AND FREQUENCY OF MEASUREMENTS USED IN THIS WORK DURING 3 JULY- 5 AUGUST 1985

| Instrument | Parameter | Frequency |
|---|---|---|
| Auto-Telephotometer (Gama Scientific 2009) | Radiance values for contrast ($b_{ext}$) | 3 measurements each hour for 8 hours per day |
| Integrating Nephelometer (MRI 5197 EPA #119) | Fine particulate $b_{sp}$ | Continuously (hourly averages of readings recorded below 60% relative humidity) |
| Manual Dichotomous Sampler (ERC 200) | Mass concentration in 2 size ranges | 9 hours each day |
| Chemiluminescence Analyzer (RTI) | Concentration of $NO_2$ gas in ppb | Every two minutes (hourly averages) |
| Integrating Plate Method (light transmission) | Light intensity $b_{ap}$ | Daily average<br>Daily average |
| Meteorological Sensors: | | |
| Temperature | | Hourly |
| Dew Point | | Hourly |

### 2.4 QUALITY ASSURANCE OF THE DATA

A quality assurance program was enforced to ensure that the measurements were consistent and reliable. Our measurements were based on the brightness contrast telephotometry, which gives a direct measurement of $b_{ext}$. To assure the quality of telephotometric measurements, the components on the right-hand side of Eq. (1) were added and compared with the direct measurements of $b_{ext}$. All test cases of cloud cover were used and a quality check was performed following the procedure outlined by Dzubay and Clubb (1981).

## 3. DISCUSSION AND RESULTS

### 3.1 BRIGHTNESS RATIO OF NONIDEAL TARGETS

In Fig. 2 is displayed the average brightness ratio (the target's brightness to that of the background sky brightness) as a function of eastern daylight saving time (EDT) from 0800 to 1800 hr for the period 3 July - 5 August 1985 for seven targets, namely a black box, gray wall, willow oak tree, white wall, green tank, red brick wall, and forest. For nonideal targets, the target to background sky brightness should be below 0.05 - 0.08 for all observation angles. It is evident from the figure that the willow oak tree, gray wall, and forest come closer to satisfying the criterion. The black box

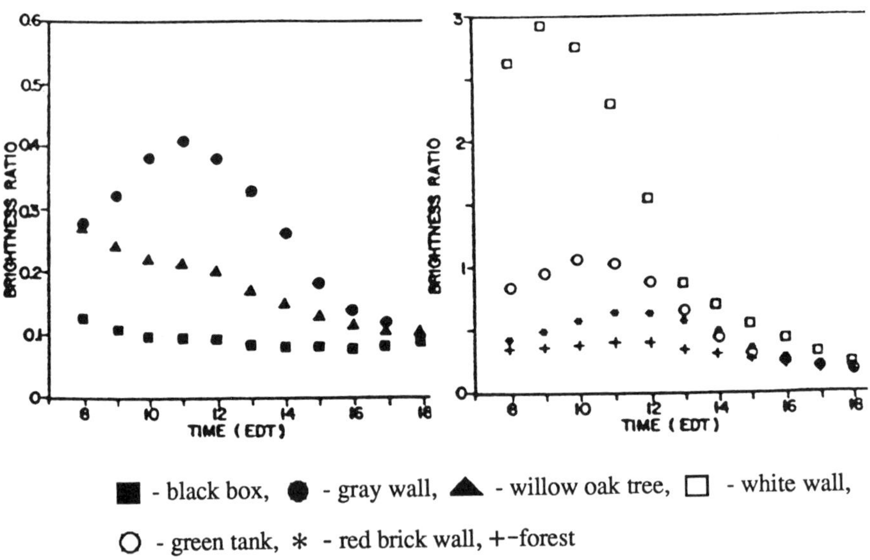


FIGURE 2. Diurnal pattern for target brightness ratio averaged over all observations from 3 July to 5 August 1985.

showed its lowest brightness at 0800 EDT and that is when the maximum contrast should occur. However, as seen in Fig. 2, the brightness ratio of the black box is not negligible compared to that of the sky due to the box orientation. At sunrise, the sun's rays were striking directly inside the box, illuminating the baffles and the back causing the brightness to increase. This example illustrates the necessity of using brightness and contrast throughout the day. The actual decrease of the brightness ratio in the late afternoon is due to the overwhelming forward scattering of solar radiation rather than the diminishing brightness of the targets themselves.

## 3.2 INTERCOMPARISON OF CONTRASTS

In Table 2 are shown the intercorrelation results for the apparent contrasts of the test targets. The impact of nonstandard conditions is evident from this table. The target with the best correlation to the black box is the willow oak tree ($r = 0.72$). The next best target, namely the forest, gives a correlation coefficient of 0.57 only. To be noted is the very high correlation of the red brick wall with the gray wall, white wall with the green tank, and gray wall with the green tank. This is due primarily to the similar diurnal variation in the apparent contrast of the test targets. The table summarizes the results of 155 observations.

## 3.3 VISUAL RANGE MODEL FOR NONSTANDARD INHERENT CONTRAST

In the derivation of Eq. (4), $\varepsilon = 0.02$ was assumed. Our computations indicate that the use of $\varepsilon = 0.05$ instead of 0.02 (Co = -1) produces a 23% change in $V_r$. The factor $\varepsilon$ depends upon the angle subtended by the object at the observer's eye, boundary sharpness of the object, and observer's physical health and is consequently difficult to assess due to its psychometric nature. Our measurements resulted in the black box inherent contrast (Co) to be -0.84 (average value based on 155 observations). If we use realistic value of $\varepsilon = 0.05$ based on Blackwell's (1946) realistic results and measured value of $C_o = -0.84$, the numerator of Eq. (4) turns out to be 2.82 leading to values of visual ranges smaller by 27 to 33% under identical environmental conditions and aerosol content of the airmass. This demonstrates that large errors in the estimation of $V_r$ are encountered if standard conditions, as outlined in the derivation of Eq. (4), are not used.

TABLE 2. INTERCORRELATION OF APPARENT CONTRAST FOR ALL TEST TARGETS.[1]

| | Black Box | Red Brick | White Wall | Gray Wall | Green Tank | Willow Oak Tree | Forest |
|---|---|---|---|---|---|---|---|
| Black Box | 1.00 | .248 | .194 | .305 | .192 | .722 | .573 |
| Red Brick Wall | .248 | 1.00 | .565 | .960 | .802 | .633 | .841 |
| White Wall | .194 | .565 | 1.00 | .674 | .907 | .732 | .605 |
| Gray Wall | .305 | .960 | .674 | 1.00 | .884 | .697 | .855 |
| Green Tank | .192 | .802 | .907 | .884 | 1.00 | .722 | .742 |
| Willow Oak | .722 | .633 | .732 | .697 | .722 | 1.00 | .855 |
| Forest | .573 | .841 | .605 | .855 | .742 | .855 | 1.00 |

[1]Computations are based on $n = 155$.

### 3.4 VARYING SKY CONDITIONS

Nonuniform illumination could be one of the major sources of error in $V_r$ as shown in Table 3. Here $V_o$ represents the visual range computed using the model as modified above. All three cases in the table consistently exemplify considerable (41-104%) increases in visual range due to the passage of a bright cloud behind the black box target. The cloud increases the horizon brightness thereby producing a higher contrast, hence a lower extinction coefficient that may be perceived arising from a cleaner airmass. For light-colored targets, the contrast would decrease and result in a lower visual range. As pointed out by Dzubay and Clubb (1981) also, smaller values of $V_r$ may result from a cloud casting a shadow over the telephotometer and target while larger values from a dark cloud behind the target.

### 3.5 RELATIONSHIP BETWEEN FINE PARTICULATE MASS AND SCATTERING COEFFICIENT

Our measurements showed that the scattering ($b_{sp}$) by aerosol particles accounted for 74% of the total extinction ($b_{ext}$). Fine particulate mass ($m_f$) for $d_p < 2.5\ \mu m$ was obtained by dichotomous sampler resulting in the average ratio $b_{sp}/m_f = (3.7 \pm 0.8) m^2 g^{-1}$. This value is comparable to the value of 3.52 $m^2 g^{-1}$ reported by Dzubay et al. (1982) on the basis of their Houston field study. As demonstrated by Eq. (2), $b_{sp}$ is dependent upon the refractive index of aerosol particles and hence on their chemical composition also.

## 4. CONCLUDING REMARKS

The major objective of this paper was to examine the impact of nonstandard conditions on visibility measurements and to assess the influence of violations of the Koschmieder visibility model on visual range estimates. To this end, we focussed on

TABLE 3. SUMMARY OF BRIGHT CLOUD CASES EFFECT $b_{ext}$

| | | | Bright Cloud Behind the Target | | | Cloudiness Sky | | | % Change in | |
|---|---|---|---|---|---|---|---|---|---|---|
| Date | Time (EDT) | B* | $b_{ext}$ † | $V_r$ | $V_o$ ≠ | $b_{ext}$ | $V_r$ | $V_o$ | $b_{ext}$ | $V_r$ |
| 8/2/85 | 10:00 | 74 | 47.6 | 82 | 61 | 66.7 | 58 | 43 | 28 | 41 |
| 8/2/85 | 10:30 | 112 | 33.4 | 117 | 86 | 68.2 | 57 | 42 | 51 | 104 |
| 8/3/85 | 11:00 | 81 | 33.0 | 118 | 88 | 61.0 | 63 | 47 | 46 | 84 |

* percent increase over normal background brightness
\+ units in $Mm^{-1}$ ($1 Mm^{-1} = 10^{-6} m^{-1}$)
≠ based upon inherent contrast of black box

the aspects of nonblackness of the target and nonuniform sky conditions that commonly occur. The former was approached by comparison of targets with different colors, sizes, and textures while the latter involved measurements being made under varying cloud cover using an ideal black target. These effects, examined separately, were used to determine the extent to which our interpretation of the actual visual range is influenced. The above discussion has made it clear that although visibility reduction is largely caused by an increase in the mass concentration of fine particulates, any inference based on the visual estimates regarding the particulate pollutants could be misleading inasmuch as the use of nonblack targets and varying sky conditions could also produce appreciable changes in visual estimates. This work also provides additional data for a rural site in the eastern U.S.A., where the visibility is relatively low (Tombach and Allard, 1983) and undergoes large seasonal variations (Trijonis, 1982).

## ACKNOWLEDGMENTS

During the tenure of this work, the authors received invaluable help from the EPA's Atmospheric Sciences Research Laboratory personnel: Dr. Robert Stevens, Dr. Charles Lewis, Dr. Thomas Dzubay, and Mr. John Bell. The Research Triangle Institute allowed the authors to use their mobile Environmental Research Monitoring Laboratory wherein the field study was based. We extend our gratitude to the following RTI personnel: Dr. James Tomerdahl, Mr. Curtis Moore, and Mr. Clifford Decker. The authors express their heartfelt thanks to the technical reviewer of this paper for giving them an opportunity to clarify and strengthen its contents.

## REFERENCES

Allard, D., and I. Tombach, 1981: The Effects of Nonstandard Conditions on Visibility Measurements, Atmos. Environ., 15, 1847-1857.

Bergstrom, R. W., 1973: Extinction and Absorption Coefficients of the Atmospheric Aerosol as a Function of Particle Size, Contrib. Atmos. Phys., 46, 223-234.

Blackwell, H. R., 1946: Contrast Thresholds of the Human Eye, J. Opt. Soc. Am., 36, 624-643.

Charlson, R J., 1969: Atmospheric Visibility Related to Aerosol Mass Concentration: A Review, Environ. Sci. Technol., 3, 913-917.

Charlson, R. J., H. Horvath and R. F. Pueschel, 1967: The Direct Measurement of Atmospheric Light Scattering Coefficient for Studies of Visibility and Pollution, Atmos. Environ., 1, 469-478.

Charlson, R. J., N. C. Ahlquist, H. Selvidge and P. B. McCready, 1969: Monitoring the Atmospheric Aerosol Parameter With the Integrating Nephelometer, J. Air Poll. Cont. Assoc., 19, 937-942.

Dzubay, T. G., and K. W. Clubb, 1981: Comparison of Telephotometer Measurements of Extinction Coefficients With Scattering and Absorption Coefficients, Atmos. Environ., 15, 2617-2624.

Dzubay, T. G., R. K. Stevens, C. W. Lewis, D. H. Hern, W. J. Courtney, J. W. Tesch and M. A. Mason, 1982: Visibility and Aerosol Composition in Houston, Texas, Envir. Sci. Technol., 16, 514-524.

Friedlander, S. K., 1977: Smoke, Dust and Haze, John Wiley & Sons, New York, 317 pp.

Griffing, G. W., 1980: Relations Between the Prevailing Visibility, Nephelometer Scattering Coefficient and Sun Photometer Turbidity Coefficient, Atmos. Environ., 14, 577-584.

Horvath, H., and K. E. Noll, 1969: The Relationship Between Atmospheric Light Scattering Coefficient and Visibility, Atmos. Environ., 3, 543-550.

Horvath, H., 1971: On the Applicability of the Koschmieder Visibility Formula, Atmos. Environ., 5, 177-184.
Horvath, H., 1981: Atmospheric Visibility, Atmos. Environ., 15, 1785-1796.
Koschmieder, H., 1924: Theorie der hoirzontelen Sichtweite, Beitr. Phys. frei. Atmos., 12, 33-53, 171-181.
Lewis, C. W., 1981: On the Proportionality of Fine Mass Concentration and Extinction Coefficient for Bimodal Size Distribution, Atmos. Environ., 15, 2639-2646.
Malm, W. C., E. G. Walther, K. O'Dell and M. Klein, 1981: Visibility in the Southwestern United States from Summer 1978 to Spring 1979, Atmos. Environ., 15, 2031-2042.
Middleton, W. E. K., 1952: Vision Through the Atmosphere, University of Toronto Press, Toronto, Ont., Canada.
Stevens, R. K., T. G. Dzubay, R. W. Shaw, W. A. McClenny, C. W. Lewis and W. E. Wilson, 1980: Characterization of the Aerosol in the Great Smokey Mountains, Environ. Sci. Technol., 14, 1491-1498.
Stevens, R. K., T. G. Dzubay, C. W. Lewis and R. W. Shaw, Jr., 1984: Source Apportionment Methods Applied to the Determination of the Origin of Ambient Aerosol That Affect Visibility in Forested Areas. Atmos. Environ., 18, 261-272.
Tombach, I., and D. Allard, 1983: Comparison of Visibility Measurement Techniques: Eastern United States, Final Report EA-3292, Electric Power Research Institute, Palo Alto, CA.
Trijonis, J., 1982: Existing and Natural Background Levels of Visibility and Fine Particles in the Rural East, Atmos. Environ., 16, 2431-2445.
U.S. EPA, 1980: Interim Guidance for Visibility Monitoring, EPA-450/2-80-082, available from Office of Air Quality Planning and Standards, Research Triangle Park, NC 27711.
Waggoner, A. P., and R. J. Charlson, 1976: Measurements of Aerosol Optical Parameters. In B.Y.H. Liu (Ed.), Fine Particles, Academic Press, New York, N.Y., 511-533.

# AEROSOLS FROM PAST AND PRESENT VOLCANIC EMISSIONS

*D.J. Hofmann*
University of Wyoming
Laramie, Wyoming 82071, USA

ABSTRACT

Our knowledge of the effects of volcanic eruptions on the stratosphere has increased dramatically since 1980. This is due mainly to the occurrence of several major eruptions and to the availability of both direct and remote techniques of measuring these effects. It has become clear that large quantities of sulfur gas injected into the stratosphere, which remain confined owing to lack of global transport, can greatly affect the aftermath of the eruption through vapor-to-liquid-droplet conversion. Thus the eruption of El Chichón in 1982 created perhaps the largest such perturbation since that of Krakatau 100 years earlier. As a result, the term "volcanic dust" has largely been replaced by "volcanic aerosol," referring to the dominant sulfuric acid - water droplets that form in the stratosphere and cause the long-term effects of such eruptions. While the critical question of whether such eruptions affect climate, or have done so in the past, has not been answered, the input to models that may someday answer this question are now available.

## 1. INTRODUCTION

The great eruption of Krakatau, which obliterated an entire island in Indonesia in 1883, created a great deal of interest in volcanic eruptions and their possible global effects. Although Benjamin Franklin, 100 years earlier, noted the "dry fog," which formed following the great eruption of Lakagigar in Iceland and Asama in Japan in 1783, and suggested that volcanic activity may have caused the severe winter that followed, it was not until the great eruption of Krakatau that one can say a science of volcanic eruptions began. Expeditions were mounted to study the aftermath of the eruption and reports of the optical effects seen round the world were recorded. Much of this information was published by the Royal Society (Symons, 1888) in London, constituting the first (and most comprehensive) report of a volcanic eruption on record. Another century passed before a similarly well-studied volcanic event took place.

Only thirty years after Krakatau, the eruption of Novarupta and Katmai in 1912 on the Alaskan Peninsula, while not of the magnitude of Krakatau, indicated what a major eruption closer to home can do in terms of visual effects. By contrast, the major eruption of Agung (8.3°S) in 1963 in the Southern Hemisphere did not have similar visual effects in the Northern Hemisphere, although some of the first stratospheric dust measurements by balloons (Rosen, 1964) and lidar (Grams and Fiocco, 1967) were made following this eruption.



ISBN 0-937194-11-5

The visual effects were aptly described as "dust veils" by Lamb (1970, 1977) in his treatise on volcanic eruptions, wherein he defined the "dust veil index." In an attempt to further classify volcanic eruption intensity, Newhall and Self (1982) defined the "volcanic explosivity index" (VEI), which combines the total volume of eruptive products, apparent height of the eruptive cloud, duration, and qualitative descriptions.

Although the eruption of Mt. St. Helens (46.2°N) in the state of Washington in 1980 cannot be classed as a great eruption in terms of stratospheric aerosol effects, its proximity to large populations and incredible explosive nature created a new awareness of this devastating phenomenon. It was not until 1982 that an eruption that can be placed in the class of Krakatau and Katmai occurred and that was the eruption of El Chichón (17.3°N) in Mexico. As will become clear in this review, more was learned about the effects of volcanic aerosols on the stratosphere following the El Chichon eruption than from all previous eruptions combined. This was due mainly to the wealth of research techniques that could be brought to bear with little delay.

Figure 1 (adapted from a compilation by Simkin et al., 1981), shows the historical record of the VEI (≥4) from 1880 to 1980. Also shown are Northern Hemisphere annual average temperature deviations from the mean over this period. The question of whether volcanic eruptions affect climate, or for that matter whether such events have been responsible for major climate changes in the past, returns periodically, generally with a new aspect, for example, altered global circulation. This question has been a driving force in this field of research and will remain so.

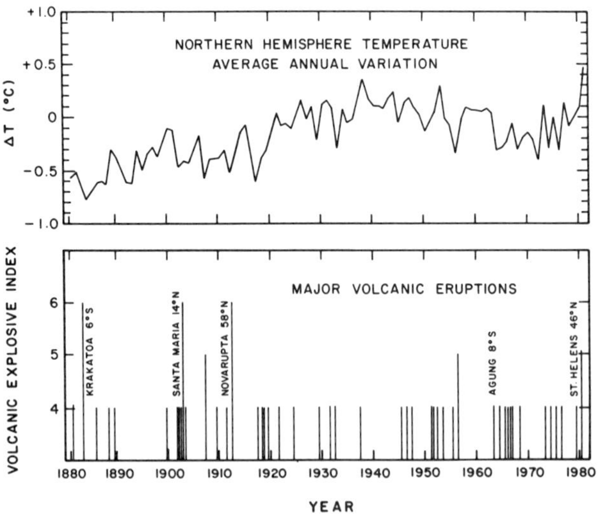


FIGURE 1. The volcanic explosivity index (VEI) of major volcanic eruptions and Northern Hemisphere average annual temperature deviations versus time during the past century.

In view of what was learned following the eruption of El Chichón, which was assigned a VEI of 4 compared to a 5 for Mt. St. Helens and a 6 for Krakatau, it appears that the VEI may not adequately measure stratospheric effects of eruptions for the following reasons. The lasting effect of an eruption apparently does not depend solely on the explosivity and related debris produced but on the amount of sulfurous gas injected into the stratosphere and on global transport, i.e., on the season of the eruption. Although it was known for some time that most stratospheric particles were volatile and had an approximate 75% $H_2SO_4$, 25% $H_2O$ composition (Rosen, 1971), the dominant component following volcanic eruptions was thought to be dust or silicate, possibly with an $H_2SO_4$ coating.

Following the Mt. St. Helens eruption in 1980, homogeneous nucleation from the gas phase was observed (Hofmann and Rosen, 1982). Composition measurements through aerosol vaporization techniques 6 months after the El Chichón eruption revealed the stratospheric aerosol to be 99% volatile and composed of various dilutions of $H_2SO_4$, depending on temperature and humidity (Hofmann and Rosen, 1983). Thus it was concluded that most of the silicate or "dust" component in an eruption consists of particles that are too large to survive very long in the stratosphere.

The growth of aerosol, owing to $H_2SO_4$ acretion, was observed in detail following El Chichón. This was due to the springtime eruption to the 25 km altitude level, which precluded meridional distribution throughout the summer of 1982, allowing the aerosol to grow unusually large (Hofmann and Rosen, 1984) and resulting in unprecedented atmospheric optical effects and stratospheric temperature perturbations at low latitudes (Labitzke et al., 1983). Figure 2 is a schematic depiction of events believed to have occurred following the El Chichón eruption. A bimodal aerosol size distribution formed owing to growth of the preeruption aerosol combined with nucleation, coagulation, and growth of new aerosol.

Perhaps the most significant measure of volcanic eruptive effects on the stratosphere is the ability of suspended particulates to block out the Sun, as has been measured by NOAA/Geophysical Monitoring for Climate Change (GMCC) at Mauna Loa, Hawaii (19.5°N) since 1958. Figure 3 (personal communication, 1 June 1987, J. DeLuisi, NOAA/ERL, Boulder, Colorado) shows the change in atmospheric transmission at a wavelength of 500 nm and indicates the relative magnitude of the 1982 eruption of El Chichón. The transmission of direct solar radiation was reduced by about 15% and did not recover for about 3 years. Since the aerosol scatters mainly in the forward direction, only a small fraction of this amount is actually lost to the earth-atmosphere system and shows up as an increase in diffuse scattering of sunlight, resulting in a milky appearance of the sky around the Sun. Figure 3 indicates that the eruption of Agung in 1963 and possibly Fuego (14.5°N) in Guatemala in 1974, caused smaller perturbations.

Figure 4 summarizes the atmospheric effects of large volcanic eruptions. The chemical effects involving the conversion of $SO_2$ to $H_2SO_4$ and the creation of aerosol, which can cause heating of the

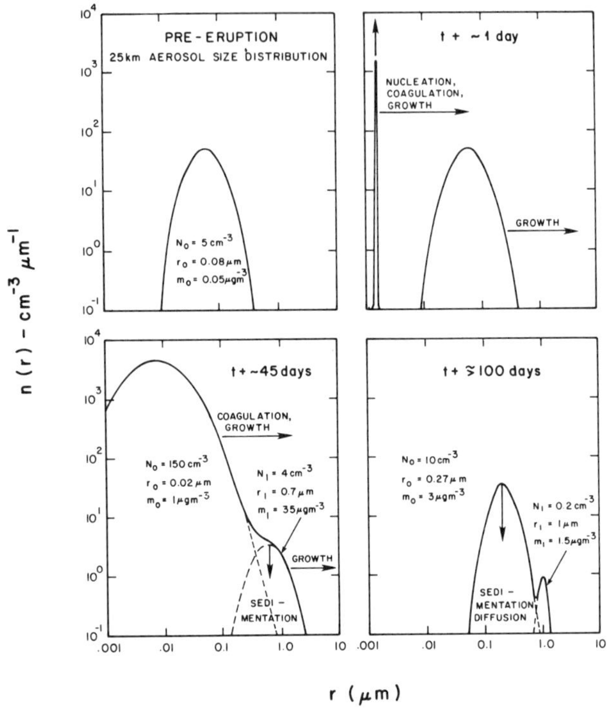


FIGURE 2. Schematic representation of the time development of the bimodal stratospheric aerosol differential size distribution, based on balloonborne particle counter measurements, following the eruption of El Chichón in 1982. $N_0$, $r_0$, $m_0$ and $N_1$, $r_1$, $m_1$ are, respectively, the total concentrations, median radii, and mass concentrations in each distribution.

stratosphere and cooling of the troposphere, are by far the most important. Other chemical perturbations including the introduction of water vapor and other trace species such as chlorine into the stratosphere with possible subsequent perturbations on the ozone layer are believed to be relatively minor but not ignorable (Hofmann, 1987a). In what follows, we will review observations related to the aerosol layer that formed in the stratosphere following major eruptions since about 1970. Although the eruption of Agung in 1963 is considered to have been major, insufficient measurements exist to discuss this Southern Hemisphere eruption in the same context as those after 1970 when balloonborne particle counters, surface- and aircraft-based lidar, aircraft filter and impactor techniques, and satellite-based extinction measurements evolved from infancy to routine measurement tools.

In a short review of this nature it is impossible to do justice to all those who contributed to our knowledge of volcanic aerosols.

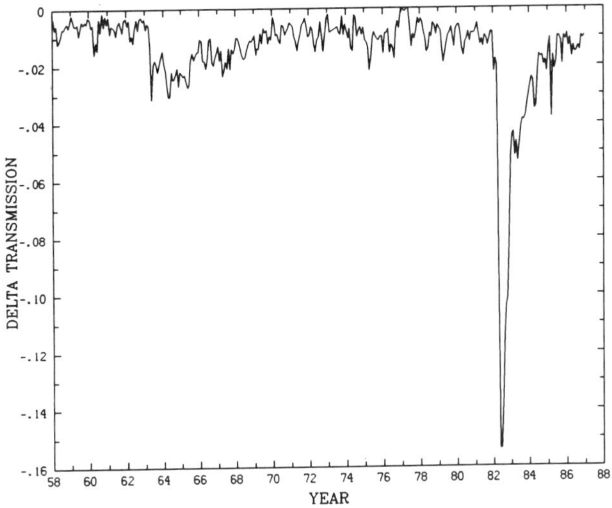


FIGURE 3. Decrease in the atmospheric transmission of direct solar radiation at a wavelength of 500 nm at Mauna Loa, Hawaii over the past 30 years.

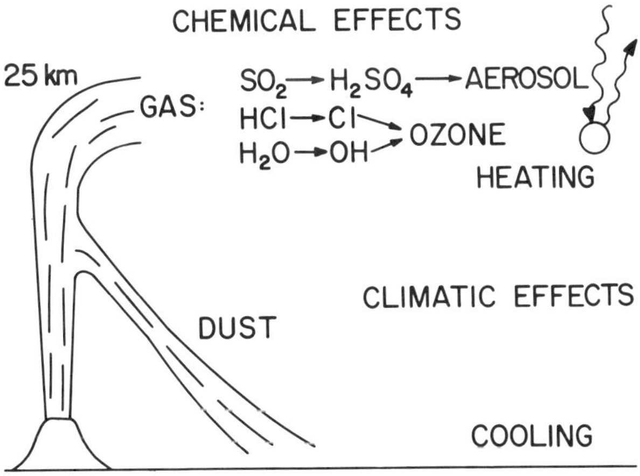


FIGURE 4. Schematic representation of the aftermath of a major volcanic eruption. The silicate dust particles fall to the surface rapidly but the sulfur gases convert to long-lived sulfuric acid aerosol and other trace gases may alter the stratospheric ozone distribution. The aerosol formed can cause stratospheric heating and surface cooling.

The more extensive data records will be emphasized and, resorting to convention, those most accessible, namely the balloonborne measurements of the University of Wyoming group of which the author is a member, will inevitably appear prominently. For this I apologize.

## 2. VOLCANIC AEROSOL OBSERVATIONS SINCE 1970

### 2.1 LONG-TERM DATA RECORDS

Figure 5 shows the time variation of the peak stratospheric aerosol mixing ratio (particles per mg ambient air) since 1971 as measured by balloonborne particle counters (Hofmann and Rosen, 1982; 1987) for 2 size ranges in 216 balloon flights at Laramie, Wyoming and 5 at several locations in southern Texas (the latter following the El Chichón eruption). Figure 6 shows the stratospheric integrated aerosol backscattering as measured by surface-based ruby (694.3 nm) lidar at 3 stations around the world. (The lidar data were supplied by M. McCormick, NASA Langley, Hampton, Virginia, T. DeFoor, NOAA/GMCC, Mauna Loa, Hawaii, and H. Jager, International Astronomical Union, Garmisch, West Germany.) The times of major volcanic eruptions, which are believed to have perturbed the stratospheric aerosol level at midlatitudes in the Northern Hemisphere, are also given in Figs. 5 and 6. General agreement between the direct and remote techniques and between various lidars distributed around the world is remarkable and underlines the global nature of volcanic aerosol effects. It is also clear from Figs. 5 and 6 that the El Chichón event of 1982 was by far the largest such event in the records.

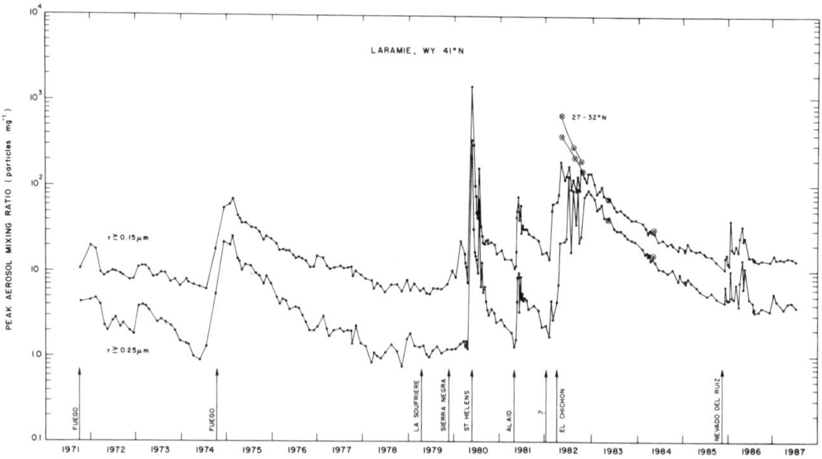


FIGURE 5. The peak stratospheric (≈20 km) aerosol mixing ratio for two size ranges since 1971 as measured by balloonborne particle counters at Laramie, Wyoming and in southern Texas (circled crosses). Times of major volcanic eruptions are indicated. (After Hofmann and Rosen, 1987.)

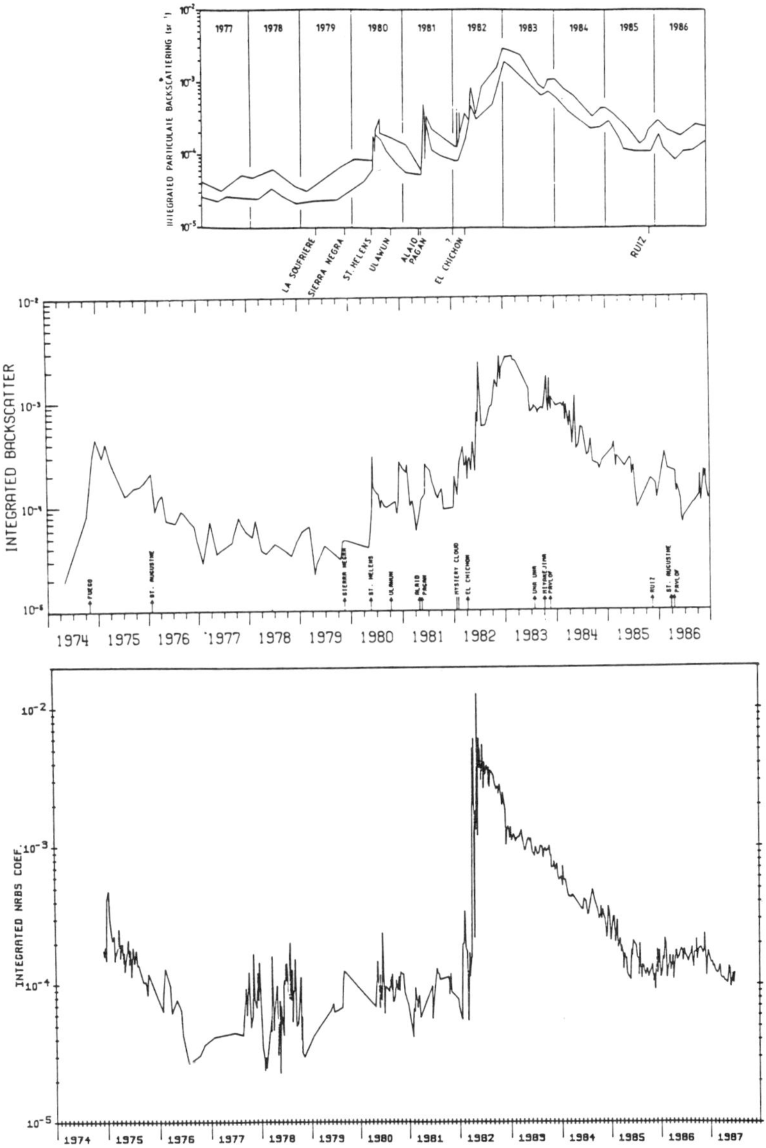


FIGURE 6. Integrated stratospheric particulate backscattering (per steradian) versus time as measured by 3 ruby lidars at Garmisch, West Germany (top), Hampton, Virginia (middle), and Mauna Loa, Hawaii (bottom). The 2 curves in the Garmisch data define uncertainty limits.

Solar extinction at 1 $\mu$m has been measured with the SAM II instrument on the satellite Nimbus 7 since 1978. It continues to function to this day and has returned a remarkable record of stratospheric extinction in the polar regions. Figure 7 (from McCormick and Trepte, 1987) shows stratospheric optical depth in the Arctic and Antarctic derived from these data. Again, the occurrence of volcanic injections are obvious in the arctic data. It appears that only the El Chichón eruption was large enough to seriously affect the antarctic stratosphere. The large mid-year increases followed by decreases in the antarctic data are due to the formation of polar stratospheric clouds during the austral winter followed by stratospheric cleansing owing to subsidence and/or gravitational settling (McCormick and Trepte, 1987). The solar-extinction satellite instrument SAGE (McCormick et al., 1979) operated from February 1979 to November 1981 and SAGE II from October 1984 to the present. The latter instrument covered midlatitude ranges and multiple wavelengths and was used to study volcanic effects in the 1980s (Kent and McCormick, 1984; Lenoble et al., 1984).

## 2.2 SOME OBSERVATIONAL MILESTONES

The first observation of the production of new aerosol from the gas phase following a volcanic eruption was made at Laramie by balloonborne condensation nuclei (CN) counters after the eruption of Mt. St. Helens in May 1980. Figure 8 (from Hofmann and Rosen, 1982), shows the abrupt increase in CN, reaching a peak about 40-50 days after the eruption, and the subsequent decay owing to coagulation and diffusion. By July 1980, most of the ash had fallen out and the remaining aerosol was observed to be mostly volatile at 150°C, i.e., a sulfuric acid - water mixture.

Aircraft were used extensively to study the effects of the eruption of Mt. St. Helens and were able to systematically sample the fresh eruption plume. This allowed direct particle sampling by filters and impactors and also allowed measurements of $SO_2$ and other trace gases. Space does not permit presentation of this large quantity of data and the interested reader is referred to the summary by Pollack et al. (1981).

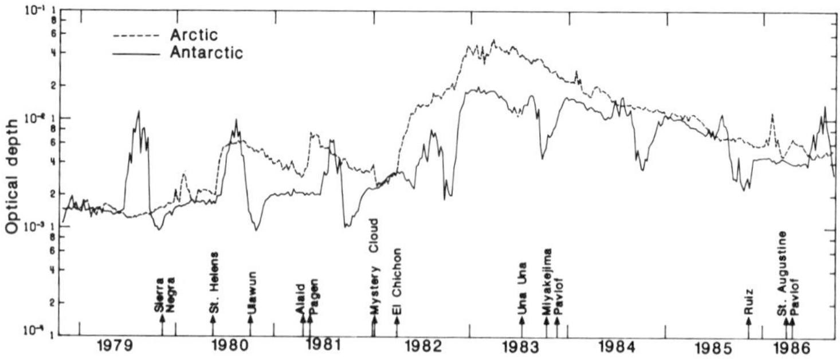


FIGURE 7. Stratospheric optical depth versus time derived from SAM II solar extinction measurements over the Arctic and Antarctic. (After McCormick and Trepte, 1987.)

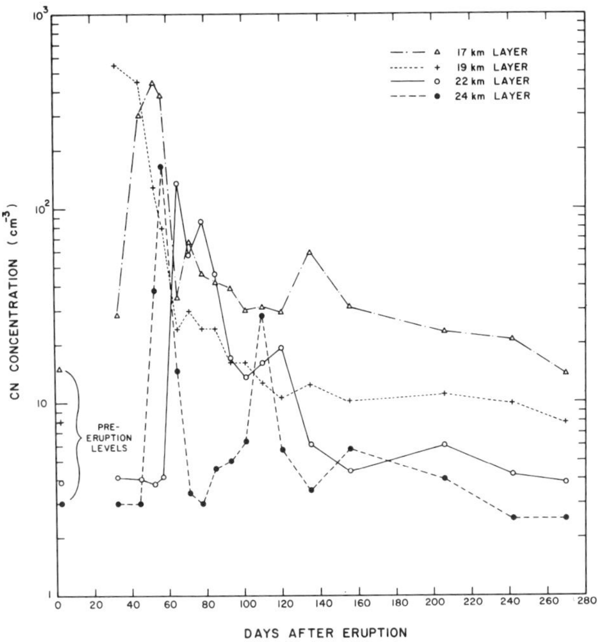


FIGURE 8. Condensation nuclei (r ≥ 0.01 μm) concentration versus time in several observed layers from balloonborne measurements at Laramie, Wyoming following the Mt. St. Helens eruption in 1980. (After Hofmann and Rosen, 1982, copyright by the American Geophysical Union.)

Airborne lidar was used effectively to define the latitudinal extent of the eruption cloud in 1982 and 1983 following the eruption of El Chichón (McCormick et al., 1984). Figure 9 (from the referenced work) shows an airborne lidar survey conducted about 6 months after the El Chichón eruption and clearly shows the confinement of the aerosol to low latitudes. It was this confinement that apparently resulted in aerosol growth to unusually large sizes (above 20 km), as shown in Fig. 10 (from Hofmann and Rosen, 1983), where we see two distinct aerosol layers, one above and one below the summer wind reversal region at about 19 km. Transport northward occurred in the lower layer soon after the April 1982 eruption but did not do so in the upper layer until fall and winter. Thus in Fig. 10 we see relatively more larger particles in the upper layer as compared to the lower layer. Aerosol growth was observed to occur with a time constant of about 40-50 days as shown in Fig. 11 (from Hofmann and Rosen, 1984), in agreement with satellite observations of the decay of $SO_2$ (Krueger, 1983), owing to conversion to $H_2SO_4$, and theoretical considerations concerning this transformation (McKeen et al., 1984).

Figure 12 shows a typical aerosol vaporization experiment (Hofmann and Rosen, 1983) after the El Chichón eruption. Nearly all of the aerosol is observed to be volatile at 130°C. From such measurements, the percentage of sulfuric acid was found to vary from about

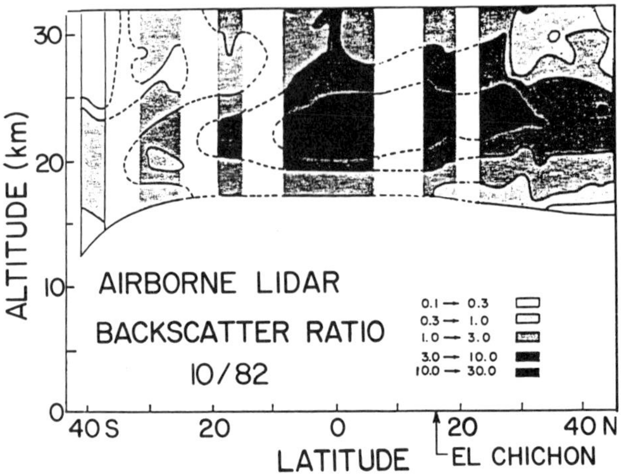


FIGURE 9. Aerosol backscatter ratio (relative to molecular scattering) as a function of altitude and latitude from an airborne lidar survey in October 1982, following the El Chichón eruption. (After McCormick et al., 1984.)

60% at the cold tropopause to about 80% in the regions of higher temperature at about 25 km. Combining such observations on composition and size distributions and the data of McCormick et al. (1984), on global extent, allowed an estimate of total mass in the aerosol. Although only 3.3 megatons of $SO_2$ were detected from space, at least 12 megatons of aerosol (about 75% of which was $H_2SO_4$) formed in the stratosphere. This apparent descrepancy in the sulfur budget exists to this day and suggests that not everything is known about these processes (Hofmann, 1987a).

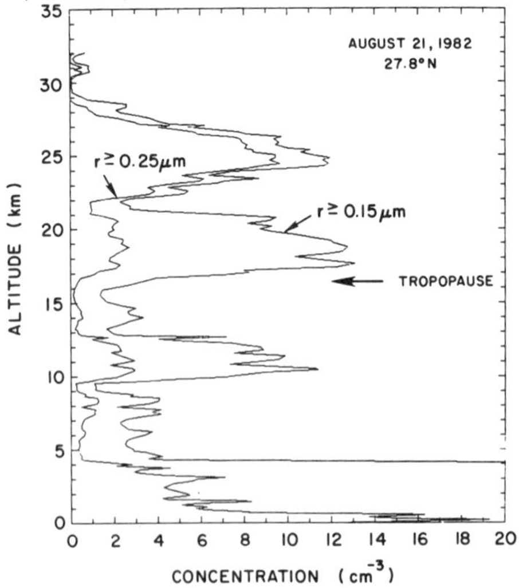


FIGURE 10. Aerosol concentration profiles measured in the El Chichón plume by balloonborne particle counters in southern Texas. (After Hofmann and Rosen, 1983.)

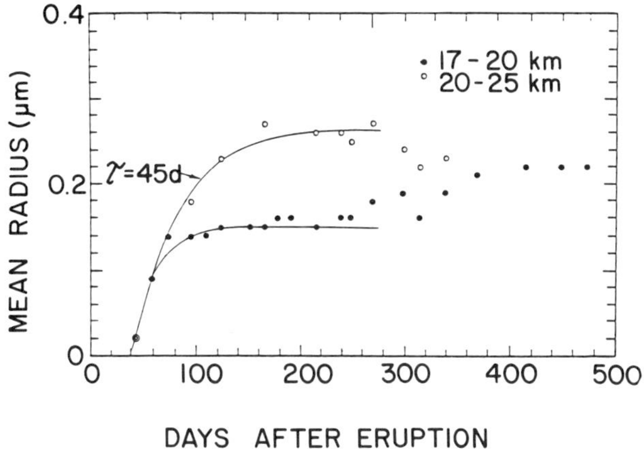


FIGURE 11. The growth of stratospheric aerosol as measured with balloonborne particle counters at Laramie, Wyoming in the two layers that formed after the El Chichón eruption in 1982. $\tau$ is the decay time of sulfuric acid vapor inferred from a model of aerosol growth by vapor acretion. (After Hofmann and Rosen, 1984.)

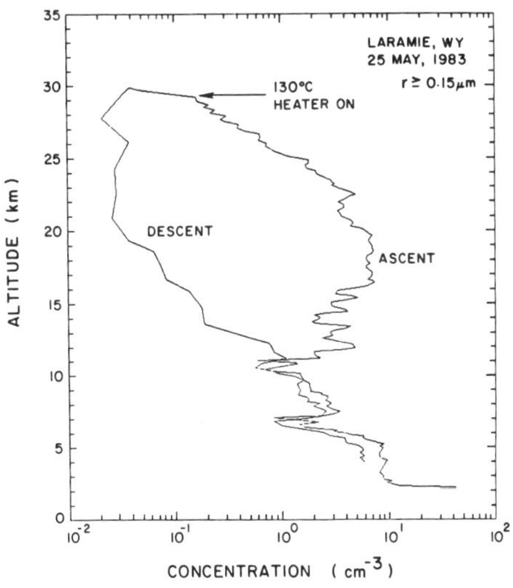


FIGURE 12. A balloonborne aerosol vaporization experiment. Near the top of the sounding, a 130°C air intake heater is turned on, which vaporizes the predominantly sulfuric acid/water aerosol during the parachute descent of the optical particle counter. (After Hofmann and Rosen, 1983.)

Comparing the rate of decay of aerosol following the eruption of Fuego in 1974 and El Chichón in 1982 with the same balloonborne particle counters led Hofmann and Rosen (1987) to conclude that simple sedimentation and diffusion are not the only factors affecting the decay of stratospheric aerosol from volcanic eruptions. Figure 13, from this work, shows the comparison. The size difference in the Fuego and El Chichón events is reflected in the ratio of the two sizes measured in Fig. 13. We see that even though the El Chichón aerosol grew considerably larger (0.2-0.3 $\mu$m versus 0.10-0.15 $\mu$m for the characteristic aerosol radius), it decayed more slowly (longer characteristic decay time). Jäger and Carnuth (1987), from ruby lidar data at Garmisch, West Germany, similarly found that the aerosol from El Chichón decayed much more slowly than that from either Mt. St. Helens in 1980 or Alaid (50.8°N) in 1981. The Garmisch lidar was not in operation at the time of the Fuego eruption.

Hofmann and Rosen (1987) attributed the longer lifetime of the El Chichón aerosol to the magnitude and altitude of the event combined with stratospheric warmings in the polar regions, which vaporize sulfuric acid aerosol at high altitude. The resulting vapor then diffuses to the 20 km region where it takes part in aerosol growth, thus prolonging the decay and maintaining the relatively larger aerosol

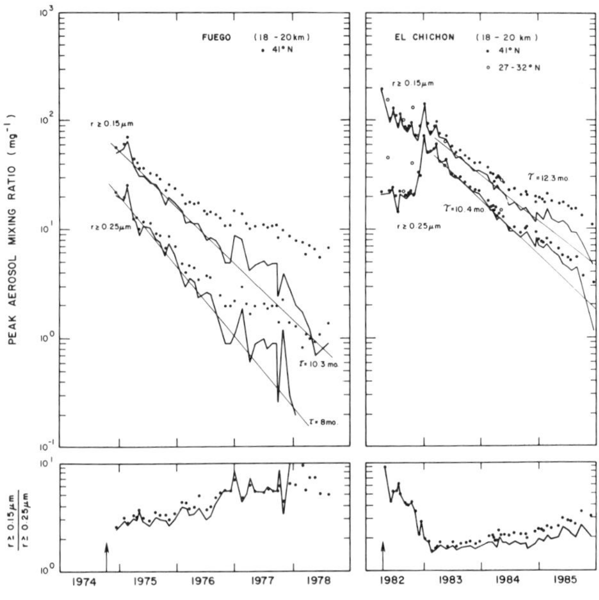


FIGURE 13. A comparison of the decay of 2 volcanic events with the same balloonborne particle counters. Points are actual measurements, the continuous lines are background corrected data, and the straight lines are exponential decay fits with indicated decay (1/e) times ($\tau$). The arrows mark the times of the eruptions. (After Hofmann and Rosen, 1987, copyright by the American Geophysical Union.)

size. Even at the present time, over 5 years after the El Chichón eruption, the stratospheric aerosol has not recovered to its pre-eruption size distribution although most of the original aerosol are now gone. It is possible that the stratospheric aerosol background, from natural and anthropogenic sulfurous gases that diffuse into the stratosphere and from minor volcanic eruptions, has also changed its nature with larger particles at the present time. The question of the present nature of the natural background was complicated by the relatively minor eruption in 1985 of Nevado del Ruiz in Colombia, and by further minor perturbations in 1987 (to be discussed later) and probably can only be resolved with several additional eruption-free years.

The El Chichón eruption not only provided a unique opportunity for measurement of the atmospheric effects of an eruption but also spawned numerous models of the possible climatic effects. These models have been outlined by Hofmann (1987a) and will not be discussed here except to say that they all predicted surface temperature reductions of about 0.5-1.0°C. Such reductions were not observed in global temperature records. This is perhaps an indication of the enormous complexity of the earth-ocean-atmosphere system involved. For example, Angell (1986) showed that if one corrects global surface and tropospheric temperatures for substantial variations in equatorial sea surface temperature, then a slight cooling (≈0.5°C) remains following both the eruptions of Agung and El Chichón. The data (from Angell, 1986), are shown in Fig. 14.

A more obvious atmospheric perturbation following El Chichón was that on extratropical stratospheric ozone in the Northern Hemisphere. Figure 15 (from Mantis et al., 1986) shows the time series of monthly departures of total ozone averaged for 15 Dobson stations located between about 30° and 60°N between 1963 and 1983. The greater than $2\sigma$ (dashed lines in Fig. 15) reduction in 1983 closely paralleled the

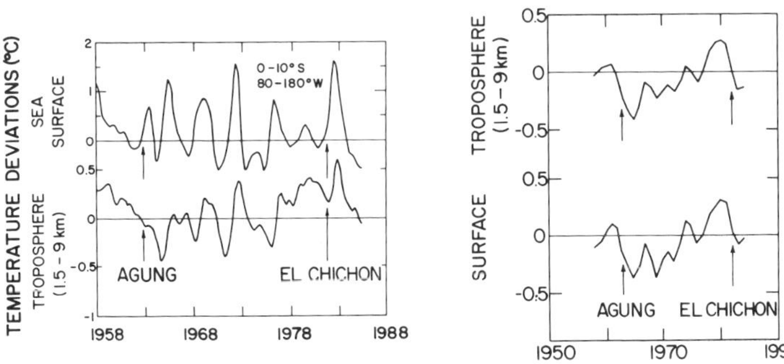


FIGURE 14. Global average tropospheric temperature deviations from the mean and equatorial sea surface temperature deviations (left) and surface and tropospheric temperature deviations after modification for sea surface temperature effects (right) versus time. (After Angell, 1986.)

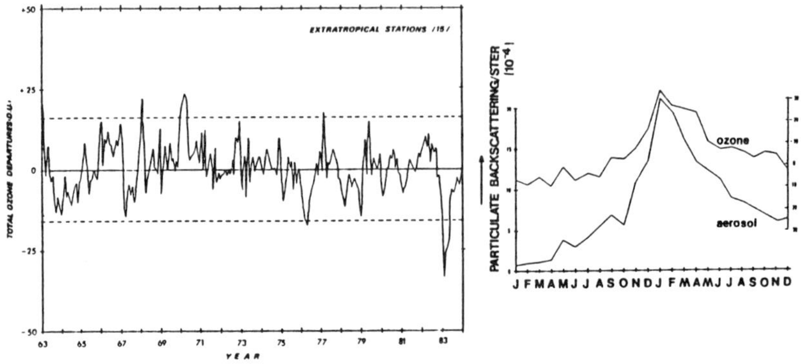


FIGURE 15. Total ozone departures from the mean for 15 extratropical Dobson stations versus time (left) and the apparent anti-correlation of these departures and stratospheric aerosol backscattering (right) as measured by lidar at Garmisch in 1982 and 1983. (After Mantis et al., 1986.)

stratospheric aerosol increase at midlatitudes as a comparison of the ozone data and the Garmisch lidar, also shown in Fig. 15, clearly indicates. Although the possibility of coincidental occurrence of low ozone and high aerosol exists, the variations are more likely related. The connection may be a direct reduction of ozone through aerosol and/or trace gas reactions associated with the eruption cloud or it may be related to modified ozone production and transport associated with a radiative imbalance induced by the eruption cloud. Quite possibly it is a combination of such effects, as discussed by Bojkov (1987). Optical contamination of the ozone observations, associated with enhanced aerosol scattering of ultraviolet radiation, should also be considered. This includes both surface-based and satellite-based remote sensing ozone measurements. At any rate, the possibility of a major volcanic eruption altering the global ozone layer is a significant observation, which, when studied in its entirety, may shed light on other anomalous ozone variations, for example as have been recently observed in Antarctica (the subject of the next section).

## 2.3 VOLCANIC AEROSOLS IN ANTARCTICA

In addition to the satellite instrument SAM II's monitoring of aerosol extinction in the polar regions mentioned earlier (see Fig. 5), particle counters have been flown on balloons in Antarctica since 1972 (Hofmann, 1987b). Figure 16 (from Hofmann et al., 1987a) shows vertical profiles of aerosol as measured annually at McMurdo Station (78°S) since the El Chichón eruption and clearly shows the decay of the volcanic aerosol. Such measurements are important as particles from the El Chichón eruption have been implicated in theories of the springtime depletion of stratospheric ozone in Antarctica (Tung et

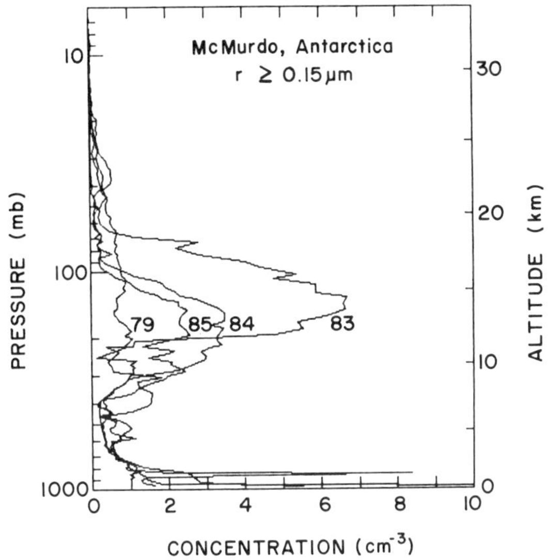


FIGURE 16. Aerosol profiles measured in Antarctica by balloonborne optical counters following the El Chichón eruption in 1983 (October 27), 1984 (November 1), and 1985 (November 5) compared to a pre-eruption profile measured on January 16, 1979. (After Hofmann et al., 1987a, copyright by the American Geophysical Union.)

al., 1986; McElroy et al., 1986). However, since the ozone deficit grew while the El Chichón aerosol was decaying, such a connection does not exist. However, there may be an indirect relationship.

Figure 17 (from data reported by Komhyr et al., 1986), shows total ozone measured from the surface at South Pole Station for the months of January and October and for the annual average since 1964. The times of the eruptions identified in Fig. 5 are also indicated. The spring (October) ozone reduction, since about 1980, is clear in these data. There is also a suggestion of a recent reduction in the summer values (January) and in the annual average. The large October values in some years prior to 1969 are due to polar vortex breakdown and ozone transport from midlatitudes earlier than in recent years. Thus, while the particles from El Chichón and possibly the other eruptions, which occurred since 1980, cannot be directly responsible for the ozone decrease, the recent late and weakened recovery of ozone in the spring in Antarctica may be a volcanic-related phenomenon. The latter could take the form of a perturbation to normal transport, owing to stratospheric heating and possible coupling to the troposphere, as witnessed by enhanced El Niño events following major equatorial eruptions that produced enhanced stratospheric aerosols (Handler, 1984). Although this cannot account for the rapid reduction in ozone between 10 and 20 km observed in ozone profiles in September 1986 at McMurdo (Hofmann et al., 1987b), volcanic eruptions may have been involved in the unusual severity of the ozone deficit in recent years.

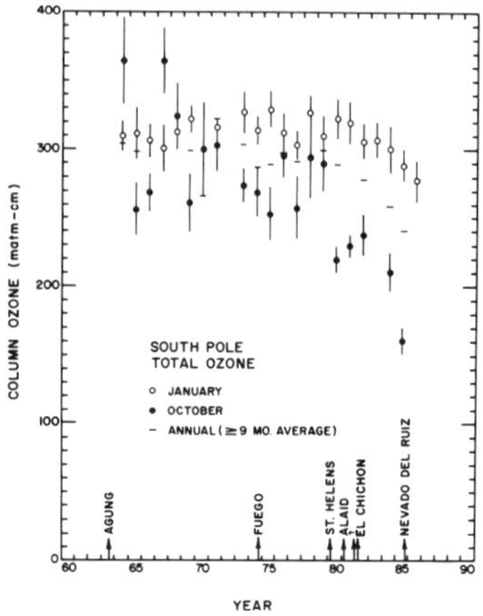


FIGURE 17. Total ozone monthly averages for January and October and the annual average at South Pole Station since 1964. The times of major volcanic eruptions are indicated by arrows.

## 2.4 RECENT STRATOSPHERIC AEROSOL PERTURBATIONS

In November 1985, Nevado del Ruiz (4.9°N) in Colombia erupted and ended the 3-year, 8-month period of stratospheric aerosol decay after the eruption of El Chichón. Figures 5, 6, and 7 show that the effect of this eruption on the stratosphere was actually rather minor. Figure 18 shows details of the eruptive effect from balloon-borne particle counter measurements at Laramie. Although a minor eruption, it appears that it was sufficient to terminate the study of the El Chichón decay and prevent the stratosphere from returning to a background condition such as in 1978-1979.

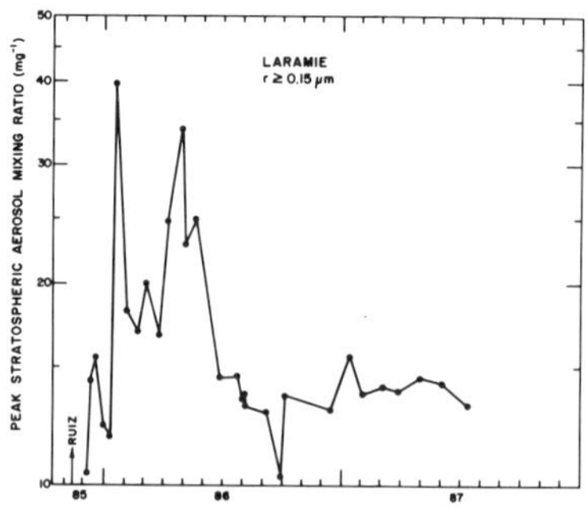


FIGURE 18. Stratospheric aerosol mixing ratio at the maximum (≈20 km) from balloonborne particle counter measurements at Laramie versus time following the eruption of Nevado del Ruiz in November 1985.

Figure 19 shows two r ≥ 0.15 μm aerosol profiles measured recently at Laramie. The sounding on 29 May 1987 indicated an unusual aerosol enhancement from the surface, through the tropopause at about 10 km and into the lower stratosphere where a layer with a concentration of about 10 $cm^{-3}$ was observed. The smaller, broad layer at about 20 km is the normal stratospheric aerosol layer with a concentration of only about 1 $cm^{-3}$ at this time. The next sounding, on 8 July, indicated an apparent stratospheric remnant layer between about 10 and 14 km with a concentration of about 3 $cm^{-3}$. The layer was not present on a sounding conducted in late July and thus does not appear to be of global extent, although lidars in Virginia, Hawaii, Germany, and Japan saw intermittent signs of a particle layer in the lower stratosphere (SEAN Bulletin, 1987). As reported in this bulletin, although there were no reported explosive volcanic eruptions or signs of such in satellite imagery, large forest fires, which burned from 6-22 May in northern China, produced extensive clouds of smoke that could be tracked for thousands of kilometers by satellites. Conceivably some of this smoke could have gotten into the stratosphere and resulted in the observations in Fig. 19. If indeed true, then this may represent an important observation for climatic studies such as those involved in the nuclear winter theory in which large quantities of smoke from nuclear attacks on major cities are envisioned to enter the stratosphere and possibly cause severe climatic change. While the nature of the smoke particles from biomass burning, presumably quite small, may be different than those from a burning city, the dynamics, driven by the intense heat, may be quite similar.

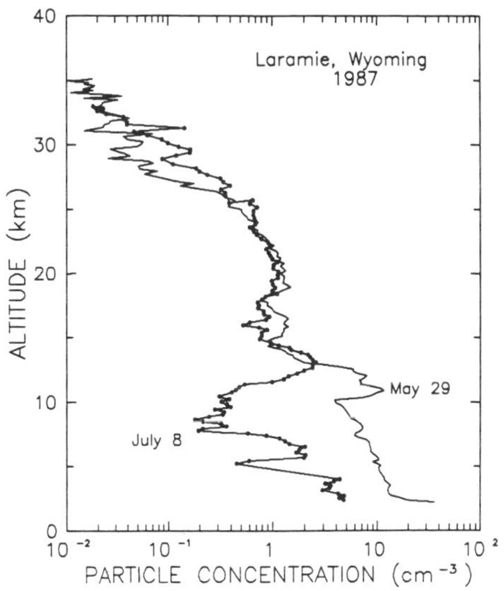


FIGURE 19. Aerosol (r ≥ 0.15 μm) concentration profiles measured at Laramie with balloonborne particle counters in 1987.

## 3. SUMMARY AND CONCLUSIONS

This brief review attempted to give the reader the flavor of the advances made recently in the study of stratospheric aerosols from volcanic eruptions. For those of us in the field it has been an incredibly exciting and satisfying time, which has not yet seen its end. We now feel that we know what happens after a volcanic eruption has injected large quantities of gas into the stratosphere. From the height and season of the eruption and some initial measurements of the amount of gas injected, as may be obtained by satellites, the severity of the aftermath may be predicted with reasonable reliability. The gas-to-particle-conversion process and subsequent particle decay is reasonably well understood, perhaps better than the dynamics involved. Extrapolation to climatic effects is wrought with uncertainty, however, relying on relatively primitive radiative, dynamical and chemical models.

Perhaps the most important aspect, and one not dwelt on at any length in this review, is the question of the nature and sources of the nonvolcanic stratospheric aerosol, i.e., the natural background. Is it increasing? If indeed it is formed from sulfurous gases as studies of volcanic eruptions have led us to believe, then it may well be increasing with the inevitable increase in the release of such gases at the surface of the earth, especially relatively inert gases such as carbonyl sulfide. It has been estimated (Hofmann and Rosen, 1981) that the background stratospheric aerosol may be increasing with a doubling time of only 8 years. If true, then this is indeed a serious problem that warrants study as the climatic effect of such an increase is completely unpredictable.

## REFERENCES

Angell, J. K., 1986: Tropospheric Effects of the 1963 Agung and 1982 El Chichon Eruptions, paper presented at the Norman D. Watkins Symposium "Environmental Impact of Volcanism," Grad. Sch. of Ocean., Univ. of Rhode Island, Narragansett, 12-14 March 1986.

Bojkov, R. D., 1987: The 1983 and 1985 Anomalies in the Ozone Distribution in Perspective, Mon. Weather Rev., 115, 2187-2201.

Grams, G., and G. Fiocco, 1967: The Stratospheric Aerosol Layer During 1964 and 1965, J. Geophys. Res., 72, 3523-3542.

Handler, P., 1984: Possible Association of Stratospheric Aerosols and El Nino Type Events, Geophys. Res. Letts., 11, 1121-1124.

Hofmann, D. J., 1987a: Perturbations to the Global Atmosphere Associated With the El Chichon Volcanic Eruption of 1982, Rev. Geophys., 25, 743-759.

Hofmann, D. J., 1987b: Balloonborne Measurements of Middle Atmosphere Aerosols and Trace Gases in Antarctica, preprint, Dept. Phys. & Astro., Univ. Wyoming, Laramie, WY 82071.

Hofmann, D. J., and J. M. Rosen, 1981: On the Background Stratospheric Aerosol, J. Atmos. Sci., 38, 168-181.

Hofmann, D. J., and J. M. Rosen, 1982: Balloonborne Observations of Stratospheric Aerosol and Condensation Nuclei During the Year Following the Mt. St. Helens Eruption, J. Geophys. Res., 87, 11039-11061.

Hofmann, D. J., and J. M. Rosen, 1983: Stratospheric Sulfuric Acid Fraction and Mass Estimate for the 1982 Volcanic Eruption of El Chichon, Geophys. Res. Letts., 10, 313-316.

Hofmann, D. J., and J. M. Rosen, 1984: On the Temporal Variation of Stratospheric Aerosol Size and Mass During the First 18 Months Following the 1982 Eruptions of El Chichon, J. Geophys. Res., 89, 4883-4890.

Hofmann, D. J., and J. M. Rosen, 1987: On the Prolonged Lifetime of the El Chichon Sulfuric Acid Aerosol Cloud, J. Geophys. Res., 92, 9825-9830.

Hofmann, D. J., J. M. Rosen, J. W. Harder and S. R. Rolf, 1987a: Observations of the Decay of the El Chichon Stratospheric Aerosol Cloud in Antarctica, Geophys. Res. Letts., 14, 614-617.

Hofmann, D. J., J. M. Rosen, J. W. Harder and S. R. Rolf, 1987b: Balloonborne Observations of the Development and Vertical Structure of the Antarctic Ozone Hole in 1986, Nature, 326, 59-62.

Jager, H., and W. Carnuth, 1987: The Decay of the El Chichon Stratospheric Perturbation, Observed by Lidar at Northern Midlatitude, Geophys. Res. Letts., 14, 696-699.

Kent, G. S., and M. P. McCormick, 1984: SAGE and SAM II Measurements of Global Stratospheric Aerosol Optical Depth and Mass Loading, J. Geophys. Res., 89, 5303-5314.

Komhyr, W. D., R. D. Grass and R. K. Leonard, 1986: Total Ozone Decrease at South Pole, Antarctica, 1964-1985, Geophys. Res. Letts., 13, 1248-1251.

Krueger, A. J., 1983: Sighting of El Chichon Sulfur Dioxide Clouds with Nimbus 7 Total Ozone Mapping Spectrometer, Science, 220, 1377-1379.

Labitzke, K., B. Naujokat and M. P. McCormick, 1983: Temperature Effects on the Stratosphere of the April 4, 1982 Eruption of El Chichon, Mexico, Geophys. Res. Letts., 10, 24-26.

Lamb, H. H., 1970: Volcanic Dust in the Atmosphere; With a Chronology and Assessment of its Meteorological Significance, Phil. Trans. R. Soc. London, 266, 425-533.

Lamb, H. H., 1977: Supplementary Volcanic Dust Veil Assessments, Clim. Monit., 6, 57-67.

Lenoble, J., P. Provost and C. Brogniez, 1984: SAGE Satellite Observations of Stratospheric Aerosols From Mt. St. Helens Eruption: A Two-Wavelength Analysis, J. Geophys. Res., 89, 11666-11676.

Mantis, H. T., C. S. Zerefos, A. Bais, I. Ziomas and A. Kelessis, 1986: The Northern Hemisphere Ozone Minimum in 1982-1983, Arch. Met. Geoph. Biocl., Ser. B 36, 135-145.

McCormick, M. P., P. Hamill, T. J. Pepin, W. P. Chu, T. J. Swissler, and L. R. McMaster, 1979: Satellite Studies of the Stratospheric Aerosol, Bull. Am. Meteorol. Soc., 60, 1038-1046.

McCormick, M. P., T. J. Swissler, W. H. Fuller, W. H. Hunt and M. T. Osborn, 1984: Airborne and Ground-Based Lidar Measurements of the El Chichon Stratospheric Aerosol From 90°N to 56°S, Geofis. Int., 23-2, 187-221.

McCormick, M. P., and C. R. Trepte, 1987: Polar Stratospheric Optical Depth Observed Between 1978 and 1985, J. Geophys. Res., 92, 4297-4306.

McElroy, M. B., R. J. Salawitch, S. C. Wofsy and J. A. Logan, 1986: Reductions of Antarctic Ozone Due to Synergistic Interactions of Chlorine and Bromine, Nature, 321, 759-762.

McKeen, S. A., S. C. Liu and C. S. Kiang, 1984: On the Chemistry of Stratospheric $SO_2$ From Volcanic Eruptions, J. Geophys. Res., 89, 4873-4881.

Newhall, C. G., and S. Self, 1982: The Volcanic Explosivity Index (VEI): An Estimate of Explosive Magnitude for Historical Volcanism, J. Geophys. Res., 87, 1231-1238.

Pollack et al. (37 authors), 1981: Measurements of the Volcanic Plumes of Mount St. Helens in the Stratosphere and Troposphere, Science, 211, 815-838.

Rosen, J. M., 1964: The Vertical Distribution of Dust to 30 Kilometers, J. Geophys. Res., 69, 4673-4676.

Rosen, J. M., 1971: The Boiling Point of Stratospheric Aerosols, J. Appl. Meteorol., 10, 1044-1045.

SEAN Bulletin, 1987: Sci. Event Alert Net. Bull., (Smith. Inst., Wash. D.C.) 12, No. 6, 15-18.

Simkin, T., L. Siebert, L. McClelland, D. Bridge, C. Newhall and J. H Latter, 1981: Volcanoes of the World, Hutchison Ross, Stroudsburg, Pennsylvania, 232 pp.

Symons, G. J. (Ed.), 1888: The Eruption of Krakatoa and Subsequent Phenomena, Report of the Krakatoa Committee of the Royal Society, Harrison and Trubner, London.

Tung, K. K., M. K. W. Ko, J. M. Rodriguez and N. D. Sze, 1986: Are Antarctic Ozone Variations a Manifestation of Dynamics or Chemistry?, Nature, 321, 811-814.

# STRATOSPHERIC EXTINCTION AND MASS VARIATIONS AFTER A MAJOR VOLCANIC ERUPTION, DERIVED FROM LIDAR MEASUREMENTS AT NORTHERN MIDLATITUDES

*H. Jäger and M. Littfass*
Fraunhofer Institute for Atmospheric Environmental Research
D-8100 Garmisch-Partenkirchen, Federal Republic of Germany

and

*D.J. Hofmann and J.M. Rosen*
University of Wyoming
Laramie, Wyoming 82071, USA

ABSTRACT

Lidar backscatter profiles of the stratospheric aerosol layer, observed at Garmisch-Partenkirchen, have been converted to extinction and mass values by utilizing in situ particle measurements of the University of Wyoming. The cumulative concentrations recorded in six channels of the balloon-borne counter system have been differentiated with a height resolution of 1 km to fit a bimodal lognormal size distribution. These size distributions, combined with Mie backscatter intensity and extinction cross-section functions, yield height- and time-resolved conversion factors, extinction/backscatter, and mass/backscatter, which have been applied to Garmisch-Partenkirchen lidar backscatter measurements of the period of dispersion and decay of the El Chichon eruption cloud, 1982 through 1985.

## 1. INTRODUCTION

Ground- or aircraft-based lidar (laser radar) systems are well established tools to observe the stratospheric aerosol layer, to detect volcanic injections, and to obtain data on the global distribution of such disturbances. These lidar systems measure the intensity of light scattered back from both air molecules and air-borne particles when a laser pulse propagates through the atmosphere. Particle backscattering itself depends on a number of parameters like particle concentration, particle size distribution, shape, and optical properties. Interpretation of lidar backscattering therefore is only possible with additional information. Molecular backscattering is calculated from radiosonde data and can be subtracted from the received lidar signal. The inversion of particle backscattering to quantities of interest, such as extinction or particle mass (e.g., McCormick and Swissler, 1983; Jäger et al., 1984), can be done via models that offer the respective conversion factors for certain stratospheric situations (e.g., Russell et al., 1981; Russell and Hamill, 1984). Another possibility is to engage simultaneous in situ measurements of particle parameters to derive the necessary conversion factors. This has been applied, e.g., to verify the quality of lidar data (Swissler et al., 1982; Hofmann et al., 1983).



ISBN 0-937194-11-5

In this work balloon-borne particle counter measurements of the University of Wyoming (UW) (41°N) are used to derive height- and time-resolved factors to convert lidar backscatter data obtained at Garmisch-Partenkirchen (GP) (47.5°N) to extinction and mass concentration values. This is done by combining particle size distributions derived from UW data with Mie scattering and extinction functions. Mie theory has been applied to stratospheric aerosol particles, since these particles remain spheres of sulfuric acid-water droplets at prevailing stratospheric temperatures (Steele and Hamill, 1981).

Latitude and longitude differences do not allow the observation of same air masses. Thus during the early phase of spreading of a fresh volcanic injection the results should be examined with some reservation. However, it has been shown that some months after an eruption homogeneity has progressed to a degree where the differences in location do not play a significant role (Hofmann et al., 1983). To reduce the influence of different air masses, average conversion factors will be applied to corresponding periods of lidar observation.

## 2. OBSERVATIONS

In this work the period 1/82 to 12/85 is under consideration. At the beginning of 1982 the stratosphere had almost recovered from the April 1981 Alaid (Kurile Islands) and May 1981 Pagan (Marianas Islands) eruptions and lidar backscattering at Garmisch-Partenkirchen was close to pre-St. Helens (Washington, May 1980) background values, when a yet unknown tropical eruption caused a disturbance in the lower stratosphere up to about 20 km. The April 1982 eruption of the Mexican volcano El Chichón then led to a global stratospheric perturbation, which became one of the largest in this century. Figure 1 shows the dramatic increase of integral particle backscattering (integrated from 1 km above the local tropopause to the top of the aerosol layer at about 30 km) in 1982 as observed by lidar at Garmisch-Partenkirchen.

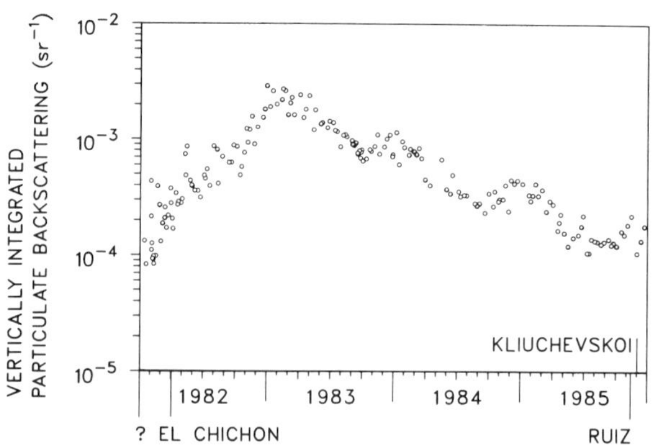


FIGURE 1. Time variation of vertically integrated particulate backscattering (tropopause + 1 km to layer top) at 694.3 nm.

The lidar system, which works at 694.3 nm, has been described elsewhere (Reiter et al., 1979). Structures of the decay beginning in early 1983 have been discussed by Jäger and Carnuth (1987). The decay was terminated by the eruptions of Nevado del Ruiz (Colombia, November 1985) and Kliuchevskoi (Kamchatka, December 1985).

The optical particle counter system employed by the University of Wyoming was recently described by Rosen and Hofmann (1986). It detects particles with radii greater than 0.01, 0.15, 0.25, 0.95, 1.2, and 1.8 μm up to more than 30 km altitude. Twenty-nine particle profiles (4/82-3/85), plus a February 1981 profile for background comparison, have been used in this study. Data were smoothed with a 1.5 km averaging window and data at each km level have been used for calculations.

## 3. CALCULATIONS

A bimodal lognormal distribution has been chosen as differential size distribution to fit the integral particle concentrations:

$$n(r)dr = (2\pi)^{-1/2}\ [N_1\ \exp(-\alpha_1^2/2)d\alpha_1 + N_2\ \exp(-\alpha_2^2/2)d\alpha_2] \qquad (1)$$

where $N_i$ are the number concentrations in the single modes, and $\alpha_i = \ln(r/r_i)/\ln\sigma_i$ with $r_i$ and $\sigma_i$ being the modal radii and modal widths, respectively. The calculation procedure leading to six distribution parameters will be described elsewhere. The 0.95 μm values have been regarded an upper limit in this channel, but have not been used in the calculations because of uncertainties discussed by Rosen and Hofmann (1986). Consequently, six parameters have been determined with the aid of five pieces of information, resulting in some interdependences, particularly between $N_2$ and $\sigma_1$. Since the same calculation procedure including certain assumptions to substitute channel 0.95 μm has been used for all data sets, the results allow a description of the stratospheric aerosol layer by these six resulting distribution parameters.

The variations with time of these height-resolved distribution parameters are demonstrated at some characteristic altitude levels in Figs. 2 and 3. In Fig. 2 radii at 18 and 21 km of both modes are presented. All radii show a maximum at the time of maximum perturbation. Whereas mode 1 radii have returned to pre-El Chichón values in early 1985, mode 2 radii remain at an enhanced level. This has already been reported and interpreted by Hofmann and Rosen (1987). In Fig. 3 the substantial contribution of the relatively large mode 2 particles to the total number concentration is apparent.

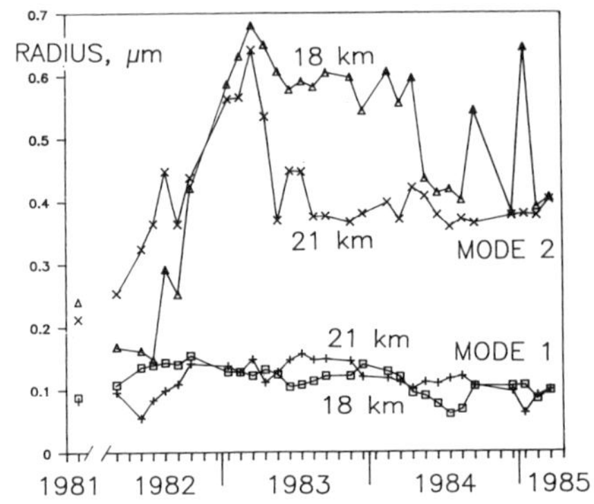


FIGURE 2. Time variations of modal radii.

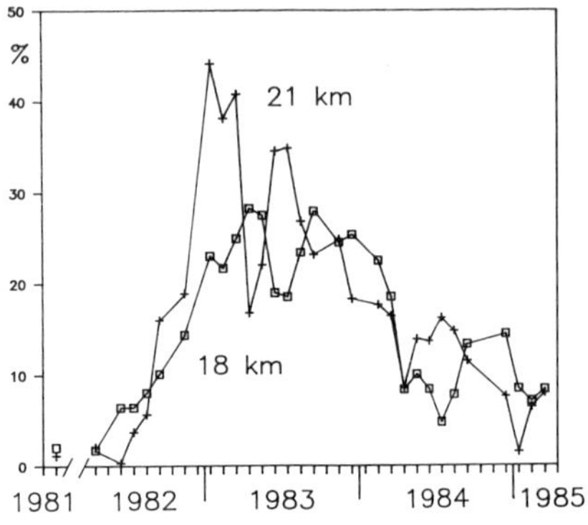


FIGURE 3. Percentage of large particles $N_2$ in $N_1+N_2$.

Figure 4 shows characteristic particle size distributions, one at the maximum of the El Chichón perturbation and another one from the decay period, both exhibiting the large particle mode.

To convert GP lidar measurements to extinction and mass concentration, conversion factors have been calculated by combining Mie calculations with the above described particle size distributions. The conversion of particle backscattering B to extinction E requires a factor E/B, which is given by

$$E/B = \pi k^2 \int_0^\infty r^2 Q_e n(r) dr \Big/ \int_0^\infty I_b n(r) dr \quad [sr] \tag{2}$$

where $Q_e$ is the Mie extinction efficiency function, $I_b$ the Mie backscatter intensity function and $k=4\pi/\lambda$. To convert particle backscattering to mass concentration M, the factor M/B has to be calculated as

$$M/B = (4\pi\rho k^2/3) \int_0^\infty r^3 n(r) dr \Big/ \int_0^\infty I_b n(r) dr \quad [g\ sr\ m^{-2}] \tag{3}$$

These calculations have been carried out at a wavelength $\lambda$ = 694.3 nm, applying a refractive index of 1.44-0i (Palmer and Williams, 1975; Steele and Hamill, 1981), and a specific aerosol gravity $\rho$ = 1.65 g $cm^{-3}$, which corresponds to a 75%/25% sulfuric acid/water solution. Variations of these properties have not yet been considered.

Figure 5 shows the product of size distribution and MIE backscatter intensity function. The effect of large particles contributing to backscattering is evident when comparing Figs. 4 and 5 curves. At 21 km in January 1983, 44% large (mode 2) particles were responsible for 98% backscattering and in July 1984, 16% large particles still produced 86% backscattering.

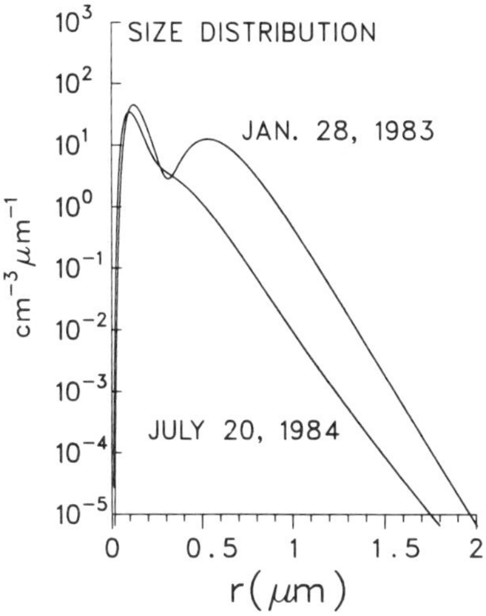


FIGURE 4. Particle size distributions at 21 km.

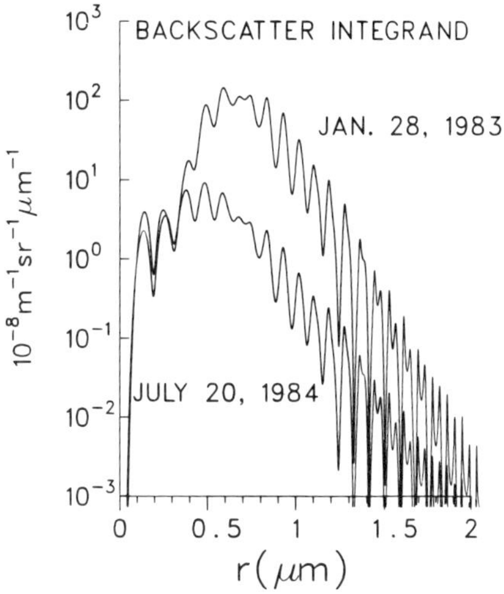


FIGURE 5. Backscattering integrands at 21 km.

For each Wyoming balloon experiment ratios E/B and M/B have been determined at 1 km height intervals. Since GP lidar measurements most probably did not probe same air masses, the conversion ratios were averaged at each height level over certain periods to eliminate local air mass variations. Figures 6 and 7 show the height and time variations of smoothed conversion ratios for six such periods.

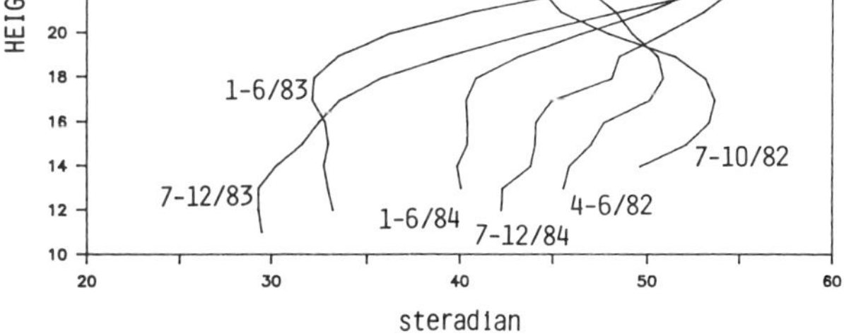


FIGURE 6. Extinction to backscatter ratios.

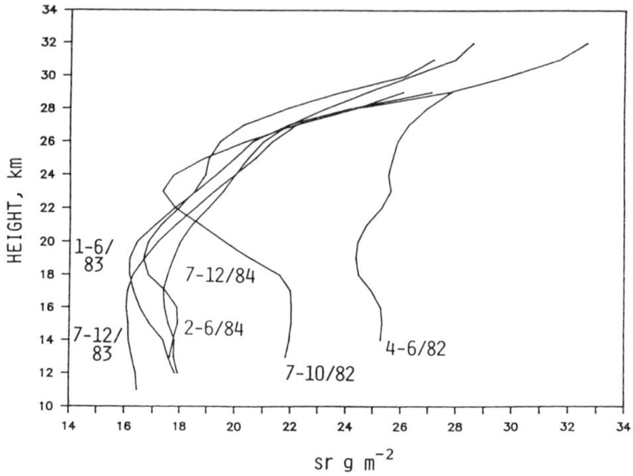


FIGURE 7. Mass concentration to backscatter ratios.

## 4. CONVERSIONS

These factors have been applied to GP lidar values. The results have been integrated from 1 km above the tropopause to the top of the stratospheric aerosol layer, resulting in optical depth (integrated extinction) and mass density (integrated mass concentration) variations, which are shown in Fig. 8 (smoothed with a 90-day running window). In Fig. 9 all three quantities have been normalized with the average January 1982 value. This figure shows that an increase in integral backscattering from January 1982 to January 1983 by a factor of 20 caused a 14-fold increase of optical depth and mass density. During the decay period the ratio of integral backscattering to optical depth did not return to the pre-El Chichón value, neither did the ratio of mass density to integral backscattering.

## 5. CONCLUSIONS

The conversion factors contain five quantities describing the size distribution: both modal radii and widths and the ratio of particle concentrations of both modes. The total particle number concentration has been eliminated by the calculation procedure. This independence of conversion factors from the absolute particle concentration complies with the necessity to use those factors independent of their place of origin.

Figures 8 and 9 show that integral particle backscattering was close to pre-El Chichón values at the end of the El Chichón decay period in 1985, whereas conversion to optical depth and mass density produced lower than the early 1982 values. This discrepancy is

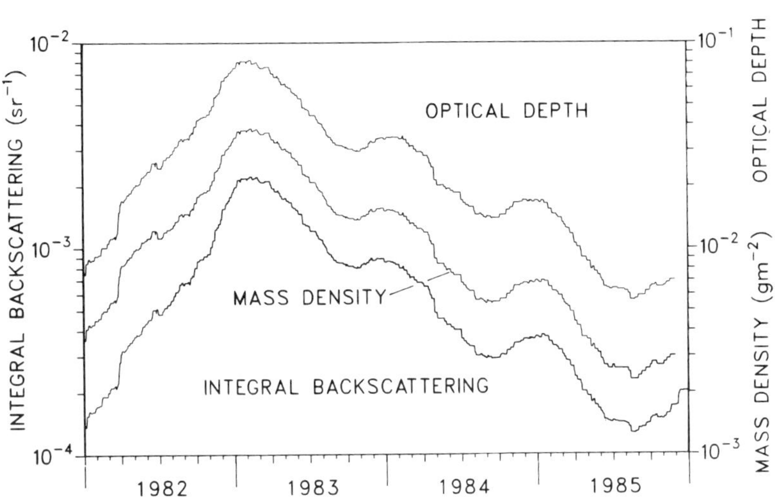


FIGURE 8. Ninety-day running mean of integral backscattering and optical depth (both at 694.3 nm) and mass density.

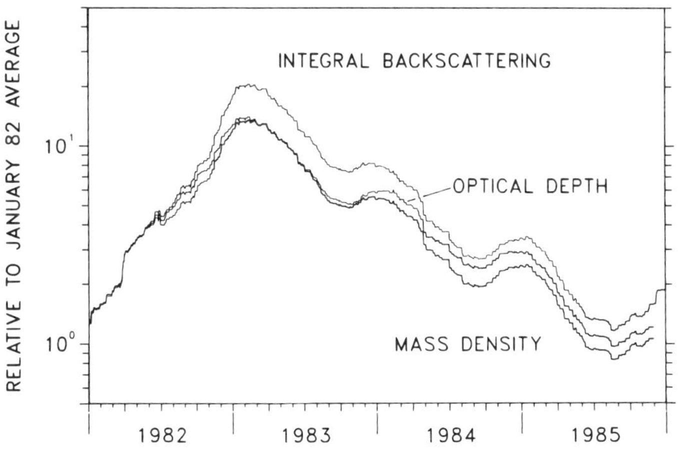


FIGURE 9. Ninety-day running mean, normalized with January 1982 average value.

explained by the continued presence of a relatively great amount of large particles. It was proposed by Hofmann and Rosen (1987) that a source for condensation nuclei and sulfuric acid vapor exists in high polar regions that supports continued growth of sulfuric acid droplets.

## REFERENCES

Hofmann, D.J., J.M. Rosen, R. Reiter, H. Jager, 1983: Lidar and Balloon-Borne Particle Counter Comparisons Following Recent Volcanic Eruptions, J. Geophys. Res., 88, 3777-3782.

Hofmann, D.J., and J.M. Rosen, 1987: On the Prolonged Lifetime of the El Chichon Sulfuric Acid Aerosol Cloud, J. Geophys. Res., 92, 9825-9830.

Jager, H., R. Reiter, W. Carnuth and W. Funk, 1984: El Chichon Cloud Over Central Europe, Geof.Int., 23-2, 243-257.

Jager, H., and W. Carnuth, 1987: The Decay of the El Chichon Stratospheric Perturbation, Observed by Lidar at Northern Midlatitudes, Geophys. Res. Letts., 14, 696-699.

McCormick, M.P., and T.J. Swissler, 1983: Stratospheric Aerosol Mass and Latitudinal Distribution of the El Chichon Eruption Cloud for October 1982, Geophys. Res. Letts., 10, 877-880.

Palmer, K.F., and D. Williams, 1975: Optical Constants of Sulfuric Acid: Application to the Clouds of Venus?, Appl. Opt., 14, 208-219.

Reiter, R., H. Jager, W. Carnuth and W. Funk, 1979: The Stratospheric Aerosol Layer Observed by Lidar Since October 1976, A Contribution to the Problem of Hemispheric Climate, Arch. Met. Geophys. Biokl., Ser. B, 27, 121-149.

Rosen, J.M., and D.J. Hofmann, 1986: Optical Modeling of Stratospheric Aerosols: Present Status, Appl. Opt., 25, 410-419.

Russell, P.B., T.J. Swissler, M.P. McCormick, W.P. Chu, J.M. Livingston and T.J. Pepin, 1981: Satellite and Correlative Measurements of the Stratospheric Aerosol, I: An Optical Model for Data Conversions, J. Atmos. Sci., 38, 1279-1294.

Russell, P.B., and P. Hamill, 1984: Spatial Variation of Stratospheric Aerosol Acidity and Model Refractive Index: Implications of Recent Results, J. Atmos. Sci., 41, 1781-1790.

Steele, H.M., and P. Hamill, 1981: Effects of Temperature and Humidity on the Growth and Optical Properties of Sulfuric Acid-Water Droplets in the Stratosphere, J. Atmos. Sci., 12, 517-528.

Swissler, T.J., P. Hamill, M. Osborn, P.B. Russell and M.P. McCormick, 1982: A Comparison of Lidar and Balloon-Borne Particle Counter Measurements of the Stratospheric Aerosol, 1974-1980, J. Atmos. Sci., 39, 909-916.

# THE ABNORMALLY COLD SUMMERS OF CENTRAL CHINA AND THEIR RELATION TO VOLCANIC ERUPTIONS

*Xu Qun*
Jiangsu Meteorological Institute
Nanjing, China

ABSTRACT

Through the analysis of comprehensive Chinese historical writings the author has found 98 abnormally cold summers since A.D. 1500, 35 of them extremely cold. Using the acidity record from the ice core of Hammer et al. (1980), in combination with the volcanic eruptions data of Simkin et al. (1981), it appears that most abnormally cold summers and extremely cold summers occurred 0-2 years after great volcanic eruptions. Such historical extremely cold summers of central China, reflecting the lowering of summer temperature by dense volcanic clouds, could exceed 20°C. Therefore, the present work strongly supports the "nuclear winter" hypothesis.

A concentration of abnormally cold summers occurred between the 17th and mid-19th centuries, then disappeared between 1913 and 1964. Recently renewed, frequent occurrences of these abnormally cold summers give warning of future possible climatic change.

## 1. INTRODUCTION

The events of "summer snow" not only appear in Chinese myths and popular verses, but occur as historical facts in Chinese ancient writings. Moreover, according to the electrical measurements for the acidity record of the Greenland ice core from Hammer et al. (1980), considered in combination with the volcanic eruption data of Simkin et al. (1981), the author has calculated each volcanic dust veil index of the northern temperate zone since A.D. 1500. Most abnormally cold summers (ACSs) in central China were found to be highly correlated with great volcanic eruptions. This relationship suggests that dense volcanic dusts in the stratosphere can cause significant lowering in air temperatures in central China, even in summer.

## 2. ABNORMALLY COLD SUMMERS

The following description in Chinese historical writings appears in four local annals of the lower Yangtze for 1601: "Snow flying in heaps and men were clothed with cotton" during July. Since, of course, the occurrence of such an extremely cold summer is a disaster, the author decided to systematize those abnormally cold summers for the past 500 years, as recorded by the Chinese provincial meteorological bureaus from more than 2200 local annals and other



ISBN 0-937194-11-5

numerous historical writings.[1] Based on compiled materials from Shanghai and 11 provinces of mideast China (east of $105^{\circ}$E, 25-$35^{\circ}$N),[2] the author sorted out climatic descriptions of abnormal coldness during each half-year of summer (May-October), combined with meteorological data since 1911.

An abnormally cold summer in central China was considered to have taken place if any one of the following occurred:

1) Two or more sources recorded "cold, chill, frost or snow" for the same summer and showed the temperature lower than normal, by more than $10^{\circ}$C.

2) A single source recorded summer coldness when the temperature was very severe, such as a hard frost or huge snow appearing between May and October.

3) The large-scale air temperature of central China for two successive summer months since 1911 was abnormally low, compared to the average temperature over a 50-year period.

According to the above criteria, 98 years of ACSs were selected from 1500 to 1984 as follows:

1509, 1524, 1530, 1546, 1554, 1566, 1576, 1577,
1586, 1587, 1588, 1589, 1598, 1601, 1602, 1603,
1612, 1620, 1624, 1625, 1626, 1628, 1631, 1632,
1640, 1642, 1647, 1650, 1651, 1652, 1653, 1655,
1660, 1661, 1663, 1664, 1667, 1668, 1669, 1670,
1678, 1679, 1683, 1693, 1695, 1697, 1698, 1699,
1702, 1705, 1724, 1727, 1729, 1730, 1743, 1746,
1755, 1756, 1764, 1769, 1783, 1792, 1794, 1808,
1809, 1812, 1813, 1817, 1831, 1832, 1833, 1834,
1837, 1839, 1843, 1848, 1851, 1856, 1860, 1862,
1863, 1865, 1867, 1870, 1876, 1877, 1880, 1882,
1892, 1896, 1901, 1907, 1912, 1965, 1974, 1976,
1980, 1982.

The 35 years that are underlined are the extremely cold summers (ECSs). Through analysis, it was found that the occurrences of most of the above ACSs were related to volcanic eruptions.

---

[1]The scope of such systematical compilation of historical climatic materials from local annals was expanded during the 1970s.

[2]The 11 provinces are: Jiangsu, Zhejiang, Anhui, Jiangxi, Hunan, Fujian, Sichuan, Quizhou, the southern part ($\leq 35^{\circ}$N) of Henan and Shanxi.

## 3. ASSESSMENT OF THE VOLCANIC DUST VEIL INDEX OF THE NORTHERN TEMPERATE ZONE ($V_{N.M}$)

Lamb (1970) first sorted out the volcanic dust veil index (DVI) of the Northern Hemisphere and the Globe since 1500, but the basis for the assessment of DVI was poor. It only included some air temperatures from a single station and the description of optical phenomena from limited writings. The author (1985) has reestablished $V_{N.M}$ for a 100-year period (1883-1982) on the basis of the yearly variation of direct radiation of clear skies and the atmospheric optical depths for the northern temperate zone. It was found that a large difference existed between the yearly DVI and $V_{N.M}$ since 1883. Simkin et al. (1981) has published a volcanic explosivity index (VEI) for historical years, but the explosive lava volume is not always directly proportional to the stratospheric volcanic dust (gases) effects. As a remedy, the acidity record of the Greenland ice core from Hammer (1980) provides us with the basic data for objectively defining the yearly $V_{N.M}$.

The Greenland icesheet is a rich source of information on past volcanism. This is not due to deposition of volcanic microparticles, for they are apparently too large to float over Greenland. However, volcanic acids in snow layers deposited shortly after a large volcanic eruption can be detected--as elevated acidities reveal by an electric current through an ice core. In this way, Hammer (1980) has measured the acidity index (AI) for the past 900 years to within an accuracy of one year. A comparison between $\Delta$AI peak and the corresponding $V_{N.M}$ during 1883-1972 is shown in Table 1.

TABLE 1. THE COMPARISON BETWEEN AI AND $V_{N.M}$ FOR DIFFERENT LATITUDE ZONES

| Latitude Zone | Volcano and Year of Eruption | $V_{N.M}$ | Value of $\Delta$AI Peak and the Peak Year | Average Lag (Yrs) of $\Delta$AI Peak* and Averaged ($V_{N.M}/\Delta$AI) in Parentheses |
|---|---|---|---|---|
| 15°N-10°S | Krakatao (1883) | 1000 | 1.4 (1885) | +1 - +2 |
| | Pelee et al. (1902) | 1200 | 0.54(1904) | |
| | Agung (1963) | 600 | 0.60(1964) | (1119) |
| | Awu (1966) | 350 | 0.65(1967) | |
| 16°N-55°N | Bandai San (1888) | 750/2 | 0.60(1888) | 0 |
| | Suwanose Zima (1889) | 750/2 | 0.50(1889) | (727) |
| | Bogoslof (1890) | 250 | 0.60(1890) | |
| | Ksudach (1907) | 480 | 0.43(1907) | |
| North of 55°N | Katmai (1912) | 750 | 2.05(1912) | 0 |
| | Katla (1918) | 170 | 0.50(1918) | |
| | Hekla (1947) | 260 | 0.52(1947) | (404) |

*Relative to corresponding volcanic eruption.

From Table 1, it can be concluded that:

1) Most peak years of ΔAI coincide well with the same years of extratropical volcanic eruption. For tropical eruptions, the lag of ΔAI peak may extend to 1-2 years.

2) ΔAI peaks can sensitively reflect volcanic eruptions in the extratropical zone. As for tropical eruptions, they are only in the larger class--at least at levels of Agung (1963) and Awu (1966).

In summary, the $\Delta AI \geq 0.4$ unit (above the background of 1.2 μm equiv. H/kg ice) can generally reflect the great volcanic activities at all latitudes north of $10^{o}$S, thus it is a critical ΔAI value of acid fallout of the large volcanic eruptions depositing on the Greenland icesheet.

The large volcanic eruption year and its $V_{N.M}$ may be defined as follows:

1) All peak years of AI($AI \geq 1.6$ unit) were initially defined with an uncertainty $\pm 1$ year since 1500. These AI data were supplied by Dr. H. B. Clausen (personal communication).

2) The actual volcanic eruptive years occurred only at 0-2 years before AI peaks were selected and defined by Simkin et al. (1981), with larger volcanic eruptive cases ($VEI \geq 3$).

3) Each yearly $V_{N.M}$ was calculated through the average ratio of ($V_{N.M}/\Delta AI$) from Table 1, according to the latitude of selected large volcanic eruptive cases and their corresponding AI peak.

In combination with the assessed $V_{N.M}$ for the recent 100-year period, we can thus fix each large volcanic eruption year and its $V_{N.M}$ since 1500 (see Fig. 1).

## 4. THE RELATIONSHIP BETWEEN ABNORMALLY COLD SUMMERS AND VOLCANIC ERUPTIONS

By reason of the significantly decreasing influence of volcanic dusts over the direct radiation of the northern temperate zone (Xu, 1986), we define the period during and after the large volcanic eruption (no more than 2 years) as the year(s) influenced by volcanic dusts (Yd). There are 225 Yd among 485 years (1500-1984), but Yd reaches 82 for the 98 abnormally cold summers as a whole. Through the Chi-square test we obtain: $\chi^2=53.28>>\chi^2.001$. This indicates the close relationship existing between the large volcanic eruption and the occurrence of the ACS, which lags 0-2 years behind a tropical eruption and 0-1 year behind an extratropical eruption.

In central China 35 extremely cold summers appeared before 1865; 33 of them occurred at the interval of 0-2 years after large volcanic eruptions. This relationship can be summarized as follows:

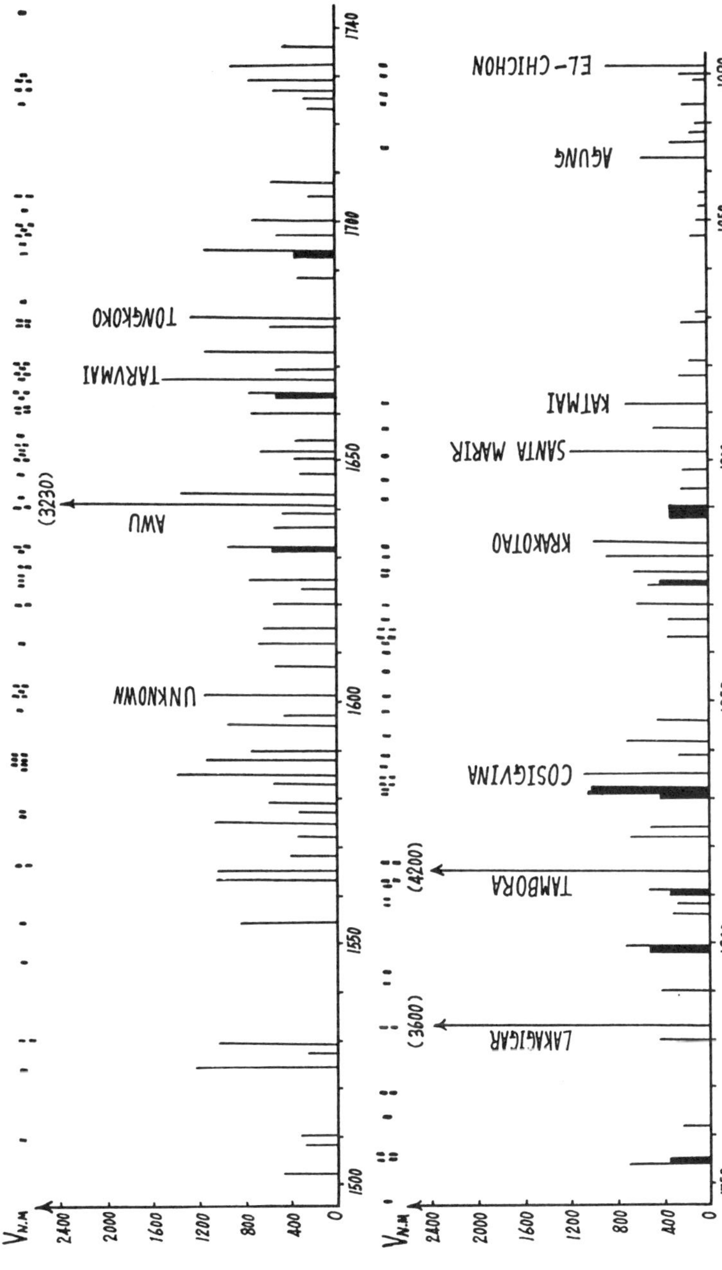


FIGURE 1. Temporal distribution of yearly $V_{N.M}$, ACSs, and ECSs (ACSs indicated by symbol ' and ECSs by ' ). (Reprinted by permission from Nature, 228, 230-235; copyright (c) 1980, Macmillan Journals Limited.)

1) The strongest AI peak of the 16th century occurred around 1586 (Fig. 1); it reflected the influence of great volcanic eruptions at temperate low latitudes (such as Kirishima 31.9°N, 130.9°E, Kelut 7.9°S, 112.3°E, etc.). In the successive 3 years (1587-1589), extremely cold summers prevailed in central China, snow and frost frequently occurred at high summer especially in 1588-1589. The large-scale summer drought of central China reflected the abnormal weakness of the summer monsoon in Asia. According to Bryson the abandonment of the great city of Fatepur Sikri (27.1°N, 77.4°E) in 1588, only 16 years after its construction, can be attributed to the failure of the water supply (Lamb, 1982).

2) The strong AI peak in 1601-1602 coincided well with Lamb's writing (1970): The Sun and Moon appeared "reddish, faint and lacked brilliance in central Europe all through the year 1601 up to the end of July 1602." Responding sensitively to this large volcanic eruption in the mid or high latitudes around 1601, abnormally cold summers prevailed in central China for 3 successive years (1601-1603); in 1601 the annals of four counties (30-31°N, 117-121°E) in lower Yangtze recorded phenomena (as described in Section 2) that suggested the air temperature in July in lower Yangtze was 25°C lower than normal. During this summer a notable frost ring event also occurred in the western U.S.A. (LaMarche et al., 1984).

3) In 1660-1661 the scope of the cold summer of central China extended far and wide, especially during 1661 when the abnormal cold extended from lower Yangtze to Hunan and Fujian provinces. Heavy frost fell like snow over three counties (located at 27°N) of Fujian province during July, revealing the extreme cold of high summer in the coastal area of southeast China.

4) Some severe cold summers of central China occurred in the drought periods of high summer, such as in 1588, 1589, 1632, 1640, and 1679. The abnormally cold and dry air mass controlling central China during high summer showed the extreme condition that the summer monsoon of China had wholly disappeared once more in the high summers of the 16th and 17th centuries.

5) The occurrences of ECS were less severe in the 19th century; the most significant ECS appeared in 1817 (just 2 years after the great eruption of Tambora). The annals of three counties (29.5-30.0°N,116.0-116.6°E) of northern Jiangxi recorded the following phenomena: "So cold was the north wind in the first half of August, especially on 11 August; the next morning snow could be seen on the hills

[3]The dates have been translated from the traditional Chinese calendar.

and most silk cotton trees were frostbitten. During mid-August, the weather was colder with gales and rain, some people warmed themselves with fires and dressed in fur coats."

6) The severe cold autumns were also notable. The river normally flowing through Xishui (30.4°N, 115.2°E) had frozen over during mid-autumn of 1566, suggesting that an abnormally low air temperature of -10°C would continue for some duration. Fish in the lakes of lower Yangtze froze to death in the autumn of 1628. Such severe cold occurring in late October suggested that an abnormal cold wave with a downward amplitude of air temperature of more than 20°C existed.

7) Significant long trends existed separately in the time variation of ACSs and ECSs. The ACSs were concentrated in two stages: from 1586 to 1705 and from 1808 to 1882. From Fig. 2, it is obvious that two main peaks, one in the 17th century and the other in the mid-19th century coincided well with the corresponding peaks of the long-trend variation of $V_{N.M}$. Correlation between these two curves (ACS-$V_{N.M}$) reaches +0.65(N=46), far exceeding the 99.9% level. A 61% concentration of ECSs occurred at the above-mentioned first stage, the central period of the recent little ice age. No ECSs have appeared since 1865, such a trend might be influenced by the recent warming since the late 19th century.

The phases of the three curves ($V_{N.M}$, ACS, and ECS) notably coincide with each other (correlations are significant at the 99.9% level), but just in antiphase with the winter temperature curve. This means that the active volcanic eruption periods not only correspond with the ACS-concentrated periods but with the periods of lower winter temperature in China. Hence, it appears that volcanic eruptions in the 17th and mid-19th centuries not only caused the frequent occurrences of ACS of central China, but also played the dominant role in forming lower winter temperatures in China during these two stages.

## 5. CONCLUSION

Since 1500, 98 abnormally cold summers have appeared in central China, 35 of them extremely cold summers. Most ACSs (or ECSs) occurred 0-2 years after a large volcanic eruption. All ECSs appeared before 1865, most of them during the central period (1587-1705) of the little ice age. During the ECSs, certain very unusual weather phenomena occurred in central China, such as snow falling in high summer (1589, 1601, 1603, 1653, 1660, 1661, 1664, 1668, 1670, 1705, 1727, 1769, and 1817) and heavy frost falling during the dog days (1530, 1587, 1589, 1628, 1640, 1661, 1679, 1693, 1775, and 1783). Hard frost during mid-autumn froze the river in Xishui (1566) and the fish of the lakes in lower Yangtze (1628). These phenomena indicate that air temperatures can be lower by 20-25°C during the summer half

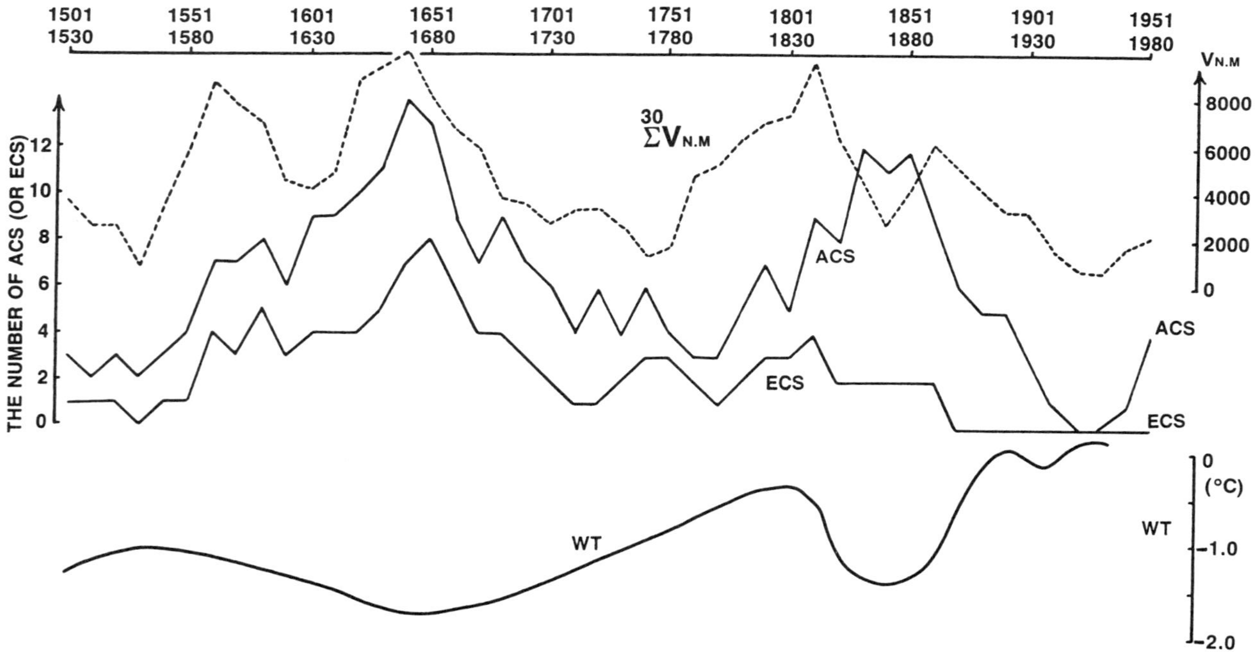


FIGURE 2. Long-trend variations (the running average sum of each 30-year period) of ACS, ECS, and $V_{N.M}$. The long-trend variation of the winter temperature in China is also depicted here. (From Chu Kochen, 1973.)

of the year in colder climatic periods under the influence of stratospheric volcanic aerosols. These findings strongly support the viewpoint of "nuclear winter."

The lower winter temperatures of the 17th and mid-19th centuries in China also are closely related to the frequent occurrences of volcanic eruptions during the same periods.

It is exceptional that since 1500 one 52-year period (1913-1964) occurred without ACSs. Only after the eruptions of Agung (1963), Mount St. Helens (1980), and El Chichon (1982), the direct solar radiation of the Northern Hemisphere decreased significantly. ACSs have occurred frequently since the mid-1960s; the summer monsoon of China has weakened significantly, and the high summer monsoon rain belt of east China has shifted southerly, corresponding well with the southward trend of the intertropical convergence zone (ITCZ) in tropical Africa and the persistent droughts of Sahel since 1968 (Xu, 1986).

We must continue to sound the warning for this notable weakening trend of summer monsoons over the Northern Hemisphere, for it could possibly mean that a new important climatic change has just commenced.

## ACKNOWLEDGMENTS

I am greatly indebted to Dr. H. B. Clausen and Dr. C. U. Hammer for supplying annual acidity records from the Greenland ice core.

## REFERENCES

Chu Kochen, 1973: A Preliminary Study on the Climatic Fluctuation During the Last 5000 Years in China, Science Sinica, 16 (2).

Hammer, C.U., H.B. Clausen and W. Dansgaard, 1980: Greenland Ice Sheet Evidence of Post-Glacial Volcanism and its Climatic Impact, Nature, 288, 230-235.

Lamb, H.H., 1970: Volcanic Dust in the Atmosphere, Phil. Trans. Roy. Soc., A.266, 426-533.

Lamb, H.H., 1982: Climate History and the Modern World, Methuen and Co., Ltd., London.

LaMarche, V.C. Jr., and Katherine K. Hirschboeck., 1984: Frost Rings in Trees as Records of Major Volcanic Eruption, Nature, 307, 121-126.

Simkin, T., L. Siebert, L. McClelland, D. Bridge, C. Newhall and J.H. Latter, 1981: Volcanoes of the World, Smithsonian Institute, Hutchinson Ross Publishing Company.

Xu Qun, 1985: The Assessment of Volcanic Dust Veil Index of Northern Temperate Zone in the Recent 100 Years, Acta Geophysica Sinica, 28, 558-568.

Xu Qun, 1986: The Main Cause of the Persistent Droughts of Sahel and the Increasing Trend of El Nino Events in Recent Years. The Proceedings of Beijing International Radiation Symposium, Beijing, August 1986.

# MODELING OF OPTICAL PROPERTIES OF POSTVOLCANIC STRATOSPHERIC LAYER

*V.E. Zuev, G.M. Krekov and R.F. Rakhimov*
Institute of Atmospheric Optics, USSR Academy of Sciences
Siberian Branch, Tomsk, USSR

## ABSTRACT

The data on the optical characteristics dynamics of the layer at the postvolcanic relaxation stage have been obtained. They are based on empirical information of the spatial-time trend of integral parameters of stratospheric aerosol.

## I. INTRODUCTION

Long-period observations have shown that the stratospheric aerosol layer (SAL) is affected by volcanic eruptions about 50% of the time. The so-called "Junge disturbed layer" varies markedly when tending to equilibrium. To determine the possible climatic consequences of this phenomenon, time series analysis is required. The development of optical models of the stratospheric layer is needed in order to predict the effects on climate of volcanic materials injected into the stratosphere.

Beginning with the application of Junge's theory (Junge et al., 1961), our concepts of the stable formation mechanism of the stratospheric aerosol layer have changed as our knowledge of the nature of atmospheric aerosol has increased. For example, when constructing SAL models either self-preserving aerosol particle size distribution is assumed or statistical mixing of heterogeneous aerosols is used (Burgmeier and Blifford, 1975). Recognition of the significance of secondary processes of aerosol formation has determined subsequent versions of the SAL model (Shibata et al., 1984; Hofmann and Rosen, 1984), where much attention has been paid to analyzing cycles of photochemical synthesis of aerosol-generating compounds ($SO_2$,$NO_2$,NO,$H_2SO_4$,$H_2O$,$HNO_3$), as well as the processes of heteromolecular nucleation and condensation.

However, the necessity of considering not only exchange processes in the dispersed SAL but also mesoscale processes, together with the requirement of reasonably simple models, has necessitated the search for generalized parameters, which can be used as predictors of the optical state of the SAL.

Development of SAL optical models requires the data selection of microphysical parameters, calculations of their optical characteristics over a wide spectral range, and comparison of the results with field observations of the SAL (Krekov and Rakhimov, 1982; Zuev and

AEROSOLS AND CLIMATE
Peter V. Hobbs and M. Patrick McCormick (Eds.)


ISBN 0-937194-11-5

1982; Zuev and Krekov, 1986; Krekov and Rakhimov, to be published). It is evident that the application of this approach to the modeling of postvolcanic variations of the optical states of the SAL is difficult, since information on the dispersed structure as a function of altitude and time is necessary. Such information is not likely in the near future.

## 2. METHOD OF CALCULATION

For the numerical prediction of light scattering by aerosols we have used a polymodal parametric structure defined as

$$f(R) = C^{(\nu)}R^{-\nu} \sum_i M_i \exp\left\{-b_i\left[\ln\left(\frac{R}{R_i^{(\nu)}}\right)\right]^2\right\} \tag{1}$$

where $R$ is the particle radius, $M_i$ is the relative portion of the i-th fraction particles with respect to the background one (usually $M_2 = 1$), $A_i^{(\nu)}$ is the normalization parameter, $\nu$ is the characteristic parameter with the values $\nu = 1,2,3$ for expressions $f(R) = dN/dR$, $dS/dR$ or $dV/dR$, respectively; $R_i^{(\nu)}$ is the geometric mean radius of particles, $b_i$ is inversely proportional to the width of the logarithmically normal distribution mode.

In this case, acceptable results (determined by the accuracy of estimates of light scattering parameters) can be obtained when determining the disperse composition of atmospheric haze (Zuev and Krekov, 1986), if instead of the measured histograms $f(R_j)$, one can confine oneself to the data on integral features of the size distribution, such as total volume $V_i$, surface area $S_i$, and concentration $N_i$ of separate fractions of the particles. The errors in the calculation, in this case, do not exceed 10% to 15% and practically all of the regularities are preserved in the functions behavior (e.g., angular ones) of aerosol light scattering.

The value of the geometric mean volume radius $R_i^{(\nu)}$ at $\nu = 3$ can be estimated from the formula (Krekov and Rakhimov, to be published):

$$R_i^{(3)} = \frac{\left(3\ V_i\right)^{2/3}}{\left(4\pi\ N_i\right)^{1/6}\ S_i^{1/2}} \tag{2}$$

The value of the parameter $b_i$ is determined by the ratio:

$$1/b_i = \ln\left[\left(4\pi\ N_i\right)^{2/3}\left(3V_i\right)^{4/3}/S_i^2\ \right] \tag{3}$$

If the values of the parameters $R_i$ and $b_i$ are known, it is not difficult to estimate the value of the parameter $F_i$, which determines the content of this fraction of particles,

$$F_i^{(3)} = C^{(3)}M_i = \frac{S_i}{\sqrt{16\pi^3/b_i}} = \frac{3V_i}{R_i\exp\left(1/4b_i\right)\sqrt{16\pi^3/b_i}} = \tag{4}$$

$$= N_iR_i^2/\left(\exp(1/b_i)\ \sqrt{\pi/b_i}\right)$$

Krekov, 1986; Krekov and Rakhimov, to be published). It is evident that the application of this approach to the modeling of postvolcanic variations of the optical states of the SAL is difficult, since information on the dispersed structure as a function of altitude and time is necessary. Such information is not likely in the near future.

## 2. METHOD OF CALCULATION

For the numerical prediction of light scattering by aerosols we have used a polymodal parametric structure defined as

$$f(R) = C^{(\nu)}R^{-\nu} \sum_i M_i \exp\left\{-b_i\left[\ln\left(\frac{R}{R_i^{(\nu)}}\right)\right]^2\right\} \tag{1}$$

where R is the particle radius, $M_i$ is the relative portion of the i-th fraction particles with respect to the background one (usually $M_2 = 1$), $A_i^{(\nu)}$ is the normalization parameter, $\nu$ is the characteristic parameter with the values $\nu = 1,2,3$ for expressions $f(R) = dN/dR$, $dS/dR$ or $dV/dR$, respectively, $R_i^{(\nu)}$ is the geometric mean radius of particles, and $b_i$ is inversely proportional to the width of the logarithmically normal distribution mode.

In this case, acceptable results (determined by the accuracy of estimates of light scattering parameters) can be obtained when determining the disperse composition of atmospheric haze (Zuev and Krekov, 1986), if instead of the measured histograms $f(R_i)$, one can confine oneself to the data on integral features of the size distribution, such as total volume $V_i$, surface area $S_i$, and concentration $N_i$ of separate fractions of the particles. The errors in the calculation, in this case, do not exceed 10% to 15% and practically all of the regularities are preserved in the functions behavior (e.g., angular ones) of aerosol light scattering.

The value of the geometric mean volume radius $R_i^{(\nu)}$ at $\nu = 3$ can be estimated from the formula (Krekov and Rakhimov, to be published):

$$R_i^{(3)} = \frac{\left(3\ V_i\right)^{2/3}}{\left(4\pi\ N_i\right)^{1/6}\ S_i^{1/2}} \tag{2}$$

The value of the parameter $b_i$ is determined by the ratio:

$$1/b_i = \ln\left[\left(4\pi\ N_i\right)^{2/3}\left(3V_i\right)^{4/3}/S_i^{\,2}\right] \tag{3}$$

If the values of the parameters $R_i$ and $b_i$ are known, it is not difficult to estimate the value of the parameter $F_i$, which determines the content of this fraction of particles,

$$F_i^{(3)} = C^{(3)}M_i = \frac{S_i}{\sqrt{16\pi^3/b_i}} = \frac{3V_i}{R_i\exp\left(1/4b_i\right)\sqrt{16\pi^3/b_i}} = \tag{4}$$

$$= N_iR_i^2/\left(\exp(1/b_i)\sqrt{\pi/b_i}\right)$$

The problem of prediction of aerosol turbidity of the SAL over a wide spectral range cannot be solved correctly with the use of data on variations of structure and content of secondary aerosols. The particles in the coarse-dispersed fraction ($R \simeq 1$ - 20 $\mu m$) play a crucial role in the attenuation of IR radiation, especially during the period of postvolcanic layer relaxation. The calculated relationships expressing kinetic interdependence of the microphysical parameters of particles of intra-atmospheric synthesis due to their origin (Rozenberg, 1982) cannot be used for predicting the content of coarse-dispersed areosols.

The suggested model estimates are based on the empirical data on time variation of integral characteristics of the aerosol size distribution of two fractions, namely, the accumulation mode ($0.03<R<1.0$ $\mu m$) and the coarse-dispersed mode (Hofmann and Rosen, 1984; McCormick et al., 1984; Hirono et al., 1984a; Chuan and Woods, 1984). As input variables, the empirical data on calculated number density $N_i$ and mass concentration $\rho V_i$ ($\rho$ is the density of aerosol substance) are used as well as those on characteristic particle sizes for every fraction $R_i$. In particular, when determining the number density at the initial stage of development of the layer, the data (Hofmann and Rosen, 1984; McCormick et al., 1984) were used according to those $N_o^A > 100$ $cm^{-3}$ (in model calculations $N_o^A = 150$ $cm^{-3}$ (Hirono et al., 1984a) for accumulated fraction and $N_o^C = 0.9$ $cm^{-3}$ for the coarse-dispersed fraction). At the same time, the volume concentration was estimated from data given in Chuan and Woods (1984), $V_o = 2$ $\mu m^3/cm^3$ (in the model $V_o^A = 0.2$ $\mu m^3/cm^3$ for the accumulated fraction and $V_o^C = 1.8$ $\mu m^3/cm^3$ for the coarse-dispersed particles).

Fraction modeling implies a relative independence of the formation and evolution of aerosol spectral composition for separate ranges of particle size. Degradation of the SAL in time in the model is simulated by an exponential decrease of number and mass density of each fraction separately. Besides, as the measurements show (Uchino et al., 1984; Hayashida and Iwasaka, 1985; Shibata et al., 1984; Hirono et al., 1984a and b; Chuan and Woods, 1984), the rate of decrease in time of number density of microdisperse fraction is about 1.5 times greater than that of mass fraction. Thus, the relative rise of aerosol particles of submicron range is considered indirectly on the basis of empirial data in model estimates.

As the observations show (Uchino et al., 1984; McCormick et al., 1984; Hirono et al., 1984b), under the influence of atmospheric circulation processes for postvolcanic cloud, the considered stratification is divided into the main layer relaxing during 1.5-2 years and some additional layers whose lifetime is some months. In the model evaluated when determining the initial profiles $N_i(h)$ and $V_i(h)$, the following parametric structure is tested:

$$N_i(h,t) = N_{i(h)}^{(b)} + \sum_j^2 N_{i0}^{(j)} \exp\left\{-\alpha_{N_i}^{(j)}(t)\left[h-h_{ij}^{(N_i)}(t)\right]^2\right\} \quad (5)$$

$$V_i(h,t) = V_{i(h)}^{(b)} + \sum_j^2 V_{i0}^{(j)} \exp\left\{-\alpha^{(j)}_{V_i}(t)\left[h-h_{ij}^{(V_i)}(t)\right]^2\right\} \quad (6)$$

Interconnection exists between the typical particle size of the separate fractions, expressed in terms of number density, that is, $R_i^{(v=o)}$, and similar parameters interpreted in terms of mass (or volume) concentration according to the spectral model (1). The above interconnection is determined by:

$$R_i^{(0)} = R_i^{(3)} \exp\left[-\frac{3}{2} b_i\right] \quad (7)$$

The Stokes settling velocity of particles in a gravitational field is determined by:

$$W_s = \frac{2R^2 \rho g}{g\eta} \left[1 + A\frac{1}{R} + Q \exp\left\{-B\frac{R}{1}\right\}\right] \quad (8)$$

where $\eta$ is the viscosity of air, A = 0.8648; Q = 0.29; B = 1.25; and 1 is the mean free path of molecules. When imitating the effect of the sedimentation in the model the spectral deformations F(R, h, t) to the first and second orders of magnitude were taken into account. Due to differences in typical particle sizes, the subdividing of profiles, not only $N_1(h,t)$ and $N_2(h,t)$, of accumulated and coarse-dispersed fractions but also the spectrum moments $N_i(h,t)$, $V_i(h,t)$ were modeled within each fraction separately. The latter at initial time (t = 0) coincided.

Peak aerosol concentrations, being scattered in the course of zonal and meridian transport, are mixed with the background concentration. When determining the components $N_i(h)$ and $V_i(h)$ in Eqs. 5 and 6, the empirical data tested earlier in Zuev and Krekov (1986) were used as initial profiles of the SAL size distribution f(h,R) of the background optical model.

## 3. THE RESULTS

The results of model estimates are given partially in Table 1. For every altitude (the values are indicated by numbers located vertically in the first column of the table) the four versions of spectral dependence of extinction coefficient are calculated. These correspond to various stages of layer development, indicated in the second column of the table (the unit of measurement in time is three months).

Figure 1 gives the model estimates of the altitude profile of extinction coefficient for three wavelengths compared to lidar measurements: 5,9 - molecular profile; 6,7 - the data (Uchino et al., 1984; Hayashida and Iwasaka, 1985); and 8,10 - the data (Shibata et al., 1984; Hirono et al., 1984a). The good agreement between the theoretical and experimental estimates enables one to state that

TABLE 1. SPECTRAL VARIATION OF AEROSOL EXTINCTION COEFFICIENT $\beta_t(\lambda,t,h)\mathrm{km}^{-1}$

| H | T | WAVELENGTH (μm) | | | | | | | |
|---|---|---|---|---|---|---|---|---|---|
| (km) | | 0.248 | 0.530 | 0.694 | 1.060 | 2.360 | 3.390 | 5.300 | 10.600 |
| | 0 | 3.961-4 | 1.913-4 | 1.316-4 | 6.565-5 | 1.690-5 | 1.728-5 | 6.742-6 | 5.824-6 |
| 18 | 1 | 4.049-4 | 1.995-4 | 1.396-4 | 7.383-5 | 2.436-5 | 2.289-5 | 1.074-5 | 9.645-6 |
| | 2 | 4.112-4 | 2.062-4 | 1.464-4 | 8.067-5 | 3.121-5 | 2.850-5 | 1.548-5 | 1.450-5 |
| | 4 | 3.978-4 | 1.945-4 | 1.347-4 | 6.872-5 | 2.059-5 | 2.002-5 | 8.875-6 | 7.965-6 |
| | 0 | 1.023-3 | 4.907-4 | 3.934-4 | 2.949-4 | 1.063-4 | 1.002-4 | 4.923-5 | 3.647-5 |
| 19 | 1 | 8.923-4 | 4.745-4 | 3.836-4 | 2.968-4 | 1.610-4 | 1.326-4 | 7.572-5 | 6.402-5 |
| | 2 | 6.929-4 | 3.589-4 | 2.779-4 | 2.000-4 | 1.086-4 | 9.160-5 | 5.267-5 | 4.565-5 |
| | 4 | 4.507-4 | 1.903-4 | 1.243-4 | 6.242-5 | 2.204-5 | 2.320-5 | 9.205-6 | 7.277-6 |
| | 0 | 6.372-3 | 3.230-3 | 2.834-3 | 2.372-3 | 8.483-4 | 7.853-4 | 4.060-4 | 2.957-4 |
| 20 | 1 | 2.876-3 | 1.545-3 | 1.300-3 | 1.040-3 | 4.684-4 | 4.005-4 | 2.199-4 | 1.717-4 |
| | 2 | 1.105-3 | 5.739-4 | 4.480-4 | 3.110-4 | 1.158-4 | 1.086-4 | 5.623-5 | 4.204-5 |
| | 4 | 2.592-4 | 1.115-4 | 7.411-5 | 3.785-5 | 1.304-5 | 1.393-5 | 5.641-6 | 4.236-6 |
| | 0 | 1.321-2 | 6.723-3 | 5.927-3 | 4.986-3 | 1.781-3 | 1.647-3 | 8.541-4 | 6.216-4 |
| 21 | 1 | 3.386-3 | 1.652-3 | 1.330-3 | 9.702-4 | 3.234-4 | 3.195-4 | 1.658-4 | 1.203-4 |
| | 2 | 7.978-4 | 3.765-4 | 2.746-4 | 1.587-4 | 3.920-5 | 4.764-5 | 2.354-5 | 1.665-5 |
| | 4 | 1.149-4 | 5.075-5 | 3.390-5 | 1.779-5 | 6.233-6 | 6.172-6 | 2.315-6 | 1.789-6 |
| | 0 | 6.243-3 | 3.176-3 | 2.800-3 | 2.356-3 | 8.416-4 | 7.785-4 | 4.035-4 | 2.937-4 |
| 22 | 1 | 9.722-4 | 4.349-4 | 3.334-4 | 2.165-4 | 5.462-5 | 6.493-5 | 3.278-5 | 2.324-5 |
| | 2 | 1.733-4 | 7.610-5 | 5.201-5 | 2.663-5 | 7.256-6 | 8.790-6 | 4.118-6 | 2.967-6 |
| | 4 | 4.732-5 | 1.950-5 | 1.279-5 | 6.997-6 | 2.739-6 | 2.572-6 | 8.730-7 | 6.983-7 |
| | 0 | 6.893-4 | 3.476-4 | 3.034-4 | 2.528-4 | 9.045-5 | 8.372-5 | 4.309-5 | 3.141-5 |
| 23 | 1 | 9.616-5 | 3.973-5 | 2.807-5 | 1.581-5 | 4.253-6 | 5.087-6 | 2.283-6 | 1.683-6 |
| | 2 | 4.306-5 | 1.696-5 | 1.102-5 | 5.925-6 | 2.185-6 | 2.187-6 | 7.891-7 | 6.276-7 |
| | 4 | 3.759-5 | 1.479-5 | 9.581-6 | 5.206-6 | 2.012-6 | 1.941-6 | 6.596-7 | 5.373-7 |

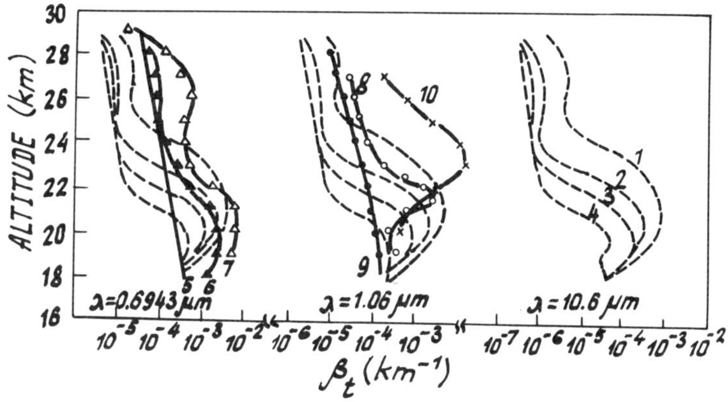


FIGURE 1. The model estimates of the altitude profile of extinction coefficient for three wavelengths compared to lidar measurements.

the suggested methodology for modeling the optical state of the SAL up to the altitudes of the middle atmosphere is of utility. The calculated data reveal an essential transformation of spectral dependence $\beta(\lambda,h,t)$ both in altitude and in time (Fig. 2a,b).

The results of model estimates with a detailed interval according to ($\Delta\ \lambda \sim 0.1\ \mu m$) radiation spectrum in the 0.248-13.6 $\mu$m range are fully given in Krekov and Rakhimov (to be published). There the results of modeling are also considered for another aerosol stratification within 0-90 km altitudes and the regional features of atmospheric haze optical characteristics in the lower 4 km tropospheric layer are analyzed.

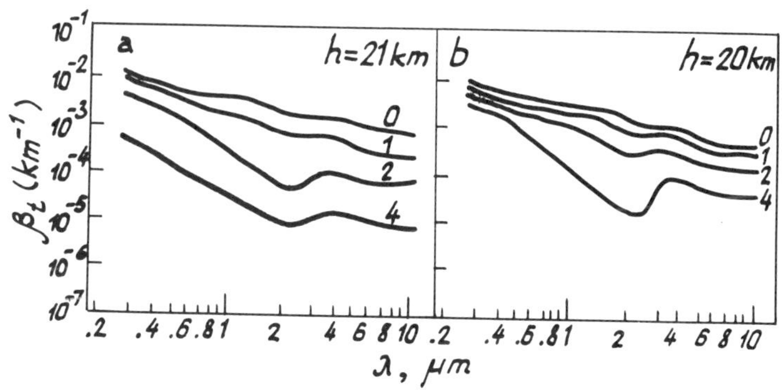


FIGURE 2. Transformation of spectral dependence $\beta(\lambda$, h, t) in altitude and in time.

## REFERENCES


Burgmeier, J.W., and I.H. Blifford, 1975: A Reinforced Coagulation Sedimentation Model for Stratospheric Aerosols, Water Air Soil Pollut., 5, 133-147.

Chuan, R.L., and D.C. Woods, 1984: Temporal Variations in Characteristics of the El Chichon Stratospheric Clouds, Geofis. Int., 23-3, 335-349.

Hayashida, S., and Y. Iwasaka, 1985: On the Long-Term Variation of Stratospheric Aerosol Content After the Eruption of Volcano El Chichon: Lidar Measurements at Nagoya, Japan, J. Meteor. Soc. of Japan, 63, No. 3, 465-472.

Hirono, M., et al., 1984a: Enormous Increase of Volcanic Clouds in the Stratosphere over Fukuoko After April 1982, Geofis. Int., 23-2, 259-276.

Hirono, M., M. Fujiwara, T. Shibata and N. Kugumiya, 1984b: Lidar Observations of Atmopheric Aerosols Following the 1980 Eruption of Mt. St. Helens: Part I, J. Atmos. and Terrestrial Phys., 46, No. 12, 1147-1157.

Hofmann, D.J., and J.M. Rosen, 1984: Balloonborne Particle Counter Observations of the El Chichon Aerosol Layers in the 0.01-1.8 μm Range, Geofis. Int., 23-2, 155-185.

Junge, C.E., C.W. Chagnon and J.E. Manson, 1961: Stratospheric Aerosols, J. Meteor., 18, 81-108.

Krekov, G.M., and R.F. Rakhimov, 1982: Optical-Sensing Model of Continental Areosol, Nauka, Novosibirsk, 196 pp.

Krekov, G.M., and R.F. Rakhimov, Optical Models of Atmospheric Aerosol, Tomsk, Edition of Tomsk Branch of SB USSR Academy of Sciences (to be published).

McCormick, M.P., T.J. Swissler, W.H. Fuller, W. H. Hunt and M. T. Osborn, 1984: Airborne and Ground-Based Lidar Measurements of the El Chichon Stratospheric Aerosol from 90°N to 56°S, Geofis. Int., 23-2, 187-221.

Rozenberg, G.V., 1982: Finely Dispersed Aerosol of the Stratosphere and Mesosphere, Izv. Akad. Nauk SSSR Fiz. Atmos. Okeana, 18, No. 6, 609-622.

Shibata, T., M. Fujiwara and M. Hirono, 1984: The El Chichon Volcanic Cloud in the Stratosphere: Lidar Observation at Fukuoka and Numerical Simulation, J. Atmos. Terrestrial Phys., 46, No.12, 1121-1146.

Uchino, O., K. Takahashi, I. Tabata, I. Akita, Y. Okada and K. Naito, 1984: Ruby Lidar Observations of El Chichon Dust Clouds at Tsukuba (36.1°N) and Comparisons with UV Lidar Measurements at Fukuoka (33.6°N), J. Meteor. Society of Japan, 62, No. 4, 679-687.

Zuev, V.E., and G.M. Krekov, 1986: Optical Models of the Atmosphere, Gidrometeoizdat, Leningrad, 256 pp.

# RADIATIVE EFFECTS OF ATMOSPHERIC AEROSOL PARTICLES

*Hartmut Grassl*
Forschungszentrum Geesthacht
D-2054, Federal Republic of Germany

ABSTRACT

The contribution of atmospheric aerosol particles to shortwave (solar) and longwave (terrestrial) radiative flux profiles in a cloudfree and a cloudy atmosphere is reviewed. The change of cloud optical properties by aerosol particle characteristics is emphasized. From radiation budget calculations, based upon aerosol particle profiles given by a two-dimensional global aerosol transport model, the possible latitude dependent manmade albedo change is derived. The main result: albedo enhancement is caused by cloud albedo increase rather than by enhanced backscattering from cloud-free parts of the atmosphere. The strong dependence of the albedo change on size distribution change and soot content underlines the need for more reliable input parameters and more sophisticated models before the sign and the relative magnitude of an aerosol climate signal as compared to the greenhouse effect by gases may be given.

## 1. INTRODUCTION

There is no considerable doubt that tropospheric as well as stratospheric aerosol particles are important for the earth's climate. These particles scatter, absorb, and emit in cloud-free parts of the atmosphere. They strongly influence stratus cloud droplet size distribution, and they may influence the droplet size distribution of convective clouds. Thus they also play a role in precipitation formation in all clouds. Moreover, they may have their strongest climatic relevance through cloud radiative properties, thereby being involved in one of the two most important interactions in the climate system: air-sea exchange and cloud-radiation interaction.

This contribution tries to cover the particles' influence on the radiative budget of the atmosphere/surface system by discussing four combinations:

1) cloud-free, solar radiation,
2 cloud-free, terrestrial radiation,
3) cloudy, solar radiation, influence on cloud optical properties, and
4) cloudy, terrestrial radiation, influence on cloud optical properties.

In the following sections aerosol particles will be discussed separately and/or in combination for the following parameters: local heating rates, and net fluxes at the surface and top of the atmosphere. This is done both for static situations, i.e., prescribed aerosol burden and aerosol load resulting from a two-dimensional global aero-

AEROSOLS AND CLIMATE
Peter V. Hobbs and M. Patrick McCormick (Eds.)  
ISBN 0-937194-11-5

sol transport model resolving latitude and height. The anthropogenic perturbation is also included. However, it should be borne in mind, that this perturbation is far from being understood so well as the additional greenhouse effect by growing concentrations of well-mixed, thus long-lived atmospheric trace gases. An understanding would mean realistic modeling of the cloud optical properties and the precipitation cycle, the hardest task for atmospheric modelers at present.

## 2. RADIATIVE EFFECTS IN A CLOUD-FREE ATMOSPHERE

Nearly all publications on aerosol radiative properties (mostly calculations) have concentrated on the particles' role in a cloud-free atmosphere and most of them were only for solar radiation. Since the dominant negative feedback in the climate system is the tendency for compensation of a solar radiation absorption increase by a terrestrial emission increase, the omission of the terrestrial or longwave radiation calculation has exaggerated the aerosol particle influence, or may even have led to the incorrect sign of an energy budget change for a given turbidity change. For an accurate local radiation budget calculation we would need the vertical profile of the following particle characteristics:

1) number density,
2) size distribution,
3) chemical composition (complex index of refraction) of individual particles or agglomerates of particles,
4) shape of the particles or agglomerates, and
5) some method of calculating cross sections.

In addition, profiles of other radiatively active constituents of the atmosphere will be needed, the most important being the water vapour profile.

### 2.1 SOLAR RADIATION

Atmospheric turbidity or reduced horizontal visibility is an obvious sign of aerosol particle scattering in the atmosphere; however, it is not as important as absorption from the point of view of local tropospheric heating rates, which are mainly due to absorption by water vapour and soot. There is a rather intriguing interplay between the parameters that determine the sign and the magnitude of a net solar radiation flux or local shortwave planetary albedo change for a given aerosol particle parameter change. These are:

1) absorption coefficient,
2) size distribution,
3) surface albedo,
4) total optical depth of aerosol particles,
5) zenith distance of the Sun, and
6) relative humidity.

There should also be a means of deriving the scattering and absorption cross sections from the above physical and chemical properties of the aerosol. Also the directionality of surface reflectivity should not be overlooked. Yamamoto and Tanaka (1972) and Eschelbach (1973) were among the first to show by calculations that distinct combinations of particle size, complex index of refraction, and surface albedo may lead to an increase or decrease of shortwave local planetary albedo. The critical point, i.e., the parameter combination with no change in local planetary albedo at a given turbidity change was roughly estimated by Chylek and Coakley (1974), as well as Coakley and Chylek (1975). More detailed calculations by Grassl and Newiger (1982) led to the following list of parameters, if ordered according to importance for local planetary albedo change at a given surface albedo:

1) absorption/backscattering ratio (determined by both the aerosol particle size distribution and the complex index of refraction and thus also relative humidity),
2) zenith distance of the Sun,
3) total optical depth of aerosol particles, and
4) directionality of surfaces.

In order to demonstrate the possible variations Fig. 1 contains the differences of critical surface albedo, $\Delta A_c$, if compared to Chylek and Coakley's (1974) first estimates. Here critical cloud albedo does mean, adding of particles has no influence on local planetary albedo.

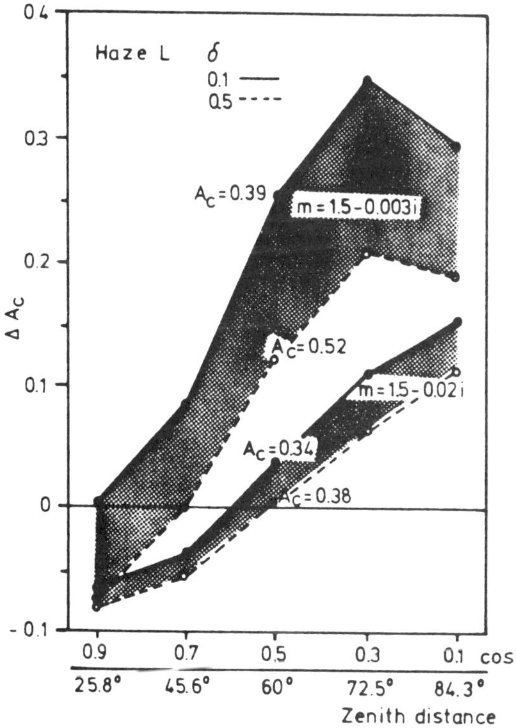


FIGURE 1. Difference of critical surface albedo $A_c$ (as compared to the Chylek and Coakley approximation) at which adding of particles has no effect on local planetary albedo as a function of solar zenith angle and for Haze L at two total optical depths $\delta$ and for two complex indices of refraction m. The critical $A_c$ values are also indicated for $\theta = 60°$.

Simplified, one could summarize: Over oceans additional particles will increase local planetary albedo in most cases even at noon and with a rather high soot content, while over fresh snow moderately absorbing particles will decrease local planetary albedo also at rather low Sun elevation.

Therefore, even the sign of an aerosol effect on radiation budget will be uncertain especially during winter and spring in high latitudes with a maximum of turbidity and soot content, but also for desert aerosols under duststorm conditions. While Cess (1983) points to an albedo reduction by arctic haze during winter and spring relying on a single measuring station, Heintzenberg (1984) has shown that remote arctic aerosol will enhance albedo during summer even over snow surfaces. As an example for the possible net solar radiation flux changes, Table 1 (drawn from Cess, 1983) lists solar net flux and changes of net flux at the surface and at the top for arctic haze. Although the single scattering albedo (scattering/extinction) $\omega_o$= 0.703 used by Cess (1983) corresponds to an urban aerosol type absorption and might be far too low on a climatic timescale, Table 1 nevertheless underlines the effects possible during strong pollution episodes.

The local heating rates will be discussed together with cooling rates caused by longwave or terrestrial radiation within the next section.

## 2.2 TERRESTRIAL RADIATION AND NET RADIATION

The discussion up to now referred to simple one-dimensional radiative transfer models of solar radiation or point measurements of optical parameters in the solar radiation wavelength region. A next step on the way to more realistic models would be the use of still one-dimensional solar and terrestrial radiation models and a second step radiative convective models, in order to relate the total radiation flux perturbation, caused by aerosol particles, to surface temperature or stratospheric temperature changes. The second step was done already in the early seventies, however, without measured absorption properties of aerosol particles. For three similar radiative convective models, two of them also accounting for heat storage and seasonal variations, Charlock and Sellers (1980) have found the same basic difficulties, i.e., the uncertainty in aerosol absorption results in a broad range of surface temperature changes from roughly 2 K cooling up to no effect at all. Surely neither the surface albedo influence nor the spatial variability of aerosol optical depth and absorption can be modeled realistically with one-dimensional radiative convective models.

Fortunately, there are two rather complete studies, from the point of view of net radiation, on the effect of Saharan desert aerosol particles by Carlson and Benjamin (1980) and Fouquart et al. (1987, a,b). Both studies try to establish the net effect of aerosol particles on the local radiation budget by radiative transfer calculations using either their own (Fouquart et al., 1987a,b) and/or earlier measurements. Besides the confirmation of the importance of the complex

TABLE 1. SOLAR RADIATION FLUX PERTURBATION DUE TO ANTHROPOGENIC ARCTIC HAZE[1]

| | Nov. | Dec. | Jan. | Feb. | March | April | May |
|---|---|---|---|---|---|---|---|
| | | | | 65° N | | | |
| $\tau_{0.55}$ | 0.035 | 0.125 | 0.115 | 0.148 | 0.089 | 0.087 | 0.035 |
| $S_N$ | 26 | 4 | 11 | 63 | 162 | 295 | 412 |
| $\Delta S_N$(top) | +1.1 | +1.0 | +2.0 | +6.4 | +7.2 | +9.4(8.4)[2] (7.7)[3] | +2.8 |
| $\Delta S_N$(surf) | -1.0 | -0.4 | -1.1 | -4.9 | -5.3 | -7.7(-10.1)[2] (-10.6)[3] | -5.4 |
| $A_S$ | 0.53 | 0.63 | 0.61 | 0.58 | 0.56 | 0.50 | 0.30 |
| | | | | 75° N | | | |
| $\tau_{0.55}$ | | | | 0.148 | 0.089 | 0.087 | 0.035 |
| $S_N$ | | | | 5 | 89 | 243 | 415 |
| $\Delta S_N$(top) | | | | +1.8 | +7.5 | +14.0(12.9)[2] (14.0)[3] | +8.2 |
| $\Delta S_N$(surf) | | | | -0.3 | -3.0 | -5.1(-6.7)[2] (-7.0)[3] | -3.8 |
| $A_S$ | | | | 0.81 | 0.72 | 0.71 | 0.64 |
| | | | | 85° N | | | |
| $\tau_{0.55}$ | | | | | 0.089 | 0.087 | 0.035 |
| $S_N$ | | | | | 16 | 228 | 428 |
| $\Delta S_N$(top) | | | | | +3.8 | +18.8(17.2)[2] (18.8)[3] | +10.7 |
| $\Delta S_N$(surf) | | | | | -0.7 | -5.7(-7.9)[2] (-8.0)[3] | -2.5 |
| $A_S$ | | | | | 0.82 | 0.77 | 0.78 |

[1]Diurnally averaged net solar radiation flux $S_N$, $\Delta S_N$ [surface] and $\Delta S_N$ [top] for given aerosol optical depth $\tau_{0.55}$ at 0.55 µm wavelength and 70% relative humidity and given mean solar radiation surface albedeo $A_S$; cloudless, single scattering albedo $\omega_o$ = 0.703.
[2]Maritime aerosol added.
[3]Continental aerosol added.

index of refraction it is well established by Fouquart et al. (1987 b), that the main uncertainty is with the particle size distribution. The large discrepancy between size distributions measured with optical particle counters and those retrieved from spectral extinction measurements immediately impacts on the net radiation flux and thus on the sign of a flux change at an aerosol optical depth change. Since most substances (liquids or solids) absorb rather strongly in the thermal infrared window from 8-14 µm wavelength, the size of a particle is most important for its absorption, thus giving the size distribution an overwhelming influence on the possible compensation of solar radiation effects by terrestrial radiation. Since the discrepancies found by Fouquart et al. (1987a, b) are due to the small particle size range, future experiments should also concentrate on the establishment of a size distribution at radii below 1 µm. Figure 2, taken from Fouquart et al. (1987b), points to three remarkable effects (two caused by particles):

1) rather strong absorption of solar radiation and thus heating within the dust layer leading to a greatly reduced solar radiation net flux at the ground amounting to more than 100 W $m^{-2}$ in agreement with observation,
2) thermal infrared heating of the lowest atmospheric layers caused by the strong superadiabatic lapse rate in the surface layer, and
3) more than a doubling of the 8-14 µm infrared cooling rate to ~ 1 K $d^{-1}$ within the dust layer.

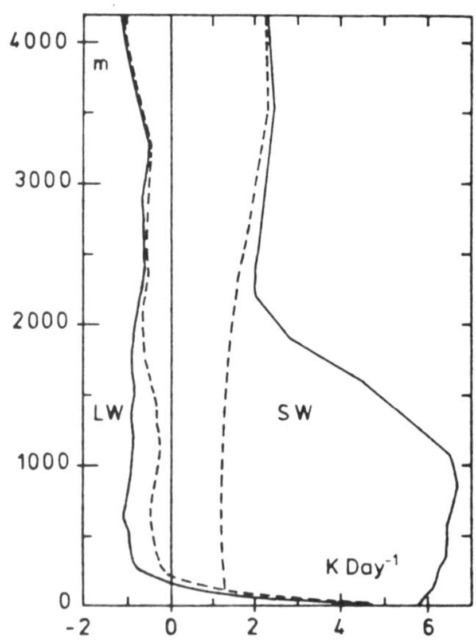


FIGURE 2. Heating by absorption of solar radiation (SW) and cooling or heating by terrestrial radiation (LW) as a function of height for a Saharan dust layer, derived from observations at 27 November and 9 December 1980 (taken from Fouquart et al., 1987b). Dashed curves: neglection of particles.

Compared to the results by Carlson and Benjamin (1980), where, over land, the desert aerosol load led to an energy gain of the atmosphere surface system, the increased number of small particles found by Fouquart et al. (1987a) will certainly reduce the energy gain or even turn it into a loss. Again the answer to the question of the sign of a climatic impact by aerosol particles can still not be given even for the most important desert aerosol. Up to now only one-dimensional radiative transfer models or radiative convective models have been discussed. What really would be needed for the aerosol particles influence on atmospheric circulation and climate is a three-dimensional global aerosol transport model including rather sophisticated radiative transfer calculations. Prior to the conference, however, only the results of three-dimensional circulation models at fixed aerosol load and unchanged cloudiness (Joseph, 1977 and 1984; Tanre et al., 1984; Randall et al., 1984) or two-dimensional global aerosol transport including the change of cloud optical properties (Rehkopf et al., 1984), and concomittant radiative transfer calculations (Newiger, 1985) were available. Although the circulation model studies did show the expected surface temperature variations, for instance a cooling over most surfaces, they were still rather close to the model's noise level. The most comprehensive study by Tanre et al. (1984) points to a net energy loss caused by the aerosol load as typically observed.

The radiative transfer results (Newiger, 1985) based upon the two-dimensional aerosol transport model results (for details see Rehkopf, 1984) are only available for the solar radiation range. However, since they are part of a rather advanced aerosol modeling including cloud effects (where longwave radiation changes are small), they will nevertheless be discussed here. Figure 3, for the possible manmade aerosol effect for particles outside clouds, indicates two dominating parameters: the amount of soot and the size distribution. Therefore both parameters were allowed to vary in order to show the possible range. Even for an exaggerated 20% soot fraction in anthropogenic particles, solar net flux decreases or local planetary albedo increases in lower and midlatitudes. The unrealistic fixation of the slope of the aerosol size distribution (dashed lines) gives a totally different latitude dependence, because the anthropogenic particle component in all three size classes (nucleus, accumulation, and coarse mode) is then equally distributed, despite the model's capability to react on precipitation, mean vertical motion, and gas-to-particle conversion from $SO_2$ into sulfate or sulfuric acid differently for different latitudes. The size distribution change causes (only for winter) a shift of the main influence southward if compared to the main particle and $SO_2$ input at 50 °N. The anthropogenic effect is stronger in summer than in winter. In both seasons high latitudes experience an albedo reduction for high soot content mainly caused by high surface albedo and not fully compensated by the low Sun. The soot particles have been handled as an external mixture for radiative transfer calculations. In view of the lack of reliable data on soot content and due to the sensitivity to size distribution changes, consequences of these model results should be discussed carefully.

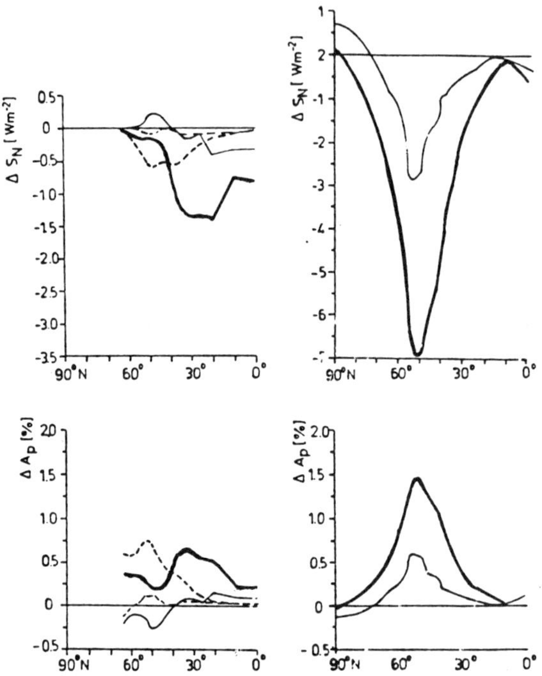


FIGURE 3. Anthropogenic aerosol particle influence on solar radiation net flux change $\Delta S_N$ or local planetary albedo change $\Delta A_p$ for summer and winter (left), for particles without soot (thick curves), and particles with 20% soot content (thin curves). Dashed curves for a fixed slope of the size distribution $\alpha = 3.5$.

## 3. AEROSOL PARTICLE INFLUENCE ON CLOUD OPTICAL PROPERTIES

The optical properties of clouds certainly have an impact on climate. If these properties vary with aerosol particle characteristics, the particles' influence would not be confined only to cloudless parts of the atmosphere. Since the number of cloud droplets is mainly determined by updraft and three aerosol parameters, namely total number, size distribution, and chemical composition, there is definitely a way for such an influence to occur. Twomey (1974 and 1977) has demonstrated the possible impact of condensation nucleus number on planetary albedo. The albedo reduction as a function of absorption by aerosol particles has been calculated by Grassl (1975). A simple calculation, connecting the above four parameters in broad agreement with more sophisticated microphysical models of clouds, gave: all four parameters are important for both stratus and convective clouds (Grassl and Levkov, 1984). The relative variation in cloud droplet number $\Delta N_c/N_c$ at relative variations in aerosol particle number $\Delta N/N$, slope $\alpha$ of the aerosol size distribution $n(r) \sim r^{-\alpha}$, and chemical composition represented by A* (where A* is the size at which a certain cloud condensation nucleus is activated at a supersaturation of 1 %, normally ranging between 0.01 - 0.025 µm) varies between $0.35 \leq \frac{\Delta N}{N}/\frac{\Delta f}{f} \leq 1.7$ for observed bandwidths in f = N, $\alpha$ or A*, except for the influence of aerosol size distribution slope $\alpha$ on stratus clouds to which $\Delta N/N$ is even more sensitive.

Relative magnitudes using the dependencies on $\alpha$ and N for the modification of cloud optical properties, according to the aerosol transported in the two-dimensional model (Rehkopf, 1984) resolving latitude and height, Newiger (1985) could show that the solar net flux change $\Delta F_{net}$ at the top of the atmosphere is larger than for the cloudless parts of the atmosphere. The cloud amount has been fixed according to a published climatology as well as the parameter $A^* = 1.4 \cdot 10^{-6}$ cm used as a measure for chemical composition. Figure 4, only for winter conditions and the Northern Hemisphere, again shows a latitudinal shift, if the radiative transfer calculations did account for the varying slope of the size distribution. Only the low stratus clouds contribute remarkably to the net flux change in Fig. 4; higher clouds are not affected so strongly since the anthropogenic particle increase is small in midtropospheric levels.

While Figs. 3 and 4, both resulting from a two-dimensional global aerosol transport model plus radiative transfer calculations, did not account for infrared radiation effects and thus are only part of the real radiative effect, our own calculations (Grassl, 1982) have pointed to negligible compensation of the increased shortwave cloud albe-

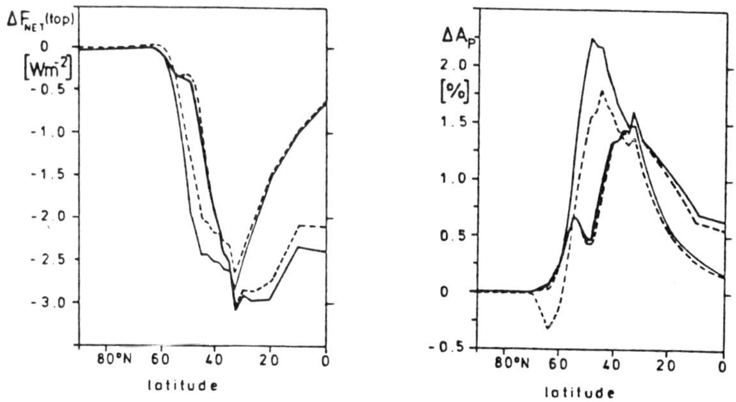


FIGURE 4. Solar net flux change at the top of the atmosphere and local planetary albedo as a function of latitude caused by changed optical properties of stratus clouds reacting on anthropogenic aerosol particles; thin curves: 20% soot content, dashed curves: fixed slope ($\alpha$ = 3.5) of the aerosol size distribution.

do by reduced longwave emission. This fact is itself due to a compensation: first, the increased optical depth at increased cloud droplet numbers causes, as expected, an upward shift of the emitting cloud layer into colder layers; second, the reduced emission, however, does not occur because of a concomittant single scattering albedo reduction or emissivity increase for smaller droplets at the thermal window wavelengths.

## 4. OVERALL EFFECT AND DISCUSSION

The combined effect, i.e., aerosol particle influences on the solar radiation budget in the cloudless and the cloudy parts of the atmosphere, as shown in Fig. 5, enhances local planetary albedo at all latitudes. Figure 5 also shows the reduction of the effect by an exaggerated value of 20% soot content of additional particles and the increase due to gas-to-particle conversion. Surely, a two-dimensional model for aerosol transport is a rather poor representation of the flow field in the troposphere. The radiative transfer model results presented should thus be seen as a first step only.

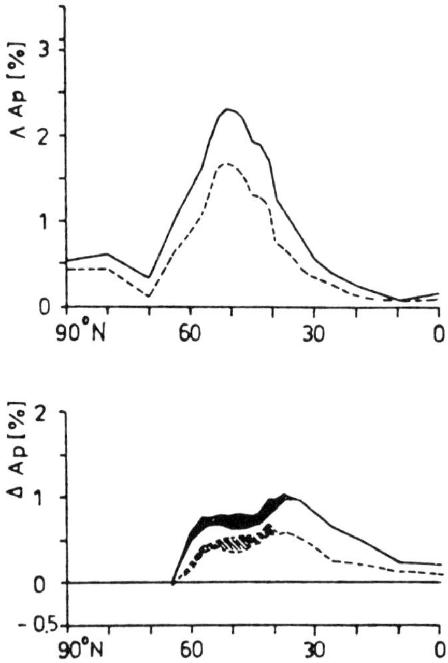


FIGURE 5. Local shortwave planetary albedo change $\Delta A_p$ caused by anthropogenic aerosol particles combining the effects of particles in cloudless parts of the atmosphere and of changed cloud optical properties. Dashed curves: 20% soot, hatched area: contribution by gas-to-particle conversion ($SO_2$ to $SO_4$) inside and outside clouds; upper panel: summer; lower panel: winter.

Any substance having different sources at the surface and being involved in the precipitation cycle has to show extreme spatial and temporal fluctuations. For such atmospheric constituents, aerosol particles being a good example, point observations do not give the mass budget sufficiently accurately and thus modeling their radiative effects is much more difficult than for well-mixed gases. Therefore, we do not have a clear understanding of the climatic impact of aerosol particles. We know, however, that two parameters besides total mass have the strongest influence, size distribution, and single scattering albedo, both depending on relative humidity through the growth of particles, with relative humidity causing a decrease (solar) or increase (thermal) absorption by a liquid water coating or the dilution of solution droplets.

Further steps should be: measurement of the relation between aerosol particle parameters and optical cloud parameters as well as the introduction of aerosol transport into atmospheric general circulation models.

## REFERENCES


Carlson, T.N., and S.G. Benjamin, 1980: Radiative Heating Rates for Saharan Dust, J. Atm. Sci., 37, 193-213.

Charlock, T.P., and W.D. Sellers, 1980: Aerosol Effects on Climate: Calculations with Time-Dependent and Steady-State Radiative-Convective Models, J. Atm. Sci., 37, 1327-1341.

Cess, R.D., 1983: Arctic Aerosols: Model Estimates of Interactive Influences Upon the Surface-Atmosphere Clear-Sky Radiation Budget, Atm. Environment, 17, 2555-2564.

Chylek, P., and J.A. Coakley, 1974: Aerosols and Climate, Science, 183, 75-77.

Coakley, J.A., and P. Chylek, 1975: The Two-Stream Approximation in Radiative Transfer: Including the Angle of the Incident Radiation, J. Atm. Sci., 32, 409-418.

Eschelbach, G., 1973: Computations of the Influence of Aerosol on the Atmospheric Radiation Balance in the Visible Spectrum, Contr. Atm. Phys., 46, 249-261.

Fouquart, Y., B. Bonnel, M. Chaoui Roquai, R. Santer and A. Cerf, 1987a: Observation of Saharan Aerosols: Results of ECLATS Field Experiment, Part I: Optical Thicknesses and Aerosol Size Distributions, J. Climate and Appl. Met., 26, 28-37.

Fouquart, Y., B. Bonnel, G. Brogniez, J.C. Buriez, L. Smith, J.J. Morcrette and A. Cerf, 1987b: Observations of Saharan Aerosols: Results of ECLATS Field Experiment, Part II: Broad Band Radiative Characteristics of the Aerosols and Vertical Radiative Flux Divergence, J. Climate and Appl. Met., 26, 38-52.

Grassl, H., and M. Newiger, 1982: Changes of Local Planetary Albedo by Aerosol Particles. In Atmospheric Pollution, Studies of Environmental Science, 20, 313-320.

Grassl, H., and L. Levkov, 1984: A Simple Model Allowing to Investigate the Activation Process in Water Clouds, Proceedings 11th Int. Conf. on Atmosph Aerosols, Condensation and Ice Nuclei, Budapest, Hungary.

Grassl, H., 1982: The Influence of Aerosol Particles on Radiative Parameters of Clouds, Idöjaras, 86, 60-74.

Grassl, H., 1975: Albedo Reduction and Radiative Heating of Clouds by Absorbing Aerosol Particles, Contr. Atm. Phys., 48, 199-210.

Heintzenberg, J., 1984: Baseline Measurements of Tropospheric Aerosol Properties in the Northern and Southern Hemispheres. In H.E. Gerber and A. Deepak (Eds.), Aerosols and Their Climatic Effects, A. Deepak Publishing, Hampton, Virginia, 95-104.

Joseph, J.H., 1977: Effect of a Desert Aerosol on a Model of the General Circulation. In H.-J. Bolle (Ed.), Proceedings of Symposium on Radiation in the Atmosphere, Princeton Science Press, 487-492.

Joseph, J.H., 1984: The Sensitivity of a Numerical Model of the Global Atmosphere to the Presence of Desert Aerosol. In H.E. Gerber and A. Deepak (Eds.), Aerosols and Their Climatic Effects, A. Deepak Publishing, Hampton, Virginia, 215-226.

Newiger, M., 1985: Einfluß anthropogener Aerosolteilchen auf den Strahlungshaushalt der Atmosphäre, Ph.D. Thesis, University of Hamburg, Hamburger Geophys. Einzelschriften, Reihe A, Heft 73, 88 pp.

Randall, D., T. Carlson, Y. Mintz, 1984: The Sensitivity of a General Circulation Model to Saharan Dust Heating. In H.E. Gerber and A. Deepak (Eds.), Aerosols and Their Climatic Effects, A. Deepak Publishing, Hampton, Virginia, 123-132.

Rehkopf, J., M. Newiger and H. Grassl, 1984: A 2-D Model of Global Aerosol Transport, Atm. Environment, 18, 2745-2752.

Rehkopf, J., 1984: Ein zweidimensionales globales Ausbreitungsmodell für Aerosole und Schwefelkomponenten, Ph.D. Thesis, University of Hamburg, Hamburger Geophys. Einzelschriften, Reihe A, Heft 71, 106 pp.

Tanre, D., J.F. Gelyn and J. Slingo, 1984: First Results of the Introduction of an Advanced Aerosol-Radiation Interaction in ECMWF Low Resolution Global Model. In H.E. Gerber and A. Deepak (Eds.), Aerosols and Their Climatic Effects, A. Deepak Publishing, Hampton, Virginia, 133-177.

Twomey, S., 1974: Pollution and the Planetary Albedo, Atmos. Environment, 8, 1251-1256.

Twomey, S., 1977: The Influence of Pollution on the Shortwave Albedo of Clouds, J. Atm. Sci., 34, 1149-1152.

Yamamoto, G., and M. Tanaka, 1972: Increase of Global Albedo Due to Air Pollution, J. Atm. Sci., 29, 1405-1412.

# EFFECT OF SHIP-STACK EFFLUENTS ON THE RADIATIVE PROPERTIES OF MARINE STRATOCUMULUS: IMPLICATIONS FOR MAN'S IMPACT ON CLIMATE[1]

*J.A. Coakley, Jr.*
National Center for Atmospheric Research[2]
Boulder, Colorado 80302, USA

*R.L. Bernstein*
Sea Space
San Diego, California 92122, USA

and

*P.A. Durkee*
Naval Postgraduate School
Monterey, California 93943, USA

## ABSTRACT

The effect of manmade aerosols on the earth's radiation budget through changes in cloud reflectivity is expected to be several times that due to the direct scattering and absorption of sunlight by the aerosols alone. Furthermore, these changes are expected to be comparable in magnitude but of opposite sign to that due to increasing $CO_2$. We use satellite observations to show that ships at sea increase the reflectivities of overlying stratus and stratocumulus cloud layers. Observations at 0.63, 3.7, and 11 $\mu m$ indicate that effluents from ships cause a decrease in droplet size and an increase in droplet number. The increase in the reflectivity at 0.63 $\mu m$ seems to be somewhat less than that expected on the basis of calculations for a plane-parallel cloud in which droplet size and droplet number change so that the liquid water content remains constant.

## 1. INTRODUCTION

Through their scattering and absorption of sunlight, manmade aerosols may alter the climate (Deepak and Gerber, 1983). By changing the amount of sunlight reflected by clouds, however, the cloud condensation nucleii (CCN) component of the aerosols may have an effect on the earth's energy budget that is several times that due to the direct scattering and absorption of sunlight by the aerosols alone (Twomey, *et al.*, 1984). Here we review theoretical estimates for the influence of increased aerosol on cloud reflectivity. We then show results for

[1] This work was supported in part by AFGL and AFOSR through transfer of funds to NASA, GLH6-6031 and by NASA Grant L-79877B.

[2] The National Center for Atmospheric Research is sponsored by the National Science Foundation.


A. DEEPAK Publishing
ISBN 0-937194-11-5

changes in the reflectivities of marine stratocumulus caused by CCN formed in the stack-exhausts of ships beneath the clouds.

## 2. THEORY

Twomey (1977) assumed that an increase in CCN leads to an increase in the number of cloud droplets and a decrease in droplet size so that the liquid water content remains unchanged. To a first approximation this assumption is consistent with condensation models for cloud droplet spectra (Mordy, 1959; Twomey, 1959) and the compensation between droplet number and droplet radius has been observed in the field (Warner and Twomey, 1967; Hudson, 1983; Noonkester, 1984).

Following Twomey's assumption, the total cross section per unit volume, which is also known as the extinction coefficient, is to a fair approximation given by

$$\sigma = KN^{1/3}\,W^{2/3} \tag{1}$$

where $K$ is a constant of proportionality; $N$ is the number of droplets per unit volume, and $W$ is the liquid water content.

Cloud reflectivity is also well approximated in either the two-stream or the Eddington approximation by

$$R = \frac{\chi\sigma\ell}{1 + \chi\sigma\ell} \tag{2}$$

where $\chi \sim 0.1$ is a weak function of droplet size and $l$ is the thickness of the cloud. Thus for a given cloud thickness the change in cloud reflectivity caused by an increase in droplet number is given by

$$\frac{\Delta R}{R} = \frac{1}{1 + \chi\sigma\ell}\,\frac{\Delta\sigma}{\sigma}. \tag{3}$$

Taking the global average cloud albedo to be 0.5, we obtain using (1) and (3)

$$\Delta R \sim \frac{1}{12}\,\frac{\Delta N}{N}. \tag{4}$$

The change in reflectivity given by (4) is consistent with those given by others (Grassl, 1982; Twomey *et al.*, 1984; Charlson *et al.*, 1987).

If we make the following assumptions:

1) The contribution of manmade CCN to the current level is in proportion to the manmade contribution to the global mass loading of aerosols,

2) The fractional increase in manmade aerosols and CCN is equal to the fractional increase in the atmospheric concentration of $CO_2$,

then a doubling of the $CO_2$ concentration will be accompanied by an increase in cloud reflectivity of 0.02, assuming that the current manmade contribution to the aerosol mass loading is about 20% (Pruppacher and Klett, 1980). The changes in the earth's radiation budget associated with this change in cloud reflectivity is given in Table 1 along with the changes associated with the direct scattering and absorption by manmade aerosols (taking the manmade contribution to be 20% that of the naturally occurring aerosol, Coakley *et al.*, 1983) and for a doubling of $CO_2$ (Ramanathan and Coakley, 1978). These estimates take global cloud coverage to be 50%. Table 1 shows the indirect effect of manmade aerosols on cloud reflectivity to be comparable in magnitude but opposite in sign of that due to the doubling of $CO_2$. The table also shows that the indirect effect is several times that due to the direct scattering and absorption by the aerosol.

The estimate given in Table 1 assumes that all clouds are affected equally. Of course, this will not be the case. Those affected will most likely be continental, and of those, most likely low level. Nevertheless, such clouds may become heavily polluted. How the degree of pollution might offset the type and extent to which cloud albedos are modified and how these changes will affect the global radiation budget and climate remain to be determined.

The estimate given in Table 1 also assumes that the CCN are non-absorbing. Any decrease in reflectivity due to the absorption added by the CCN is expected to be small compared with the increase that results from the increase in droplet number and decrease in droplet size (Twomey, 1977).

## 3. OBSERVATIONS

Kondrat'yev *et al.* (1981) seem to provide the only observations of cloud albedos affected by pollution. They reported aircraft observations of cloud albedos for an industrial region of the Soviet Union. There, contrary to the theoretical discussions of the preceding section, the pollution lowered the cloud albedo. Here we report on how ships at sea affect the reflectivity of overlying stratocumulus. The observations are made with $(1\ km)^2$ Advanced Very High

TABLE 1. CHANGE IN GLOBAL RADIATION BUDGET

| PERTURBATION | CHANGE IN EARTH'S NET RADIATION BUDGET |
|---|---|
| Change in cloud albedo due to doubling of manmade aerosols | -3.4 $W\ m^{-2}$ |
| Scattering and absorption due to doubling of manmade aerosols | -0.8 $W\ m^{-2}$ |
| Doubling of $CO_2$ | 3.9 $W\ m^{-2}$ |

Resolution Radiometer (AVHRR) data from the National Oceanic and Atmospheric Administration (NOAA) polar orbiters. Figure 1 shows such a ship track for a stratus deck off the coast of California observed during the FIRE (First International Satellite Cloud Climatology Project Regional Experiment) marine stratocumulus field experiment. This particular track is by no means an outstanding example of the tracks observed thus far (Coakley *et al.*, 1987). During the experiment many ship tracks were found in the satellite imagery data and some were penetrated by aircraft that made *in situ* microphysics and radiation observations.

The presence of the ship underneath the cloud is detected in satellite observations as an increase in the 3.7-$\mu m$ radiance (Channel 3 in Fig. 1). This increase is due to an increase in reflected solar radiation. Observations at 11 $\mu m$ indicate that there are no changes in cloud temperature associated with ship tracks.

At 3.7 $\mu m$ water is a moderate absorber and the cloud droplets have radii comparable to the wavelength of radiation. While the scattering cross section of a droplet scales like its geometric cross section, the absorption cross section scales like its volume. Consequently, as the droplets decrease in size the ratio of scattering to absorption increases. Stratocumulus clouds are typically opaque at 3.7 $\mu m$ so that cloud reflectivity and emissivity are insensitive to droplet concentration (Coakley and Davies, 1986). So, the increase in reflected solar radiation at 3.7 $\mu m$ indicates a decrease in droplet size.

To determine the extent to which the visible reflectivity of the cloud is altered, we isolate the satellite fields of view contaminated by the ship. We do this by taking for a mesoscale-sized region of approximately (60 km)$^2$ those pixels having 3.7 $\mu m$ radiances greater than the 90th percentile and which also are linearly connected to at least 11 other fields of view that also have 3.7 $\mu m$ radiances that exceed the 90th percentile. In Fig. 1 the mesoscale-sized region is indicated by the boxes and the contaminated fields of view are indicated by the highlighted line. For comparison, fields of view that are free of contamination are chosen at random from the remainder of the mesoscale region so that: 1) the fields of view are between 10 km and 30 km from the contaminated fields of view, and 2) the distribution of 11 $\mu m$ radiances for the noncontaminated fields of view closely matches that of the contaminated fields of view. The second restriction ensures that the contaminated and noncontaminated fields of view have similar distributions of fractional cloud cover and cloud height. Cloud cover and often cloud height affect the visible reflectivity of clouds. For the case shown in Fig. 1 all of the fields of view were overcast and there was little variability in cloud height. Figure 2 shows the distribution of 0.63 $\mu m$ reflectivities, and 3.7 and 11 $\mu m$ radiances for the contaminated and noncontaminated fields of view. The contaminated fields of view show higher 0.63 $\mu m$ reflectivities. These higher reflectivities are also perceptible in Fig. 1. The higher reflectivities are consistent with an increase in droplet number.

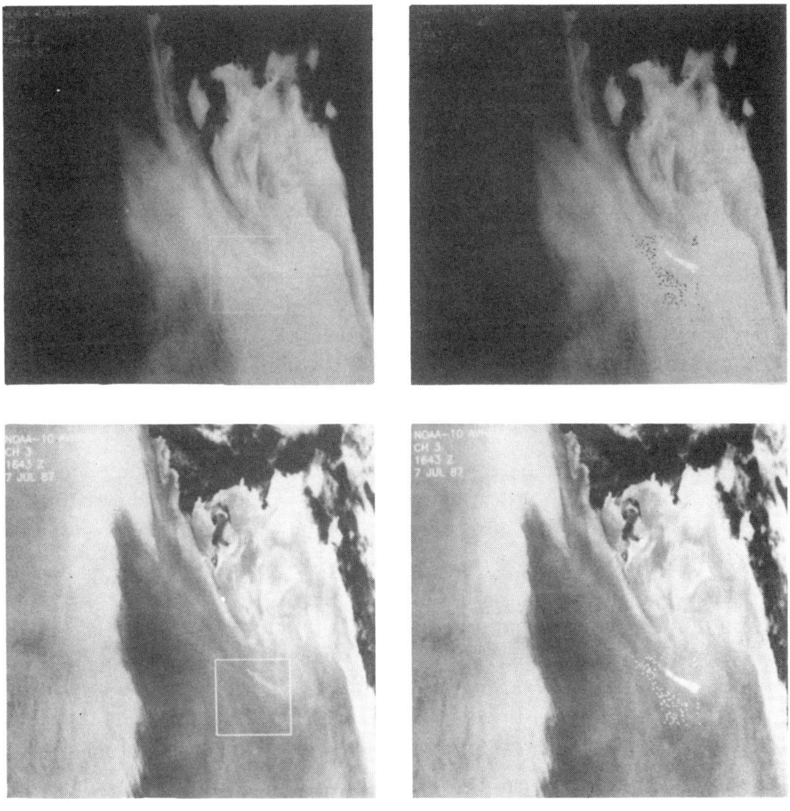


FIGURE 1. Images showing a ship track observed at 0.63 $\mu m$ (top) and 3.7 $\mu m$ (bottom) by satellite during FIRE. On the left the track is isolated in the boxed region. On the right it is highlighted. Randomly selected noncontaminated fields of view that serve as a control are indicated by dots.

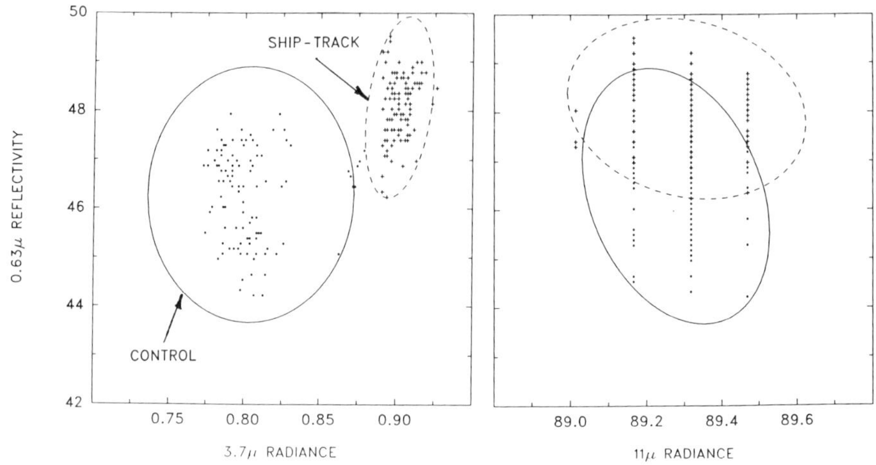


FIGURE 2. **Distributions of 0.63 $\mu m$ reflectivity and 3.7 and 11 $\mu m$ radiances for contaminated (crosses) and noncontaminated (dots) fields of view.** The ellipses are at two standard deviations from the means of the distributions, which are taken to be joint Gaussian distributions.

We add this case to an ensemble of 15 ship tracks studied previously (Coakley *et al.*, 1987). All of the tracks are wholly contained within large-scale cloud systems. For this collection we find a significant increase in the reflectivity at 0.63 $\mu m$. The results are summarized in Table 2.

In addition to the changes in reflectivities and radiances the table also gives the ratio of the change in 0.63 $\mu m$ reflectivity to that at 3.7 $\mu m$. To a first approximation this ratio is insensitive to the fractional cloud cover that affects the magnitudes of the individual reflectivity changes. The range of this ratio is also given for the simple model in which the clouds are taken to be

TABLE 2. CHANGES IN REFLECTIVITIES AND RADIANCES FOR STRATUS AND STRATOCUMULUS CLOUDS CONTAMINATED BY SHIPS

| | | |
|---|---|---|
| $\Delta R_{(0.63)}$ | 0.016 ± 0.007 | |
| $\Delta R_{(3.7)}$ | 0.039 ± 0.004 | |
| $\Delta I_{(11)}$ $(mWm^{-2}\,sr^{-1}\,cm)$ | 0.03 ± 0.05 | |
| | OBSERVATIONS | THEORY |
| $\Delta R_{(0.63)}/\Delta R_{(3.7)}$ | -0.2 - 1.2 | 0.6 - 2.6 |

plane-parallel, the liquid water content is held fixed, and changes in droplet size compensate changes in droplet number. The ratio is a function of droplet size distribution and cloud thickness. The observed changes in reflectivity seem to fall short of those expected from simple theory. The short fall may be due to absorption added by the CCN, to changes in cloud top geometry, or a decrease in liquid water content for the polluted clouds.

## 4. CONCLUSIONS

Ships at sea generally increase the reflectivities of overlying stratus and stratocumulus. Satellite observations would suggest that they do so by adding CCN, which lead to a decrease in droplet size and an increase in droplet number. The observations are thus in qualitative agreement with the mechanism proposed by Twomey (1977) and which had been used to explain the occurrence of ship-generated cloud streaks in otherwise cloud-free regions that appeared in the earliest satellite imagery data (Conover 1966; Twomey *et al.*, 1968). The observations indicate, however, that the increase in reflectivity is somewhat less than that obtained theoretically by a decrease in droplet radius and an increase in droplet number that leaves liquid water content unchanged. This study has dealt with changes in a maritime environment. Changes in a continental environment may well differ given the different populations of naturally occurring CCN and the different relative humidities.

## REFERENCES

Charlson, R. J., J. E. Lovelock, M. O. Andreae and S. G. Warren, 1987: Oceanic Phytoplankton, Atmospheric Sulphur, Cloud Albedo and Climate, *Nature*, *326*, 655-661.

Coakley, J. A., Jr., R. D. Cess and F. B. Yurevich, 1983: The Effect of Tropospheric Aerosols on the Earth's Radiation Budget: A Parameterization for Climate Models, *J. Atmos. Sci.*, *40*, 116-138.

Coakley, J. A., Jr., and R. Davies, 1986: The Effect of Cloud Sides on Reflected Solar Radiation as Deduced From Satellite Observations, *J. Atmos. Sci.*, *43*, 1025-1035.

Coakley, J. A., Jr., R. L. Bernstein and P. A. Durkee, 1987: Effect of Ship-stack Effluents on Cloud Reflectivity, *Science*, *237*, 1020-1022.

Conover, J. H., 1966: Anomalous Cloud Lines, *J. Atmos. Sci.*, *23*, 778-785.

Deepak, A., and H. E. Gerber (Eds.), 1983: *Report of the Experts Meeting on Aerosols and Their Climatic Effects*, WCP-55, World Meteorological Organization, Geneva, Switzerland.

Grassl, H., 1982: The Influence of Aerosol Particles on Radiation Parameters of Clouds, *Időjárás*, *86*, 60-75.

Hudson, J. G., 1983: Effects of CCN Concentrations on Stratus Clouds, *J. Atmos Sci.*, *40*, 480-486.

Kondrat'yev, V. I. Binenko and O. P. Petrenchuk, 1981: Radiative Properties of Clouds Influenced by a City, *Izv. Acad. Sci. USSR Atmos. and Oceanic Phys.*, (Engl. Transl.) *17*, 122-127.

Mordy, W., 1959: Computations of the Growth by Condensation of a Population of Cloud Droplets, *Tellus*, *11*, 16-44.

Noonkester, V. R., 1984: Droplet Spectra Observed in Marine Stratus Cloud Layers, *J. Atmos. Sci.*, *41*, 829-845.

Pruppacher, H. R., and J. D. Klett, 1980: *Microphysics of Clouds and Precipitation*, Reidel, Boston.

Ramanathan, V., and J. A. Coakley, Jr., 1978: Climate Modeling Through Radiative-convective Models, *Rev. Geophys. Space Phys.*, *16*, 465-489.

Twomey, S. A., 1959: The Nuclei of Natural Cloud Formation Part II: The Supersaturation in Natural Clouds and the Variation of Cloud Droplet Concentration, *Geofis. Pura Appl.*, *43*, 243-249.

Twomey, S. A., H. B. Howell and T. A. Wojciechowski, 1968: Comments on "Anomalous Cloud Lines," *J. Atmos. Sci.*, *25*, 333-335.

Twomey, S. A., 1977: The Influence of Pollution on the Shortwave Albedo of Clouds, *J. Atmos. Sci.*, *34*, 1149-1152.

Twomey, S. A., M. Piepgrass and T. L. Wolfe, 1984: An Assessment of the Impact of Pollution on Global Cloud Albedo, *Tellus*, *36B*, 356-366.

Warner, J., and S. A. Twomey, 1967: The Production of Cloud Nuclei by Cane Fires and the Effect on Cloud Droplet Concentration, *J. Atmos. Sci.*, *24*, 704-706.

# A DESERT AEROSOL MODEL FOR RADIATIVE TRANSFER STUDIES

*David R. Longtin, Eric P. Shettle,[1] John R. Hummel and James D. Pryce*
OptiMetrics, Incorporated
Burlington, Massachusetts 01803, USA

**ABSTRACT**

A desert aerosol model that can be used to predict aerosol radiative properties during background and severe dust storm conditions is presented. The model is based upon measurements taken at many of the world's desert and arid locations. The model separates the desert aerosol into its three main compositional components, each with its own size distribution and set of indices of refraction. The model also incorporates mass loading during dust storms.

The radiative properties of the present model are compared with other desert aerosol formulations. The comparisons suggest that the absorption of the solar radiation is less than that calculated by previous models.

## 1. INTRODUCTION

One of the major sources of natural atmospheric aerosols is wind-blown dust and sand. These predominantly originate from arid and semi-arid regions that make up approximately one-third of the earth's land area. The aerosols in the Standard Radiation Atmospheres of the International Association of Meteorology and Atmospheric Physics Radiation Commission do not include an explicit desert aerosol model. A tentative desert aerosol model has been developed by Shettle (1984) for use with the Air Force Geophysics Laboratory (AFGL) model atmospheres. The desert aerosol model presented in this paper is an extension of the Shettle model.

Compositional data (e.g., Levin and Lindberg, 1979, and Pinnick et al., 1987) indicate that the composition of the desert aerosol is a function of particle size. Small particles ($r < 0.5\ \mu m$) tend to be associated with the generally observed background aerosol. Larger particles are primarily made up of crustal material representative of loose and finely grained soils. During dust storm conditions, additional less-absorbing, locally produced dust becomes airborne, which changes the radiative effects of desert aerosols. Therefore, a model for desert aerosols needs to include a different size distribution and set of indices of refraction for each type of particle because they contribute differently to the total radiative impact.

1 AFGL/OPA, Hanscom AFB, MA 01731.



ISBN 0-937194-11-5

Measurements made in the Saharan region by Jaenicke (1984) and Fouquart et al. (1983) suggest that for calm winds, the size distribution of the desert aerosol is similar to that of a remote continental aerosol. Furthermore, local windspeed controls the concentration of large particles and therefore, must be accounted for when developing a desert aerosol model.

## 2. THE DESERT AEROSOL MODEL

### 2.1. OVERVIEW

The present model separates the desert aerosol into three components and uses a different lognormal size distribution and set of indices of refraction for each component. The three components represent carbonaceous particles, water soluble particles, and sand. Mie calculations were performed for each component and then weighted according to their fraction by volume of the total aerosol. The calculations were performed for 61 wavelengths between 0.2 and 40 $\mu$m.

### 2.2. WINDSPEED DEPENDENCE ON AEROSOL LOADING

The mass loading equation of Jaenicke (1984) is the basis for the windspeed dependence

$$c = 52.77 \exp(0.30\, u) \qquad (1)$$

where c is the mass concentration in $\mu g/m^3$ and u is the windspeed in m/s at 10 meters above the ground. This equation is an average based on values in the literature gathered by Jaenicke. It is meant to represent a generic desert aerosol and is only valid within the mixing layer.

The present model assumes that the volume (or mass) of the water-soluble and carbonaceous components remains the same as that for background conditions. Consequently, all extra mass loading due to the wind is from sand. Also, the mode radius and standard deviation of the sand component increase as the windspeed increases. The standard deviation is increased so that the smaller sand particles are retained. Based on Eq. (1), the aerosol fractions by volume of the three components are calculated and the optical parameters for each component weighted according to these values. Table 1 summarizes the dependency of the size distributions on windspeed for the aerosol components, and Figs. 1a and 1b give the number density and surface area distributions, respectively, as a function of windspeed. The magnitudes of the number density distributions in Fig. 1a differ from those reported by d'Almeida (1987) because they have been calculated such that the total mass (or volume) of the aerosol is consistent with Eq. 1. The present model has adopted this approach instead of using the observed size distributions given by d'Almeida because his size distributions do not behave consistently as a function of windspeed; that is, in going from background to wind-blown dust conditions, the number density for the middle mode of the d'Almeida model decreases, which is not realistic.

TABLE 1. WINDSPEED DEPENDENCE ON THE DESERT AEROSOL SIZE DISTRIBUTION PARAMETERS

| AEROSOL COMPONENT | WIND SPEED (m/s) | AEROSOL FRACTION BY VOLUME | SIZE DISTRIBUTION PARAMETERS $r_i$ ($\mu$m) | $\mathrm{Log}(\sigma_1)$ | COMMENTS |
|---|---|---|---|---|---|
| Carbonaceous | 0 | 0.0100* | 0.0118 | 0.301 | Deepak and Gerber (1983b) |
| | 10-30 | small | | | |
| Water Soluble | 0 | 0.2900* | 0.0285 | 0.350 | Deepak and Gerber (1983b) |
| | 10-30 | small | | | |
| Sand | 0 | 0.7000* | 6.24 | 0.277 | Coarse Mode of the Remote Continental Model of Jaenicke (1984) |
| | 10 | 0.9835 | 7.76 | 0.331 | Linear Interpolation of the Values at 0 and 30 m/s |
| | 20 | 0.9992 | 9.28 | 0.384 | |
| | 30 | 0.9996 | 10.80 | 0.438 | Coarse Mode of the Dust Storm Model of Jaenicke (1984) |

*Deepak and Gerber (1983a).

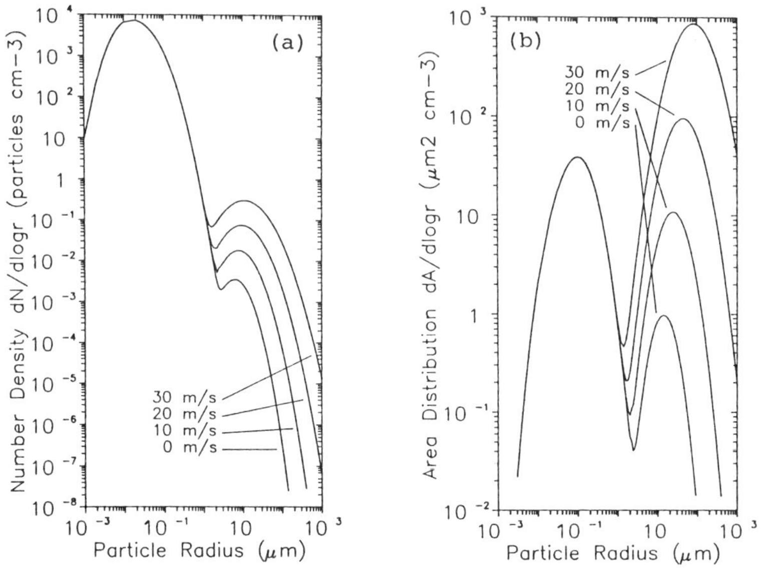


FIGURE 1. (a) Number density and (b) area distributions as a function of particle radius for different windspeeds.

### 2.3 INDICES OF REFRACTION

The indices of refraction for the carbonaceous component are from the work of Shettle and Fenn (1979). These values represent various types of carbon blacks and soots. The water soluble component utilizes measurements made by Toon et al. (1976) for pure crystalline ammonium sulfate. The sand component is assumed to be pure crystalline quartz. The measurements of Gray (1963) are used in the visible and near infrared (IR), those of Drummond (1936) in the region 2.5 to 7.0 μm, and the measurements of Spitzer and Kleinman (1961) for the region 7.0 to 40.0 μm. Both the O-ray and E-ray have been included because quartz is highly anisotropic in the region from about 7 to 30 μm. That is, the optical properties of the sand component are calculated as the sum of two-thirds of the O-ray values plus one-third of the E-ray values since two of the three principal axes of quartz are the same. (Strictly speaking, this approach is only valid for birefringent particles that are small compared with the wavelength. Since our quartz particles approach the geometric optics limit, the approach may not be truly applicable.)

The use of a separate set of optical constants for each mode of the size distribution is the main difference between the present model and those of Shettle (1984) and d'Almeida (1987) who both use only one set of optical constants for all particles making up the aerosol. The distinction between the two approaches is subtle, but important: in the present model, the indices of refraction for each component are based on measurements of pure substances while the Shettle and d'Almeida models use average indices of refraction that attempt to characterize the complex mixture of discrete particles comprising the desert aerosol. Since they have different physical meanings, it is difficult to make a direct comparison between the average indices of refraction used by Shettle and by d'Almeida, which are based on those in the literature, and the ones used in the present model; however, we would expect the present model to give rise to less absorption in the visible regime at high winds since our sand component is transparent at these wavelengths.

The present method of dealing with the index of refraction has been adopted because the average optical constants cited in the literature do not accurately describe the radiative properties of desert aerosols. The problem with using average indices of refraction in radiative transfer calculations is that each particle is treated as though it has the same optical properties, which is clearly not true. In the case of desert aerosols, most of the absorption in the visible is due to the small discrete carbon-based particles from the background global aerosol. In high wind conditions when more of the less-absorbing, local dust and sand are lofted, the relative contributions of the highly absorbing carbon particles to the total radiative impact are lessened. Also, it can be shown through a simple argument that the average values for the imaginary index cited in the literature are too high in the visible (Bohren and Huffman, 1983). That is, as a visible light wave travels through a single desert particle having an imaginary index on the order of 0.001, enough absorption will occur to make the individual particle appear black, which is clearly not the case.

## 3. RESULTS

### 3.1 GENERAL DISCUSSION

The extinction, scattering, and absorption coefficients are shown in Figs. 2a and 2b for 0 m/s (background) and 30 m/s (severe dust storm) windspeed conditions, respectively. The absorption in visible wavelengths changes little as the windspeed changes. This absorption is due to carbonaceous particles and their abundance does not change with windspeed. The absorption peaks near 3, 7, and 9 $\mu$m, for 0 m/s windspeed are primarily due to the strong absorption bands of ammonium sulfate. At windspeeds of 30 m/s, the absorption in the IR is dominated by the sand component. Interestingly, there are minima in the absorption and maxima in the scattering near 8 and 20 $\mu$m, corresponding to the centers of the strong crystal lattice absorption bands of quartz. It is believed that this phenomenon is a result of the sand component acting as a reflector. (This may appear contradictory, but it can be shown that a material in a nonabsorbing medium is highly reflective when its real index is much less than one and its imaginary index is much greater than one; consequently, the absorption drops because the incident wave cannot penetrate the material to be absorbed.) Finally, the extinction is wavelength dependent

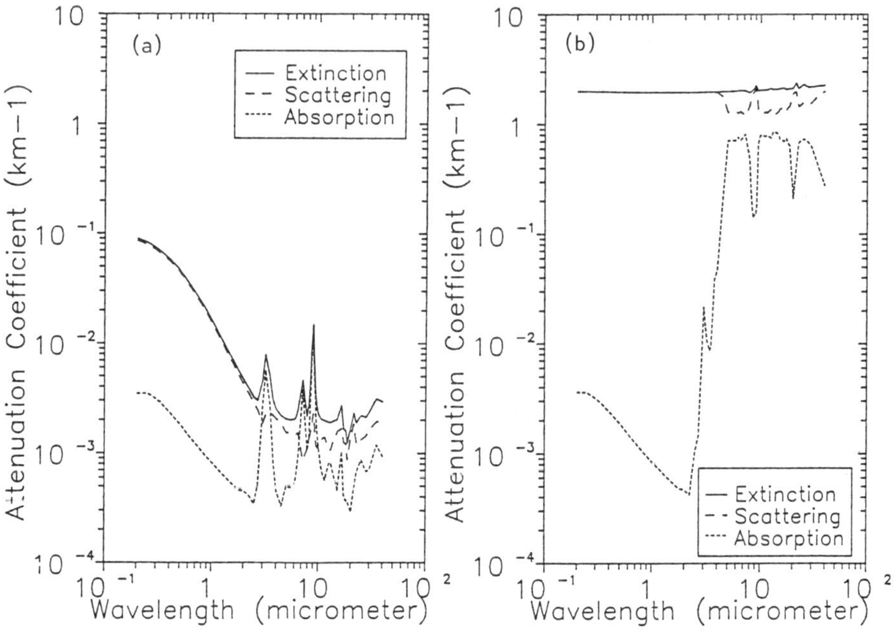


FIGURE 2. Attenuation coefficients versus wavelength for windspeeds of (a) 0 m/s and (b) 30 m/s.

for winds of 0 m/s, but nearly constant at 30 m/s. These differences are driven by the relative contributions of the three aerosol components. For winds of 0 m/s, only the smaller carbonaceous and water soluble particles dominate the extinction at visible and near-IR wavelengths and, therefore, a wavelength dependence exists. On the other hand, in high windspeed conditions, the extinction is dominated by the much larger sand particles that approach the geometric optics regime.

### 3.2 COMPARISONS WITH OTHER RESEARCHERS

Figures 3a and 3b compare the extinction coefficients against those of Shettle (1984) and d'Almeida (1987), respectively. The present model gives similar values as those of Shettle, as expected, because both models use similar size distributions. Figure 3b suggests that the values of d'Almeida for background conditions are much greater than ours. The differences appear to result from d'Almeida defining his background conditions in terms of "clear" visibility and not windspeed, and some mass loading due to the wind may be occurring. The extinction calculated by d'Almeida for sandstorm conditions show strong variations with wavelength while ours do not. The structure beyond about 1 $\mu$m arises because much of the mass loading in the d'Almeida model is from particles that are smaller than ours and not in the geometric optics limit.

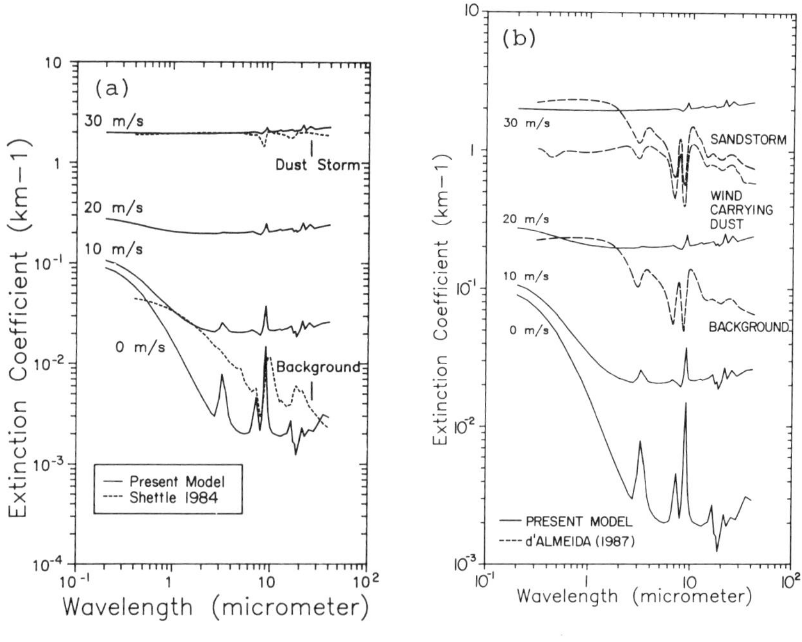


FIGURE 3. Comparison of extinction coefficients against (a) Shettle (1984) and (b) d'Almeida (1987).

In order to compare the three models with observed data, the extinction at 0.55 μm can be converted to visual ranges using the formula, visual range = 3.912/extinction at 0.55 μm. Both the present and Shettle background models correspond to visual ranges of about 80 km while the d'Almeida background model yields a value close to 20 km. Measurements of visual range at White Sands, New Mexico (Pitchford et al., 1981) and Big Bend, Texas (Malm et al., 1981) indicate that visibilities of 80 km occur frequently in desert and arid regions, which adds credibility to the present model, and supports our suggestion that d'Almeida's "background" model is representative of moderate winds and not calm conditions.

Figures 4a and 4b compare the values of single scattering albedo for background and dust storm conditions, respectively, against those of Shettle and d'Almeida. At visible and near-IR wavelengths, Fig. 4a suggests that the values for 0 m/s windspeed conditions are similar to those for background conditions of the Shettle model, but greater than those for the d'Almeida model. Perhaps the most striking differences between the present model and Shettle and d'Almeida models are the values for the single scattering albedo in dust storm conditions (Fig. 4b). Our values approach 1.0 at visible and near-IR wavelengths for 30 m/s wind conditions whereas the values for the two other models are actually less than their respective background values. The differences are directly related to the way the present model deals with the index of refraction. The comparisons of the asymmetry parameters are given in Figs. 5a and 5b. Interestingly, there is little agreement between the three models for background conditions, especially at visible and near-IR wavelengths. In dust storm conditions, values for the asymmetry parameter are similar for the three models.

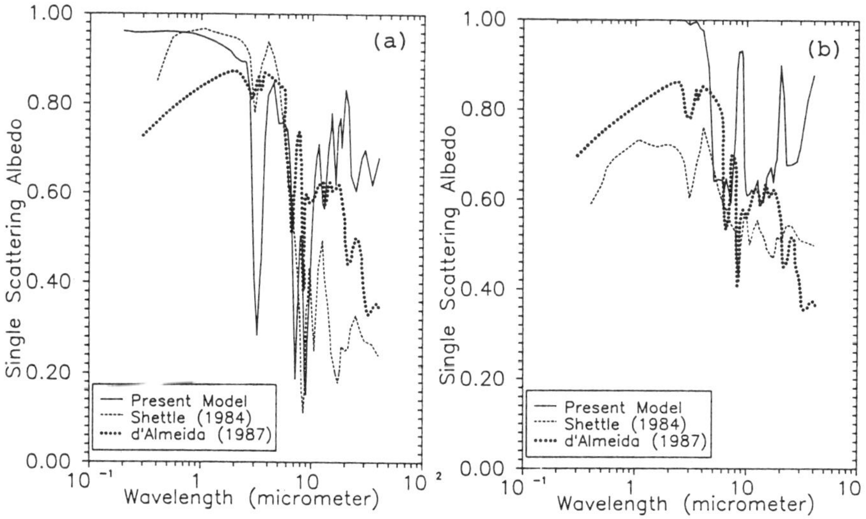


FIGURE 4. Comparison of single scattering albedos against those of Shettle (1984) and d'Almeida (1987) for (a) background and (b) dust storm conditions.

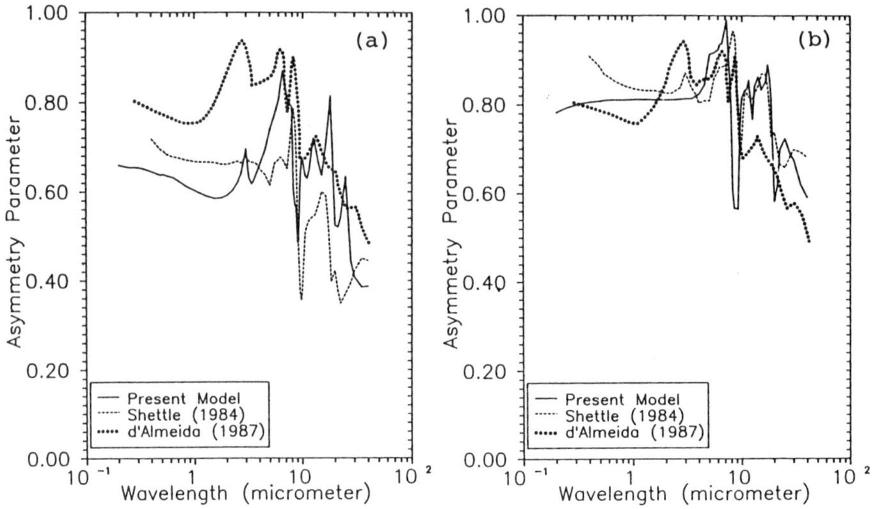


FIGURE 5. Comparison of asymmetry parameters against those of Shettle (1984) and d'Almeida (1987) for (a) background and (b) dust storm conditions.

## 4. SUMMARY

A desert aerosol model has been presented that treats the desert aerosol as a three-component mixture. Each component has its own size distribution and set of indices of refraction. Mass loading during dust storms has been incorporated. The absorption of solar radiation is less than that calculated by previous desert aerosol models although this result would be modified by the inclusion of iron oxides into the sand component.

## REFERENCES


Bohren, C., and D. Huffman, 1983: Absorption and Scattering of Light by Small Particles, Wiley-Interscience, New York, 279 pp.

d'Almeida, G. A., 1987: On the Variability of Desert Aerosol Radiative Characteristics, J. Geophys. Res., 92, 3017-3026.

Deepak, A., and H. E. Gerber (Eds.), 1983: Report of the Experts Meeting on Aerosols and Their Climatic Effects (WCP-55), World Meteorological Organization, Geneva, Switzerland, (a) 83; (b) 31.

Drummond, D.G., 1936: Absorption Coefficients of Crystal Quartz in the Infrared, Proc. Roy. Soc. (London)-Series A, 153, 328-338.

Fouquart, Y., B. Bonnell, A. Cerf, M. Chaoui, L. Smith, and J.C. Vanhoutte, 1983: Size distribution and optical properties of Saharan aerosols during ECLATS. In H. E. Gerber and A. Deepak (Eds.), Aerosols and Their Climatic Effects, A. Deepak Publishing, Hampton, Virginia, 35-62.

Gray, D. C., 1963: American Institute of Physics Handbook, 2nd Edition, McGraw-Hill, New York, New York, 6-97.

Jaenicke, R., 1984: Aerosol Physics and Chemistry, Meteorology Volume, Landolt-Boernstein Numerical Data and Functional Relationships in Science and Technology, to be published by Springer-Verlag.

Levin, Z., and J. D. Lindberg, 1979: Size Distribution, Chemical Composition, and Optical Properties of Urban and Desert Aerosols in Israel, J. Geophys. Res., 84, 6941-6950.

Malm, W. C., E. G. Walther, K. O'Dell and M. Kleine, 1981: Visibility in the Southwestern United States from Summer 1978 to Spring 1979, Atmos. Environ., 15, 2031-2042.

Pinnick, R. G., S. G. Jennings and G. Fernandez, 1987: Volatility of Aerosols in the Arid Southwestern United States, J. Atmos. Sci., 44, 562-576.

Pitchford, A., M. Pitchford, W. Malm, R. Flocchini, T. Cahill and E. Walther, 1981: Regional Analysis of Factors Affecting Visual Air Quality, Atmos. Environ., 15, 2043-2054.

Shettle, E. P., and R. W. Fenn, 1979: Models for the Aerosols of the Lower Atmosphere and the Effects of Humidity Variations on Their Optical Properties, AFGL-TR-79-0214, 20 Sept 1979 (available from NTIS AD A0859).

Shettle, E. P., 1984: Optical and radiative properties of a desert aerosol model. In G. Fiocco (Ed.), IRS '84: Current Problems in Atmospheric Radiation, A. Deepak Publishing, Hampton, VA, 74-77.

Spitzer, W. G., and D. A. Kleinman, 1961: Infrared Lattice Bands of Quartz, Phys. Rev., 121, 1324-1335.

Toon, O. B., J. B. Pollack and B. N. Khare, 1976: The Optical Con stants of Several Atmospheric Aerosol Species: Ammonium Sulfate, Aluminum Oxide, and Sodium Chloride, J. Geophys. Res., 81, 5733-5748.

# RADIATION STUDIES IN THE ARCTIC

*Francisco P.J. Valero, Thomas P. Ackerman, Warren J.Y. Gore and Melinda L. Weil*
NASA Ames Research Center
Moffett Field, California 94035, USA

## ABSTRACT

The interaction of the arctic winter aerosol (arctic haze) with solar radiation produces changes in the radiation field that result in the enhancement of scattering and absorption processes, which alter the energy balance and solar energy distribution in the arctic atmosphere-surface system. During the second Arctic Gas and Aerosols Sampling Project (AGASP II) field experiment in 1986, we measured radiation parameters using the NOAA WP-3 research aircraft as a platform. State of the art instrumentation was employed to measure in situ the absorption of solar radiation by the arctic atmosphere during severe haze events. Simultaneously with the absorption measurements, we determined optical depths and total-diffuse radiation fields. The experimentally determined parameters were used to define an aerosol model that was employed to integrate the absorption by the aerosols over the full solar spectrum and also to calculate atmospheric heating rate profiles. Changes in the heating rate profiles ranged up to about 0.15 K/day. Besides, the increased absorption (30% to 40%) and scattering of radiation by the atmosphere produces a reduction in the absorption of solar energy by the surface amounting to values around 6% to 10% and in a reduction of 3% to 6% in the effective planetary albedo over ice surfaces.

## 1. INTRODUCTION

The pollution of the Arctic by anthropogenic aerosols and gases has recently become the subject of great scientific and geopolitical interest. Mitchel (1957) noted the arctic haze for the first time; Shaw (1975) confirmed that the atmosphere became very turbid during haze events, although the origin of the haze was still uncertain at that time. More recently, additional studies (Rahn and McCaffrey 1980; Shaw and Stamnes 1980; Barrie et al. 1981; Heintzenberg et al. 1982, etc.), encouraged the organization of several international research projects, notably the Arctic Gas and Aerosols Sampling Projects in 1983 (AGASP I) and 1986 (AGASP II). The results of the AGASP I research were mostly published in Geophys. Res. Letts. 1984) and in Arctic Air Pollution (1986) and indicate that large concentrations of aerosols were found in the arctic atmosphere. Results of these studies suggest that their impact on the ecology may be significant.

Characteristically, the haze increases in density from November to April, with a maximum between late winter and early spring. Rahn and McCaffrey (1980) conclude that the winter arctic pollution originates in industrialized regions of Europe, the United Kingdom, and the Soviet Union, with minor contributions from North America.

The interaction of the haze with solar and infrared radiation and the associated effects on the atmosphere-surface system have the potential to be climatically significant because they affect the radiative energy balance between the different components of the system. The Arctic is thought to be particularly sensitive to such changes.

In this paper we report measurements of the absorption of solar radiation and optical depths obtained during AGASP II in April 1986. These measurements were made using


A. DEEPAK Publishing
ISBN 0-937194-11-5

techniques similar to the ones used by us during AGASP I. At that time, we measured in situ the absorption of solar radiation in several narrow-spectral bandpasses as well as the integrated broadband solar spectrum. During AGASP II a multichannel total-direct-diffuse radiometer was added. With this addition we were able to determine optical depths simultaneously with the radiation absorption measurements.

## 2. INSTRUMENTATION

To carry out the measurements required to determine heating rate profiles we used specially designed radiometric systems that were installed in the NOAA WP-3D research aircraft. The instrumentation includes four hemispherical field-of-view narrow-spectral-bandpass (0.01 μm) solar radiometers, one broad bandpass (0.25 to 3.0 μm) solar flux radiometer, and a five-channel total-direct-diffuse radiometer that was used to determine optical depths (Valero et al.,1982, 1983, 1984; Ackerman and Valero,1984). The spectral bandpasses for four of the five channels of the total-direct-diffuse radiometer were identical to the bandpasses of the four narrow-spectral-bandpass radiometers. The fifth channel had a bandpass centered at 380 nm. Detailed instrument descriptions will be published elsewhere.

The narrow bandpass and broadband radiometers were mounted in the aircraft's wing tip and the total-direct-diffuse radiometer in a zenith-looking window in the WP-3D fuselage. All systems were computer controlled by a central system that also collected navigational data together with radiometer measurements and housekeeping information.

## 3. EXPERIMENTAL

In order to measure the absorption of radiation by different atmospheric layers, we measured upwelling and downwelling radiative flux densities at several altitudes and over the same geographical area to ensure constant ground conditions. By differencing the net fluxes at different altitudes we determined the absorption of radiation by the atmospheric layers defined by the flight altitudes, which were chosen to include the various layers of haze with the highest measuring level selected to be above the highest haze layer. The technique employed to measure net radiative fluxes is described by Valero et al., 1982.

Since the important quantities that we intended to measure were the absorption of radiation and optical depth associated exclusively with the haze, we selected the spectral bandpasses to avoid absorption by gaseous components of the atmosphere. The wavelengths chosen for both optical depth and absorption measurements were 412, 500, 675 and 778 nm with a bandwidth of ±5 nm. Using the measured absorption and optical depths at these discrete wavelengths the absorption by the aerosols, integrated over the full solar spectrum, can be calculated. From the total amount of energy absorbed by the aerosols, we can determine the contribution to the atmospheric heating rate profiles, planetary albedo, and surface absorption parameters that can then be compared to their values under aerosol-free conditions.

## 4. RESULTS

Figures 1 to 4 and Table 1 show the values measured for absorption and optical depths during flights 201 and 203. To deduce heating rate profiles from the measured values it was required that model simulations of the measurements be conducted. This was done in three steps. First, the measurements of aerosol optical depth as a function of wavelength were used to infer an aerosol size distribution. Second, the measured optical depth and absorption per layer in the 500 nm channel was used to constrain the fraction of absorbing aerosol in each layer. Calculations were also carried out at the other radiometer wavelengths

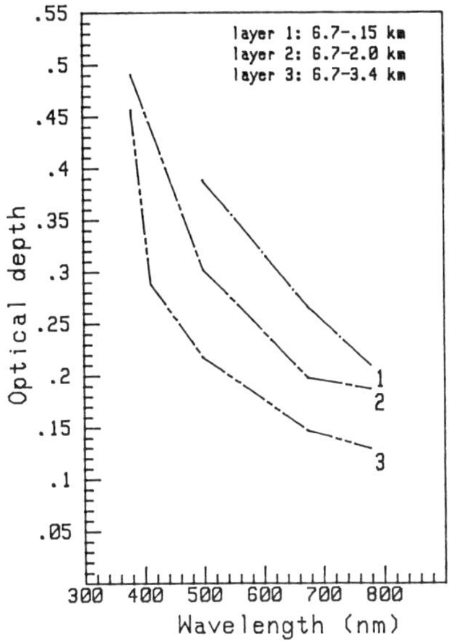


FIGURE 1. Averaged aerosol optical depth vs. wavelength. Flight 201.

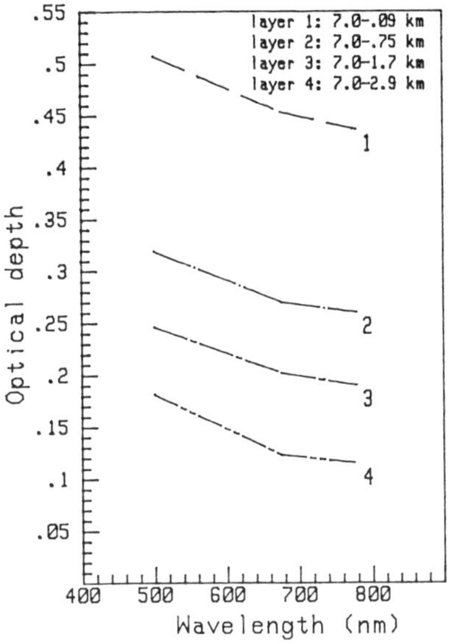


FIGURE 2. Averaged aerosol optical depth vs. wavelength. Flight 203.

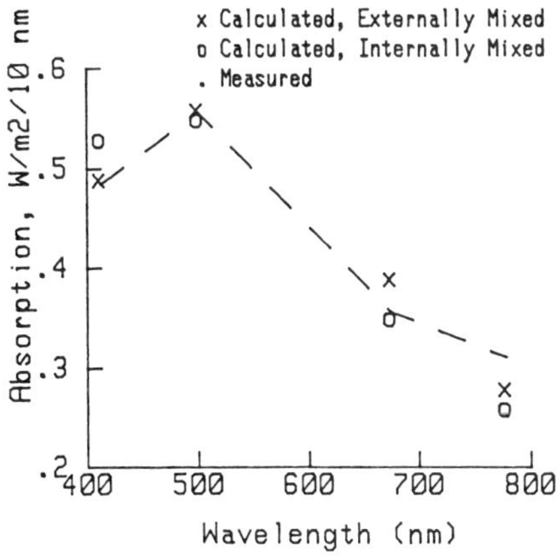


FIGURE 3. Measured and calculated values for absorption. Flight 201.

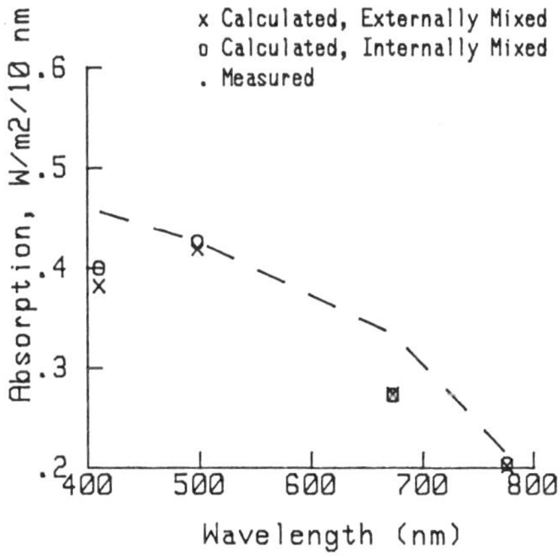


FIGURE 4. Measured and calculated values for absorption. Flight 203.

TABLE 1. OPTICAL DEPTHS AND ENERGY ABSORBED

| AGASP Flight | Channel $\Delta\lambda$=10nm $\lambda_o$(nm) | Aerosol Optical Depth $\tau_a$ | Energy Absorbed ($wm^{-2}$) Measured | Calculated Internal Mix | Calculated External Mix |
|---|---|---|---|---|---|
| | 412 | | 0.483 | 0.533 | 0.491 |
| | 500 | 0.388 | 0.555 | 0.551 | 0.555 |
| 201 | 675 | 0.266 | 0.356 | 0.349 | 0.275 |
| | 778 | 0.210 | 0.311 | 0.255 | 0.280 |
| | 412 | | 0.456 | 0.402 | 0.384 |
| | 500 | 0.507 | 0.425 | 0.428 | 0.421 |
| 203 | 675 | 0.453 | 0.333 | 0.275 | 0.276 |
| | 778 | 0.436 | 0.211 | 0.206 | 0.204 |

(412, 675, 778 nm) in order to verify the estimated absorbing fraction. Finally, using the inferred size distribution and absorbing fraction, calculations of the total solar absorption by the haze and the associated atmospheric heating rates were made using a multilayer version of the δ4-stream model described by Cuzzi et al. (1982). Calculations were made for both internal and external mixtures of the aerosol components. The results are shown in columns 5 and 6 of Table 1. The diurnally averaged heating rate profiles deduced using the values for the absorbed energy, are shown in Figs. 5 and 6. Broken traces correspond to the haze-free (clear) atmosphere; continuous traces represent the result of adding the haze to the clear atmosphere.

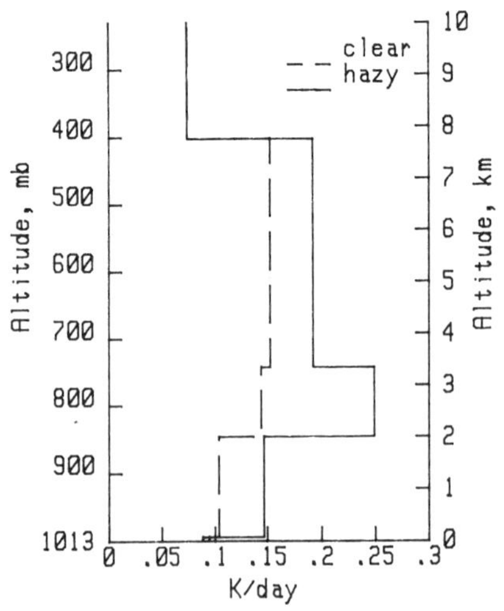


FIGURE 5. Diurnally averaged heating rate. Flight 201 over ice at 75N 156W.

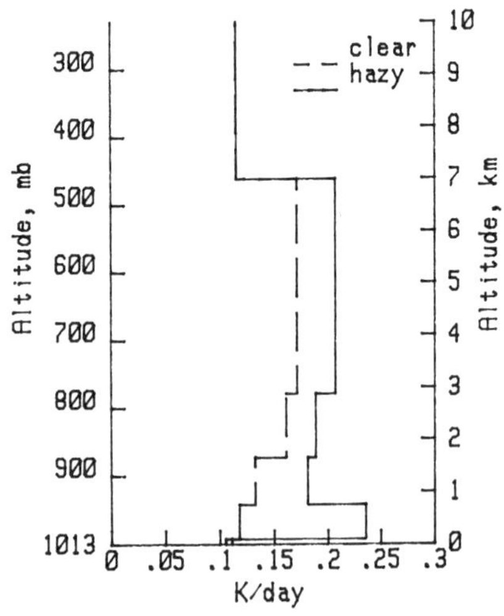


FIGURE 6. Diurnally averaged heating rate. Flight 203 over ice at 71N 156W.

An additional and important result is that aerosols change the effective planetary albedo, which modify (through absorption and scattering of radiation) the amount of energy that penetrates to the lower layers of the atmosphere and the surface. For the AGASP II experiment, our analysis indicates that together with the heating rate changes described above, the following variations in the distribution of solar energy occur:

1) Increase of the total energy absorption by the atmosphere, in the 0.1 to 7 km layer, by about 30% to 40%.
2) Decrease of the amount of energy absorbed by the ice surface by 6% to 10%.
3) Increase of the total absorption of solar energy by the atmosphere-solar system by 10% to 15%.
4) Reduction of the effective planetary albedo by 3% to 6%.

The above results refer to AGASP flights 201 and 203 that took place over ice surfaces and are qualitatively similar to the results obtained during the AGASP I experiment also for ice surfaces.

## 5. DISCUSSION

The above measurements and calculations indicate that changes in climatologically important parameters (i.e., heating rate profiles, planetary albedo) take place as a consequence of the interaction of solar radiation with the haze. It is not clear,however, if the magnitude of the changes is large enough to originate any significant climatic and ecological effects. To be able to give an answer to this question we need to further our understanding of the energetic balance in the arctic basin as well as any modifications that may result from the cumulative effects of the haze on the surface, including optical properties, and the interaction of radiation with the modified ice surface.

A comparison of the results reported here with those obtained during AGASP I show some notable differences. The 1983 experiment indicates higher solar energy absorption by the aerosol and lower optical depths than those measured in 1986. Assuming constant aerosol composition, we would normally associate higher absorption optical depths with higher total extinction optical depths. However, the comparison of the AGASP I and II results suggests that the absorption optical depth was higher in 1983 while the scattering optical depth was higher in 1986. This may be the result of the presence of a nonabsorbing but scattering aerosol during the 1986 deployment.

No infrared measurements have yet been performed and, although the haze effects in the infrared are not expected to be large, they occur over a long period of time and may result in contributions to the overall radiative balance.

## REFERENCES

Ackerman, T.P., and F.P.J. Valero, 1984: The Vertical Structure of Arctic Haze as Determined from Airborne Net-Flux Radiometer Measurements, Geophys. Res. Letts., 11, 469-472.

Barrie, L.A., R.M. Hoff and S.M. Daggupaty, 1981: The Influence of Mid-Latitudinal Pollution Sources on the Haze in the Canadian Arctic, Atmos. Environ., 15, 1407-1419.

Cuzzi, J.N., T.P. Ackerman and L.C. Hemle, 1982: The Delta-Four-Stream Approximation for Radiative Flux Transfer, J. Atmos. Sci., 39, 917-925.

Geophys. Res. Letts., 11, 359-472.

Heintzenberg, J., 1982: Size-Segregated Measurements of Particulate Elemental Carbon and Aerosol Light Absorption at Remote Arctic Locations, Atmos. Environ., 16, 2461-2469.

Mitchel, J.M., Jr., 1957: Visual Range in the Polar Regions with Particular Reference to the Alaskan Arctic, Jour. Atmos. and Terrest. Phys. Special Supplement, Pt. I, 195-211.

Rahn, K. and R.J. McCaffrey, 1980: On the Origin and Transport of the Winter Arctic Aerosol, Ann. N.Y. Acad. Sci., 338.

Shaw, G.E., 1975: The Vertical Distribution of Atmospheric Aerosols at Barrow, Alaska, Tellus, 27(1), 39-50.

Shaw, G.E., and K. Stamnes, 1980: Arctic Haze: Perturbation of the Polar Radiation Budget, Ann. N.Y. Acad. Sci., 338, 533-539.

Stonehouse, B. (Ed.), 1986: Arctic Air Pollution, Cambridge University Press, Great Britain.

Valero, F.P.J., W.J.Y. Gore and L.P. Giver, 1982: Radiative Flux Measurements in the Troposphere, App. Opt., 21, 831-838.

Valero, F.P.J., T.P. Ackerman and W.J.Y. Gore, 1983: Radiative Effects of the Arctic Haze, Geophys. Res. Letts., 10, 1184-1187.

Valero, F.P.J., T.P. Ackerman and W.J.Y. Gore, 1984: The Absorption of Solar Radiation by the Arctic Atmosphere During the Haze Season and its Effects on the Radiation Balance, Geophys. Res. Letts., 11, 465-468.

# CHARACTERISTICS OF AEROSOLS IN THE LOWER STRATOSPHERE

*David C. Woods*
NASA Langley Research Center
Hampton, Virginia 23665, USA

and

*Raymond L. Chuan*
Brunswick Corporation
Costa Mesa, California 92692, USA

ABSTRACT

In situ sampling of aerosols in the stratosphere in recent years, starting at the end of a quiet period in 1979 and extending through 1985, including a volcanically active period between 1980 and 1983 and another relatively quiet period from 1983 to 1985, has revealed some interesting features of the aerosol. The particles sampled ranged in size from less than 0.05 μm to greater than 2.0 μm aerodynamic diameter. Hydrated sulfuric acid droplets were dominant in the narrow size band between about 0.2 μm and 0.8 μm. When the sulfuric acid mass, as measured by a quartz crystal microbalance cascade impactor (QCM), was converted to sulfate mixing ratios they "compared favorably" with results obtained by filter measurements. A large fraction of the samples collected contained an abundance of carbonaceous material and some iron oxide that may have climatic implications. The source of carbonaceous particles has not been determined, but it is suspected that they are meteoritic.

## 1. INTRODUCTION

Integrated backscatter data from lidar records (McCormick et al., 1984) and aerosol concentrations from balloon-borne measurements (Hofmann and Rosen, 1983) suggest that the aerosol concentration in the stratosphere was at a minimum in 1979 after several years of relatively low volcanic activity. The series of eruptions between late 1979 and 1982, including Sierra Negra (1979), Mount St. Helens (1980), Alaid (1981), and El Chichon (1982), contributed to a substantial increase in aerosols that caused more than an order of magnitude increase in integrated backscatter in the stratosphere. This increase was followed by a decay in aerosol loading starting about 1983 and continuing to 1985. In this report, we examine aerosol data collected in 1979 when the loading was at a minimum or near "background level," in 1980 and 1982 following the Mount St. Helens and El Chichon eruptions, and in 1985 well after the aerosol had started to decay towards background level. The year 1979 represents a case in which the stratosphere is likely to have been free of volcanic influences, whereas the stratosphere was influenced greatly by volcanic emissions between


A. DEEPAK Publishing
ISBN 0-937194-11-5

1980 and 1982. Thus, there is an opportunity to compare the aerosols under background conditions with aerosols under volcanically perturbed conditions and perhaps assess possible influences of various kinds of aerosols on climate.

## 2. SAMPLING PROCEDURE

The aerosol particles were collected and sampled in the stratosphere by a multistage cascade impactor aboard a NASA U-2 aircraft (Woods and Chuan, 1983). The impactor, referred to as a quartz crystal microbalance (QCM) (Chuan, 1975), classifies aerosol particles by aerodynamic size by means of inertial impaction and senses mass by frequency changes in piezoelectric crystals. The resulting data may be presented in the form of mass loading versus particle size. The size segregated samples were retained for post-flight elemental analyses by scanning electron microscopy (SEM) and x-ray energy spectroscopy (XES). Aerosols suspected of containing elements with atomic numbers less than 11 (carbon for example), were analyzed by Auger spectroscopy to identify the low atomic number elements. The sampling and analysis tehniques are more fully described by Chuan and Woods (1984a).

A number of sampling flights aimed at characterizing the stratospheric aerosol and determining possible climate effects were conducted between 1979 and 1985 aboard a NASA U-2 research aircraft. These flights varied in location but were confined to the Northern Hemisphere. In July 1979 and August 1985, flights were conducted near Fairbanks, Alaska (≈68°N, 148°W) in conjunction with satellite validation experiments (Russell et al., 1981). After the Mount St. Helens eruption in 1980, flights were designed primarily for measuring eruption material and were conducted along the western U.S.A. coast from northern California to Washington (Chuan et al., 1981). After the El Chichon eruption in southern Mexico in 1982, flights were again focused on eruption material and were conducted mostly along the California coast between Moffett Field (37°N, 122°W) and the tip of Baja California (23°N, 111°W). Several flights were conducted between Moffett Field and Laramie, Wyoming (41°N, 106°W). Additional latitudinal flights were conducted aboard a NASA WB-57 aircraft in 1981 as a part of Project Airstream, a Department of Energy (DOE) high altitude sampling program.

## 3. RESULTS

Selected data from flights covering the entire time span (1979-1985) are presented in the form of aerosol mass loading as a function of latitude in Fig. 1. Each individual plot represents a flight segment at a fixed altitude and is identified by a date and altitude. For flights that were generally localized and did not cover a long latitudinal path, the mass loading is indicated as a single point at the approximate latitude. The open and closed stars near 40°N for the May 1980 Mount St. Helens flights and the large and small asterisks for the measurements near Fairbanks, Alaska represent such cases.

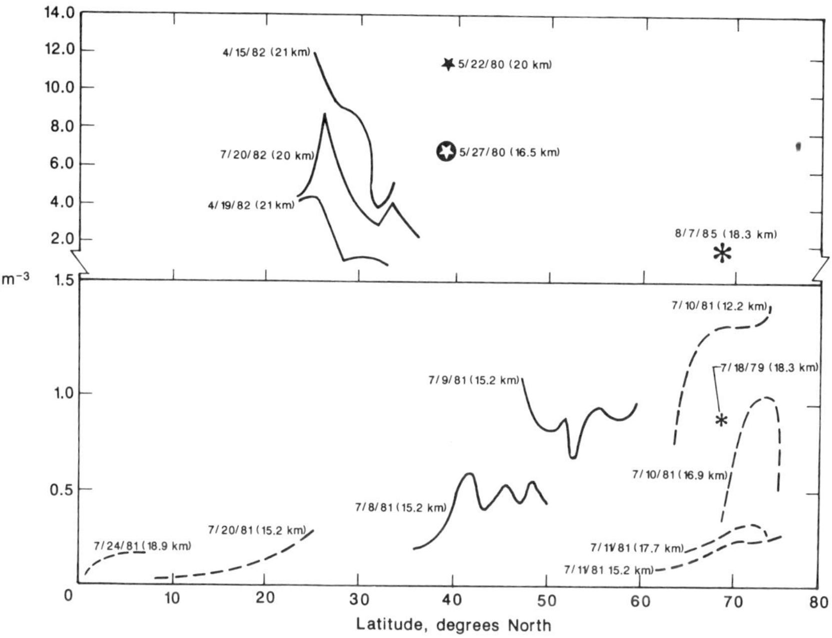


FIGURE 1. Aerosol mass loading vs latitude measured by a QCM during aircraft sampling flights conducted between 1979 and 1985. Solid lines represent measurements from the NASA U-2 aircraft and the broken lines represent measurements from the NASA WB-57 aircraft.

The data show a wide variability of mass loadings both in time and location. On 22 and 27 May 1980, the aircraft flew directly into the stratospheric cloud formed from the Mount St. Helens eruptions, which accounts for the very high mass loadings. Except for the May 1980 measurements, all of the measurements prior to the El Chichon eruption in April 1982 show mass loadings on the order of 1 μg $m^{-3}$ or less. Other measurements following the eruption were made in or near the eruption clouds and were on the order of several μg $m^{-3}$, indicating the vast amount of material injected into the stratosphere by the eruptions (Chuan and Woods, 1984b). The estimated uncertainty in the measured mass loading is 0.2 μg $m^{-3}$.

The aerosol mass loading as a function of altitude on 18 July 1979, and 7 and 9 August 1985, are shown in Fig. 2. Each data point

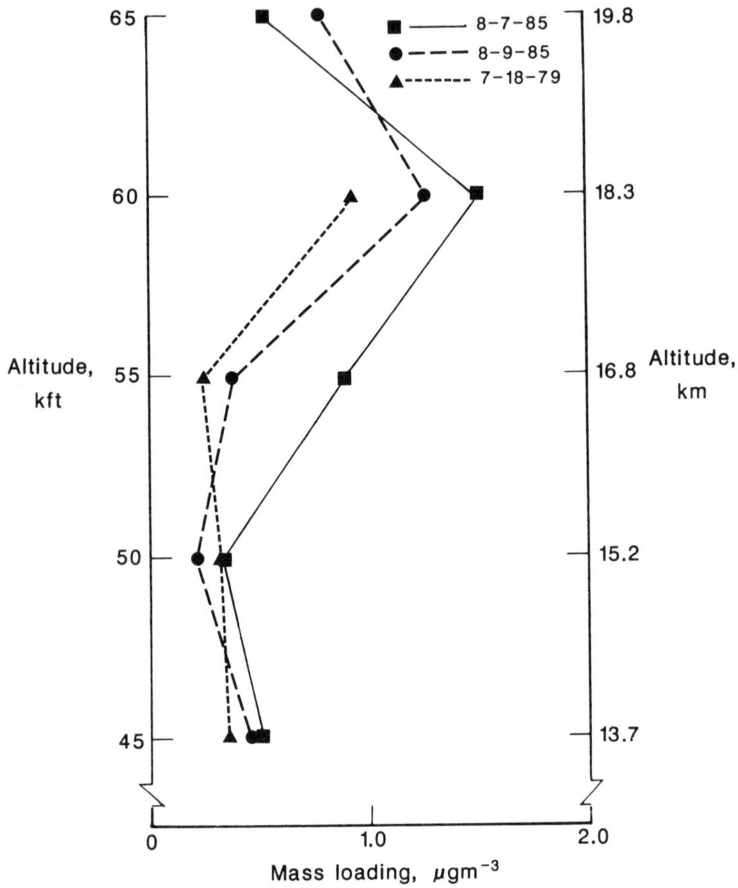


FIGURE 2. Vertical profiles of aerosol mass loadings measured near Fairbanks, Alaska on 18 July 1979, and 7 and 9 August 1985. Each data point represents an average over a 20-minute flight segment at a constant altitude.

represents an average integration time of 20 minutes at a fixed altitude. The peak concentration near 18.3 km represents the stratospheric aerosol layer. A NASA airborne uplooking lidar system that was operating at the same time from a lower altitude also indicated a peak aerosol concentration at about the same altitude. At the lower altitudes, 13.7 km just above the tropopause, and 15.2 km, the mass loadings for the three data sets are very nearly the same, but at higher altitudes, moving closer to the peak layer, the differences are most pronounced. The lower mass loading in 1979 is a reflection of the period following low volcanic activity, while in 1985 the aerosol loading was still decaying from the series of eruptions since 1979.

In Figs. 3a and 3b, the aerosol size distributions (mass loadings as a function of diameter) are plotted for 18 July 1979 and 9 August

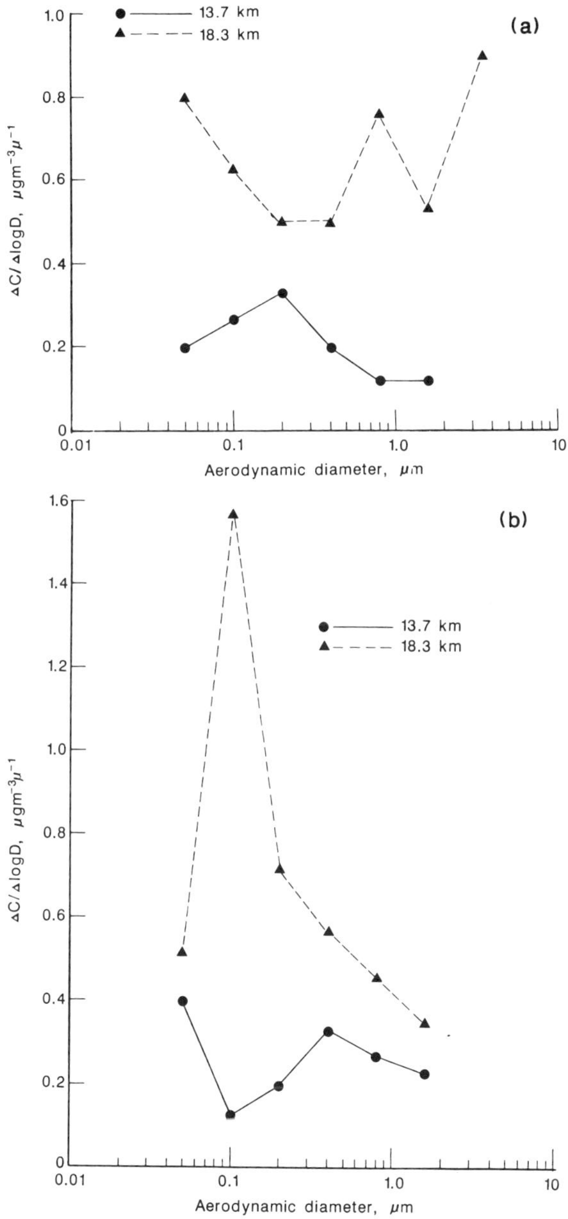


FIGURE 3. Aerosol particle size distributions (mass loading vs particle aerodynamic diameter) at altitudes of 13.7 and 18.3 km near Fairbanks, Alaska on 18 July 1979(a) and 9 August 1985(b).

1985, respectively. In each case, size distributions are presented for altitudes of 13.7 km just above the tropopause but clearly out of the peak aerosol layer, and at 18.3 km near the center of the peak aerosol layer. At 13.7 km, the size distributions on both days have no special features but the mass loading in each impactor stage is very small, approaching the resolution of the instrument. At 18.3 km in 1979 the distribution is irregular with notable amounts of mass in the size range larger than 1 μm. The distinctive feature to be noted is the very sharp peak near 0.1 μm diameter at the 18.3 km altitude level on 9 August 1985, indicating a high concentration of material collected in impactor stage 9.

A scanning electron photomicrograph of the particles collected in stage 9 is shown in Fig. 4. They are very similar in appearance to particles in the same size range collected in 1979 at 13 km and 18 km (Chuan and Woods, 1984a). X-ray energy spectroscopy and Auger spectroscopy revealed that these particles consisted of iron and carbon. We have not determined the source of these particles, but we believe it is possible that they are meteoritic (Chuan and Woods, 1984a). Cadle (1972), Rosen (1969), and Shedlovsky and Paisly (1966) have suggested the possibility of extra-terrestrial particles in the stratosphere. Brownlee (1978) has collected particles in the stratosphere and confirmed their origin to be extra-terrestrial. Our efforts have not been directly aimed at studying meteoritic particles but rather at the climatic effects of aerosols (Woods and Chuan, 1983) and in some cases validation of satellite measurements (Russell et al., 1981). The presence of carbonaceous and iron particles in the stratosphere has interesting implications relating to possible climatic effects, since many forms of carbon and iron are highly absorbent optically (Rosen et al., 1978).

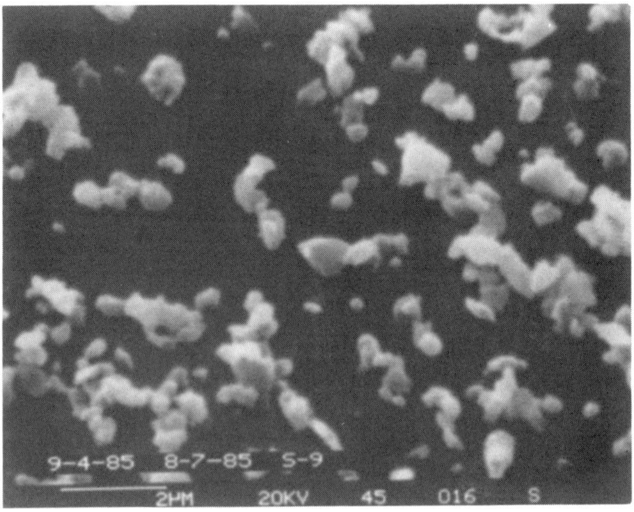


FIGURE 4. Scanning electron micrograph of particles collected on QCM impactor stage 9 on 7 August 1985, near Fairbanks, Alaska. XES indicated the presence of iron in the particle and Auger spectroscopy indicated large amounts of carbon.

As indicated earlier, the aerosol mass loadings measured within the eruption clouds were at least several times higher than those measured outside the clouds. The size distributions and relative amounts of various materials within the clouds differ also. In the case of Mount St. Helens (stars in Fig. 1) the size distributions were very narrow and sharp with peaks near 0.2 μm (27 May 1980) and 0.4 μm (22 May 1980), as shown in Fig. 5. In both cases, the material in each peak consisted mainly of liquid $H_2SO_4$ droplets (Chuan et al., 1981). This represents a considerable increase in $H_2SO_4$ compared to the background.

The size distributions in the El Chichon cloud were bimodal and much broader than those in the Mount St. Helens cloud. Within the first few months after the eruptions, a large percent of the aerosol mass consisted of very large magmatic particles extending to larger than 20 μm (Woods and Chuan, 1983). However, there was a submicron mode ranging between about 0.2 and 0.8 μm that consisted primarily of liquid $H_2SO_4$ corresponding to the sulfate mode in the Mount St. Helens

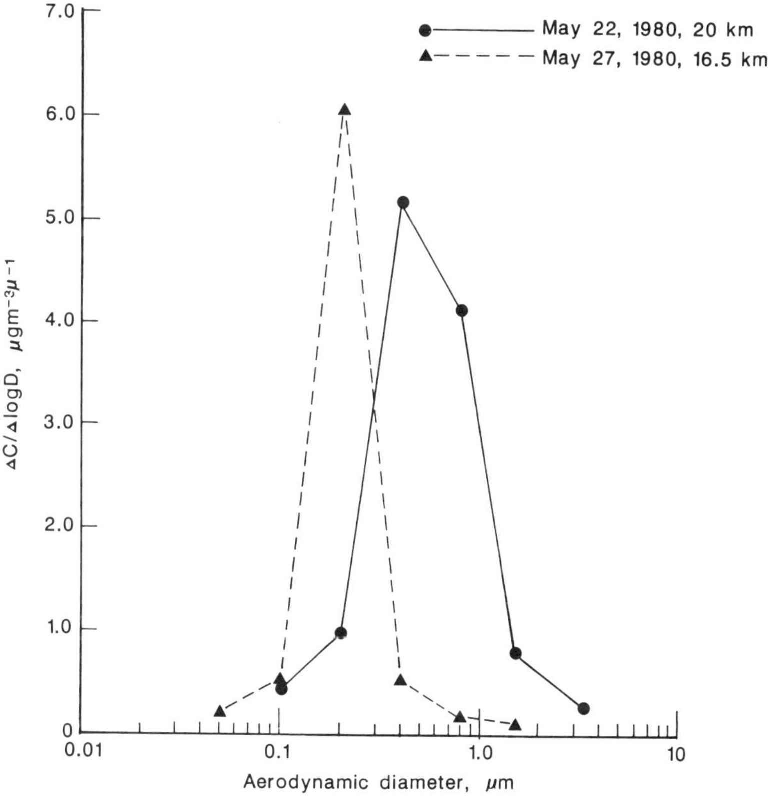


FIGURE 5. Aerosol particle size distributions measured in the Mount St. Helens eruption clouds 22 and 27 May 1980.

cloud. These sulfate particles are preferentially collected in impactor stages 6 through 9 indicating aerodynamic diameters in the range of 0.1 to 0.8 μm. In most cases they are most abundant in stages 7 and 8 corresponding to 0.4 μm and 0.2 μm.

Since the sulfate particles account for nearly all of the mass over the narrow size band in eruption clouds (usually 0.2 to 0.8 μm), which can be determined from Scanning Electron Microscope (SEM) photographs, we suggest that it may be possible to estimate sulfate mixing ratios in volcanic eruption clouds from QCM data. We converted QCM-derived sulfate mass to sulfate mixing ratios and compared them with filter measurements obtained on the same flight in the El Chichon cloud on 5 May 1982, (Gandrud et al., 1983). This comparison is shown in Fig. 6. Data are compared over fixed altitude segments along the flight path between 23° and 34° latitude, north. The trends are similar for the two measurement techniques and they agree to better than 3 parts per billion, with the exception of the one usually high filter measurement at 20.4 km altitude. These results should encourage future efforts to make comparisons of this kind.

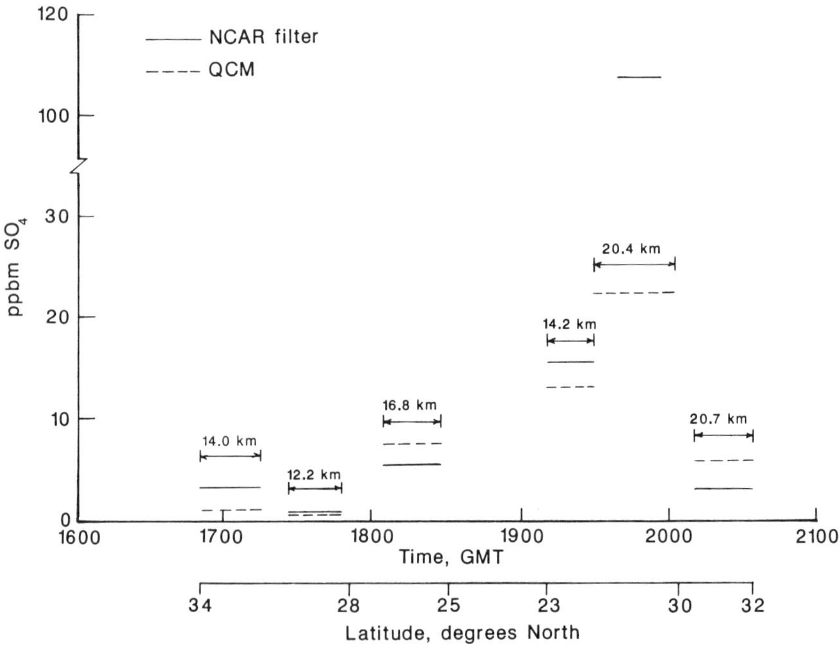


FIGURE 6. Comparison of sulfate mixing ratios obtained on 5 May 1982, by converting $H_2SO_4$ mass measured by the QCM to mass mixing ratio with measurements obtained by filters. Line segments span the latitudinal range over which the measurements were made. The distances are indicated over each segment.

## 4. CONCLUSIONS

The abundance of sulfuric acid droplets found in the stratosphere during periods following major volcanic eruptions may have a potential impact on the global climate. Particles in this size band (0.2 to 0.8 μm diameter) are efficient scatterers of visible radiation. They also have long residence times (several years) in the stratosphere because of their low settling rates. Thus they may interact with solar and terrestrial radiation and possibly alter temperatures at the surface and or in the stratosphere. Also, the presence of the very fine iron and carbonaceous particles in the stratosphere, regardless of origin, may cause temperature perturbations in the stratosphere since many forms of both iron and carbon are highly absorbent.

## REFERENCES

Brownlee, D. E., 1978: Microparticle Studies by Sampling Techniques. In J.A.M. McDonnell (Ed.), Cosmic Dust, John Wiley & Sons, New York, New York, 295-336.

Cadle, R. D., 1972: Composition of the Stratospheric Sulfate Layer, EOS Trans. American Geophysical Union, 53, 9.

Chuan, R. L., 1975: Rapid Measurement of Particulate Size Distribution in the Atmosphere, In B.Y.H. Liu (Ed.), Fine Particles, Aerosol Generation, Measurement, Sampling and Analysis, Academic Press, New York, 763.

Chuan, R. L., D. C. Woods and M. P. McCormick, 1981: Characterization of Aerosols From Eruptions of Mt. St. Helens, Science, 211, 830-832.

Chuan, R. L., and D. C. Woods, 1984a: The Appearance of Carbon Aerosol Particles in the Lower Stratosphere, Geophys. Res. Letts., 11, 5, 553-556.

Chuan, R. L., and D. C. Woods, 1984b: Temporal Variations in Characteristics of the El Chichon Stratospheric Cloud, Geof. Int., 23-3, 335-349.

Gandrud, B. W., M. A. Kritz, and A. L. Lazrus, 1983: Balloon and Aircraft Measurements of Stratospheric Sulfate Mixing Ratio Following the El Chichon Eruptions, Geophys. Res. Letts., 10, 1037-1040.

Hofmann, D. J., and J. M. Rosen, 1983: Lidar and Balloon-borne Particle Counter Comparisons Following Recent Volcanic Eruptions, J. Geophys. Res., 88, C6, 3777-3782.

McCormick, M. P., T. J. Swissler, W. H. Fuller, W. H. Hunt and M. T. Osborn, 1984: Airborne and Ground-based Lidar Measurements of the El Chichon Stratospheric Aerosol From 90°N to 56°S, Geof. Int., 23-2, 187-221.

Rosen, J. M., 1969: Stratospheric Dust and Its Relationship to the Meteoric Influx, Space Science Rev., 9, 58-59.

Rosen, H., A. D. A. Hansen, L. Gundel and T. Navakov, 1978: Identification of the Optically Absorbing Component in Urban Aerosols, Appl. Opt., 17, 3859.

Russell, P. B., M. P. McCormick, T. J. Swissler, W. P. Chu, J. M. Livingston, W. H. Fuller, Jr., J. M. Rosen, D. J. Hofmann, L. R. McMaster, D. C. Woods and T. J. Pepin, 1981: Satellite and Correlative Measurements of the Stratospheric Aerosol II: Comparison of Measurements Made by SAM II, Dustsondes, and on Airborne Lidar, J. Atmos. Sci., 38, 6, 1295-1312.

Shedlovsky, J. P., and S. Paisly, 1966: On the Meteoric Component of Stratospheric Aerosols, Tellus, 18, 499-503.

Woods, D. C., and R. L. Chuan, 1983: Size-Specific Composition of Aerosols in the El Chichon Volcanic Cloud, Geophys. Res. Letts., 10, 1041-1044.

# NITRIC ACID HAZE FORMATION IN THE LOWER STRATOSPHERE: A MAJOR CONTRIBUTING FACTOR TO THE DEVELOPMENT OF THE ANTARCTIC "OZONE HOLE"

*P.J. Crutzen, C. Brühl and U. Schmailzl*
Max-Planck-Institute for Chemistry
D-6500 Mainz, Federal Republic of Germany

and

*F. Arnold*
Max-Planck-Institute for Nuclear Physics
D-6900 Heidelberg, Federal Republic of Germany

ABSTRACT

We show here that large ozone depletions result, following conversion of HCl to $ClO_X$ during periods when $HNO_3$ is frozen out as polar stratospheric haze or cloud particles. Gravitational settling of $HNO_3/H_2O$ particles may separate HCl from $HNO_3$ and so produce similar chemical conditions in narrow layers even after evaporation of the particles. This may especially activate ClOX catalyzed ozone destruction and it is conceivable that some of these conditions are transferred to regions outside the polar vortex by atmospheric transport.

The inorganic chlorine originates mostly from industrial chlorofluorocarbons. Large amounts of ClO, OClO and $ClONO_2$ can form during the springtime warming phase in the lower stratosphere, as observed. The results are very temperature dependent, indicating the potentially great influence of meteorological variability.

## 1. INTRODUCTION

From 1979 to 1985, increasingly drastic ozone reductions occurred during September and October in the Antarctic lower stratosphere (Farman et al., 1985; Stolarski et al., 1986; Chubachi and Kajiwara, 1986; Komhyr et al., 1987). For its explanation both dynamic (Tung, 1986, Mahlman and Fels, 1986) and chemical (Solomon et al., 1986a; McElroy et al., 1986a,b; Crutzen and Arnold, 1986) causes have been invoked. Polar stratospheric clouds (PSCs) have been proposed to play a decisive role in promoting the formation of the Antarctic "ozone hole". Several processes involving PSC aerosols were considered, including aerosol surface-catalyzed trace gas reactions and trace gas scavenging by aerosols. A common feature of all of these mechanisms is their potential to liberate active chlorine from the chlorine reservoir species $ClONO_2$ and HCl, thereby promoting halogen-catalyzed ozone destruction. Another



ISBN 0-937194-11-5

common feature of these ozone-hole-formation mechanisms is their critical dependence on the PSC characteristics, particularly chemical composition, physical state, and spatial and temporal extent. Until recently, PSC particles were thought to be composed of water ice (Steele et al., 1983), formed by heterogeneous monomolecular deposition of stratospheric water vapor. This may occur at temperatures below about 191 K near 17 km, being only a few degrees higher than the lowest weekly averaged temperatures (about 188 K) occurring in the winter south polar stratosphere (McCormick and Trepte, 1986).

Building on this view, two different ozone-hole-formation mechanisms involving water ice particles were proposed. McElroy et al. (1986a) invoked reactions of $N_2O_5$ and $ClONO_2$ with $H_2O$ on the surface of the PSC particles. Solomon et al. (1986a) proposed surface-catalyzed reactions of HCl with $ClONO_2$ and of $ClONO_2$ with $H_2O$.

More recently, however, it was recognized that the stratospheric particles may also be composed of nitric acid and water with roughly equal mass fractions. These could be formed by heterogeneous heteromolecular deposition of stratospheric nitric acid and water vapor on background $H_2SO_4/H_2O$ aerosols (Toon et al., 1986; McElroy et al., 1986b; Crutzen and Arnold, 1986). Interestingly, $HNO_3/H_2O$ nucleation can take place at temperatures that are significantly higher than those required for pure $H_2O$ vapor deposition, so that in this case it might be much more appropriate to speak of polar stratospheric haze (PSH), since these particles are probably much smaller than water ice particles and, therefore, optically much thinner. While Toon et al. (1986) just mentioned a potential role of $HNO_3/H_2O$ particles in promoting ozone depletion, McElroy et al. (1986b) and Crutzen and Arnold (1986) carried out model calculations of the ozone depletion induced by $HNO_3/H_2O$ particle formation. Moreover, Crutzen and Arnold explicitly pointed out that the main effect of $HNO_3/H_2O$ particle formation on ozone is the depletion of gaseous $NO_X$ and $HNO_3$.

This sets the stage for a rapid buildup of enhanced hydroxyl radical concentrations, which are amplified by the methane-photooxidation chain after the end of the polar night. In the presence of $NO_2$ and $HNO_3$, OH reacts preferably with these gases, setting up a catalytic OH destruction chain:

$$(OH+NO_2 \longrightarrow HNO_3)+(HNO_3+OH \longrightarrow H_2O+NO_3) + (NO_3+h\nu \longrightarrow NO_2+O) = (2\ OH \longrightarrow H_2O + O).$$

In the absence of $NO_2$ and $HNO_3$, OH reacts preferably with $CH_4$ and HCl. Reactions of OH with $CH_4$ amplify odd hydrogen production; those with HCl liberate active chlorine, which destroys ozone via catalytic cycles.

In contrast to other mechanisms, that proposed by Crutzen and Arnold (1986) does not require surface catalyzed reactions involving the chlorine reservoir species HCl and $ClONO_2$ since active chlorine is liberated from HCl by a gas phase reaction involving enhanced OH. The gas phase concentrations of $NO_X$ and $N_2O_5$ are likewise supposed to be low in the Crutzen and Arnold mechanism, which can be accomplished by $N_2O_5$ formation from $NO_X$ during the long polar night and surface reactions of $N_2O_5$ on the polar stratospheric particles. Ion-catalyzed $N_2O_5$-hydrolysis reactions may also be important (Crutzen and Arnold, 1986).

The ozone-hole-formation mechanism by Crutzen and Arnold (1986) has several interesting implications: 1) the altitude range of the ozone hole is restricted to about 10-25 km, 2) nitric acid vapor abundances are low also after the end of the polar night, and 3) there is a critical dependence of the ozone destruction efficiency on atmospheric temperatures, so that low temperatures promote ozone destruction.

In the meantime, additional information on trace gases in the winter polar stratosphere and surface-catalyzed reactions was obtained from polar expeditions and laboratory studies. In particular, measurements of the height distribution of ozone (Hofmann et al., 1987a) and the column densities of $O_3$ and $NO_2$ (Farmer et al., 1987; Mount et al., 1987), NO, $HNO_3$, HCl, $ClONO_2$ and HF (Farmer et al., 1987), OClO (Solomon et al., 1987b) and ClO (deZafra et al., 1987; Solomon et al., 1987a) were obtained within the framework of the National Ozone Expedition (NOZE), which was conducted from August through November 1986 by U.S. scientists at the McMurdo Station (78°S, Antarctica). Also, detailed balloon-borne measurements of the height distribution of nitric acid vapor (Knop and Arnold, 1987) were performed during the CHEOPS (Chemistry of Ozone in the Polar Stratosphere) expedition, which took place from December 1986 through February 1987, mostly at the Esrange Station in Kiruna, Sweden (70° N).

Concerning surface-catalyzed reactions, the one of $ClONO_2$ with $H_2O$ was studied in the laboratory at room temperature on $H_2SO_4$ surfaces, containing only 4.4 weight percent water. In these experiments, a reaction probability of $\gamma=3.2\times10^{-4}$ per collision was measured (Rossi et al., 1987). Since stratospheric $H_2SO_4/H_2O$ aerosol contains 25 weight percent $H_2O$ at temperatures near 220 K and 55-60 percent at 190-195 K (Rosen, 1971; Hofmann et al., 1987b), $\gamma \simeq 2\text{-}5\times10^{-3}$ may apply for stratospheric conditions. In this study, however, only reactions of $N_2O_5$ and $ClONO_2$ on $HNO_3/H_2O$ and ice particles are considered. In the present paper we extend our previous ideas about the ozone-hole-formation mechanism, involving particle formation (Crutzen and Arnold, 1986) and discuss them in the light of the new information on trace gas distributions in Antarctica and surface-catalyzed reactions that recently has become available.

## 2. POLAR STRATOSPHERIC HAZE (PSH) FORMATION

According to our theory (Crutzen and Arnold, 1986), during sunlit conditions, precipitous loss of ozone will take place when nitric acid is frozen out in polar stratospheric particles (see also Toon et al., 1986; McElroy et al., 1986b). This causes a rapid rise in OH concentrations, conversion of HCl to ClOX ($Cl+ClO+ClOH+2Cl_2O_2+OClO$) via

$$HCl+OH \longrightarrow Cl+H_2O \qquad (1)$$

and rapid catalytic loss of ozone through the following reaction cycles (Solomon et al., 1986a; Crutzen and Arnold, 1986; Molina and Molina, 1987; Hayman et al., 1986; McElroy et al., 1986a), all having as net result $2O_3 \longrightarrow 3O_2$:

$(Cl+O_3 \longrightarrow ClO+O_2) + (ClO+HO_2 \longrightarrow ClOH+O_2) + (ClOH+h\nu \longrightarrow Cl+OH) + (OH+O_3 \longrightarrow HO_2+O_2) = (2O_3 \longrightarrow 3O_2)$;

$2x(Cl+O_3 \longrightarrow ClO+O_2) + (ClO+ClO+M \longrightarrow Cl_2O_2+M) + (Cl_2O_2+h\nu \longrightarrow 2\,Cl+O_2) = (2O_3 \longrightarrow 3O_2)$;

$(Cl+O_3 \longrightarrow ClO+O_2) + (Br+O_3 \longrightarrow BrO+O_2) + (BrO+ClO \longrightarrow Br+Cl+O_2) = (2O_3 \longrightarrow 3O_2)$.

Freeze-out of $HNO_3$ and $H_2O$ takes place whenever both the $H_2O$ ($a_1$) and $HNO_3$ ($a_2$) atmospheric vapor pressures are larger than the equilibrium vapor pressures of $H_2O$ ($e_1$) and $HNO_3$ ($e_2$), respectively. The equilibrium vapor pressures of $H_2O$ and $HNO_3$ are functions of the temperature and composition of the $HNO_3/H_2O$ mixture. Laboratory measurements of $e_1$ and $e_2$, which are only available for temperatures from 0°C to 70°C, are reviewed by Clavelin and Mirabel (1979). The logarithms of the experimental $e_1$ and $e_2$ values exhibit a linear variation with $T^{-1}$, according to the Clausius-Clapeyron equation:

$$\ln e = \frac{(\Delta H)_V}{RT} + C \qquad (2)$$

where $(\Delta H)_V$ is the molar heat of vaporization, R the gas constant, and C a constant. Estimations of $e_1$ and $e_2$ at temperatures below 0°C may be obtained by extrapolating Eq. (2) towards lower temperatures assuming constant $(\Delta H)_V$. Such extrapolations would apply to liquid $HNO_3/H_2O$ mixtures. However, at the stratospheric temperatures of interest, the particles may be frozen. Two forms of solid

$HNO_3/H_2O$ aerosol exist, namely $HNO_3 \cdot H_2O$ (monohydrate, melting point: 235.48 K, heat of fusion 4184 cal mole$^{-1}$) and $HNO_3 \cdot 3H_2O$ (trihydrate, melting point: 254.63 K, heat of fusion: 6954 cal mole$^{-1}$) (Forsythe and Giauque, 1942). Since $e_1$ and $e_2$ for solid $HNO_3/H_2O$ mixtures are much lower than for liquid mixtures, solid particle formation is favored over liquid droplet formation, except if supercooling occurs. We will discuss this complication below.

Estimations of $e_1$ and $e_2$ for solid $HNO_3/H_2O$ mixtures may be obtained by approximating the heat of sublimation by the sum of the heats of condensation and fusion. Hence, at the melting point, the slope of ln e versus $T^{-1}$ becomes steeper, leading to much lower $e_1$ and $e_2$ compared to those for supercooled liquids. For solid $HNO_3 \cdot 3H_2O$ ($HNO_3$ mass fraction about 54%), extrapolations of $e_1$ and $e_2$ toward lower temperatures are given in Fig. 1. In comparison, $e_1$ and $e_2$ for solid $HNO_3 \cdot H_2O$ (monohydrate) are much larger (Crutzen and Arnold, 1986). Also shown in Fig. 1 are $e_1$ values for pure $H_2O$ ice ($f_2$ = 0%), being much larger than $e_1$ for solid $HNO_3 \cdot 3H_2O$.

The $H_2O$ and $HNO_3$ equilibrium vapor pressures over solid $HNO_3 \cdot 3H_2O$ mixtures, according to our analysis (see Fig. 1; for more details see Arnold and Crutzen, 1987), are given by the expressions:

$$e_1 = 9 \times 10^{12} \exp(-7944/T) \text{ hPa} \quad (3a)$$

$$e_2 = 1.56 \times 10^{15} \exp(-10050/T) \text{ hPa} \quad (3b)$$

Equilibrium vapor pressures over pure ice are equal to

$$e_0(H_2O) = 7.5 \times 10^{9} \exp(-5790/T) \text{ hPa} \quad (3c)$$

Temperatures below which both $e_1$ and $e_2$ become lower than the respective atmospheric vapor pressures $a_1$ and $a_2$ can be estimated from Fig. 1. Around 17 km, the central height of the Antarctic ozone hole, for solid $HNO_3 \cdot 3H_2O$ particles, $e_1$ becomes equal to $a_1$ at T ≈ 211 K (for a typical $H_2O$ vapor volume mixing ratio of $5 \times 10^{-6}$; point C). However, at this temperature $e_2$ is still larger than $a_2$ ($\sim 5 \times 10^{-9}$), so that no $HNO_3$-trihydrate can form. Consequently, the critical parameter, the atmospheric vapor pressure of $HNO_3$. $HNO_3$-trihydrate can start to form at about 205 K (point E, Fig. 1). Since $a_1 \gg a_2$, below this critical temperature nitric acid will be drastically removed from the gas phase without much effect on water vapor, which is much more abundant. $H_2O$ ice can start forming only at T ≈ 191 K (point A, Fig. 1). We propose, therefore, that $HNO_3 \cdot 3H_2O$ particles are present in the lower stratosphere at temperatures between 191 K and 205 K, forming what we most appropriately should call polar stratospheric haze (PSH). Only the predominantly $H_2O$ containing ice particles that

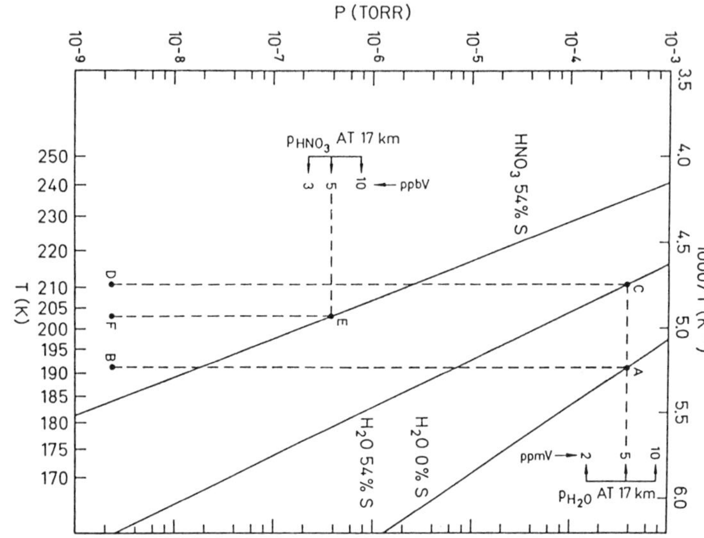


FIGURE 1. Equilibrium saturation vapor pressures for $HNO_3$ ($HNO_3$ 54% S) and $H_2O$ ($H_2O$ 54% S) over solid $HNO_3 \cdot 3H_2O$ ($HNO_3$ mass fraction 54%). Also shown are equilibrium saturation vapor pressures for $H_2O$ ($H_2O$ 0% S) over solid $H_2O$. Typical atmospheric partial pressures (at 17 km altitude) for $HNO_3$ and $H_2O$ are indicated by the arrows.

are formed below 191 K should be called polar stratospheric clouds or cirrus (PSC) (Steele et al., 1983; Heymsfield, 1986).

The nitric acid haze particles most probably form on preexisting $H_2SO_4/H_2O$ aerosol. It is quite possible that only a fraction of the particles, the largest ones in the size range (Poole, 1987; Poole and McCormick, 1987a,b), may act as deposition nuclei. The $H_2SO_4/H_2O$ aerosol may be supercooled liquid droplets or frozen particles. If the former were the case, solid $HNO_3 \cdot 3H_2O$ could not form around the nuclei (a ternary $H_2SO_4/H_2O/HNO_3$ solution would then form). However, when temperatures have once reached below about 191 K, it has been observed (Poole and McCormick, 1987a; Heymsfield, 1986) that cirrus-like ice crystals definitely do form. At this stage also the $H_2SO_4$ will become solid, if it was not yet frozen before. It will stay solid until evaporation at temperatures above about 240 K (Hamill and McMaster, 1984). Therefore, once an air parcel has been cooled to temperatures below 191 K, the $H_2SO_4/H_2O$ aerosol will most likely be solid. On such $H_2SO_4/H_2O$

aerosol $HNO_3 \cdot 3H_2O$ can crystallize in the temperature range 191-205 K. Since, during the winter, temperatures in the lower stratosphere do normally go below 191 K, the formation of solid $HNO_3 \cdot 3H_2O$ particles should occur readily in the Antarctic. In the Arctic this is not so obvious. Nevertheless, according to Poole and McCormick (1987a), even there, solid $HNO_3/H_2O$ particles can form at temperatures above 191 K.

It seems likely that during the very cold and dark Antarctic winter period, with temperatures that can go below 190 K, most NOX is converted to $HNO_3$ by ion-catalyzed (Crutzen and Arnold, 1986) or heterogeneous reactions of $N_2O_5$ on polar stratospheric cirrus particles.

At temperatures below about 190 K, together with $HNO_3$, also $H_2O$ and HCl may freeze and settle out (Toon et al., 1986). In the temperature range 191-205 K, mainly $HNO_3$ and $H_2O$ will be taken up in the particles. Gravitational settling can, therefore, lead to partial separation of $HNO_3$ from HCl, probably leading to a large vertical variability of chemical composition and reaction environment and ozone breakdown (see Hofmann et al., 1987a). This potentially very important process has not yet been taken into account in any theoretical study. Gravitational settling may be particularly important if only a small fraction of the available particles do nucleate. For cirriform ice particles in the tropics this fraction may be only $10^{-4}$ to $10^{-3}$ (Heymsfield, 1986). Low values may also be possible for PSH particles (Poole, 1987; Poole and McCormick, 1987b).

## 3. MODEL CALCULATIONS

We have used our height and time dependent model (Gidel et al., 1983) to simulate photochemistry and vertical exchange in the latitude belt between 60°S and the South Pole from the beginning of April to the end of October. For the springtime lower stratosphere, we have adopted the temperatures measured by Hofmann et al. (1987a), at McMurdo (78°S) at 18±1 km altitude to derive gas phase equilibrium $HNO_3$ concentrations (Eq. 3b), assuming the remaining $HNO_3$ is included in the particles, when condensation conditions are satisfied. At higher temperatures we allow for reevaporation of $HNO_3$.

Only the effects of vertical transport were considered, with vertical advection taken from Rosenfield et al. (1987). Vertical eddy diffusion was approximated by adopting a small $K_Z$ value of $1 \times 10^3 cm^2 s^{-1}$, in agreement with modern dynamical theories (McIntyre, 1980; Holton, 1981; Solomon et al, 1986b). Horizontal transport of tracers in and out of the 60-90°S region was thus neglected. This assumption implies that our calculations may underestimate the tracer concentrations above about 50 km and below 35 km, where zonally averaged winds blow towards the pole, and overestimate those between 35 and 50 km.

Time dependent transport equations, including chemical source and sink terms, were solved for the "families" of odd oxygen ($O+O_3$), NOX ($N+NO+NO_2+NO_3+2N_2O_5+HNO_4+ClONO_2$), $HNO_3$, odd hydrogen ($H+OH+HO_2$), $H_2O_2$, ClOX ($Cl+ClO+OClO+2Cl_2O_2+ClOH+ClONO_2$), HCl, $CH_4$, $H_2O$, $H_2$ and $N_2O$. Within the chemical families the partitioning of the individual gases, was based on photochemistry alone. Standard reaction coefficients were used (DeMore et al., 1985). For initial conditions, the concentrations calculated with our two-dimensional model were adopted. Subsequently, the volume mixing ratios at the upper boundary (65 km), μ(top), were extrapolated from those calculated at levels 1 and 2 km below through the relation

$$\frac{\mu(top)}{\mu(top-1)} = \frac{\mu(top-1)}{\mu(top-2)} \qquad (4)$$

For odd nitrogen two different upper boundary mixing ratios were adopted, viz.,3 ppbv and 20 ppbv (of NO) for sensitivity analysis. Lower boundary concentrations (at ground level) were kept constant or treated by flux boundary conditions.

Calculated vertical distributions of $CH_4$, $N_2O$, ClOX, HCl, $O+O_3$ and OH (noon time) are presented in Fig. 2 for early May and October 1986. For the region above about 25 km large temporal decreases in $N_2O$ and $CH_4$ are calculated due to downward transport (Solomon et al., 1986b; Jones and Pyle, 1984). Due to the depletion of $CH_4$, ClOX concentrations grow substantially at the expense of HCl. This in turn leads to large ozone depletions with maxima near 40 km (Figs. 3 and 4).

In the lower stratosphere, the effects of the adopted vertical advection are less important. It may be that other advection schemes than those used here could have shown a larger effect, but the uncertainties are great, and it is not the main purpose of this study to explore the sensitivities of model results to these kinds of unresolved questions.

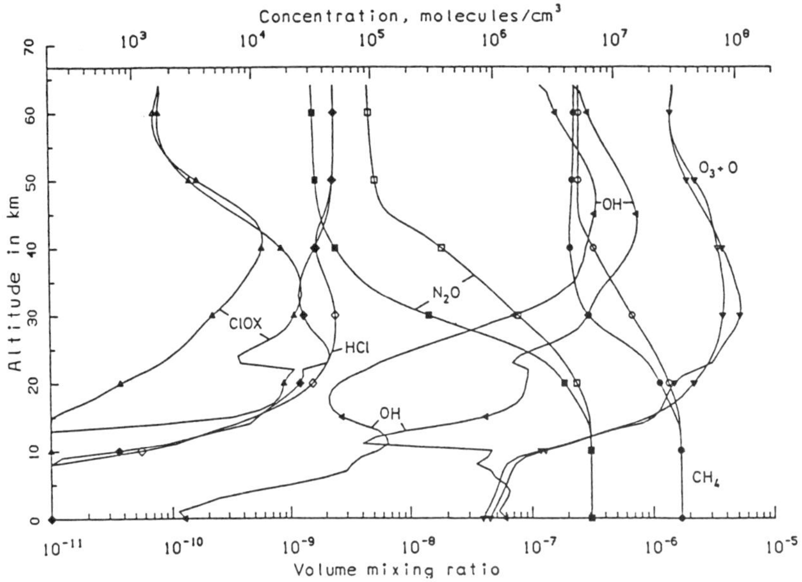


FIGURE 2. Calculated vertical profiles of OH concentrations and the volume mixing ratios of ClOX, HCl, $N_2O$, $CH_4$ and odd oxygen at 70°S in early May (open symbols), and October (filled symbols), 1986.

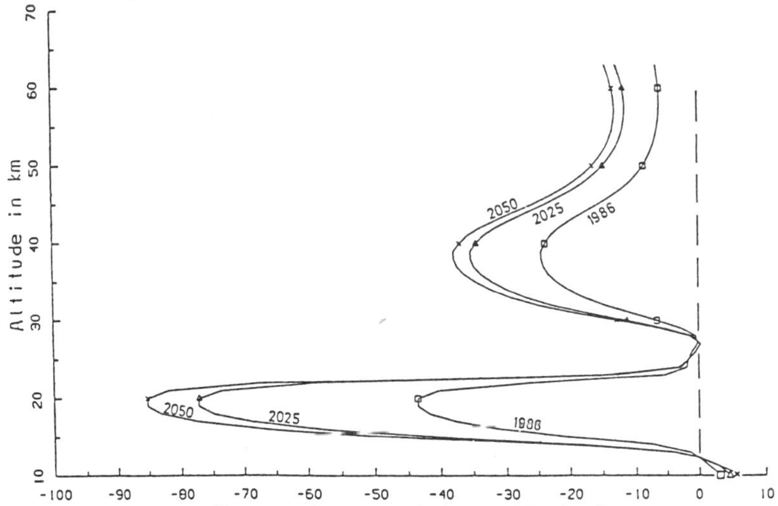


FIGURE 3. Calculated ozone depletions for October 1986, 2025, and 2050 compared to 1960, for reduced emissions of chlorofluorocarbons after 1990, as proposed by United Nations Environment Program (see Johnson, 1987). The assumed NO volume mixing ratio at the upper boundary value was 3 ppbv.

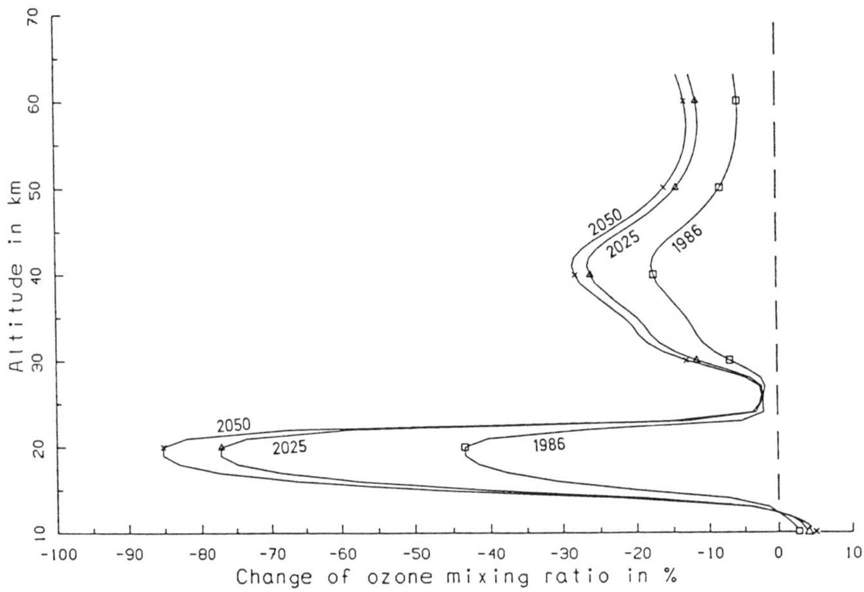


FIGURE 4. As Fig. 3, for an upper boundary NO volume mixing ratio of 20 ppbv.

With an assumed upper boundary NO volume mixing ratio of 3 ppbv, Fig. 3 shows calculated ozone changes for 1 October for the time periods 1960 to 1986, 1960 to 2025, and 1960 to 2050, under a future emission scenario for fully halogenated chlorofluorocarbons that starts with a 20% reduction from current levels in 1990, followed by another 30% cut in 1995, according to international proposals (Johnson, 1987). Total upper stratospheric ClX (= ClOX+HCl) concentrations for this scenario were estimated at 0.77 ppbv in 1960, 2.60 ppbv in 1986, 4.00 ppbv in 2025, and 4.53 ppbv in 2050. Calculated ozone depletions in the upper stratosphere are especially marked (about 1% per year) for this assumed upper boundary NO volume mixing. For 20 ppbv of NO, the calculated ozone depletions near 40 km are significantly lower, as can be seen in Fig. 4. This indicates the importance of knowing the mesospheric NO distribution at high latitudes.

During gas phase depletion of $HNO_3$, increasingly faster ozone depletions are calculated from year to year in the lower stratosphere, assuming the same temperatures as measured in 1986 (see Fig. 4). According to some analyses (Angell, 1986; Newman and Schoeberl, 1986; Sekiguchi, 1986), average stratospheric temperatures near 100 mb have, however, shown a cooling trend by 5 to 10 °C during October and November between 1978 and 1985. Increasing the temperatures by 7°C, we do not at all calculate such dramatic ozone depletions as shown in Fig. 4. Realistic predictions of ozone depletions are, therefore, also quite dependent on climatic conditions in the lower stratosphere, which may exhibit long-term trends. We must, however, point out that the above authors do not present temperature data for September. For this month, during which most ozone depletion occurs, Gardiner (1987) and Austin et al. (1987), do not find average temperature changes over the past three decades in the lower stratosphere of Antarctica.

Figure 5 shows calculated temporal changes during Antarctic winter and spring in the chemical composition for current conditions at 18 km. In particular, we note a sharp rise in OH concentrations as long as most $HNO_3$ is present in the condensed phase, because of low temperatures. The rise in OH leads to the conversion of HCl to ClOX via reaction (1), while reconversion of ClOX to HCl through reactions

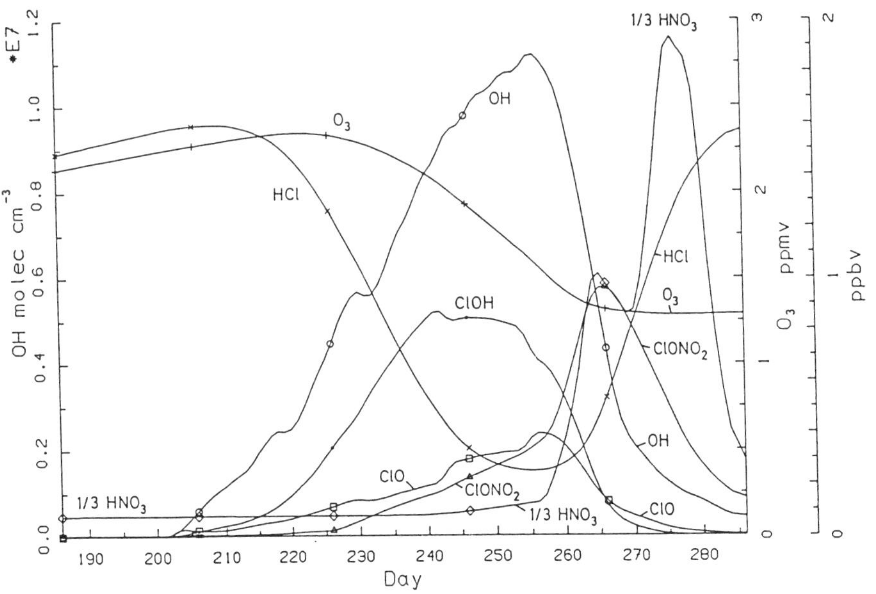


FIGURE 5. Chemical developments at 18 km from day 185 (5 July) to day 285 (15 October) in 1986. Left-hand scale: OH concentrations in units of $10^7$ molecules $cm^{-3}$; right-hand scales: volume mixing ratios of $O_3$ (ppmv) and for ClO, $HNO_3$, $ClONO_2$, ClOH and HCl (ppbv) for the temperature sequence measured by Hofmann et al. (1987a) at McMurdo (78°S) in Antarctica. The small bumps in the curves indicate the times when daylight hours changed discontinuously in the program.

$$NO + ClO \longrightarrow Cl + NO_2 \quad (5)$$
$$Cl + CH_4 \longrightarrow HCl + CH_3 \quad (6)$$

is hindered due to very low concentrations of NO (not shown in Fig. 5). The rise in ClOX leads to a strong depletion of ozone. The calculations indicate also that following reevaporation of $HNO_3$ near day 260, OH concentrations decline rapidly. Initially much $ClONO_2$ is produced by the reaction of ClO with the $NO_2$ that is produced following $HNO_3$ photolysis and reaction with OH. Paralleling this rise in $ClONO_2$ to about 1 ppbv, there is also a strong decrease in OH and an increase in HCl and a decrease in ClO, so that ozone depletion slows down.

At cold temperatures below about 190 K, HCl can be incorporated in PSC particles and reactions with $ClONO_2$ can take place (Molina et al, 1987; Tolbert et al, 1987):

$$HCl + ClONO_2 \longrightarrow HNO_3 + Cl_2(gas) \quad (7a)$$

As long as there is some polar stratospheric $HNO_3 \cdot 3H_2O$ haze, some reaction of $ClONO_2$ on the particles may take place in a transition period:

$$ClONO_2 + H_2O \longrightarrow HNO_3 + ClOH(gas) \quad (7b)$$

However, we do not consider this to contribute importantly to the ozone depletion, as $ClONO_2$ formation occurs only while the particles are evaporating. Also, reaction (7a) can not occur, because HCl can not be incorporated in the PSH particles. In any case, surface reactions of HCl and ClOH on the PSC particles at $T \leq 190K$:

$$HCl + ClOH \longrightarrow H_2O + Cl_2(gas) \quad (7c)$$

are probably much more important.

As long as the concentrations of NOX remain smaller than those of ClOX, $ClONO_2$ will have a relatively long lifetime. As shown in Fig. 5, when this is no longer the case, following enough conversion of $HNO_3$ to NOX, HCl formation occurs rapidly via reactions (5) and (6). During this phase there is a strong increase in Cl atom concentrations, which, along with $CH_4$, will react very rapidly with other hydrocarbons (e.g., $C_2H_6$). As OH concentrations are likewise high during the polar stratospheric haze period, we predict that the ozone hole regions should be devoid of nonmethane hydrocarbons. They should also contain low concentrations of CO, about 3-4 ppbv, because of the strong temperature dependence of the reaction between OH and $CH_4$, which leads to CO formation.

Several observations of the chemical composition of the springtime stratosphere, especially of ClO, OClO and

$ClONO_2$, can be explained by our reaction mechanism (Mount et al., 1987; Solomon et al., 1987b; deZafra et al., 1987; Farmer et al., 1987). However, interpretation is complex due to interfering dynamic processes.

Since removal of odd nitrogen compounds and their inclusion in solid $HNO_3 \cdot 3H_2O$ polar stratospheric haze particles is the essential element in our theory and because this process is very temperature dependent, there is potentially a strong influence of interannual meteorological variability on lower stratospheric chemistry. (For instance our calculations show that an increase in temperatures by 6 K is sufficient to wipe out much of the effects shown in Fig. 5). However, changing ozone concentrations may influence stratospheric temperatures and circulation. These processes can be positively coupled in some parts of the stratosphere, such as the polar regions. The analysis by Austin et al. (1987), indicates that this may indeed be the case. These authors did not find any evidence of average temperature declines in September in the ozone hole region over a 30-year period, but they did find a temperature drop in October, which could well have been caused by the increasing ozone loss during September and the resulting reduction in solar heating (Shine, 1986).

## 4. CONCLUSIONS

Polar stratosphere haze particles can start forming at higher temperatures ($T < 200$-$205$ K) than can PSC ice particles, which form below about 190 K. Both kinds of particles probably grow on available $H_2SO_4/H_2O$ nuclei. The PSH particles consist mainly of solid $HNO_3 \cdot 3H_2O$ crystals. A requirement for PSH formation is that the $H_2SO_4/H_2O$ nuclei are frozen. This will most likely be the case for particles in any air masses that have experienced temperatures below 190 K. This happens frequently in south polar regions. In the Arctic this is less likely. However, also here PSH formation may be commonplace, whenever temperature conditions are favorable (Poole and McCormick, 1987a).

Whenever PSH particles exist, the gas phase is depleted in nitric acid and other oxides of nitrogen ($NO_X$, $NO_3$, $N_2O_5$, $ClONO_2$). The latter are converted to $HNO_3$ especially during the polar night, because in the cold and dark lower stratosphere $NO_X$ is oxidized to $N_2O_5$ by well-known reactions and $N_2O_5$ is hydrolyzed to $HNO_3$ by surface- or ion-catalyzed reactions on PSC and PSH particles.

After the removal of NOX and $HNO_3$ from the gas phase and the return of ultraviolet solar radiation during early spring, hydroxyl concentrations grow rapidly (amplified by reactions with $CH_4$ and $CH_4$ oxidation products). At this stage HCl is efficiently converted to active gaseous chlorine, ClOX (= $Cl+ClO+ClOH+OClO+Cl_2O_2$). This leads to rapid ozone destruction by catalytic reactions involving ClOX and BrOX (= Br+BrO+BrOH) radicals. The ozone depletion continues as long as an appreciable fraction of the odd nitrogen molecules is captured in the PSH particles. When $HNO_3$ is again released to the gas phase by evaporation at higher temperatures, the ozone decay slows down and formation of $ClONO_2$ and HCl occurs. It is only during this stage that heterogeneous reactions of $ClNO_3$, $H_2O$, and HCl may occur on the particles, but we consider this to be rather insignificant (in fact, reactions of HCl with ClOH are much more relevant). Initially $ClONO_2$ is more abundant than HCl, but as soon as there is some free $NO_X$ ($NO+NO_2$), the formation of HCl accelerates and $ClONO_2$ is rapidly lost from the stratosphere.

Gravitational settling, especially of PSC particles, may be important in bringing HCl and $HNO_3$ down into the lower stratosphere. Downward settling of PSH particles will affect $HNO_3$, but not HCl. Although fall velocities of PSH particles may be small (~ 0.01 cm/s at 18 km; Hamill et al., 1986), after reevaporation of the PSH particles, the gases may separate into discrete layers with strong ozone destruction properties, because layers may develop with high NOX and low ClOX, or with low NOX and high ClOX concentrations. It is conceivable that these conditions are transferred to regions outside the polar vortex by atmospheric transport processes. These potentially very important processes should be studied by detailed observations and modeling. They may explain observed ozone depletions in regions surrounding the Antarctic.

Finally, we would again like to emphasize the great importance of the branching ratio of the reactions

$$ClO+OH \longrightarrow HO_2+Cl \quad (+1.7\ \text{kcal mol}^{-1}) \qquad (8a)$$
$$ClO+OH \longrightarrow HCl+O_2 \quad (+55.6\ \text{kcal mol}^{-1}) \qquad (8b)$$

in ozone depletion calculations. So far, kinetic studies have shown reaction (8a) to be by far the most important (≤ 85%; Burrows et al., 1984; Hills and Howard, 1984) or the only channel for reaction (8) (Poulet et al., 1985), despite the much larger exothermicity of channel (8b). More studies are needed to establish the branching ratios unequivocally, because even a relatively small contribution (~ 10%) from reaction (8b) will substantially moderate calculated ozone depletions.

## REFERENCES

Angell, J.K., 1986: The Close Relation Between Antarctic Total-Ozone Depletion and Cooling of the Antarctic Low Stratosphere, Geophys. Res. Letts., 13 (12),1240-1243.

Arnold, F., and P.J. Crutzen, 1987: Stratospheric $HNO_3/H_2O$ Aerosol Formation: Implications for Ozone and Other Trace Gas Chemistry (in preparation).

Austin, J., R.L. Jones, T.N. Palmer and A.F. Tuck, 1987: Circulation Changes and Chemistry: Implications for the Destruction of Ozone in Antarctica, Nature (submitted).

Burrows, J.P., T.J. Wallington and R.P. Wayne, 1984: Kinetics of the Reaction of OH and ClO, J. Chem. Soc. Faraday Trans. 2, 80, 957-971.

Chubachi, S., and R. Kajiwara, 1986: Total Ozone Variations at Syowa, Antarctica, Geophys. Res. Letts., 13, 1197-1198.

Clavelin, J.L., and P. Mirabel, 1979: Determination des pressions partielles du melange eau-acide nitrique, J. de Chimie Physique, 76, 533-537.

Crutzen, P.J., and F. Arnold, 1986: Nitric Acid Cloud Formation in the Cold Antarctic Stratosphere: A Major Cause for the Springtime "Ozone Hole", Nature, 324, 651-655.

DeMore, W.B., J.J. Margitan, M.J. Molina, R.T. Watson, D.M. Golden, R.F. Hampson, M.J. Kurylo, C.J. Howard and A.R. Ravishankara, 1985: Chemical Kinetics and Photochemical Data for Use in Stratospheric Modeling, Evaluation Number 7, JPL Publication 85-37, NASA, Jet Propulsion Laboratory, California Institute of Technology, Pasadena, CA.

DeZafra, R.L., M. Jaramillo, A. Parrish, P. Solomon, B. Connor and J. Barrett, 1987: High Concentrations of Chlorine Monoxide at Low Altitudes in the Antarctic Spring Stratosphere: Diurnal Variation, Nature, 328, 408-411.

Farman, J.C., B.G. Gardiner and J.D. Shanklin, 1985: Large Losses of Total Ozone in Antarctica Reveal Seasonal $ClO_x/NO_x$ Interaction, Nature, 315, 207-210.

Farmer, C.B., G.C. Toon, P.W. Schaper, J.-F. Blavier and L.L. Lowes, 1987: Ground-Based Infrared Measurements of the Composition of the Antarctic Atmosphere During the 1986 Spring Season: I. Stratospheric Trace Gases, Nature (in press).

Forsythe, W.R., and W.F. Giauque, 1942: The Entropies of Nitric Acid and Its Mono- and Tri-hydrates, J. Am. Chem. Soc., 64, 48-61.

Gardiner, B.G., 1987: Stratospheric Ozone and Temperature Variations in the Antarctic Spring (this symposium).

Gidel, L.T., P.J. Crutzen and J. Fishman, 1983: A Two-Dimensional Photochemical Model of the Atmosphere: 1. Chlorocarbon Emissions and Their Effect on Stratospheric Ozone, J. Geophys. Res., 88, 6622-6640.

Hamill, P., and L.R. McMaster (Eds.), 1984: Polar Stratospheric Clouds, NASA Conference Publication 2318, 72 pp.

Hamill, P., O.B. Toon and R.P. Turco, 1986: Characteristics

of Polar Stratospheric Clouds During the Formation of the Antarctic Ozone Hole, Geophys. Res. Letts., 13 (12), 1288-1291.

Hayman, C.D., J.M. Davies and R.A. Cox, 1986: Kinetics of the Reaction ClO+ClO ---> Products and Its Potential Relevance to Antarctic Ozone, Geophys. Res. Letts., 13: 1347-1350.

Heymsfield, A.J., 1986: Ice Particles Observed in a Cirriform Cloud at -83°C and Implications for Polar Stratospheric Clouds, J. Atmos. Sci., 43, 851-855.

Hills, A.J., and C.J. Howard, 1984: Rate Coefficient Temperature Dependence and Branching Ratio for the OH+ClO Reaction, J. Chem. Phys., 81, 4458-4465.

Hofmann, D.J., J.W. Harder, S.R. Rolf and J.M. Rosen, 1987a: Balloon-borne Observations of the Temporal Development and Vertical Structure of the Antarctic Ozone Hole in 1986, Nature, 326, 59-62.

Hofmann, D.J., J.M. Rosen and J.W. Harder, 1987b: Aerosol Measurements in the Winter/Spring Antarctic Stratosphere: I. Correlative Measurements With Ozone, J. Geophys. Res. (in press).

Holton, J.R, 1981: An Advective Model for Two-Dimensional Transport of Stratospheric Trace Species, J. Geophys. Res., 86, 11989-11994.

Johnson, K., 1987: Agreement Nears on Protection of the Earth's Ozone Layer, Nature, 328, 7.

Jones, R.L., and J.A. Pyle, 1984: Observations of $CH_4$ and $N_2O$ by the NIMBUS 7 SAMS: A Comparison With In-Situ Data and Two-Dimensional Numerical Model Calculations, J. Geophs. Res., 89, 5263-5279.

Knop, G., and F. Arnold, 1987: Stratospheric Trace Gas Detection Using a New Balloon-Borne ACIMS Instrument: Acetonitrile, Acetone and Nitric Acid, Geophys. Res. Letts. (submitted).

Komhyr, W.D., P.R. Franchois, B.C. Halter and C.C. Wilson, 1987: ECC Ozonesonde Observations at South Pole, Antarctica, During 1986, NOAA Data Report ERL ARL-11, Silver Spring, Maryland.

Mahlman, J.D., and S.B. Fels, 1986: Antarctic Ozone Decreases: A Dynamical Cause?, Geophys. Res. Letts., 13, 1316-1319.

McCormick, M.P., and C.R. Trepte, 1986: SAM II Measurements of Antarctica PSC's and Aerosols, Geophys. Res. Letts., 13 (12), 1276-1279.

McElroy, M.B., R.J. Salawitch S.C. Wofsy and J.A. Logan, 1986a: Antarctic Ozone: Reductions Due to Synergistic Interactions of Chlorine and Bromine, Nature, 321, 759-762.

McElroy, M.B., R.J. Salawitch and S.C. Wofsy, 1986b: Antarctic $O_3$: Chemical Mechanisms for the Spring Decrease, Geophys. Res. Letts., 13, 1296-1299.

McIntyre, M.E., 1980: Towards a Lagrangian-Mean Description of Stratospheric Circulations and Chemical Transports, Phil. Trans. Roy. Soc. (London), A296,

129-148.
Molina, L.T., and M.J. Molina, 1987: Production of $Cl_2O_2$ From the Self-Reaction of the ClO Radical, J. Phys. Chem., 91, 433-436.
Molina, M.J., T.-L. Tso, L.T. Molina and F.C.-Y. Wang, 1987: The Chemical Interaction of $ClONO_2$, HCl and Ice, and the Release of Active Chlorine in the Antarctic Stratosphere, Science (submitted).
Mount, G., R.W. Sanders, A.L. Schmeltekopf and S. Solomon, 1987: Visible Spectroscopy at McMurdo Station, Antarctica, 1. Overview and Daily Variations of $NO_2$ and $O_3$, Austral Spring 1986, J. Geophys. Res., 92 (D7), 8320-8328.
Newman, P.A., and M.R. Schoeberl, 1986: October Antarctic Temperature and Total Ozone Trends From 1979-1985, Geophys. Res. Letts., 13 (12), 1206-1209.
Poole, L.R., 1987: Airborne Lidar Studies of Arctic Polar Stratospheric Clouds, Ph.D. Dissertation, University of Arizona.
Poole, L.R., and M.P. McCormick, 1987a: Airborne Lidar Observations of Arctic Polar Stratospheric Clouds: Indications of Two Distinct Growth Stages, Geophys. Res. Letts. (submitted).
Poole, L.R., and M.P. McCormick, 1987b: Polar Stratospheric Clouds and the Antarctic Ozone Hole, J. Geophys. Res. (submitted).
Poulet, G., G. Laverdet and G. Le Bras, 1985: Rate Constant and Branching Ratio for the Reaction of OH with ClO, J. Phys. Chem., 90, 159-165.
Rosen, J.M., 1971: The Boiling Point of Stratospheric Aerosols. J. Appl. Meteor., 10, 1044-1045.
Rosenfield, J.E., M.R. Schoeberl and M.A. Geller, 1987: A Computation of the Stratospheric Diabatic Circulation Using an Accurate Radiative Transfer Model, J. Atmos. Sci., 44, 859-876.
Rossi, M.J., R. Malhotra and D.M. Golden, 1987: Heterogeneous Chemical Reaction of Chlorine Nitrate and Water on Sulfuric-Acid Surfaces at Room Temperature, Geophys. Res. Letts., 14, 127-130.
Sekiguchi, Y., 1986: Antarctic Ozone Change Correlated to the Stratospheric Temperature Field, Geophys. Res. Letts., 13 (12), 1202-1205.
Shine, K.P., 1986: On the Modelled Thermal Response of the Antarctic Stratosphere to a Depletion of Ozone, Geophys. Res. Letts., 13 (12), 1331-1334.
Solomon, S., R.R. Garcia, F.S. Rowland and D.J. Wuebbles, 1986a: On the Depletion of Antarctic Ozone, Nature, 321, 755-758.
Solomon, S., J.T. Kiehl, R.R. Garcia and W. Grose, 1986b: Tracer Transport by the Diabatic Circulation Deduced From Satellite Observations, J. Atmos. Sci., 43, 1603-1617.
Solomon, P.M., B. Connor, R.L. de Zafra, A. Parrish, J. Barrett and M. Jaramillo, 1987a: High Concentrations

of Chlorine Monoxide at Low Altitudes in the Antarctic Spring Stratosphere: Secular Variation, Nature, 328, 411-413.

Solomon, S., G.H. Mount, R.W. Sanders and A.L. Schmeltekopf, 1987b: Visible Spectroscopy at McMurdo Station, Antarctica, 2. Observations of OClO, J. Geophys. Res., 92(D7), 8329-8339.

Steele, H.M., P. Hamill, M.P. McCormick and T.J. Swisler, 1983: The Formation of Polar Stratospheric Clouds, J. Atmos. Sci., 40, 2055-2067.

Stolarski, R.S., A.J. Krueger, M.R. Schoeberl, R.D. McPeters, P.A. Newman and J.C. Alpert, 1986: Nimbus 7 SBUV/TOMS Measurements of the Springtime Antarctic Ozone Hole, Nature, 321, 808-811.

Tolbert, M.A., M.J. Rossi, R. Malhotra and D.M. Golden, 1987: Reaction of Chlorine Nitrate with Condensed $H_2O$ and HCl at Antarctic Stratospheric Temperatures, Science (submitted).

Toon, O.B., P. Hamill, R.P. Turco and J.P. Pinto, 1986: Condensation of $HNO_3$ and HCl in the Winter Polar Stratospheres, Geophys. Res. Letts., 1986: 13 (12), 1284-1287.

Tung, K.K., 1986: On the Relationship Between the Thermal Structure of the Stratosphere and the Seasonal Distribution of Ozone, Geophys. Res. Letts., 13, 1308-1311.

# SIZE DISTRIBUTION OF STRATOSPHERIC AEROSOLS FROM SAGE II MULTIWAVELENGTH EXTINCTIONS

*C. Brogniez and J. Lenoble*
Laboratoire d'Optique Atmosphérique
Université des Sciences et Techniques de Lille Flandres Artois
59655 Villeneuve d'Ascq Cedex, France

ABSTRACT

Measurements of aerosol extinction profiles in the stratosphere in four channels are used to retrieve two parameters of the particle size distribution. Good agreement between these two parameters and those deduced from balloon data has been found.

## 1. INTRODUCTION

The Stratospheric Aerosol and Gas Experiment I (SAGE I) has provided aerosol extinction profiles in two channels at 1.00 μm and 0.45 μm during two and a half years. The ratio of the two channel extinctions allows the retrieval of one parameter of the particle size distribution, i.e., a mode radius $r_m$ or an effective radius $r_{eff}$ (Yue and Deepak, 1983; Lenoble and Brogniez, 1985). Alternatively, most of the aerosol radiative characteristics can be retrieved from this two-channel ratio, unless the distribution is bimodal (Lenoble and Brogniez, 1985). Variations of the aerosol size with latitude, season, and altitude have been studied (Yue and Deepak, 1984; Brogniez and Lenoble, 1987) for the unperturbed atmosphere of 1979, and the influence of the Mount St. Helens eruption on the aerosol size has been demonstrated from SAGE I data (Lenoble et al., 1984).

Since October 1984, SAGE II provides profiles of the aerosol extinction $\sigma(\lambda)$ in four channels: 1.02 μm, 0.525 μm, 0.45 μm, and 0.385 μm. Therefore, at least, the same information can be retrieved from SAGE II as from SAGE I, using the 1.02 μm and 0.45 μm channels. The quality from SAGE II is probably a little better, due to a smaller error on the extinction, especially at 0.45 μm, where the two narrow-band channels allow a better separation of $NO_2$ and aerosol contributions.

The issue addressed here concerns what further information about the size distribution can be obtained using four channels instead of two. We do not seek an operational algorithm to retrieve a size distribution from SAGE II aerosol channels, but rather try to clarify what can be expected from the physical considerations of the data with their error bars and from comparison with aerosol models. We will limit ourselves to seeking two parameters of a monomodal size distribution; this proves difficult enough to show that the search for three parameters is absolutely unrealistic. Various combinations of the four channels (in sets of two) can be tried to achieve this goal.



ISBN 0-937194-11-5

Another approach consists of using an analytical fit that smoothes the spectral distribution of extinction $\sigma(\lambda)$ given by the four channels. If the size distribution happens to be bimodal, there would be too many parameters to expect their retrieval from SAGE II data; however, we will see that qualitative information about the bimodality can probably be achieved.

## 2. SPECTRAL VARIATION OF EXTINCTION FROM SAGE II DATA

Figure 1 shows, for example, the spectral variation $\sigma(\lambda)$ observed by SAGE II (28 November 1984 at 1645 GMT) at various altitudes between 16.5 km and 24.5 km, in relative value $(\sigma(\lambda)/\sigma(1.02))$. The dots correspond to the measured values at the three wavelengths 0.525 μm, 0.45 μm, and 0.385 μm; the curves represent a rms best fit, using the analytical expression

$$\ln \sigma(\lambda) = \ln \sigma(1.02) - a\ln(\lambda/1.02) - b(\ln(\lambda/1.02))^2 \qquad (1)$$

For comparison Fig. 2 represents the spectral variation of the extinction for aerosol models with a lognormal size distribution (LND), expressed by

$$n(r) = \frac{1}{\sqrt{2\pi}r \ln\sigma} \exp\left\{-\frac{\ln^2 r/r_m}{2\ln^2\sigma}\right\} \qquad (2)$$

where $r_m$ and $\sigma$ are the mode radius and the variance. An alternative convenient choice of parameters (Hansen and Hovenier, 1974) are the effective radius and effective variance defined by

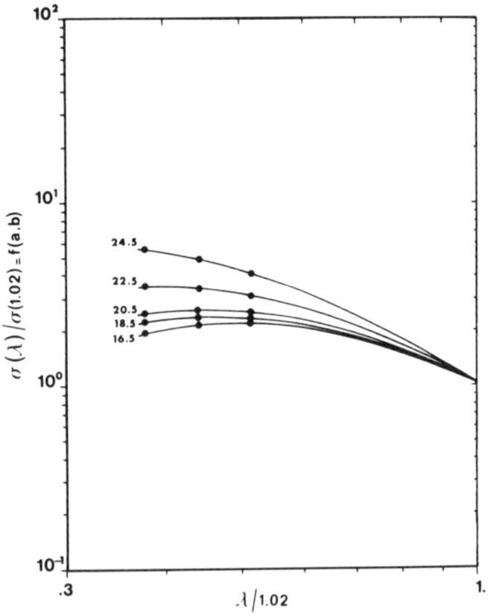


FIGURE 1. Spectral variation of the extinction coefficient $\sigma(\lambda)/\sigma(1.02)$. The dots are the measured values from SAGE II on 28 November 1984, 1645 GMT. The curves represent a rms best fit.

$$r_{eff} = r_m \exp(2.5 \ln^2 \sigma) \quad (3)$$

$$v_{eff} = \exp(\ln^2 \sigma) - 1 \quad (4)$$

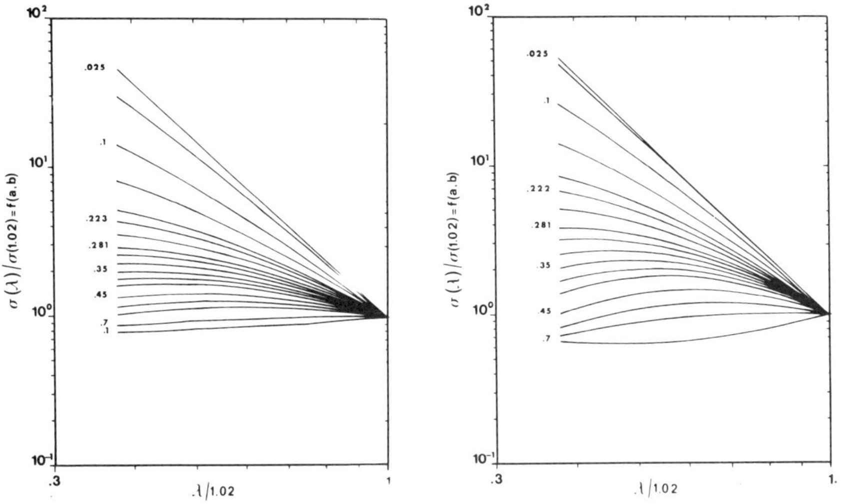


FIGURE 2. Spectral variation of the extinction coefficient $\sigma(\lambda)/\sigma(1.02)$ fitted by a rms for several aerosol models with a LND size distribution (a/ $v_{eff}$ = .25, b/ $v_{eff}$ = .05). The values on the curves are $r_{eff}$ in µm.

Figure 2a is for $v_{eff}$ = .25 ($\sigma$ = 1.60) and Fig. 2b for $v_{eff}$ = .05($\sigma$= 1.25); the set of curves corresponds to different values of $r_{eff}$. A qualitative comparison between Figs. 1 and 2 shows that the aerosols observed by SAGE II have an effective variance generally close to .05 (curvature of the curves) and an effective radius decreasing with altitude from about 0.40 µm to 0.25 µm.

A more quantitative comparison between SAGE II data and the aerosol models is possible using the expansion (1), which fits almost perfectly the spectral variation of the extinction for the LND models. Figure 3 is a diagram (a-b) for the LND models with various $r_{eff}$ and $v_{eff}$. The values of a and b for size distributions with two LND modes have also been drawn; they correspond to very small or even negative values of b (inversion of the curvature of the spectral curves from monomodal to bimodal size distributions already noted (Lenoble and Brogniez, 1985)). The points corresponding to three successive SAGE II observations in October 1985 (12 at 0502 GMT and 0639 GMT, 13 at 0651 GMT) for altitudes between 16.5 and 24.5 km are plotted on the (a-b) diagram of Fig. 3. The error rectangles are reported for two points. The points for the same altitude cluster reasonably and the shift with altitude confirms the qualitative observations of the size distribution variation. The values of $r_{eff}$

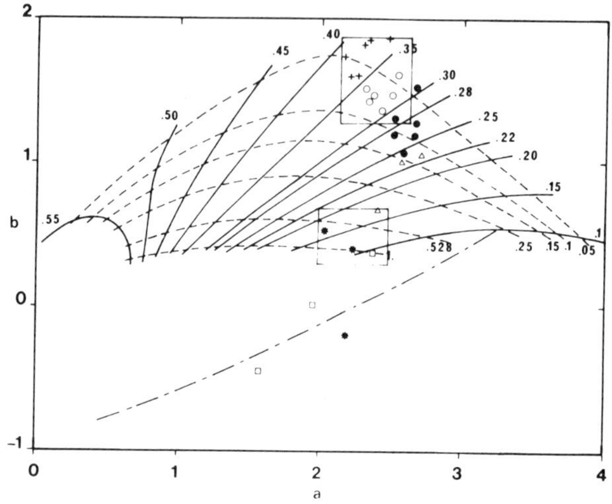


FIGURE 3. Diagram (a-b), where a and b are the coefficients of the rms fit (Eq. (1) in the text), for LND models. Full lines are for constant $r_{eff}$ (in μm), dashed lines are for constant $v_{eff}$, and dot-dashed line is for bimodal size distributions ($v_{eff}$ = .1, $r_{eff}$ = 1 μm; $v_{eff}$ = .25, $r_{eff}$ = .1 μm). The points are deduced from SAGE II observations on 12 October, 1985 (0502 GMT and 0639 GMT) and on 13 October, 1985 (0651 GMT) at different altitudes :

| | | |
|---|---|---|
| + 16.5–17.5 km | o 18.5–19.5 km | • 20.5–21.5 km |
| Δ 22.5 km | * 23.5 km | □ 24.5 km |

and $v_{eff}$ for a LND model "equivalent" to the aerosol by SAGE II, from the point of view of spectral extinction between 1.02 μm and .385 μm, can be read in Figure 3. The results are summarized in Table 1.

Up to 21 km the variance is surprisingly small; above, it increases and the possible presence of two modes cannot be rejected. However, the SAGE II data are of poor quality above 23 km, especially at short wavelengths.

TABLE 1. Value of $v_{eff}$ and $r_{eff}$ in m retrieved from the (a-b) diagram for 12 October, 1985 (0639 GMT)

| Z, km | 16.5 | 17.5 | 18.5 | 19.5 | 20.5 | 21.5 | 22.5 | 23.5 |
|---|---|---|---|---|---|---|---|---|
| $v_{eff}$ | <.05 | .05 | .07 | .05 | .05 | .10 | .40 | bimo- |
| $r_{eff}$, μm | .375 | .35 | .325 | .325 | .30 | .25 | .18 | dal? |

## 3. USE OF SAGE II CHANNEL RATIOS

Another, somewhat more direct approach to the size distribution study, is the utilization of various SAGE II channel extinction ratios

$$R_{\lambda_1 \lambda_2} = \sigma(\lambda_1) / \sigma(\lambda_2) \tag{5}$$

or of the related average Angström coefficient for the spectral intervals $(\lambda_1\lambda_2)$, defined as

$$\alpha_{\lambda_1\lambda_2} = - \ln R_{\lambda_1\lambda_2} / \ln(\lambda_1/\lambda_2) \tag{6}$$

Figure 4 shows the diagram $(\alpha_{.45/1.02}-\alpha_{.385/1.02})$ for the LND models; all the points are very close to the diagonal and the same behavior is observed for the other combinations using the ratios of two short wavelengths (.385 µm, .45 µm or .525 µm) to the channel 1.02µm. This forbids the retrieval of $r_{eff}$ and $v_{eff}$ from the SAGE II data.

Figure 5 is the diagram $(\alpha_{.45/1.02}-\alpha_{.385/.525})$ that looks much more promising for an analysis of SAGE II data, although the ratio of two short wavelength channels $(\sigma(.385)/\sigma(.525))$ has a larger error than the ratios of a short wavelength to 1.02 µm channel. In Fig. 5 is also drawn the curve for two LND mode size distributions and the points for SAGE II observations in October 1985 (same data as in Fig. 3); the error rectangles are given, as examples for two points. The retrieved values of $r_{eff}$ and $v_{eff}$ agree almost exactly with the values retrieved from the previous analysis on the (a-b) diagram (Fig. 3), and the uncertainties are of the same order.

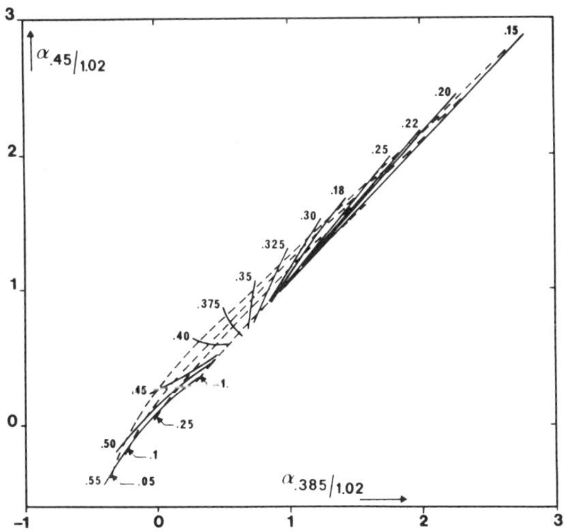


FIGURE 4. Diagram $(\alpha_{.45/1.02}-\alpha_{.385/1.02})$ for LND models. Full lines are for constant $r_{eff}$ (in µm) and dashed lines are for constant $v_{eff}$.

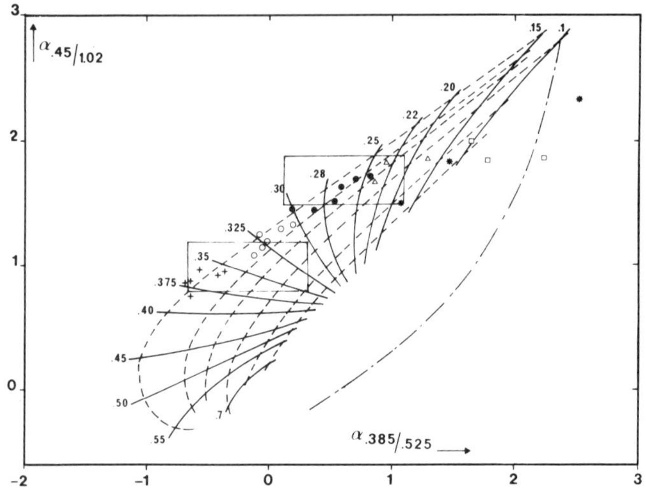

FIGURE 5. Diagram ($\alpha_{.45/1.02}$–$\alpha_{.385/.525}$) for LND models. Full lines are for constant $r_{eff}$ (in μm), dashed lines are for constant $v_{eff}$, dot-dashed line is for bimodal size distributions, and the points are for SAGE II observations as defined in Fig. 3.

## 4. COMPARISON WITH BALLOON DATA

The aerosol size distribution was deduced from balloonborne infrared polarimetry measurements on 12 October, 1985, during an European SAGE II validation experiment (Herman et al, 1986). The profiles of $r_{eff}$ and $v_{eff}$, shown in Fig. 6, compare reasonably with the values given in Table 1; a more detailed error analysis remains to be done.

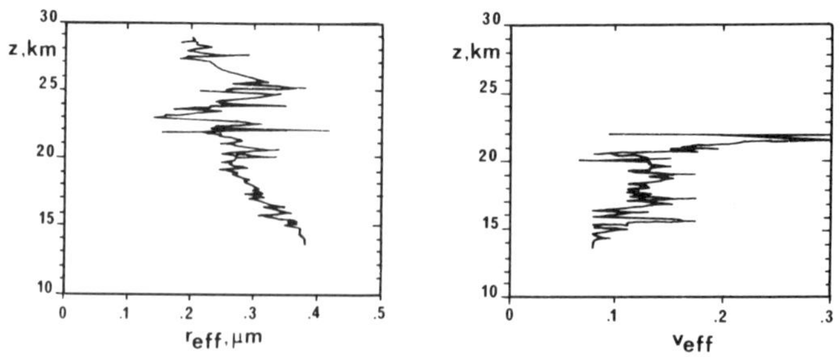

FIGURE 6. Profiles $r_{eff}$ (in μm) and $v_{eff}$ deduced from balloon experiment on 12 October, 1985.

## 5. CONCLUSION

The present analysis, although preliminary, proves that the SAGE II four-channel extinction should allow the retrieval of two parameters of a monomodal aerosol size distribution, although the error bars on the retrieved values are rather large, especially for the variance. Furthermore, it appears that the presence of a bimodal size distribution could be deduced, unless the relative contribution of one mode is too small.

## ACKNOWLEDGMENTS

The SAGE II data were kindly provided to us by the NASA Langley Research Center. This study has been supported by the Centre National d'Etudes Spatiales (Contrat CNES 1259).

## REFERENCES

Brogniez, C., and J. Lenoble, 1987: Modeling of the Stratospheric Background Aerosols from Zonally Average SAGE Profiles, J. Geophys. Res., 92, D3, 3051-3060.

Hansen, J.E., and J.W. Hovenier, 1974: Interpretation of the Polarization of Venus, J. Atm. Sci., 31, 1137-1160.

Herman, M., J.Y. Balois, L. Gonzalez, P. Lecomte, J. Lenoble, R. Santer and C. Verwaerde, 1986: Stratospheric Aerosol Observations from Balloonborne Polarimetric Experiment, App. Opt., 25, 3573-3584.

Lenoble, J., P. Pruvost and C. Brogniez, 1984: SAGE Satellite Observations of Stratospheric Aerosols from Mount St. Helens Eruption: A two Wavelength Analysis, J. Geophys. Res., 89, 11666-11667.

Lenoble, J., and C. Brogniez, 1985: Information on Stratosphere Aerosol Characteristics Contained in the SAGE Satellite Multiwavelength Extinction Measurements, App. Opt., 24, 1054-1063.

Yue, G.K., and A. Deepak, 1983: Retrieval of Stratospheric Aerosol Size Distribution from Atmospheric Extinction of Solar Radiation at Two Wavelengths, App. Opt. 22, 1639-1645.

Yue, G.K., and A. Deepak, 1984: Latitudinal and Altitudinal Variation of Size Distribution of Stratospheric Aerosols Inferred From SAGE Aerosol Extinction Coefficient Measurements at Two Wavelengths, Geophys. Res. Letts., 11, 999-1002.

# TEN YEARS OF $CFCL_3$ MEASUREMENTS IN THE STRATOSPHERE: MODEL COMPARISONS FOR THE NORTHERN HEMISPHERE

*Robert Leifer*
Environmental Measurements Laboratory, U.S. Department of Energy,
New York, New York 10014, USA

ABSTRACT

More than 800 samples of stratospheric air have been analyzed for $CFCl_3$. These samples were collected for the U.S. Department of Energy's High Altitude Sampling Program from 1974 to 1983. These samples represent the largest single data base of concentrations for the Northern Hemisphere stratosphere available to the scientific community for analysis. Presented in this paper are direct comparisons of the observed distributions with predictions of distributions of $CFCl_3$ by one-dimensional and two-dimensional models. Some agreement of observations with the predictions of the two-dimensional models is apparent, especially in the slope of isopleths, but at the present time these models do not accurately predict the distribution of $CFCl_3$ during the spring and summer periods below 20 km. The one-dimensional models for various latitudes also fail to predict the observed vertical distribution of $CFCl_3$ at altitudes below 20 km.

## 1. INTRODUCTION

It has been more than 10 years since Rowland and Molina (1975) first identified the possible consequences of the continued release of chlorofluorocarbons (CFCs) to the atmosphere. Estimates using one-dimensional models (1-D) of the stratospheric total column ozone loss from these emissions have ranged from 2% to 19% (National Research Council, 1983). These estimates have changed over the years because of our improved understanding of the chemical reactions and the rate constants that are involved in stratospheric ozone chemistry. One-dimensional models incorporate atmospheric radiative, chemical, and dynamical systems to approximate the real atmosphere for prediction of the vertical distribution of $CFCl_3$. One-dimensional models can provide insight into the response of ozone to various perturbations. These models must parameterize transport in the vertical; horizontal motions are averaged out. Feedback mechanisms involving dynamical changes are not included in these models.

Various 1-D models have been developed to represent specific latitudes and seasons by incorporating into the model the change in solar zenith angle with latitude and season. Two-dimensional (2-D) models describe latitude gradients and seasonal changes of the distribution of various trace gases. Incorporated into these models is atmospheric time-dependent meridional transport. But even these

 A. DEEPAK Publishing
ISBN 0-937194-11-5

models suffer from an inability to truly simulate the actual circulation and diffusion in the stratosphere, which is often asymmetric. Only a limited set of data has been available to validate 1-D models. Verification of 2-D models with **in situ** measurements has not been reported.

Presented in this paper are 10 years of stratospheric measurements of $CFCl_3$ in the Northern Hemisphere obtained for the U. S. Department of Energy's High Altitude Sampling Program (HASP). HASP data are compared to predictions by a DuPont 2-D model for the spring and summer seasons (Fisher, 1987) and by Harvard 1-D models reported in the National Academy Ozone update (National Research Council, 1983).

## 2. SAMPLING PROCEDURE

During the late 1950s, a high pressure gas sampling system was designed and used in the HASP program to characterize radioactive gases injected into the stratosphere from nuclear weapons testing. Early in 1973, HASP was modified to include sampling of the stratosphere for trace gases, including $CFCl_3$ (Krey et al., 1976; 1977). Over the next few years, trace gas mixing ratios of $CFCl_3$, $CF_2Cl_2$, $N_2O$, $SF_6$, and $CCl_4$ were measured. An ultra-clean small volume gas sampling system (known as CLASS) was designed in 1980. This system replaced the original high-pressure sampling system and provides measurements of CO, $CO_2$, $CH_4$, COS, and $CH_3CCl_3$, in addition to the original compounds (Leifer et al., 1981). Sampling was carried out three times a year (spring, summer, and fall) primarily in the Northern Hemisphere stratosphere. Ambient air was pressurized to 2 atmospheres in 3-liter stainless steel bottles. The samples were chemically analyzed at the Environmental Measurements Laboratory, Washington State University or the Oregon Graduate Center (Leifer et al., 1980, 1982, 1984).

Data from Schmidt et al. (1986) are compared with HASP data at similar latitude, altitude, and time of year. Differences of $< 10\%$ in absolute concentrations were found. The Schmidt data are traceable to Ale standards (Rasmussen and Lovelock, 1983).

## 3. MODEL-OBSERVATION COMPARISON: 2-D

Since the early 1970s, the concentrations of $CFCl_3$ in the troposphere and stratosphere have been increasing (Brice et al., 1982, Leifer et al., 1982). Model predictions of the latitudinal distribution of $CFCl_3$ have usually been published for one particular year. The 2-D model provides an instantaneous picture of the stratospheric distribution of $CFCl_3$ and incorporates the best state-of-the-art techniques for model calculations, but does not take into account the changing concentration with time. These model predictions are compared to measurements for a specific set of **in situ** measurements.

When a data base as extensive as HASP, which spans 10 years, is used for comparison with model predictions, the concentration data must first be corrected for time trends to make them comparable to models. In order to compare a 10-year average of the HASP data to these models, the increasing trend of $CFCl_3$ with time must be removed. The measured concentration at each sampling location in the stratosphere has been converted to a ratio that is normalized to the tropospheric concentration (stratospheric concentration divided by the tropospheric concentration). The tropospheric concentrations used for this calculation were measured at ground level at Harwell, England (Brice et al., 1982). All those samples collected and analyzed during the HASP program that were located within a grid element of ± 2 degrees latitude and ± 1 km altitude were averaged to form a 10-year average profile. This profile does not reflect the winter period because no samples were collected then. Once the average ratio profiles were obtained, they were normalized to the 1980 tropospheric value of 180 pl $l^{-1}$ (matching the tropospheric concentration used in the DuPont model) for comparison with the DuPont model. Specifically, for this study attention was paid to the comparison of $CFCl_3$ concentrations in the stratosphere, with the predictions of the 2-D model of DuPont for the spring and late summer periods (Fisher, 1987). The results of these calculations are shown in Figs. 1a - 1d, and represent an average profile for the spring, summer, fall, and total data sets for the years 1974 through 1983, respectively. Included in these graphs, as dotted lines, are the predictions of the DuPont 2-D models. Since a composite DuPont model for the three seasons was not available, only the summer 2-D model is plotted with the fall HASP data set (Fig. 1c). Similarly, the HASP total data set (Fig. 1d) contains the late-summer prediction of the DuPont model.

The differences in concentration gradients between seasons of the HASP data (solid lines, Figs. 1a - 1d) are evident. The steep concentration gradients seen at the high latitudes for the spring and fall plots reflect dynamical changes in the stratosphere during or prior to the sampling cycle. The summer plot (Fig. 1b) represents a period of greater stability in the stratosphere. The concentration gradients are smaller in the polar region, and the slope of the isopleths in the polar region are less steep during the summer period. The concentrations in the high latitudes for the spring and fall are much lower than in the summertime, reflecting either the advection of polar air into this region or air that has subsided.

Qualitatively both the HASP data and DuPont model predictions show stratospheric concentration gradients larger in the spring period than in the summer period. The HASP concentration data fall off less rapidly with height than do the model predictions at most latitudes, for all seasons. The slope of the isopleths in the polar region for the HASP data and model predictions are in reasonable agreement for the spring and summer periods. The fall months show steeper gradients in the polar region for the HASP data than are predicted by the DuPont model. The cause of this difference may be due to the fact that the DuPont model does not represent the same time period as the HASP data, but rather an earlier year. Since no DuPont prediction model is available in order to match the HASP total data set (Fig. 1d), it is difficult at this time to draw any conclusions about the composite data set or the fall set. The summer

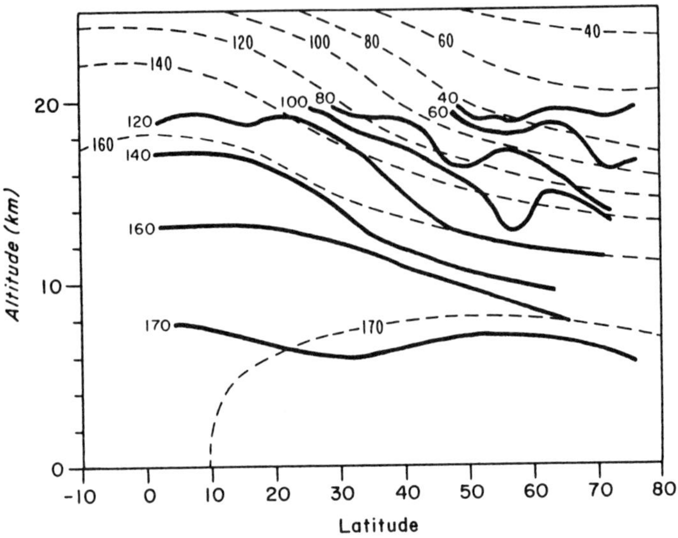


FIGURE 1a. Plot of HASP data averaged over the spring (solid lines, HASP data; dotted lines, spring 2-D DuPont model, Fisher, 1987).

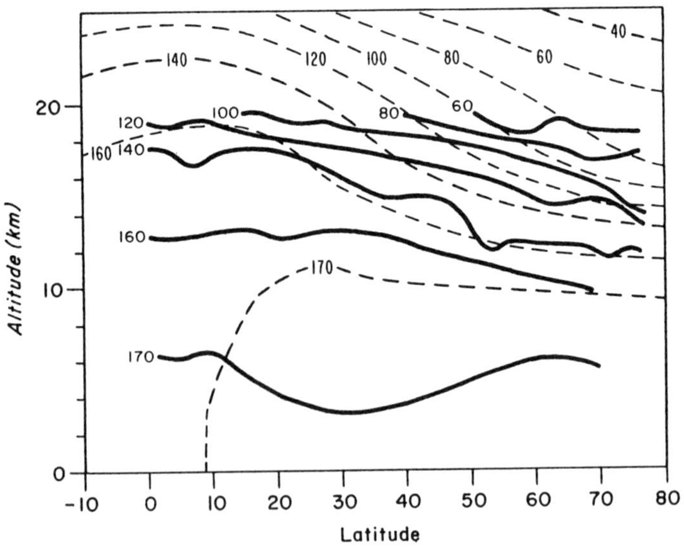


FIGURE 1b. Plot of HASP data averaged over the summer (solid lines, HASP data; dotted lines, late summer 2-D DuPont model, Fisher, 1987).

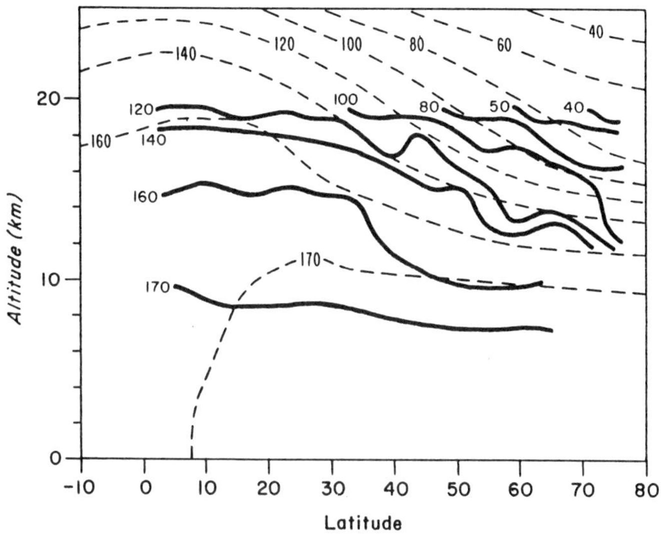


FIGURE 1c. Plot of HASP data averaged over the fall (solid lines, HASP data; dotted lines, late summer 2-D DuPont model, Fisher, 1987).

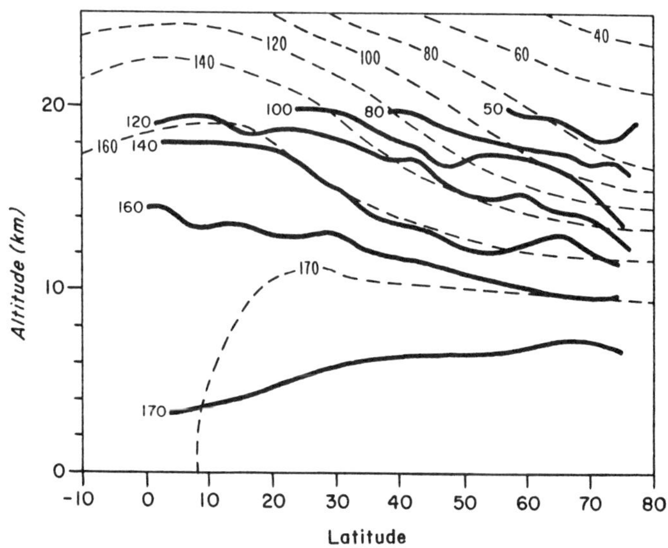


FIGURE 1d. Composite of HASP data, spring, summer, and fall (solid lines, HASP data; dotted lines, late summer 2-D DuPont model, Fisher, 1987).

DuPont model is included only for reference.

The differences seen between the HASP data and the DuPont model predictions probably reflect inaccuracies in the treatment of transport by the models. Further studies will be carried out with researchers at DuPont, focusing on the use of an objective analysis for drawing isopleths for the HASP data, and obtaining model predictions for additional data sets. HASP measurements of $CF_2Cl_2$, $CCl_4$, and $N_2O$ will be compared to similar models.

## 4. MODEL-OBSERVATION COMPARISON: 1-D

To transform our data into equivalent 1-D profiles, special normalizations were carried out. Sample collection altitudes were rounded off to the nearest whole kilometer. The total data set for the years 1974 to 1983 was reduced and plotted in two formats (Figs. 2a and 2b). These plots incorporate results for $>$ 800 samples. The actual sample altitude is plotted against a normalized ratio (the same ratio that was discussed in the previous section) in a box formatted plot (Fig. 2b). The upper and lower points of the vertical lines extend to the maximum and minimum values. The bottom and top of the boxes represent the ratio of the lowest 25% and 75% of the data, respectively. The central dot in the box is the 50% value. The width of the box is proportional to the number of samples in the analysis. Whereas the tropopause height is a function of the latitude, it is difficult to see the transition from the troposphere to stratosphere in Fig. 2a. The decrease of the ratio with altitude is observable in this figure, but is not well defined. Transforming the abscissa of Fig. 2a from "sample" altitude to a "normalized" altitude (NA) radically changes the appearance of the plotted data (Fig. 2b). The range in ratios for a given NA was reduced and the rapid decrease of the ratio with increasing higher NA is clearly seen. This transformation is accomplished by subtracting from the sample altitude the tropopause height at the latitude of sample collection. Sometimes the actual tropopause height for a given latitude and date was not available, and an extrapolated value was used instead. This occurred in $<$ 5% of the samples.

Shown in Fig. 3a is a boxplot of the total $CFCl_3$ data set. Also included in the figure is an exponential fit of the data in the stratosphere and predictions by two 1-D models from Harvard (National Research Council, 1983). In order to study the latitudinal dependence of $CFCl_3$ in the stratosphere, the data base was divided into three latitude bands, covering a low-latitude (0-25$^\circ$N), mid-latitude (26-50$^\circ$N), and high-latitude (51-75$^\circ$N) regions, and into three seasons, spring, summer, and fall.

Utilizing the subdivision of the data into latitudinal and seasonal subsets, calculations were carried out to evaluate a "least squares fit" to the data using the National Bureau of Standards "Dataplot" computer program (version 86/10) written by James Filliben. Rounding the normalized altitudes to the nearest kilometer and calculating the average mixing ratio for each altitude produced the best exponential fit. The data used for fitting were limited to

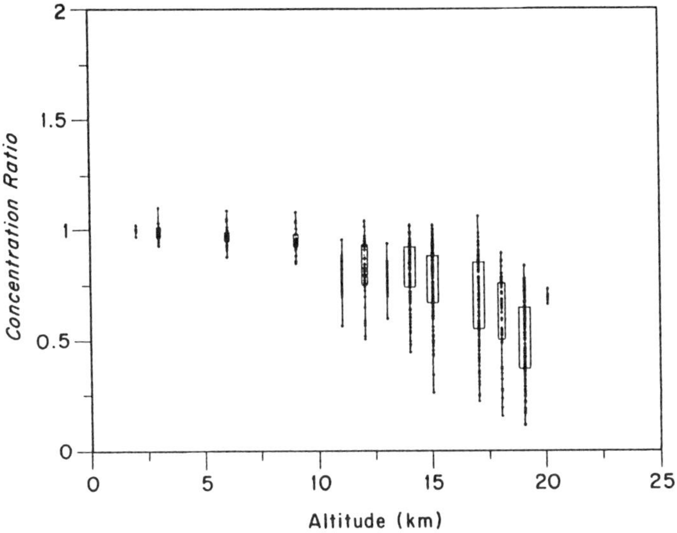


FIGURE 2a. An altitude ratio boxplot of $CFCl_3$ for the years 1974 through 1983. The top of the box is the 50% ratio; the bottom of the box is the 25% ratio. The median value is in the center of the box and width of the box is proportional to the number of samples. The upper and lower limits of the vertical line extend to maximum and minimum ratios, respectively.

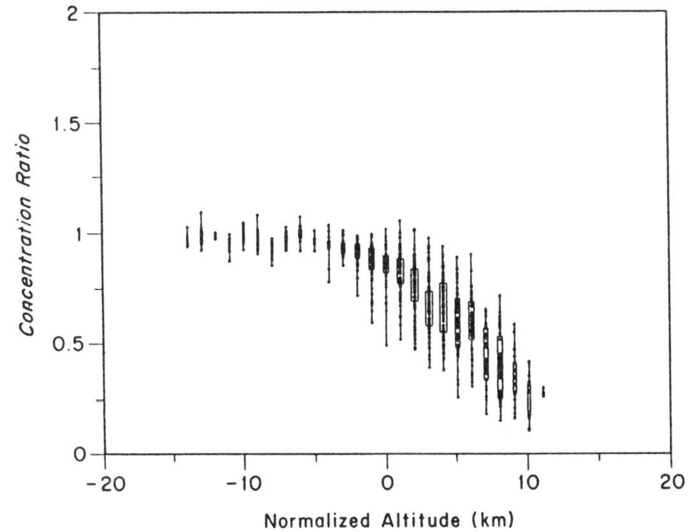


FIGURE 2b. A normalized altitude ratio boxplot of $CFCl_3$ for the years 1974 through 1983. (For definition of boxplot, see Figure 2a.)

those altitudes equal to or greater than 1 km above the tropopause. This reduces the possibility of inadvertently including tropospheric data in the stratospheric fit program. The fact that the smoothed data provide an improved fit may reflect the inaccuracy in identifying the tropopause height. Coefficients were obtained for an exponential fit of the form:

$$\text{ratio} = A_0 e^{-B_0 \text{ normal}}$$

Where ratio is the stratospheric mixing ratio divided by the tropospheric mixing ratio, $A_0$ and $B_0$ are constants, and normal is the altitude relative to the tropopause in kilometers. Other models were tried (linear power law) with no improvement in fit. The results of these analyses, including the calculated correlation coefficients, are included in Table 1 and Figs. 3a through 3c.

For this paper, the composite or total (Fig. 3a), midlatitude (Fig. 3b), and spring (Fig. 3c) data sets are discussed. The Harvard models included in these figures were obtained from the National Research Council (1983) document on ozone change. The concentration profiles in the Harvard model were converted to a normalized ratio profile by dividing each concentration at a given altitude by the tropospheric value shown in the original plot. The altitude was converted to a normalized altitude by identifying the tropopause, (the altitude where the concentration starts to decrease: in the 15°N plot at 14 km, in the 45°N plot at 11 km). The result of the transformation from plotting concentration against altitude-to-plotting ratio against normalized altitude is evident from the figures. For each of the plots shown, more than 50% of the HASP data are within ~ 10% of the mean for each normalized altitude. The stratospheric ratio falls to one half the tropospheric value within 7 km of the tropopause for all plots. Neither of the two Harvard models, at 15°N or 45°N, are able to simulate the HASP data for the total-data, summer or midlatitude sets. Concentrations predicted by the 45°N Harvard model fall off too fast with altitude, compared to those for the HASP 51-75°N data plot, while for the 15°N model the opposite occurs. The way transport in the vertical is parameterized in a model will determine to a great extent the shape of the profile. Further investigations will be carried out to quantify these differences, and we will work with modelers to better fit our data.

## 5. SUMMARY

Variations in the latitudinal distribution of the measured concentrations from one season to another are evident in the HASP data base. Average concentration and ratio profiles were obtained for 10 years of analyzed samples and have been compared to various models. Both 1-D and 2-D models, discussed in this paper for comparison with actual measured stratospheric concentrations, fail to predict accurately the seasonal and latitudinal stratospheric $CFCl_3$ profiles for the years 1974 through 1983.

TABLE 1. EXPONENTIAL FIT TO HASP DATA
COEFFICIENTS AND CORRELATIONS

| Data set | $A_0$ | $B_0$ | Correlation Coefficient |
|---|---|---|---|
| All data | 0.92 ± 0.02 | 0.09 ± 0.00 | -0.73 |
| 26 to 50°N | 0.92 ± 0.02 | 0.08 ± 0.01 | -0.68 |
| spring | 0.86 ± 0.03 | 0.09 ± 0.01 | -0.70 |

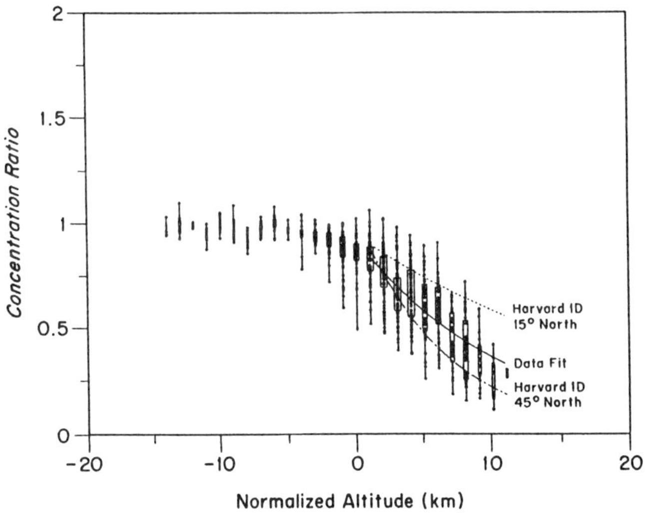


FIGURE 3a. A normalized altitude ratio boxplot of $CFCl_3$ for the years 1974 through 1983. The solid line is an exponential fit to the data. The dotted line is the Harvard 1-D model at 15°N (National Research Council, 1983). The dashed line is the Harvard 1-D model for 45°N (National Research Council, 1983). (For definition of boxplot, see Figure 2a.)

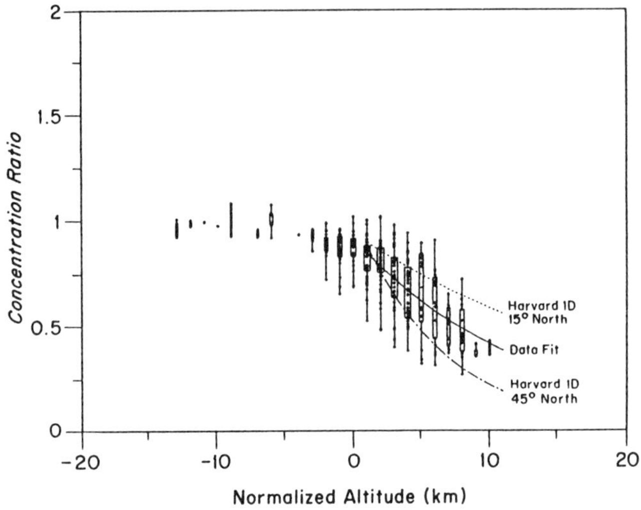


FIGURE 3b. A normalized altitude ratio boxplot of $CFCl_3$ for the years 1974 through 1983 for the latitude band 26°N to 50°N. The solid line is an exponential fit to the data. The dotted line is the Harvard 1-D model at 15°N (National Research Council, 1983). The dashed line is the Harvard 1-D model for 45°N (National Research Council, 1983). (For definition of boxplot, see Figure 2a.)

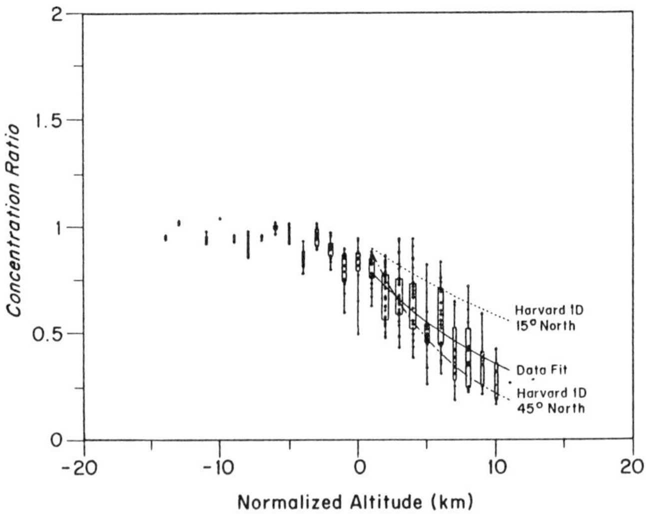


FIGURE 3c. A normalized altitude ratio boxplot of $CFCl_3$ for the years 1974 through 1983 for the spring period. The solid line is an exponential fit to the data. The dotted line is the Harvard 1-D model at 15°N (National Research Council, 1983). The dashed line is the Harvard 1-D model for 45°N (National Research Council, 1983). (For definition of boxplot, see Figure 2a.)

ACKNOWLEDGMENTS

I would like to thank the flight crew and support staff of NASA, Johnson Space Center for making this program possible. Also I would like to thank the Los Alamos National Laboratory staff for its support and cooperation in field operations; Z.R. Juzdan of EML for his help with CLASS and the data analyses; and S. Silvestri of EML for her diligent help with various data base computations. The samples were analyzed by R. Lagomarsino of EML, D. Cronn of Washington State University, and R. A. Rassmussen of Oregon Graduate Center.

REFERENCES

Brice, K. A., R. G. Derwent, A. E. J. Eggleton and S. A. Penkett, 1982: Measurements of $CCl_3F$ and $CCl_4$ at Harwell Over the Period January 1975-June 1981 and the Atmospheric Lifetime of $CCl_3F$, Atmos. Environ., 16, 2543-2554.

Fisher, D., E. I. DuPont De Nemours & Company, Wilmington, Delaware (this model available through the NASA Data Center), 1987.

Krey, P. W., R. J. Lagomarsino and J. J. Frey, 1976: Stratospheric Concentrations of $CCl_3F$ in 1974, J. Geophys. Res., 81, 1557-1560.

Krey, P. W., R. J. Lagomarsino and L. E. Toonkel, 1977: Gaseous Halogens in the Atmosphere in 1976, J. Geophys. Res., 82, 1753-1766.

Leifer, R., L. Toonkel, R. Larsen and R. J. Lagomarsino, 1980: Trace Gas Concentrations in the Stratosphere of the Northern Hemisphere During 1976, J. Geophys., 85, 1069-1072.

Leifer, R., K. Sommers and S. F. Guggenheim, 1981: Atmospheric Trace Gas Measurements With a New Clean Air Sampling System, Geophys. Res. Letts., 8, 1079-1081.

Leifer R., R. Larsen and L. Toonkel, 1982: Trends in Stratospheric Mixing Ratios of Trace Gases in the Northern Hemisphere During The Years 1974-1979, Geophys. Res. Letts., 9, 755-758.

Leifer, R., Z. R. Juzdan and R. Larsen, 1984: COS in the Stratosphere: El Chichon Observations, Geophys. Res. Letts., 11, 549-552.

National Research Council, 1983: Causes and Effects of Changes in Stratospheric Ozone: Update 1983, Committee on Causes and Effects of Changes in Stratospheric Ozone: Update 1983, National Academy Press., Washington, D.C.

Rasmussen, R. A., and E. Lovelock, 1983: The Atmosphere Lifetime Experiment 2: Calibration, J. Geophys. Res., 88, 8369-8378.

Rowland, F. S., and M. J. Molina, 1975: Chlorofluoromethane in the Environment, Rev. Geophys. Space Phys., 13, 1-35.

Schmidt, U., C. Jebsen, F. J. Johnen, A. Khedim, E. Klein, D. Knapska, G. Kulessa, J. Rudolph, G. Schumacher and E. Schunck, 1986: Stratospheric Observations of Long-Lived Trace Gases at Midlatitudes 1982-1985 (Data Report), Kernforschungsanlage Julich GmbH, F.R.G., ISSN 0343-7639.

# STRATOSPERIC AEROSOL CONDITIONS OVER MAUNA LOA DURING RECENT QUIESCENT VOLCANIC PERIODS

*Elmer Robinson and Thomas E. DeFoor*
Mauna Loa Observatory, NOAA
Hilo, Hawaii 96721, USA

ABSTRACT

Lidar observations of stratospheric aerosol conditions have been made at the Mauna Loa Observatory for more than 10 years. Plumes from large-scale volcanic eruptions, the usual causes of enhanced aerosol backscattering in the stratosphere, and the impacts of major volcanic events such as Fuego in 1974 and El Chichon in 1982 are readily evident in the lidar record. However, in the intervals between major eruptions over the past 10 years the apparent residual or "background" stratospheric aerosol optical depth, as calculated from the Mauna Loa ruby lidar at 694 nm, has gradually increased at an annual rate of about 4.5 x $10^{-4}$, or about 10% per year. The most likely explanation for this long-term increase in residual aerosol backscattering is the gradual accumulation of volcanic eruption materials in the stratosphere both from the large, well-publicized eruptions and from the more numerous, less intense events. These features of the residual stratospheric aerosol conditions and the impact of the more frequent moderate-scale eruptions are examined in this discussion with particular reference to recent Mauna Loa lidar observations.

## 1. INTRODUCTION

Investigators of long-term global climatic change have frequently considered the impacts of volcanism and the resulting persistent and widespread volcanic plumes, which penetrate the stratosphere, as an important factor in the causation of climatic anomalies. Humphreys (1940), Budyko (1974), and Lamb (1966) are prominent investigators in the field of volcanic links to climate. In the present report we have made an assessment, not of the periods of extreme volcanic activity such as the El Chichon plume period (DeLuisi et al., 1983), but rather of the interregnum or generally quiescent volcanic periods between major eruptions when, in many cases, it has been assumed that the atmosphere has returned to some sort of hard-to-define "background" state (e.g., Hofmann and Rosen, 1981; Wendler and Kodama, 1986). This time to return to a normal background has been estimated to be from a few months for small volcanic episodes to a number of years for very large eruptions.

## 2. OBSERVATIONAL DATA

This analysis of stratospheric background aerosol conditions is based on more than 10 years of regular lidar observations made at the Mauna Loa Observatory (MLO) in Hawaii (19°32'N, 155°35'W, 3400 m MSL).


A. DEEPAK Publishing
ISBN 0-937194-11-5

The ruby laser lidar system at MLO has been a regular part of the observatory program since 1973 (Miller, 1974; DeLuisi et al., 1984, 1985). Briefly, the MLO lidar system is composed of a Q-switched ruby laser (3-J @ 694 nm) transmitter, a 40 cm Cassegrainian telescope receiver, a photomultiplier tube (PMT) detector, and a Biomation model 8100 waveform recorder operating as a signal digitizer. Data transfer and coding from the Biomation to a DEC TU58 tape is performed by a PDP11/23 microcomputer system. A typical observation is performed by sequencing through three system configurations over a total of 100 laser shots. Each configuration operates at a unique PMT gain and Biomation sample delay combination designed to optimize the signal and dynamic range of the system. Later, using a PDP11/23 microcomputer system, the TU58 tapes are read and decoded, bad shots and bad data are eliminated, and aerosol backscatter profiles are calculated for the consolidated data set.

The basic features of the computer program used to process the MLO lidar data have been given by DeLuisi et al. (1984). Within this processing program the system background signal is generally determined from the highest altitude levels of the lidar profile and the basic Rayleigh backscattering signal is determined from a level in the upper troposphere near the tropopause where aerosol concentrations are considered to be negligible. In this present study the lidar data are presented in the form of aerosol backscattering integrated between the altitudes of 16 and 33 km. This is noted as the integrated non-Rayleigh backscattering coefficient, or NRBS, and has the units of $sr^{-1}$. A conversion of NRBS values to estimates of optical depth can be made by the use of a relationship proposed by DeLuisi et al. (1984, 1985), namely: (optical depth)/(NRBS) = 50. In other studies the MLO lidar data have been presented as the aerosol backscattering ($B_{sp}$) ratio: (Rayleigh + aerosol)/(Rayleigh). For profile analyses the lidar signal is integrated over 300 m vertical intervals.

## 3. DATA PRESENTATIONS

The basic data used for this analysis are the MLO integrated non-Rayleigh backscatter data, from the beginning of observations in December 1974 through some of the most recent observations of June 1987. Figure 1 shows the time-plot of the individual observation values of the NRBS coefficient integrated from 16 to 33 km, or to the top of the observation if data were not obtained at the higher altitudes. Since the tropopause in the vicinity of Hawaii is in the 15- to 17-km range these integrated NRBS values are applicable generally to stratospheric conditions. Since the MLO lidar is a ruby laser system, these data, obviously, are strictly applicable only to the ruby laser wavelength of 694 nm. As is apparent from the pattern of the plotted points, the density of observations was not uniform over the 13-year period. In the most recent period, and especially since the eruption of Mount St. Helens in May of 1980, the usual routine has been weekly lidar observations. Figure 1 shows that integrated NRBS values were relatively high early in the observation period 1974-1976, the Fuego eruption period, and again for several years beginning in mid-1982, the El Chichon era.

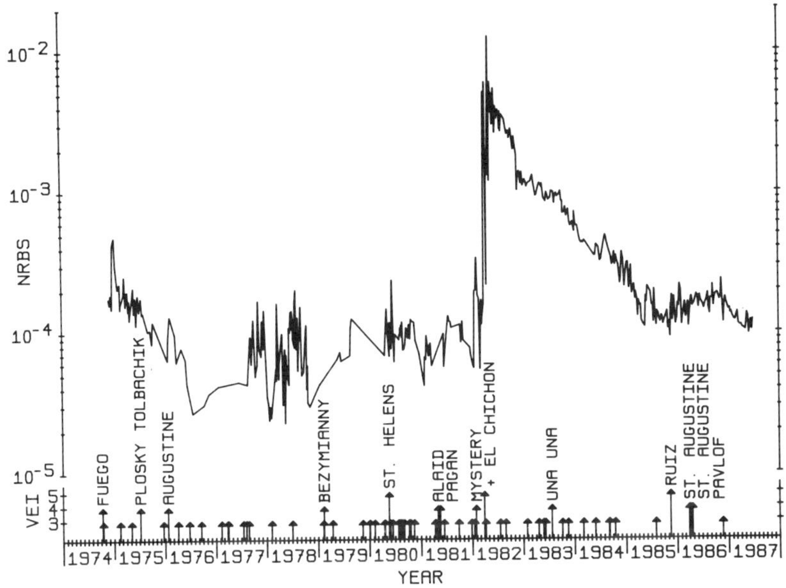


FIGURE 1. Mauna Loa lidar integrated NRBS coefficient (16 to 33 km) and volcanic explosive index (VEI $\geq$ 3) for 1974-1987.

Figure 1 also shows the eruptions of major volcanoes during this period. These volcanic eruptions have been classified on the basis of the Smithsonian Volcanic Explosive Index (VEI) to show the probability of the eruption plume entering the stratosphere (Simkin et al., 1981 and 1984). This probability has been considered to be a VEI value of 3 or larger. The record of volcanic eruptions with a VEI of 3 or greater totals 73 in the period from late 1974 through June 1987. Of these, 13 have a VEI of 4 or greater, which corresponds to an estimated plume column height of over 10 km by definition of the VEI and thus to "definite" stratospheric injection. VEI values of 3 correspond to a "possible" stratospheric injection and a plume column 3 to 15 km in height (Simkin et al., 1981). Eruptions with a VEI value of 4 or larger are shown by name in Fig. 1. Major eruptions that apparently produced stratospheric clouds that were identifiable in the MLO record include Fuego in 1974 at the beginning of MLO lidar observations, Augustine in 1976, Mount St. Helens in 1980, El Chichon in 1982, and, most recently, Ruiz in 1985. Table 1 lists by name and assigned VEI all of the eruptions indicated in Fig. 1. More details on eruptions

TABLE 1. Large[1] Volcanic Eruptions and VEI Values: 1974 - 1987

| YEAR/MONTH/DAY | | NAME | VEI |
|---|---|---|---|
| 1974 | 0103 | Sakura-Jima | 3 |
| | 0122 | Ngauruhoe | 3 |
| | 0211 | Api Siau | 3 |
| | 0219 | Great Sitkin | 3 |
| | 0304 | Manam | 3 |
| | 0408 | Kliuchevskoi | 3 |
| | 0715 | Trident | 3 |
| | 1010 | Fuego | 4 |
| | 1015 | Pavlof | 3 |
| 1975 | 0212 | Ngauruhoe | 3 |
| | 0504 | Karymsky | 3 |
| | 0706 | Plosky Tolbachik | 4 |
| | CONT | Trident | 3 |
| 1976 | 0122 | Augustine | 4 |
| | 0402 | Sakura-Jima | 3 |
| | 0621 | Poas | 3 |
| | 0915 | Api Siau | 3 |
| 1977 | 0209 | Santa Maria | 3 |
| | 0325 | Bezymianny | 3 |
| | 0330 | Ukinrek Maars | 3 |
| | 0716 | Etna | 3 |
| | 0807 | Usu | 3 |
| | 0825 | White Island | 3 |
| 1978 | 0204 | Westdahl | 3 |
| | 07xx | Dukono | 3 |
| 1979 | 0208 | Westdahl | 3 |
| | 0211 | Bezymianny | 4 |
| | 0413 | Soufriere of St. Vincent | 3 |
| | 1113 | Negra, Sierra | 3 |
| 1980 | 01xx | Kliuchevskoi | 3 |
| | 0130 | Nyamuragira | 3 |
| | 0207 | Bulusan | 3 |
| | 0415 | Lopevi | 3 |
| | 0418 | Bezymianny | 3 |
| | 0518 | St. Helens | 5 |
| | 0525 | St. Helens | 3 |
| | 0612 | St. Helens | 3 |
| | 0722 | St. Helens | 3 |
| | 0723 | Ambrym | 3 |
| | 0728 | Gorley Khrebet | 3 |
| | 0807 | Gareloi | 3 |
| | 0807 | St. Helens | 3 |
| | 0817 | Hekla | 3 |
| | 0901 | Etna | 3 |
| | 1006 | Ulawun | 3 |
| | 1016 | St. Helens | 3 |
| | 1111 | Pavlof | 3 |

TABLE 1. (Concluded)

| YEAR/MONTH/DAY | | NAME | VEI |
|---|---|---|---|
| 1981 | 0409 | Bulusan | 3 |
| | 0409 | Hekla | 3 |
| | 0427 | Alaid | 3 |
| | 0430 | Alaid | 4 |
| | 0515 | Pagan | 4 |
| | 0612 | Bezymianny | 3 |
| | 0925 | Pavlof | 3 |
| | 1225 | Nyamuragira | 3 |
| 1982 | 0115 | Gareloi | 3 |
| | 0121 | Mystery | 4 |
| | 0328 | El Chichon | 5+ |
| | 0404 | El Chichon | 5+ |
| | 0405 | Galunggung | 3 |
| | 0714 | Galunggung | 3 |
| | 0718 | Raung | 3 |
| | 0813 | Galunggung | 3 |
| | 0826 | Soputan | 3 |
| 1983 | 0126 | Sakura-Jima | 3 |
| | 0418 | Langila | 3 |
| | 0522 | Sakura-Jima | 3 |
| | 0522 | Bezymianny | 3 |
| | 0604 | Veniaminof | 3 |
| | 0721 | Una Una | 3 |
| | 1003 | Miyake-Jima | 3 |
| | 1114 | Pavlof | 3 |
| 1984 | 0302 | Home Reef | 3 |
| | 0526 | Soputan | 3 |
| | 0831 | Soputan | 3 |
| | 0913 | Mayan | 3 |
| | 1013 | Bezymianny | 3 |
| 1985 | 0730 | Sangeang Api | 3 |
| | 1113 | Nevado del Ruiz | 4 |
| 1986 | 0327 | St. Augustine | 4 |
| | 0403 | St. Augustine | 4 |
| | 0418 | Pavlof | 4 |
| | 1119 | Chikurachki | 3 |
| 1987 | 02xx | Kliuchevskoi | 3 |

[1]VEI $\geq$ 3.
(Simkin et al., 1981, 1984; SEAN Bulletin, 1985, 1986, 1987).

and the individual volcanoes can be found in Simkin et al. (1981, 1984) and the SEAN Bulletin (1985, 1986, 1987). It should be noted that although the VEI for the El Chichon eruption was given as 4 (Simkin et al., 1984), we have arbitrarily assigned the El Chichon eruption a scale value of 5+ because of its obvious global stratospheric impacts. For volcanic eruptions since the 1984 tabulation by Simkin and colleagues we have used plume altitudes reported by the SEAN Bulletin (1985, 1986, 1987) to estimate VEI values.

Although Fig. 1 clearly shows the stratospheric plume signals of a number of eruptions, it also seems to indicate that the "clean" periods between major eruption events show increasingly larger values of NRBS as time progresses. This is emphasized by the eight points of relative minimum values of integrated NRBS shown in Fig. 2 and listed in Table 2. In the period 1976 to 1979 low NRBS values were in the 0.3 to 0.4 x $10^{-4}$ $sr^{-1}$ range while in 1985 and 1987 the low values were about 1 x $10^{-4}$ $sr^{-1}$, a twofold to threefold increase. Figure 3 shows these eight relative minimum points on a time plot and the resulting regression line for this set of points. Assuming an exponential relationship, the correlation for this regression and the eight points is 0.93.

This increase in the stratospheric residual aerosol values over the 12-year period amounts to an average increase of about 14% per year in the residual value of the integrated NRBS. The period of increasing

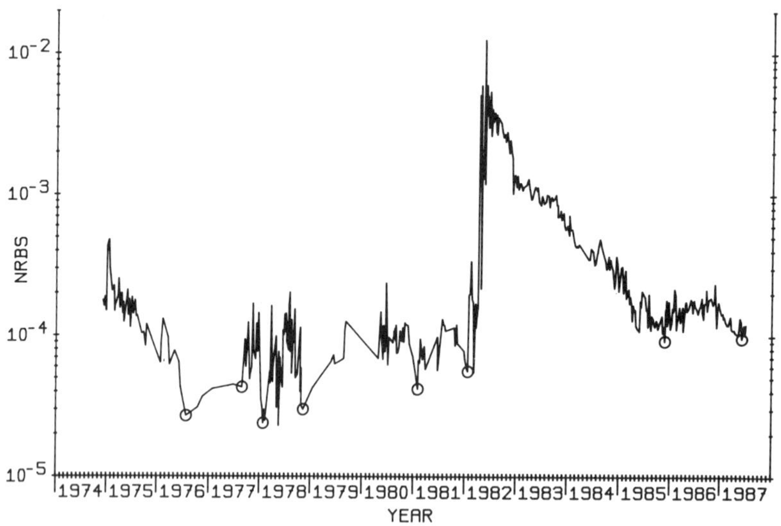


Figure 2. Similar to Fig. 1 with NRBS minimum residual values indicated.

TABLE 2. MAUNA LOA OBSERVATORY NON-RAYLEIGH BACKSCATTER RELATIVE MINIMUM POINTS

| DATE | NRBS |
|---|---|
| AUG 76 | $0.27 \times 10^{-4}\ sr^{-1}$ |
| SEP 77 | 0.43 |
| FEB 78 | 0.24 |
| NOV 78 | 0.30 |
| FEB 81 | 0.42 |
| JAN 82 | 0.56 |
| NOV 85 | 0.92 |
| JUN 87 | 0.96 |

$(NRBS \times 10^5) = 2.41 \exp(0.0105x)$ where x is months, Jan 76 = 1.

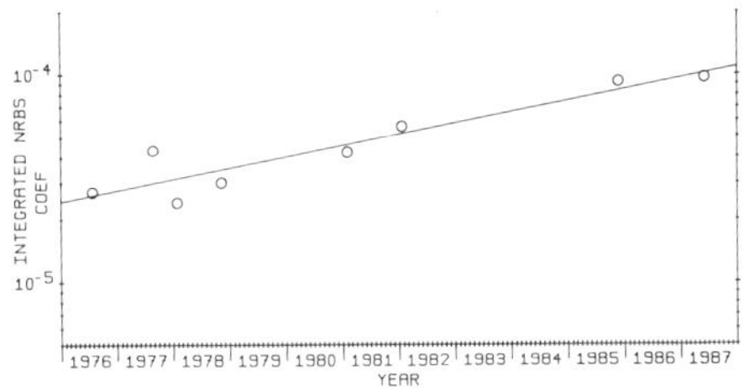


FIGURE 3. Mauna Loa lidar residual values versus time and the exponential regression line.

values also seem to follow the pattern of increasing stratospheric aerosol concentrations proposed by Hofmann and Rosen (1981) on the basis of dustsonde aerosol measurements in the period 1978-1979 and their comparison with Junge's stratospheric aerosol data from 1959 (Junge et al., 1961; Chagnon and Junge, 1961). Hofmann and Rosen assumed that both 1959 and 1977-1978 periods were times with minimal introduction of volcanic plumes into the stratosphere. For the 19-year period between Junge's 1959 data and the 1978 dustsonde data, Hofmann and Rosen estimated an annual 9% increase in background aerosol concentration.

Figure 4 shows the 1976 to 1987 MLO regression expression from Figure 3 and an approximation of Hofmann and Rosen's 1960 to 1979 relationship. Here the 1979 point for Hofmann and Rosen's expression has been placed, arbitrarily, so as to coincide with the January 1979 value on the MLO regression. The corresponding value for the Junge data was estimated from the 11.5 year exponential time constant and other information presented by Hofmann and Rosen (1981) and assigned a date of 1 January 1960 for plotting purposes. There is about a 14-fold

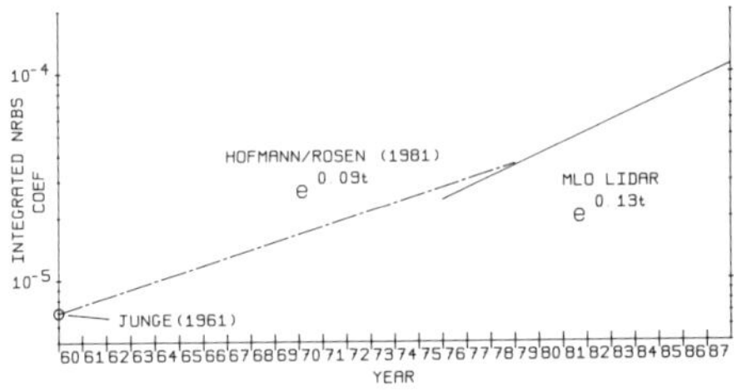


FIGURE 4. NRBS minimum residual aerosol regressions for the period 1960-1979 and 1976-1987.

change shown in Fig. 4 between the 1960 Junge value and the 1987 MLO results.

## 4. DISCUSSION

If the upward trend in residual stratospheric aerosol conditions is approximated in a reasonable fashion by Fig. 4, then there has been a 25-to 30-year trend of increasing stratospheric aerosol background concentrations averaging about 10% per year. It appears that the stratosphere, over Mauna Loa at least, has failed to recover completely from recent major volcanic plume injections and may be responding to the smaller scale but more numerous eruption plumes that fall in VEI class 3. Hofmann and Rosen (1981) estimated an increase in the background aerosol albedo from approximately zero in 1960 to values of 0.1 to 0.15% in 1979. Extending these estimates to the present according to Fig. 4 leads to an estimated background aerosol albedo increase of 0.2 - 0.3%.

Solar radiation measurements made at the ground surface have been used frequently to evaluate the impacts of volcanic eruption clouds, and data from Mauna Loa has been used for this purpose by a number of investigators (Mendonca, 1985; De Luisi et al., 1983). The fact that solar radiation at the surface is affected by tropospheric conditions as well as the stratosphere causes the record to be more complex than the lidar data record, which is applicable just to the stratosphere. The Mauna Loa solar radiation transmission record presented by Mendonca (1985) and shown in Fig. 5 for the period 1958 - 1985 indicates generally uniform "clean" background conditions from 1958 until the 1963 eruption of Agung. Following this event there was a very gradual recovery toward pre-Agung conditions. Other eruptions slowed the recovery process. A few months of 1976 or 1977 may have had some periods with close to the pre-Agung transmission but with much greater variability. Thus this 1976 to 1977 period was probably still affected by residual volcanic aerosol clouds. Since the mid-1970s there has been a very gradual decrease in transmission at Mauna Loa punctuated by the massive impact in 1982 of the El Chichon plume. As of 1987 the Mauna Loa solar radiation values have not yet made a full recovery to pre-Agung conditions.

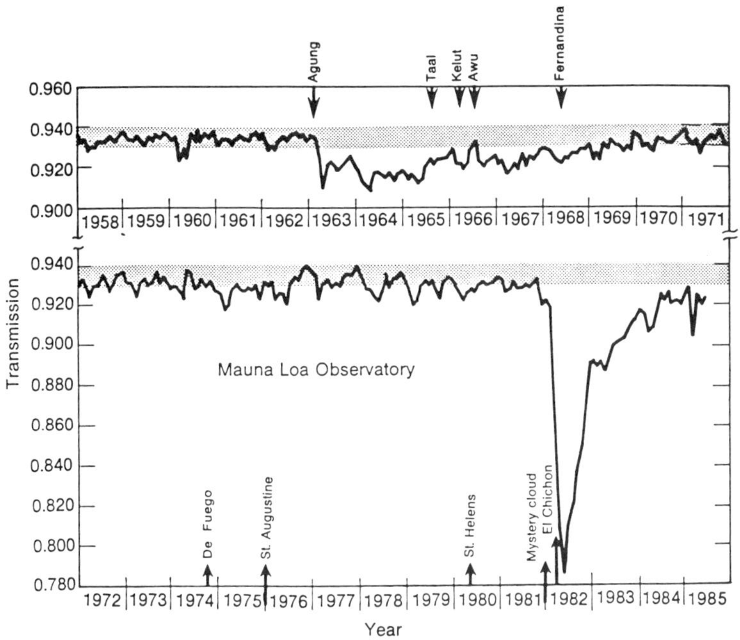


FIGURE 5. Solar radiation transmission factor at Mauna Loa Observatory, 1958-1985 (Mendonca, 1985).

## 5. CONCLUSIONS

This analysis of more than 10 years of Mauna Loa Observatory ruby lidar data indicates that there has been an apparent steady increase in the residual or background aerosol concentrations in the stratosphere over Hawaii. This conclusion is based on an assessment of the minimum aerosol concentration values observed during periods without obvious volcanic eruption impacts. The rate of increase of the Mauna Loa residual stratospheric aerosol backscatter values over the period from 1976 to 1987 has been about 10% per year. Although these observations relate only to conditions in Hawaii, the summary of lidar data from other operational lidar facilities reported by Hofmann (this volume) indicates that this increase in residual aerosol values is widespread and not localized over Hawaii.

We believe that this gradual build up of the residual stratospheric aerosol concentrations has two general causes. There doubtless is a residual aerosol concentration from the major eruption plumes that persists for an extended period after the more obvious plume characteristics have dissipated. This situation could occur with the 13 VEI-4 and -5 class eruptions shown by Table 1 to have occurred in the period 1974 to 1987. In addition to these eruptions, there were 60 eruptions classed as VEI-3 eruptions with the potential of affecting the stratosphere. Many of these seem to attract little attention, but they still may be responsible for stratospheric aerosol layers that become part of the general stratospheric aerosol concentration.

Although these long-term changes in stratospheric aerosol conditions are apparently not yet of sufficient magnitude to affect climatic conditions, they could be important in the evaluation of remote sensing satellite instrumentation and ground-based Dobson solar spectrophotometer observations.

## REFERENCES

Budyko, M. I., 1974: Climate and Life, Academic Press, N.Y., 508 pp.

Chagnon, C. W., and C. E. Junge, 1961: The Vertical Distribution of Sub-Micron Particles in the Stratosphere, J. Meteor., 18, 746-752.

DeLuisi, J., T. DeFoor, K. Coulson, F. Fernald and K. Thorne, 1984: Lidar Observations of Stratospheric Aerosol Over Mauna Loa Observatory: 1974-1981, NOAA Data Report ERL ARL-4, Air Resources Laboratory, ERL, NOAA, Rockville, MD.

DeLuisi, J., T. DeFoor, K. Coulson and F. Fernald, 1985: Lidar Observations of Stratospheric Aerosol Over Mauna Loa Observatory: 1982-1983, NOAA Data Report ERL ARL-5, Air Resources Laboratory, ERL, NOAA, Rockville, MD.

DeLuisi, J. J., E. G. Dutton, K. L. Coulson, T. E. DeFoor and B. G. Mendonca, 1983: On Some Radiative Features of the El Chichon Volcanic Stratospheric Dust Cloud and a Cloud of Unknown Origin Observed at Mauna Loa, J. Geophys. Res., 88, (C11), 6769-6772.

Hofmann, D. J., and J. M. Rosen, 1981: On the Background Stratospheric Aerosol Layer, J. Atmos. Sci., 38, 168-181.

Humphreys, W. J.,1940: Physics of the Air, 3rd ed., McGraw-Hill, N.Y.

Junge, C. E., C. W. Chagnon and J. E. Manson, 1961: Stratospheric Aerosols, J. Meteor., 18, 81-108.

Lamb, H. H., 1966: The Changing Climate, Methuen, London.

Mendonca, B. G., 1985: Measurements in the Geophysical Monitoring for Climatic Change Division of NOAA, USA: A First Step in an Integrated Global Monitoring Effort, Third International Symposium on Integrated Global Monitoring of the State of the Biosphere, Tashkent, USSR, October 13-20.

Miller, J. (Ed.), 1974: Geophysical Monitoring for Climatic Change, Summary Report No. 2, 1973, ERL, NOAA, Boulder, CO, 38-45.

Simkin, T., L. Seibert, L. McClelland, D. Bridge, C. Newhall and J. H. Latter, 1981: Volcanoes of the World, Smithsonian Institution, Washington, DC.

Simkin, T., L. Seibert, L. McClelland, D. Bridge, C. Newhall and J. H. Latter, 1984: Volcanoes of the World, Supplement, Smithsonian Institution, Washington, DC.

SEAN Bulletin, 10, 11, 12, 1985, 1986, 1987: Smithsonian Institution, Washington, DC.

Wendler, G., and Y. Kodama, 1986: Effect of El Chichon Volcanic Cloud on Surface Radiative Regime in Central Alaska, J. Clim. App. Meteor., 25, 1687-1694.

# AEROSOLS IN CLIMATE MODELING

*Thomas P. Ackerman*[1]
NASA Ames Research Center
Moffett Field, California 94035, USA

## ABSTRACT

The direct effect of aerosols on climate occurs primarily through their interaction with solar radiation and, to a lesser degree, through their interaction with infrared radiation. While the direction of this effect on climate is a function of aerosol optical depth, single-scattering albedo, and location in the atmosphere, some general rules can be developed. These general rules are illustrated by reference to one-dimensional and two-dimensional simulations of the effect of volcanic aerosols, arctic haze, and catastrophic aerosol loadings.

## 1. INTRODUCTION

Benjamin Franklin was perhaps the first person to link atmospheric aerosols with climatic perturbations when in 1784 he postulated that the exceptionally cold winter of 1783-84 was related to the diminution of sunlight caused by aerosols. In more recent times, the link between aerosols and climate has motivated a significant number of scientific studies. The sources of the aerosols considered in these studies include volcanic eruptions, pollutant emissions, biogenic emissions, wind-blown dust, asteroid and meteor impacts, biomass burning, and fires ignited by nuclear weapons. Not surprisingly, the results of these studies exhibit considerable variability. However, there are some general conclusions on aerosol and climate interactions that can be drawn from them. These conclusions are described in the next section. Section 3 then illustrates them through consideration of three specific aerosol problems.

The first of these problems, and the one with the longest history and the most voluminous reference list, is volcanic (or stratospheric) aerosols. The second is the more recent and more regional problem of arctic haze. The third is the most speculative of the aerosol-climate problems and deals with the climatological consequences of optically thick aerosol layers produced by, for example, a large asteroid impact or nuclear war. These topics were picked for special treatment not only because of their current interest, but also because they represent a wide range of possible aerosol impacts, varying from aperiodic natural events

[1]Now affiliated with Department of Meteorology, Pennsylvania State University, University Park, PA 16802.

 A. DEEPAK Publishing
ISBN 0-937194-11-5

to anthropogenic effects to potentially catastrophic events. There are other subjects, such as the effects of dust storms or regional pollutant hazes, that are also of interest but which have been omitted here in the interest of brevity. A review of aerosol-related climate effects has been prepared by Charlock (1986). It contains more detailed discussions of some of the problems mentioned here as well as a comprehensive bibliography.

## 2. GENERAL THEORY OF AEROSOL INTERACTIONS

Aerosol layers can both scatter and absorb solar radiation. The relative amount of each, as well as the reflectivity of the layer, is controlled by the aerosol single-scattering albedo ($\varpi_0$) and asymmetry factor ($g$), the layer extinction optical depth ($\tau_e$), and the albedo of the underlying surface and/or atmosphere. As a general rule, weakly absorbing aerosols ($\varpi_0 \sim 1$), which predominantly scatter radiation, increase the amount of solar radiation backscattered to space. Since this reduces the energy input to the earth-atmosphere system, it must result in a net cooling of the system. The magnitude of this increase in planetary albedo is a function of particle size, total optical depth, and the location of the aerosol layer. For fixed aerosol composition and at a given wavelength, $g$ increases with increasing particle size, implying greater forward scattering. Therefore, smaller particles tend to increase the planetary albedo more than large particles for the same optical depth. For a given aerosol distribution, an increase in optical depth will increase the albedo until some limiting value is reached as $\tau_e \rightarrow \infty$. However, since aerosol layers in the atmosphere rarely reach such large optical depths, this limiting value is of little practical importance.

The interaction between the albedo of the underlying surface and atmosphere and the aerosol layer is more complex. The normal planetary albedo of the earth is about 30%. This is caused primarily by reflection from clouds and from bright surfaces such as ice or deserts and, to a lesser degree, by Rayleigh scatter. Weakly absorbing aerosol layers over bright surfaces or clouds have only a minimal impact of the reflectivity, which is already high, and may, in fact, slightly reduce the reflectivity over very bright surfaces. However, these same layers over low albedo surfaces such as the ocean can substantially increase the albedo. Furthermore, high altitude layers tend to increase the albedo more than low altitude layers because the backscatter occurs before attenuation of the incident radiation by the atmosphere.

Absorbing aerosols ($\varpi_0 <\sim 0.8$) tend to reduce the solar energy lost from the earth-atmosphere system by absorbing solar radiation that otherwise would be reflected from clouds and from the surface. This tends to increase the energy input to the system and thereby produces a net warming. This case is more complicated than the preceding one, in that not only the magnitude but also the sign of the effect can be affected by particle size, composition, and optical depth, as well as by layer location. Obviously, the smaller the value of $\varpi_0$, the more likely the layer is to decrease planetary albedo. Hansen et al. (1980) calculated that 0.85 represents a critical value of $\varpi_0$, such that tropospheric aerosols with $\varpi_0$ less than this value would warm the system while, conversely, aerosols with $\varpi_0$ greater than this would cool the system.

However, the optical depth and asymmetry factor of the aerosols also affect the radiative transfer. For example, an increase in optical depth results in an increase in absorption within the aerosol layer but, at the same time, generally results in an increase in layer reflectivity. Also, since large particles tend to be more strongly forward scattering than small particles, they tend to have a smaller reflectivity for a given optical depth, which enhances absorption over reflection. Since aerosol optical properties are functions of aerosol size and composition, it is not easy to generalize the relationship between the change in a single aerosol optical property and the consequent change in the planetary radiation balance.

The situation is further complicated by the albedo of the underlying surface. If absorbing aerosols overlie clouds or a bright surface, they will certainly reduce the amount of reflected radiation and cause a net warming. If the underlying surface is very dark, the backscatter of the aerosols will cause an increase in the reflectivity, which will result in cooling. In intermediate cases the net effect of the aerosol layer can not be predicted a priori since it depends on the characteristics of both the aerosol and the surface. Hence, in general, the warming or cooling effects of aerosols can not be determined from a single, simple criterion such as that suggested by Hansen et al., (1980).

It should be noted that the warming or cooling discussed above applies to the system as a whole and not specifically to the planetary surface. If absorbing aerosols are placed in the stratosphere, it is possible to increase the absorption of radiation in the system and simultaneously reduce the surface temperature. A similar phenomenon has been demonstrated in the nuclear winter calculations for an optically thick aerosol layer placed in the troposphere (see Section 3.3).

These effects of aerosols on solar radiation are illustrated in Fig. 1 where we have calculated the planetary albedo of a single aerosol layer as as function of extinction optical depth. The calculations were carried out using the $\delta$4-stream model of Cuzzi et al. (1982),assuming $g=0.75$ and a cosine of the solar zenith angle of 0.5. The two values of $\varpi_0$, 0.99 and 0.70, correspond to weakly and strongly absorbing aerosols, respectively. The three values of surface albedo cover an approximate range of possible values (although sea surface values may be as low as 0.05 and fresh snow may reach values of 0.85). As discussed above, the weakly absorbing aerosols always increase the planetary albedo due to aerosol backscatter. For strongly absorbing aerosols, the planetary albedo is reduced for bright surfaces but increased for dark surfaces. Note that these

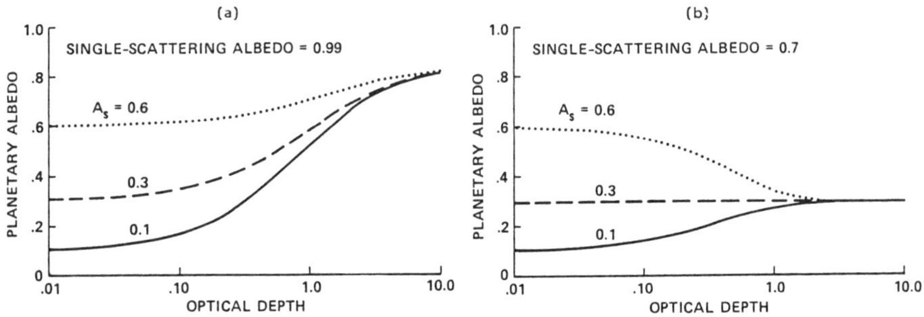


FIGURE 1. Planetary albedo as a function of optical depth for single aerosol layers having the single-scattering albedo indicated. Curves are labeled with the assumed value of the surface albedo.

calculations do not include any Rayleigh scatter or cloud effects so they do not explicitly account for the effect of vertical placement of the aerosol layer within the atmosphere.

As a general rule, atmospheric aerosols interact poorly with infrared radiation due to the small size of the particles relative to the wavelength of the radiation. This leads to small values of $\tau_e$ for typical atmospheric aerosol layers. Furthermore, since most materials exhibit moderate to strong absorption at infrared wavelengths, $\varpi_0$ tends to be small as well. This combination leads to an approximation at infrared wavelengths in which aerosols are treated as absorbers only and their scattering properties are neglected. Therefore, it follows that the primary effect of aerosols on infrared radiation is to increase somewhat the total emissivity of the atmospheric layer in which they reside. This causes local heating of the aerosol layer and may increase the temperature of the underlying atmosphere and surface through increased downward radiation by the layer. The increased emissivity and associated local heating is most important for aerosols located near an atmospheric temperature minimum where the radiative exchange is maximized.

Before moving on to discuss specific applications of the general theory, a word of warning is in order. The generalizations above and the one-dimensional calculations described in subsequent sections are based on radiative transfer theory. They do not account for dynamical or hydrological processes in the system. (However, one-dimensional models typically attempt to crudely include hydrology by making the approximation that relative humidity is constant, which increases the atmospheric water vapor concentration with increasing temperature.) Consequently, predictions of warming and cooling of the system and/or surface must be tempered with the realization that dynamical transports and hydrological processes can modify these changes.

# 3. SPECIFIC APPLICATIONS

## 3.1 VOLCANIC AEROSOLS

As mentioned above, an extensive literature exists on the effects of volcanic aerosols on climate and it cannot be adequately summarized in the limited space available here. The first studies that quantified the relationship between stratospheric aerosols and atmospheric temperatures were carried out by Pollack et al. (1976) and Hansen et al. (1978) using radiative-convective models. Pollack et al. attempted to match Mitchell's (1970) observed trend in Northern Hemisphere surface temperature through variations in stratospheric aerosol loading and increasing $CO_2$ concentration. They suggested that the decline in volcanic activity during the early part of this century made a major contribution to the observed warming. They also concluded that the increase in $CO_2$ concentration had relatively little impact through about 1960 but would dominate volcanic effects by the turn of the century. Hansen et al. (1978) looked specifically at the effects of the eruption of Mount Agung on the island of Bali in 1963. They were able to match quite well observed increases in stratospheric temperature over Australia and decreases in tropical tropospheric temperature with computed temperatures from a one-dimensional radiative-convective model (Fig. 2).

While the results shown in Fig. 2 suggest a clear relationship between stratospheric aerosol loading and climate, both Coakley (1981) and Charlock (1986) have pointed out that these results have a significant uncertainty. Following a detailed assessment of the observational constraints, Coakley concluded that the Agung layer could have caused 2 to 3 times the tropospheric cooling predicted by the models. However, he also added that, due to observational uncertainties, none of the models could be considered unrealistic. Charlock notes that the magnitude and phasing of the tropospheric temperature change calculated by Hansen et al. (1978) is critically dependent on their assumption of a ocean mixed layer depth of 70 m. Given the sensitivity of the calculation to this assumption and the spatial and temporal variability of the mixed layer depth, it is possible that the agreement between observation and theory is fortuitous. Finally, even the observed change in stratospheric temperatures is suspect because of the difficulty of separating out normal stratospheric temperature trends due to the quasi-biennial oscillation.

The eruption of El Chichon in April 1982 provided another case study of the effects of volcanic aerosols. A great deal of data on particle distributions and radiative perturbations was acquired (see, for example, DeLuisi et al., 1983; Dutton and DeLuisi, 1983; Oberbeck, 1983; Thomas et al., 1983; Pollack et al., 1985; Valero et al., 1984b). In addition to the radiative data, observations of increases in stratospheric temperatures were reported by Labitzke et al. (1983).

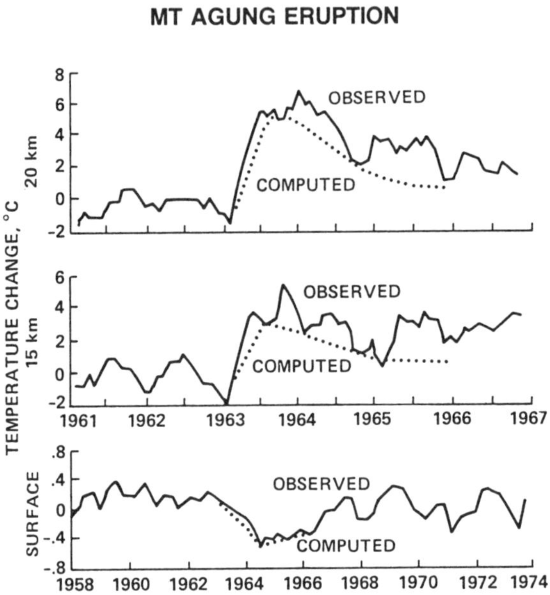


FIGURE 2. The upper two panels show the observed change (solid curves) and computed change (dots) in stratospheric temperatures from 1961 to 1967 at altitudes of approximately 20 and 15 km. The lower panel shows observed and computed (dots) tropospheric temperature change from 1958 to 1974. (Redrawn from Hansen et al., 1978.)

Given this wealth of data, one would expect that the question of the impact of stratospheric aerosols on climate could be resolved, at least for this one specific case. Unfortunately, the eruption of El Chichon was coincident with a major El Niño event (Rasmusson and Wallace, 1983) that produced large anomalies in sea surface temperature and outgoing longwave radiation. (The latter quantity is often used as a proxy indicator of precipitation over the ocean.) As a result, it is difficult to untangle the tropical tropospheric response to the combined influences of the eruption of El Chichon and the El Niño event.

However, since the stratosphere is relatively unperturbed by El Niño events, the temperature anomalies deduced by Labitzke et al. (1983) are indicative of volcanic effects. Furthermore, because the volcanic cloud spread rapidly throughout the tropics following the April eruption and then remained virtually unchanged until September, the lower stratosphere was in quasi-equilibrium radiatively during the summer of 1983. Using a radiative-convective equilibrium model, Pollack and Ackerman (1983) were able to reproduce the observed solar radiation reaching the troposphere and to simulate the observed stratospheric heating. They showed that the heating is primarily due to the absorption of upwelling terrestrial radiation by the aerosol layer and that the heating tends to be the largest in the coldest stratospheric layers.

Subsequently, MacCracken and Luther (1984) applied a two-dimensional zonally-averaged model to investigate the effects of the El Chichon eruption. They placed a stratospheric layer with optical depth 0.255 between 5°N and 35°N and found that, for an annual average model run to equilibrium, the Northern Hemisphere surface air temperature dropped by about 0.9 K. At the same time, the model produced a southward shift in the intertropical convergence zone (ITCZ) and a strengthening of the Northern Hemisphere Hadley circulation. While MacCracken and Luther are quite cautious in their interpretation of this modification to the tropical troposphere, Handler (1986) cites it as support for his theory of stratospheric aerosol and climate interaction. He postulates that tropical stratospheric aerosols tend to decrease the equator-to-pole temperature gradient while midlatitude stratospheric aerosols tend to increase the equator-to-pole gradient, and this in turn affects the global circulation as shown by MacCracken and Luther. Handler presents a statistical analysis of both sea surface temperature anomalies and variations in Indian monsoon precipitation correlated with volcanic eruptions to substantiate his hypothesis. Unfortunately, the derived relationships are somewhat suspect due to the limited data and the uncertainty of the measurements going back to the early part of this century and the latter part of the 19th century.

In summary, the effect of volcanic aerosols on climate remains uncertain. It is reasonably well documented that they cause warmings in the stratosphere primarily due to the absorption of upwelling terrestrial radiation. It is also well documented that stratospheric aerosol layers reduce the amount of direct solar radiation reaching the troposphere. They also reduce the amount of total solar radiation reaching the troposphere, but the extent to which this occurs is less certain because downward diffuse solar radiation is increased by the layer. A variety of circumstantial evidence leads to the conclusion that volcanic aerosols cause tropospheric cooling and this is consistent with radiative-convective model results. Unfortunately, unambiguous evidence of tropospheric cooling due to volcanic eruptions remains lacking. Furthermore, the relationship between tropospheric cooling and global circulation changes is not well understood either. Theories such as those proposed by Handler are thought-provoking but not solidly established.

## 3.2 ARCTIC HAZE

Arctic haze is a pollutant aerosol transported to the Arctic from predominantly midlatitude sources. It has a urban, industrial composition, including a black carbonaceous component, and consists of small particles (for detailed information on its composition and origin, see the special issue of *Atmospheric*

*Environment* on Arctic Air Chemistry, 1985, V. **19**, No. 12). In terms of an aerosol-climate problem, arctic haze stands in considerable contrast to volcanic aerosols because the haze is tropospheric, regional, and absorbing.

Surface measurements of haze properties prompted a number of radiative model studies of the possible effects of haze on the deposition of solar energy. Then, in the spring of 1983, measurements of solar flux deposition in the arctic atmosphere were carried out by aircraft as part of the Arctic Gas and Aerosol Sampling Program (Valero et al., 1984a; Valero and Ackerman, 1986). The upward and downward solar flux, at both broad-band and several narrow-band spectral intervals, was measured by aircraft at several altitudes ranging from 7 km to about 80 m. The measured fluxes were differenced to obtain heating rates and were also modeled to infer aerosol properties and compute heating rates. Best fits to the data produced aerosol optical depths on the order of 0.2 to 0.3 at 0.5 $\mu$m. The measured broad-band absorption in the haze layer of about 80 $Wm^{-2}$ converts to instantaneous heating rates of 1.1 to 1.6 $Kday^{-1}$. This is a factor of 2 to 2.5 times greater than heating in the absence of the haze. In a simultaneous experiment, Wendling et al. (1985) acquired aircraft and surface measurements of aerosol size distributions, black carbon concentrations, and vertical profiles (by lidar). From this data, they computed the radiative flux deposition. Due to much smaller optical depths during their observations (about 0.04 at 0.59 $\mu$m), the effect on solar deposition was much less pronounced.

The climatic effects of arctic haze have been modeled by MacCracken et al. (1986) using the version of their two-dimensional model that incorporates a seasonal cycle. They introduced a seasonally-varying aerosol concentration into the model, ran it out 6 years to obtain equilibrium, and compared the results to an unperturbed control simulation. The change in monthly-mean surface temperature between 65°N and 90°N latitude due to the aerosol is shown in Fig. 3. The results show that, as expected, the haze leads to an increase of surface temperature with a maximum value of about 1°C in May. What is surprising upon a closer examination of the radiative fluxes is that the increase results predominantly from an increase in downward infrared radiation. The haze layer absorbs solar radiation and heats the atmosphere, which in turn radiates increased infrared radiation to the surface.

Ackerman et al. (1986) adopted a different technique to assess the climatic impact of arctic haze. They attempted to determine the energetics of the atmosphere in the arctic basin (which they defined as the area north of 70°N latitude) both annually and seasonally. Then, using the measurements of solar deposition, they estimated the change in the energy budget due to the haze. They found that, during the spring season (March, April, and May), an average of 94 $Wm^{-2}$ of solar radiation is absorbed and 176 $Wm^{-2}$ of infrared radiation is lost. Poleward heat transport by the atmosphere supplies 66 $Wm^{-2}$ and oceanic heat transport supplies approximately another 10 $Wm^{-2}$. These

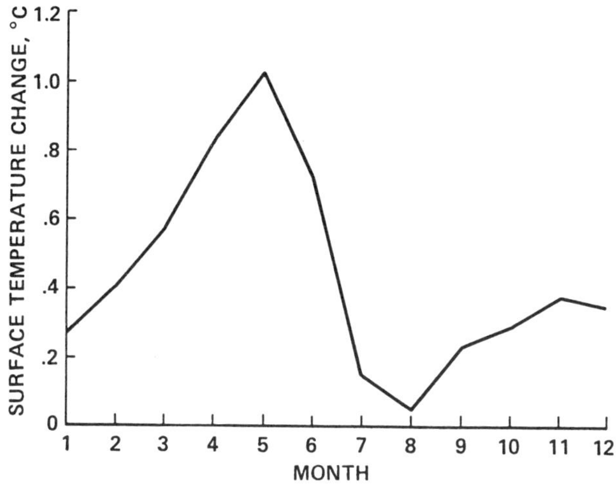


FIGURE 3. Monthly-mean changes (perturbed - control) in the surface temperature of a two-dimensional model due to the inclusion of arctic haze. (From MacCracken et al., 1986)

figures do not balance exactly due to uncertainties in the estimates and also due to seasonal heat storage effects. The measurements of solar absorption by arctic haze layers suggest that the haze might increase solar absorption by an average of 7 to 10 $Wm^{-2}$ during this spring season. (Note that this absorption may already be included in the figure given above for solar absorption by the system since that figure was derived from satellite observations.) From these figures, they conclude that the haze is climatically important on a regional and seasonal basis, but less likely to have an annual or global impact. Of course, given the simplicity of this approach, the conclusion needs to be substantiated with more sophisticated models.

Due to the scarcity of data and the uncertainties in models of polar climate, it is difficult to assess accurately the impact of arctic haze. Two different studies suggest that a climatic impact is possible, but both suffer from inadequate data and uncertain assumptions. In addition, there are haze effects which are not incorporated into the models. For instance, Clarke and Noone (1985) have suggested that sufficient aerosol deposition has occurred on the ice surface to affect the surface albedo and increase the deposition of solar radiation. It is also possible that the optical properties of the arctic stratus are modified by the haze, particularly in the spring. Also, changes in the surface energy budget will affect the rate of snow and ice melt, which will in turn affect the climate. Given the sensitivity of the Arctic to small perturbations, it is clear that more attention needs to be given to arctic haze and its possible climatic consequences.

### 3.3 CATASTROPHIC EVENTS

The potential climatic consequences of an asteroid impact (Toon et al., 1982) or a large-scale nuclear war (Turco et al., 1983) have received a great deal of attention in the literature and those discussions will not be reviewed here. However, these problems are manifestations of a limiting case of possible aerosol-climate interactions that has been much misunderstood, *i. e.*, the case of optically thick aerosol layers. The general discussion of this problem,which follows,owes much to the work of Cess et al. (1985) and Pittock et al. (1986).

The composition of the earth's atmosphere is such that most of the solar radiation absorbed by the planet is absorbed at the earth's surface rather than in the atmosphere (by about a factor of 2 to 1). This heat is transferred to the atmosphere by latent heat transfer and net infrared radiation and, to a much smaller degree, by sensible heat transfer. If an aerosol layer with an appreciable absorption optical depth at solar wavelengths is introduced into the atmosphere, solar radiation can no longer penetrate to the surface but is absorbed by the aerosol layer. This tends to produce a warming of the aerosol layer leading to local convection and to induced large-scale motions. It also leads to a cooling of the surface when the atmosphere below the aerosol becomes stable and convection stops. If the solar energy reaching the surface is reduced in a stepwise fashion, one sees an initial reduction in the transport of latent and sensible heat from the surface to the atmosphere and a slight drop in the infrared radiation lost by the surface as the surface temperature decreases. As the solar radiation is further reduced, the latent and sensible heat transfer actually changes sign because the atmosphere is warmer than the surface. However, this reverse transfer is small because the atmosphere near the now cold surface is stably stratified. Since the atmosphere is heated directly, the downward infrared radiation increases somewhat, but not nearly enough to compensate for the loss of all the solar energy, especially in view of the fact that the downward infrared is quite large even when no aerosol is present. Furthermore, because the aerosol layer radiates the absorbed solar energy both upwards and downwards, about half the solar energy is lost with respect to the surface. Finally, in the limit of an approximately isothermal, optically thick aerosol layer, the surface temperature actually approaches the radiative equilibrium temperature of the planet.

This process is shown quantitatively in Fig. 4. The calculations were carried out using the radiative-convective model described in Turco et al., (1983). The aerosol parameters used in the model are approximated such that the aerosol optical depth at 10 $\mu$m is 1/10 the value at 0.55 $\mu$m (the latter being the value plotted on the abscissa). The model grid resolution is 2 km and the surface is assumed to be in instantaneous contact with the bottom layer. For very small optical depths, the surface tends to heat slightly. This is due to a reduction in planetary albedo by the aerosol layer and to the convective coupling

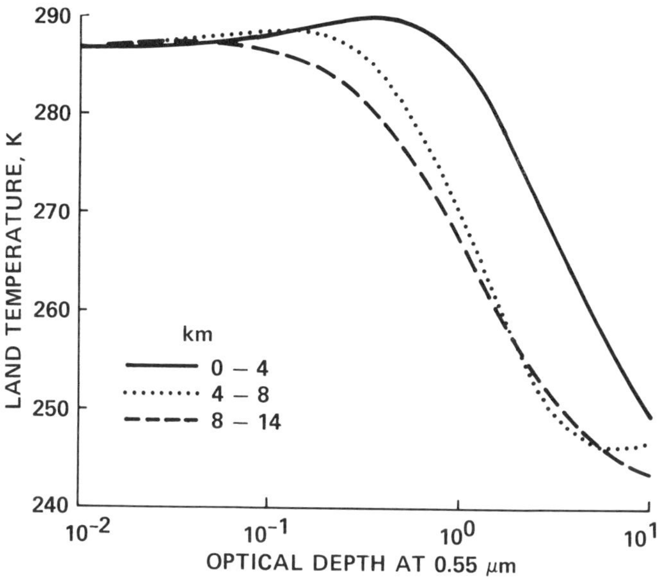


FIGURE 4. Surface temperature from a radiative-convective model as a function of aerosol visible optical depth. The aerosol in the model is assumed to have an optical depth at 10 $\mu$m that is 1/10 the visible optical depth and scales inversely with the wavelength. The three curves are for the aerosol layer at different heights in the model atmosphere as shown.

between surface and troposphere that causes both to heat simultaneously. As the optical depth is increased, the surface temperature decreases, at first slowly and then very abruptly for optical depths on the order of 1. At very large optical depths, the temperature actually increases slightly from the minimum temperature as the layer becomes optically thick in the infrared.

While it is unlikely that optically thick aerosol layers would occur, these models illustrate the sensitivity of the planetary surface temperature to the convective coupling with the atmosphere. They also show the importance of correctly modeling heat transfer at the surface-atmosphere interface if diurnal cycles, polar climates, and other phenomena where stable atmospheric conditions are encountered are to be studied and understood.

## 4. CONCLUSIONS

From the discussion above, it is apparent that the issue of aerosol-climate interactions has not yet been resolved. As a general rule, high altitude aerosol layers tend to reduce the input of solar radiation into the system and cause cooling of the surface and the atmosphere below the layer. Model studies of both volcanic and optically thick aerosol layers substantiate this conclusion. Unfortunately, the available data on volcanic effects on climate are largely statistical and inferential and, hence, are not definitive. Aerosol layers also tend to heat the atmosphere in which they are located due to absorption of solar radiation and/or absorption of upwelling thermal radiation.

The climatic effects of regional aerosols are more difficult to quantify. Issues such as horizontal and vertical transport, varying surface albedo, seasonal and temporal variability, and boundary layer interactions all need to be considered. Arctic haze was used as an example of this type of problem, but regional pollutant hazes or blowing desert sand and dust could also be used to illustrate the complexity of the issues. Most midlatitude tropospheric aerosols tend to increase solar reflectivity. However, their climatic impact can not be stated unequivocally but needs to be addressed within the context of the situation under investigation. Polar aerosols such as arctic haze tend to increase the amount of solar absorption when they are located over bright snow and ice surfaces.

Finally, it should be noted that the discussion above has been in the context of direct effects of aerosols on the radiation budget. It has been proposed that aerosols can affect climate through their interaction with cloud microphysical and optical properties. Since aerosols serve as cloud condensation nuclei, changes in aerosol amount or composition could affect cloud optical properties. It has been suggested that increasing aerosol concentrations may: 1) increase the planetary albedo by increasing the available cloud condensation nuclei (leading to more numerous but smaller cloud droplets), or 2) decrease the planetary albedo by increasing the solar absorption within clouds. This subject has not received a great deal of attention in climate modeling because of the difficulty of modeling clouds in general and aerosol-cloud interactions in particular. However, these processes are potentially so important that considerably more effort must be expended to incorporate them into climate models and understand their effects.

## REFERENCES

Ackerman, T. P., J. M. Stenback and F. P. J. Valero, 1986: The importance of arctic haze for the energy budget of the Arctic. In B. Stonehouse (Ed.), *Arctic Air Pollution*, Cambridge University Press, Cambridge, UK, 151-158.

Cess, R. D., G. L. Potter, S. J. Ghan and W. L. Gates, 1985: The Climatic Effects of Large Injections of Atmospheric Smoke and Dust: A Study of Climate Feedback Mechanisms With One and Three-dimensional Climate Models, *J. Geophys. Res.*, **90**, 12937-12950.

Charlock, T. P., 1986: Review of recent research on the climatic effect of aerosols. Prepared for Working Group VIII of the US/USSR Committee on the Environment.

Clarke, A. D., and K. J. Noone, 1985: Soot in the Arctic Snowpack: A Cause for Perturbations in Radiative Transfer, *Atm. Env.*, **19**, 2045-2053.

Coakley, J. A., Jr., 1981: Stratospheric Aerosols and the Tropospheric Energy Budget: Theory Versus Observations, *J. Geophys. Res.*, **86**, 9761-9766.

Cuzzi, J. N., T. P. Ackerman and L. C. Helmle, 1982: The Delta-Four-Stream Approximation for Radiative Flux Transfer, *J. Atmos. Sci.*, **39**, 917-925.

DeLuisi, J. J., E. G. Dutton, K. L. Coulson, T. E. DeFoor and B. G. Mendonca, 1983: On Some Radiative Features of the El Chichon Volcanic Stratospheric Dust Cloud and a Cloud of Unknown Origin Observed at Mauna Loa, *J. Geophys. Res.*, **88**, 6769-6772.

Dutton, E., and J. DeLuisi, 1983: Spectral Extinction of Direct Solar Radiation by the El Chichon Cloud During December 1982, *Geophys. Res. Letts.*, **10**, 1013-1016.

Handler, P., 1986: Possible Association Between the Climatic Effects of Stratospheric Aerosols and Sea Surface Temperatures in the Eastern Tropical Pacific Ocean, *J. Climat.*, **6**, 31-41.

Hansen, J. E., A. A. Lacis, P. Lee and W. C. Wang, 1980: Climatic Effects of Atmospheric Aerosols, *Ann., New York Acad. Sci.*, **338**, 575-587.

Hansen, J. E., W. C. Wang and A. A. Lacis, 1978: Mt. Agung Provides Test of a Global Climate Perturbation, *Science*, **199**, 1065-1068.

Labitzke, K., B. Naujokat and M. P. McCormick, 1983: Temperature Effects on the Stratosphere of the April 4, 1982 Eruption of El Chichon, Mexico, *Geophys. Res. Letts.*, 10, 24-26.

MacCracken, M. C., and F. M. Luther, 1984: Preliminary Estimates of the Radiative and Climatic Effects of the El Chichon Eruption, *Geofisica Internacional*, **23**, 385-401.

MacCracken, M. C., R. D. Cess and G. L. Potter, 1986: Climatic Effects of Anthropogenic Arctic Aerosols: An Illustration of Climate Feedback Mechanisms With One- and Two-dimensional Climate Models, *J. Geophys. Res.*, **91**, 14445-14450, copyright by the American Geophysical Union.

Mitchell, J. M., 1970: A preliminary evaluation of atmospheric pollution as a cause of of the global temperature fluctuation of the past century. In S. F. Singer (Ed.), *Global Effects of Environmental Pollution*, Springer Verlag, New York, 139-155.

Oberbeck, V. R., E. F. Danielsen, K. G. Snetsinger, G. V. Ferry, W. Fong and D. M. Hayes, 1983: Effect of the Eruption of El Chichon on Stratospheric Aerosol Size and Composition, *Geophys. Res. Letts.*, **10**, 1021-1024.

Pittock, A. B., T. P. Ackerman, P. J. Crutzen, M. C. MacCracken, C. S. Shapiro and R. P. Turco, 1986: *Environmental Consequences of Nuclear War. Volume I. Physical and Atmospheric Effects.*, John Wiley & Sons, Chichester, 342 pp.

Pollack, J. B., and T. P. Ackerman, 1983: Possible Effects of the El Chichon Volcanic Cloud on the Radiation Budget of the Northern Tropics, *Geophys. Res. Lett.*, **10**, 1057-1060.

Pollack, J. B., M. P. McCormick, E. Dutton and J. DeLuisi, 1985: A climatology of the El Chichon volcanic cloud for use in climate models. Presented at the IAMAP/IAPSO Joint Assembly, Honolulu, Hawaii, August, 1985.

Pollack, J. B., O. B. Toon, A. Summers, B. Baldwin, C. Sagan and W. Van Camp, 1976: Stratospheric Aerosols and Climatic Change, *Nature*, **263**, 551-555.

Rasmusson, E. M., and J. M. Wallace, 1983: Meteorological Aspects of the El Niño/Southern Oscillation, *Science*, **222**, 1195-1202.

Thomas, G. E., B. M. Jakosky, R. A. West and R. W. Sanders, 1983: Satellite Limb-scanning Thermal Infrared Observations of the El Chichon Stratospheric Aerosol: First Results, *Geophys. Res. Letts.*, **10**, 997-1000.

Toon, O. B., J. B. Pollack, T. P. Ackerman, R. P. Turco, C. P. McKay and M. S. Liu, 1982: Evolution of an Impact Generated Dust Cloud and Its Effects on the Atmosphere, *Geological Society of America Special Paper 190, Geological Implications of Large Asteroids and Comets on the Earth*, 187-201.

Turco, R. P., O. B. Toon, T. P. Ackerman, J. B. Pollack and C. Sagan, 1983: Nuclear Winter: Global Consequences of Multiple Nuclear Explosions, *Science*, **222**, 1283-1292.

Valero, F. P. J., and T. P. Ackerman, 1986: Arctic haze and the radiation budget. In B. Stonehouse (Ed.), *Arctic Air Pollution*, Cambridge University Press, Cambridge, UK, 121-134.

Valero, F. P. J., T. P. Ackerman and W. J. Y. Gore, 1984a: The Absorption of Solar Radiation by the Arctic Haze During the Haze Season and Its Effects on the Radiation Budget, *Geophys. Res. Letts.*, **11**, 465-468.

Valero, F. P. J., T. P. Ackerman and W. L. Starr, 1984b: Radiative effects of the El Chichon volcanic cloud. *IRS '84: Current Problems in Atmospheric Radiation*, Proceedings of the International Radiation Symposium, Perugia, Italy, 21-28 August, 1984, A. Deepak Publishing, Hampton, VA.

Wendling, P., R. Wendling, W. Renger, D. S. Covert, J. Heintzenberg and P. Moerl, 1985: Calculated Radiative Effects of Arctic Haze During a Pollution Episode in Spring 1983 Based on Ground-Based and Airborne Measurements, *Atm. Env.*, **19**, 2181-2194.

# AEROSOL RADIATIVE FORCING AND MODEL RESPONSES

*V. Ramaswamy*
Atmospheric and Oceanic Sciences Program
Princeton University
Princeton, New Jersey 08540, USA

## ABSTRACT

The effects on the surface-atmosphere system due to aerosols are surveyed, with an emphasis on the radiative aspects of aerosols generated by noncatastrophic, nonepisodic events. Solar absorption and scattering by one type of "nominal" tropospheric aerosols is chosen as an example to illustrate the nature of the perturbations in the radiative fluxes at the surface, at the top of atmosphere, and in the atmospheric layers. The range in the magnitude of the perturbations can be quite large when catastrophic or episodic events are also considered. The effects of the aerosol perturbations in the solar spectrum are analyzed by considering the compensation effects due to radiative and radiative-convective mechanisms. These constitute idealistic atmospheric responses, wherein all the mechanisms are confined in a local vertical column and changes in the hydrologic cycle are ignored. Estimates of these responses are compared with those reported from global three-dimensional general circulation model simulations. The results suggest that an aerosol-induced radiative forcing can cause changes in the local energy balance and in the circulation, besides causing changes in the atmospheric thermal profile.

The coupling of the aerosol microphysics and radiative and dynamical mechanisms has been achieved thus far only in the study of catastrophic and episodic events. Future investigations would need to focus more on these aspects for all types of aerosols. Robust estimates of the sources and optical properties as well as a better understanding of the microphysical and transport processes are needed to assess the aerosol-induced radiative-dynamical-microphysical interactions unambiguously.

## 1. INTRODUCTION

The extent to which aerosols interact with and/or modify the general circulation of the atmosphere and climate has been addressed by the use of models of varying complexity. These include one- and two-dimensional models of the energy balance and radiative-convective types, limited-area and mesoscale models and, ultimately, completing the hierarchy, the three-dimensional global general circulation models (GCMs). A general perspective on the methodology and results of model studies can be had from the collection of papers in Gerber and Deepak



ISBN 0-937194-11-5

(1984) and the Proceedings of the 1984 International Radiation Symposium (Fiocco, 1984).

The typical atmospheric aerosol (radii > 0.001 $\mu$m) is a heterogeneous entity comprised of several different components. By virtue of its size, shape, and composition, the aerosol is capable of exerting chemical and physical effects. The aerosol forcing on the atmosphere can be classified as: 1) microphysical forcing involving the transformation (including cloud drop nucleation, coalescence, aggregation, etc.) and removal of the aerosol, and 2) radiative forcing involving the scattering and absorbing of radiation. An initial microphysical forcing may eventually lead to a radiative forcing, e.g., formation of clouds. Both types of forcing can perturb the atmospheric energy budget and, thus, the climate.

Since there already exists a collection of recent research work on aerosol-related issues, as referenced earlier, a narrower aspect of the subject constitutes the focus of this review. More specifically, this review concerns itself with the general nature of the aerosol radiative forcing and the assessments of the response to this forcing. The discussions are presented with some particular numerical examples; some recent investigations into the various aspects of the subject are also mentioned, although these are necessarily limited and brief.

The radiative forcing, in general, depends on the aerosol optical properties, their vertical distribution, solar zenith angle, and presence of other scatterers (surface clouds and gas molecules; see Ackerman, 1987). Owing to the temporal and spatial variations in aerosol properties, it is not an easy task to formulate the precise nature of the aerosol radiative forcing. The lack of knowledge of the global distribution of aerosol characteristics has meant that assumptions have to be made in estimating the aerosol-induced forcing. This is unlike the more precisely defined radiative forcing of well-mixed, homogeneously distributed species, such as carbon dioxide.

The nature of the aerosol radiative perturbations is discussed first. To illustrate this point, a sensitivity study is performed with respect to one characteristic of the aerosols, namely, their absorptivity in the visible spectrum. For this purpose, "nominal" tropospheric aerosols with a scale-height dependence and located above a low albedo surface are assumed. "Nominal" here implies aerosols arising due to noncatastrophic, nonepisodic processes; mixtures of aerosols with water drops are also excluded. The range of the perturbations for various types of aerosols are also reviewed.

The response of the atmosphere to a "nominal" aerosol radiative forcing is discussed next. From a theoretical standpoint, the response to the perturbations in the radiative fluxes can involve, in ascending order of complexity, purely radiative or radiative-convective or radiative-convective-dynamical responses. A one-dimensional and a three-dimensional model are considered to investigate the change in the atmospheric temperatures and the readjustment of the energy budget. In the present text, the response of the one-dimensional model is derived on the basis of two different but idealistic assumptions: these are, respectively, the radiative (wherein radiative processes alone compensate for the perturbation) and the radiative-convective (wherein radiative and convective processes compensate for the perturbation) solutions. For the radiative response solution, aerosol perturbation (absorption) is assumed only for a particular layer while, for the radiative-convective response solutions, a purely absorbing or a purely scattering aerosol is assumed whose forcing occurs throughout the atmospheric column. Results arising out of these two different assumptions are then compared with the response from a GCM experiment performed by Coakley and Cess (1985). The comparison highlights the relative roles played by the radiative and the nonradiative components of the energy budget in the one-dimensional and the three-dimensional models, respectively. Results from other GCM studies that illustrate the nature of the model atmosphere response to aerosol perturbations are also surveyed. The various GCM studies serve to identify the mechanisms that are important in different regions of the atmosphere and for various types of aerosol radiative forcing.

Finally, the direction in which the aerosol-related modeling efforts are proceeding is surveyed; some of the studies that have coupled aerosol microphysical and radiative processes are mentioned. Problems in the determination of the aerosol-related effects, that are unsolved or whose solutions remain considerably uncertain, are stated at the end.

## 2. AEROSOL RADIATIVE FORCING

There are three ways in which aerosols can exercise a radiative perturbation to an atmosphere in radiative-convective-dynamical equilibrium:

1) change the absorption or emission of radiation at the levels where they occur in the atmosphere (Deepak and Gerber, 1984),
2) alter the radiative fluxes at levels other than where the aerosols occur (Deepak and Gerber and references therein) and
3) alter the optical properties of the atmosphere through interaction with water vapor, e.g., formation of haze drops, external or internal mixtures with cloud drops. (Twomey, 1976, Chylek et al., 1984, Charlson et al., 1987, Coakley et al., 1987).

The radiative influences can occur in both the solar and the longwave spectrum (Carlson and Benjamin, 1980; Grassl, 1987; Pollack and Ackerman (1983)). For the commonly occurring submicrometer aerosols, however, the principal influence is in the solar spectrum. Based on our present knowledge of optical properties (Shettle and Fenn, 1979), most aerosols scatter in both the visible and the near-infrared spectrum while absorbing only in the near-infrared spectrum. Graphitic carbon and particles derived from iron oxides possess absorption bands in the visible as well.

An example is given to illustrate the general nature of the aerosol-induced radiative forcing. A 1-km-scale height profile for the aerosol number concentration is assumed; the aerosol sizes are lognormally distributed (mode radius is 0.1 μm, standard deviation is 2), with refractive indices in the near infrared as given by Shettle and Fenn (1979) for water soluble aerosols. In the visible, the refractive index is assumed to be a variable. The aerosol single scattering parameters are employed in a one-dimensional radiative transfer model (Ramaswamy and Kiehl, 1985) of the clear tropical atmosphere. The cosine of the solar zenith angle is taken to be 0.6, the surface albedo is 0.07 (representative of oceans), and the calculations are performed for diurnally averaged conditions. The aerosol optical depth at a wavelength of 0.5 μm is assumed to be 0.1.

The aim is to examine the changes in the solar heating rate and the changes in the shortwave fluxes at the atmospheric boundaries (viz., top of atmosphere and surface) due to changes in the visible spectrum characteristics of the aerosols (e.g., due to presence of carbonaceous compounds or iron oxides). Specifically, changes in the imaginary part of the refractive index in the visible spectrum (Im(m)) are considered, while the real part is held fixed: these are, respectively, Im(m) of 0 (mildly absorbing, i.e., absorbing only in the near infrared), 0.01(moderately absorbing), and 0.1 (strongly absorbing). It may be noted that these values are within the range measured at rural and urban locations.

The solar heating rate in the clear tropical lower atmosphere (0-6 km.) is shown in Fig. 1a (layers are 1 km thick). The heating increases with decreasing altitude, owing to the increase in water vapor concentration and the water vapor absorption bands in the near infrared. The changes in these heating rates when aerosols are introduced into the surface-atmosphere system are shown in Fig. 1b. Due to the assumed aerosol vertical profile, the strongest perturbations occur in the lower troposphere. For the mildly absorbing aerosol, the increase is $< 0.1$ K/d; - and this is due solely to the aerosol characteristics in the near infrared. The perturbation increases with Im(m) and a value of up to 0.5 K/d is attained for an Im(m) of 0.1. While the difference in absorption between the mildly absorbing and the clear sky cases is a consequence of near-infrared aerosol characteristics, that between the various aerosol cases is due to the properties in the visible spectrum.

The changes in the fluxes at the top of the atmosphere and at the surface for the three aerosol cases are listed in Table 1. Also listed are the absolute values of the fluxes for the clear sky case and the value of the single scattering albedo at a wavelength of 0.5 μm. Increasing Im(m) decreases the value of the single-scattering albedo, resulting in increasing flux convergence in the atmosphere. Correspondingly, the flux at the surface decreases monotonically. The reflected flux at the top of the atmosphere increases with respect to the clear sky value for the mildly absorbing case; this is because the presence of the aerosols increases the backscattering. With an

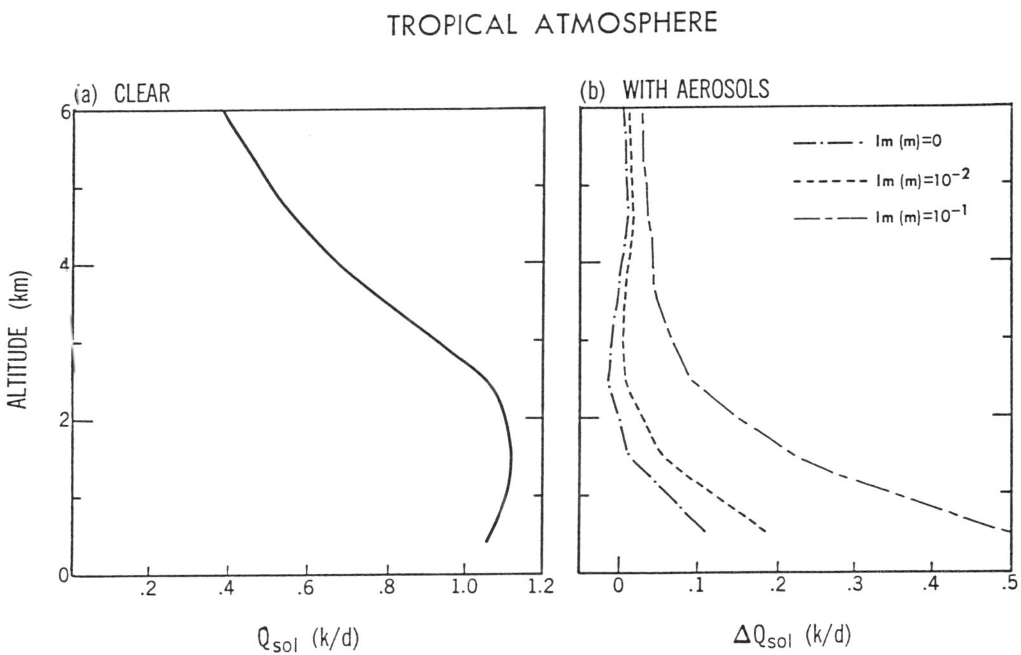


FIGURE 1. a) Solar heating rate, $Q_{sol}$ (in degrees per day), in a diurnally-averaged clear tropical atmosphere (McClatchey et al., 1972) and b) change in the solar heating rate, $\Delta Q_{sol}$, due to aerosols with different values of the imaginary part of refractive index, Im(m), in the visible spectrum. The values chosen are 0(i.e., nonabsorbing in the visible), 0.01 (moderately absorbing), and 0.1 (strongly absorbing), respectively.

TABLE 1. AEROSOL OPTICAL PROPERTIES AND BOUNDARY FLUXES

| Aerosol Absorption | Im(m) | $\omega$ | Reflected Flux ( $W/m^2$ ) | Surface Flux ( $W/m^2$ ) |
|---|---|---|---|---|
| CLEAR | - | - | 41.9 | 322.8 |
| | | | Change in Reflected Flux | Change in Surface Flux |
| Mild | 0 | 1.0 | + 3.4 | - 7.3 |
| Moderate | 0.01 | 0.7 | + 3.0 | - 8.5 |
| Strong | 0.10 | 0.6 | - 0.2 | -15.0 |

increase in the aerosol absorption, the change in the reflected flux becomes less (moderately absorbing case). In fact, for the strongly absorbing case, there is a decrease in the reflected flux with respect to the clear sky case.

For the conditions assumed, the surface always loses some solar flux. The atmosphere may or may not experience an increased solar flux convergence depending on the aerosol absorption and scattering characteristics. The net flux into the surface-troposphere system can increase or decrease, again depending on the aerosol properties. This feature of loss of energy by the surface and the gain by the surface-troposphere system is discussed in the context of absorbing aerosols by Cess et al. (1985). McCracken et al. (1986) have studied the case of aerosols in the polar regions where the surface-atmosphere system can gain energy in the presence of aerosols.

The foregoing example is just one aspect of the sensitivity of the aerosol radiative forcing. As pointed out in Section 1, the magnitude of the radiative perturbation depends on several factors. We discuss them below briefly based on earlier investigations. Coakley et al. (1983) show that while a nonabsorbing aerosol layer (optical depth 0.1) causes an increase in the planetary albedo over all types of surfaces, a moderately absorbing aerosol reduces the planetary albedo for surface albedos greater than 0.3. Their study also reveals that a decrease in solar zenith angle for a nonabsorbing aerosol layer (optical depth 1.0) leads to a decrease in the reflectivity of the layer. As shown by Ramaswamy and Kiehl (1985) in the context of smoke aerosols, the distribution of the heating perturbation in the atmosphere depends on the aerosol vertical profile.

A summary of the maximum heating rate perturbations that have been reported in the literature for various aerosol types (either based on model assumptions or observations) is presented in Table 2. These values may be compared with the clear tropical sky heating due to water vapor absorption (1.1 K/d at 0.5 km.; see Fig. 1a). Also listed are the altitudes of the assumed perturbation. The layer thicknesses,

TABLE 2. MAXIMUM PERTURBATION HEATING RATES (SOLAR) FOR VARIOUS AEROSOL TYPES

| Aerosol Type | Perturbation ( K/d ) | Altitude ( Km ) | Source |
|---|---|---|---|
| 1 km. Scale Height | 0.5 | 0.5 | This study |
| Desert Dust | 4.0<br>7.0 | 3.5<br>1.0 | Carlson and Benjamin (1980); Fouquart et al. (1984) |
| Arctic Haze | 0.4<br>1.1 | 5.5<br>0.8 | Blanchet and List (1984); Valero et al. (1984) |
| Volcanic | 0.1 | 28.0 | Valero et al. (1984) |
| Water/Ice Cloud | > 1 | Troposphere | Ramaswamy and Kiehl (1985) |
| Nuclear War | > 5 | Troposphere | Ramaswamy and Kiehl (1985) |

the averaging period, and the geographical locations considered vary for the different investigations (the intent here is to merely point out the range). The wide range of values emerging from this tabulation is a measure of the possible extent of the aerosol forcing in the Earth's atmosphere. It also points out the difficulty of formulating a global forcing that would incorporate the different types of aerosols, since each type has its own distinct signature. Toon and Pollack (1976) and Deepak and Gerber (1983) present unified models that present the average aerosol properties under nominal, non-catastrophic circumstances. One characteristic involving the aerosols, not categorised in Table 2, which is manifested in cloud optical properties, is the effect of aerosols present as external or internal mixtures with water drops or ice crystals. This perturbation can be substantially higher than that posed by the dry aerosol alone (Chylek et al., 1984; Coakley et al., 1987).

A summary can also be made about the aerosol-induced perturbations in the solar fluxes at the boundaries of the atmosphere. In this case, too, the range of possible values is large; the depletion of the flux at the surface can range from small values (few $W/m^2$ for background tropospheric aerosol) through moderate values (e.g., 10-50 $W/m^2$ for desert duststorms) to extreme values (e.g., 150 $W/m^2$ for catastrophic events). The change in the net solar flux at the top of the atmosphere can range from small negative values (for scattering aerosols) to very high positive values (for large loadings of absorbing aerosols).

The extremities in the aerosol radiative forcing imply that there would be a large variability in the surface-atmosphere response for different aerosol types. This is indeed borne out by comparing the GCM simulations of "nominal" tropospheric aerosol effects (Coakley and Cess, 1985; Geleyn and Tanre, 1984) with the GCM simulations of catastrophic aerosol effects (Covey et al., 1984; Malone et al., 1986).

Returning to the focus in this paper on the "nominal" aerosol forcing, the response to such a forcing is evaluated next. Before beginning this investigation and as a final point to the present section, the study by Coakley and Cess (1985) is cited to illustrate a global model of the background aerosol solar radiative forcing (diurnal averages for July conditions). This is shown in Fig. 2 as 1) change in the radiative flux at the top of the atmosphere, 2) change in the radiative flux at the surface, and 3) change in the atmospheric solar heating rate. In this particular study, the aerosol scattering dominates the absorption; the nature of the forcing at all latitudes involves changes as discussed earlier. The globally-averaged change at the top of the atmosphere is -3.5 $W/m^2$ and that at the surface -5.0 $W/m^2$, so that the increase in the atmospheric absorption is 1.5 $W/m^2$. The GCM response to this forcing will be discussed in Section 4.

## 3. RADIATIVE AND RADIATIVE-CONVECTIVE RESPONSES

The response of the surface-atmosphere system to the aerosol perturbations is investigated first using a one-dimensional model and by making some simple assumptions about the atmospheric physical processes. For an atmosphere in radiative-convective-dynamical equilibrium, the rate of change of temperature at any level is given by

$$\frac{dT}{dt} = Q_{sol} + Q_{lw} + Q_{conv} + Q_{dyn} = 0 \qquad (1)$$

where $Q_{sol}$, $Q_{lw}$, $Q_{conv}$ and $Q_{dyn}$ denote, respectively, the heating due to solar, longwave, convective, and dynamical processes. The convective heating is the sum of the sensible and latent heat while the dynamical heating is the sum of diffusion, and mean and eddy circulations.

### 3.1 RADIATIVE RESPONSES

Consider the response of the atmosphere when both convective and dynamical heating are assumed to be fixed, i.e., $Q_{conv}$ and $Q_{dyn}$ are

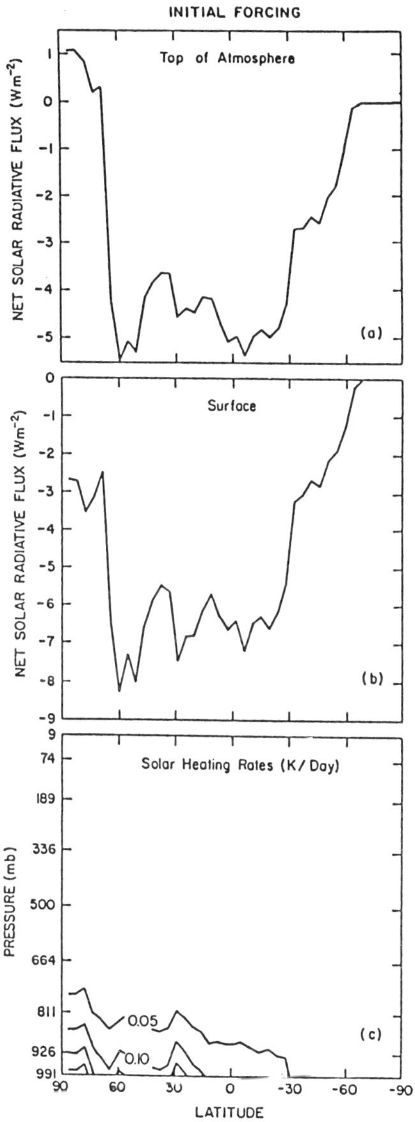


FIGURE 2. Zonally-averaged solar radiative forcing due to the background tropospheric aerosol illustrates: a) change in the net solar flux at the top of the atmosphere, b) change at the surface and c) change in the solar heating (degrees per day). (From Coakley and Cess, 1985.)

unaltered by the perturbation. Then, if the aerosol-induced increase in solar heating is $\epsilon$ and $\Delta$ denotes the change in the quantities, the response to this heating would be an increased longwave cooling, i.e.,

$$\Delta Q_{\ell w} = -\epsilon \tag{2}$$

Both the perturbation and the response involve the radiative processes only in this case. Equation (3) is essentially the fixed dynamical heating concept introduced by Fels and Kaplan (1976), Ramanathan and Dickinson (1979) and Fels et al.(1980). For an atmosphere whose properties are invariant across the longwave spectrum (i.e., "gray" atmosphere),

$$\Delta Q_{lw} \propto (-\Delta(T^4)), \tag{3}$$

hence,

$$\Delta T \propto (\epsilon/4T^3)$$

Thus, under this assumption, the result of an aerosol-induced heating is an increase in the temperature of the layer where the heating occurs. It is also seen that the magnitude of the temperature response increases with decreasing temperature so that upper levels in the troposphere would experience a higher temperature change than lower levels for the same magnitude of perturbation.

Numerical values of the temperature change in the clear tropical atmosphere under the radiative response assumption are calculated when a perturbation heating of 0.2 K/d is applied to any layer (note from Table 2 that this is a "nominal" value, well below the upper end of the range). This is performed using the radiative-convective model of Ramaswamy and Kiehl (1985) and the McClatchey (1972) tropical atmosphere profile with fixed surface temperature and fixed water vapor profile. The rationale for fixed surface temperature arises out of consideration of the large heat capacity of the oceans. The rationale for the static water vapor profile is to avoid the complicated feedback effects involving the hydrologic cycle (Ramanathan, 1981), which are not adequately represented in a one-dimensional model. In the simulations to be described below, the sum of the convective and dynamic heating is assumed to be the negative of the net radiative (solar + longwave) heating, according to Eq. (1).

Three layers in the troposphere and one in the stratosphere are considered for the perturbation experiments; these are listed in Table 3 (note that the layer configurations differ from that used to describe the aerosol radiative forcing in Section 2). Four different simulations are performed, with each layer being perturbed in turn. In each simulation, the solar heating rate is increased by 0.2 K/d in the layer under consideration. Then, the temperature profile of the model atmosphere is time-marched to a new equilibrium solution, with the longwave radiative process responding to the imposed perturbation, such that the condition given by Eq. (1) is established again. For the numerical solutions, equilibrium is deemed to occur when the left-hand side of Eq. (1) is less than 0.001 K/d. The resulting temperature

TABLE 3. LAYER SPECIFICATIONS, HEATING PERTURBATION, AND CORRESPONDING FLUX PERTURBATIONS FOR THE RADIATIVE AND RADIATIVE-CONVECTIVE RESPONSE EXPERIMENTS

| Perturbed Layer Location ( Km ) | Heating Perturbation ( K/d ) | Flux Perturbation ( $W/m^2$ ) |
|---|---|---|
| 0.1 - 0.5 | 0.2 | 0.7 |
| 3.0 - 4.2 | 0.2 | 2.1 |
| 7.2 - 9.2 | 0.2 | 2.0 |
| 19.4 - 21.3 | 0.2 | 0.1 |

change in the four different simulations for the four specific layers are listed in Table 4. Although the magnitude of the perturbation chosen is arbitrary, it is within the range that can be encountered in noncatastrophic situations. As will be seen later, were the three-dimensional model to respond in this manner, the magnitude of the temperature response, as obtained from the experiments in this section, would be well above the normal variability.

The radiative response can be expected to hold in the stratosphere where the atmosphere tends to be in radiative equilibrium on an annual basis (Ramanathan and Dickinson, 1979; Fels et al. 1980). By analogy, the radiative responses derived here should be regarded as limiting temperature changes that occur only when radiative processes are important for the compensation. In the troposphere, however, convective processes play a major role in distributing heat energy and, with the possible exception of the upper troposphere, it is not justified to assume fixed convective heating. Hence, it is more meaningful to consider the radiative-convective response of the troposphere to an aerosol radiative forcing.

TABLE 4. CHANGE IN THE RADIATIVE RESPONSE TEMPERATURE OF THE LAYER DUE TO AN AEROSOL-INDUCED INCREASE IN THE SOLAR HEATING RATE.

| Perturbed Layer Location ( Km ) | $\Delta T$ (Perturbed Layer) ( K ) |
|---|---|
| 0.1 - 0.5 | 0.31 |
| 3.0 - 4.2 | 0.97 |
| 7.2 - 9.2 | 1.58 |
| 19.4 - 21.3 | 1.97 |

3.2 RADIATIVE-CONVECTIVE RESPONSES

The radiative-convective response is evaluated by parameterizing the convective heat transfer as a diffusive process (see Ramaswamy and Kiehl, 1985); in this formulation, the convective flux exchange between two adjacent layers is linearly related to the temperature gradient. Surface fluxes follow the prescription of Ramanathan (1981). A critical lapse rate of 6.5 K/km is assumed for the simulations in this section. The McClatchey (1972) tropical atmosphere profile is again taken as the initial equilibrium profile.

As before, the heating perturbation is applied to each of the four layers individually. It is assumed first that the aerosol is absorbing only. Further, in order to simulate the reduction in solar flux below the perturbed layer (owing to aerosol absorption in the layer), the net solar flux below the layer is reduced by an amount equal to that absorbed by the layer. The amount of flux absorbed, corresponding to each of the four layers, is listed in Table 3. The simulations for the radiative-convective response solution are in contrast to the simulations for the radiative response solution where the heating of the layer was the only perturbation; in the present case, less solar flux is incident on the layers below the perturbed layer. As before, the surface temperature and the water vapor profile are held fixed.

For the simulations in this section, since convective heating is calculated explicitly, the dynamical heating is assumed to be the negative of the sum of the convective and the net radiative heating, again according to Eq. (1). The solar flux perturbation is applied according to the details in the preceding paragraph and the temperature profile is time-marched to a new equilibrium solution. In contrast to the previous section, the assumptions in this section allow both the radiative and the convective mechanisms to participate in the compensation process. For the numerical solutions, equilibrium requires, besides the heating rate criterion mentioned in Section 3.1, the additional constraint that the lapse rate be less than the critical lapse rate everywhere in the column.

The magnitude of the change in the temperature in the case of perturbation in each of the four layers is listed in Table 5. The changes are less than for the radiative response experiments. This can be explained as follows. Analogous to Eq. (2), the new balance in the heating rate in the concerned layer can be wrtten as

$$\Delta Q_{\ell w} + \Delta Q_{conv} = -\epsilon \tag{4}$$

In contrast to Section 3.1, there are now two readjustment mechanisms to compensate for the heating perturbation in the layer. The presence of convective heat transport diminishes the influence of the longwave adjustment and leads, thus, to a lesser increase in temperature. For the layers below the perturbed layer, the decrease in flux necessitates a readjustment whereby the temperature is reduced (the balance equations for the layers below are similar to Eq. (4)). The cooling below the perturbed layer is illustrated by the decreases in the surface-air temperature, as listed in Table 5.

TABLE 5. CHANGE IN THE RADIATIVE-CONVECTIVE RESPONSE TEMPERATURE ($\Delta T$ (LAYER)) DUE TO AN AEROSOL-INDUCED PERTURBATION

| Layer ( Km ) | Absorption | | Scattering | |
|---|---|---|---|---|
| | $\Delta T$ (K) (layer) | $\Delta T$ (K) (surface) | $\Delta T$ (K) (layer) | $\Delta T$ (K) (surface) |
| 0.1-0.5 | 0.08 | -0.20 | -0.51 | -1.07 |
| 3.0-4.2 | 0.17 | -2.13 | -0.69 | -1.71 |
| 7.2-9.2 | 0.67 | -2.28 | -0.17 | -2.85 |
| 19.4-21.3 | 2.32 | -0.26 | -0.02 | -0.03 |

Another comparison can be made with the radiative response solutions. The heating in any layer, unaccompanied by flux changes in layers below (as in the radiative response experiments in Section 3.1) leads to, by virtue of the longwave radiation, increased longwave flux convergence in other layers as well. This, in turn, gives rise to an additional longwave flux convergence in the layer that was perturbed directly. When the solar flux decreases in the layers below the perturbed layer (which occurs for the experiments in this subsection), there is a cooling below that counteracts the increased radiating effect of the perturbed layer. Thus, the net change in the temperature at any layer is dependent on the redistribution by the longwave radiative exchanges in the column.

Finally, the radiative-convective response when the aerosols are assumed to be nonabsorbing is considered; the perturbation in this case is one of a reduction of solar flux in the entire surface-atmosphere system, including the layers considered in Table 3. Here, the aerosol absorption is ignored and the solar flux in the entire column is depleted by an amount corresponding to a heating of 0.2 K/d. Again, the same four layers are considered individually and four simulations are performed. The magnitude of the changes in temperature are also listed in Table 3. As should be expected, the entire column cools; this is indicated by the decrease in temperature in the layer where heating was applied in the foregoing cases, and in the surface-air layer.

The conclusions to be drawn from the radiative-convective experiments are that while aerosol-induced solar absorption tends to increase the temperature of the entire column, the aerosol-induced scattering tends to cool the column. The net effect at any level in the column is a function of the degree of the aerosol absorption-to-scattering. The longwave radiative effects are felt throughout the column. Convective heat transfer, as modeled here on the assumption of a diffusive mechanism, competes with the longwave radiation and tends to limit the temperature response, especially when compared with the predictions from the radiative response solutions.

The only regions where the convective processes would not be directly effective in the compensating process would be the lower stratosphere and above (although indirect effects due to changes in the troposphere cannot be ruled out).

Two major points, which are beyond the focus of simple models and whose effects cannot be gauged by experiments with the one-dimensional model used in this section, concern the hydrologic cycle and the nature of the large-scale dynamical adjustments. The former is crucial not only for the longwave radiative changes but also for the latent heat adjustments. Both the moisture-related effects and the large-scale dynamics require the numerical solution to the primitive equations and their response can be properly addressed only with the three-dimensional general circulation models. As seen in the next section, important physical adjustments occur in GCMs, all of which cannot be anticipated by the radiative and the radiative-convective response experiments performed here.

## 4. GENERAL CIRCULATION MODEL RESPONSES

The GCM experiments reported in the literature have addressed the problem of either small perturbations such as those arising due to tropospheric and stratospheric aerosols (Coakley and Cess, 1985; Geleyn and Tanre, 1984; Tanre and Geleyn, 1984) or large perturbations following catastrophic events (Covey et al., 1984). Here, we focus mainly on the small perturbations due to tropospheric aerosols; we will mention some of the results concerning other types of aerosols at the end of the section.

Coakley and Cess (1985) have studied the response of the National Center for Atmospheric Research General Circulation Model (called the Community Climate Model or CCM) to the forcing shown in Fig. 2 and which was described in Section 2. The aerosol-induced changes in the Northern Hemisphere mean temperature profile (July conditions) from their GCM simulation with fixed sea-surface temperature is shown in Fig. 3. Also shown in that figure by the gray shading is the normal variability of the model. The GCM response was below the normal variability of the model at any level in the troposphere. It would appear that the cooling in the troposphere (Fig. 3) is the result of the aerosol scattering dominating over its absorption. While this would be consistent with the radiative-convective response solution for the scattering only case (Section 3.2), there is a more logical reason for the GCM behaviour.

The explanation can be had from Coakley and Cess' (1985) examination of the energy balance between the physical processes in the control and the perturbed cases. They analyzed the North African region in this connection and this is illustrated in Fig. 4. The control in Fig. 4a shows the tropospheric values of the net radiative heating rate (sum of solar and longwave), $Q_{rad}$, the convective heating, $Q_{conv}$, the dynamical heating, $Q_{dyn}$, and the heating due to large-scale ascending motions, $Q_{\omega}$. There is a balance essentially

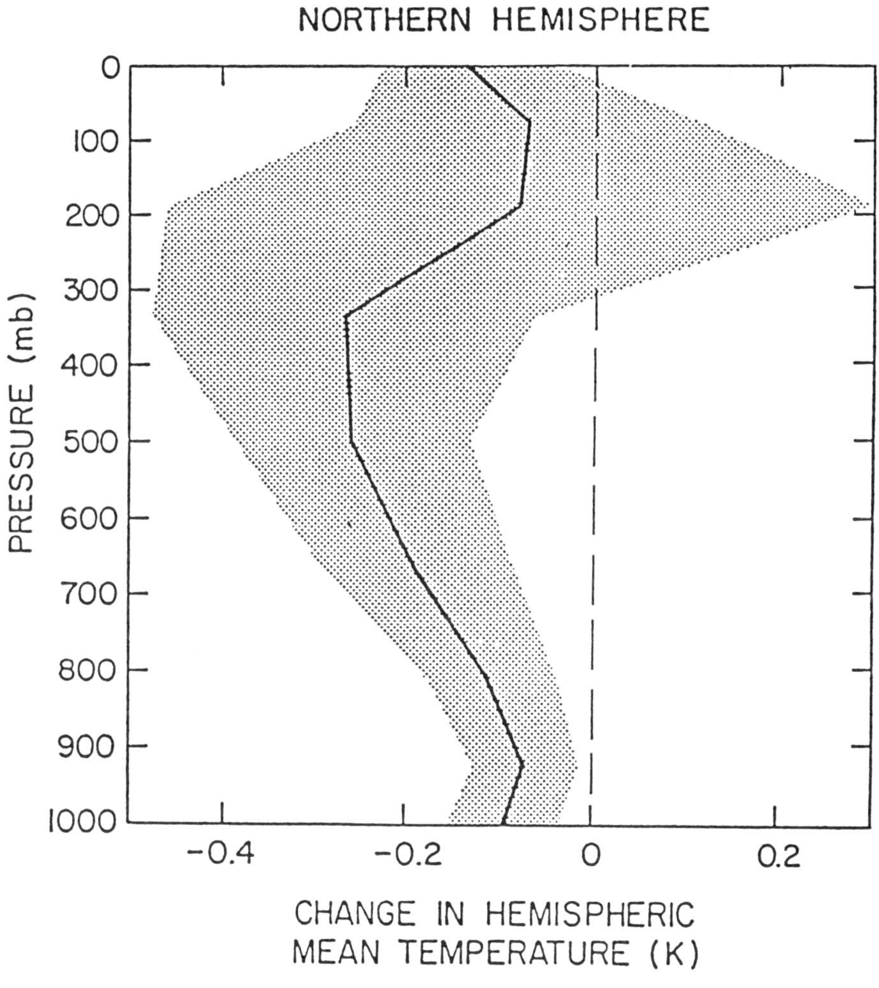


FIGURE 3. Change in the Northern Hemisphere mean temperature of the NCAR Community Climate Model due to the aerosol forcing shown in Figure 2. The gray area represents the "noise" level of the model. (From Coakley and Cess, 1985.)

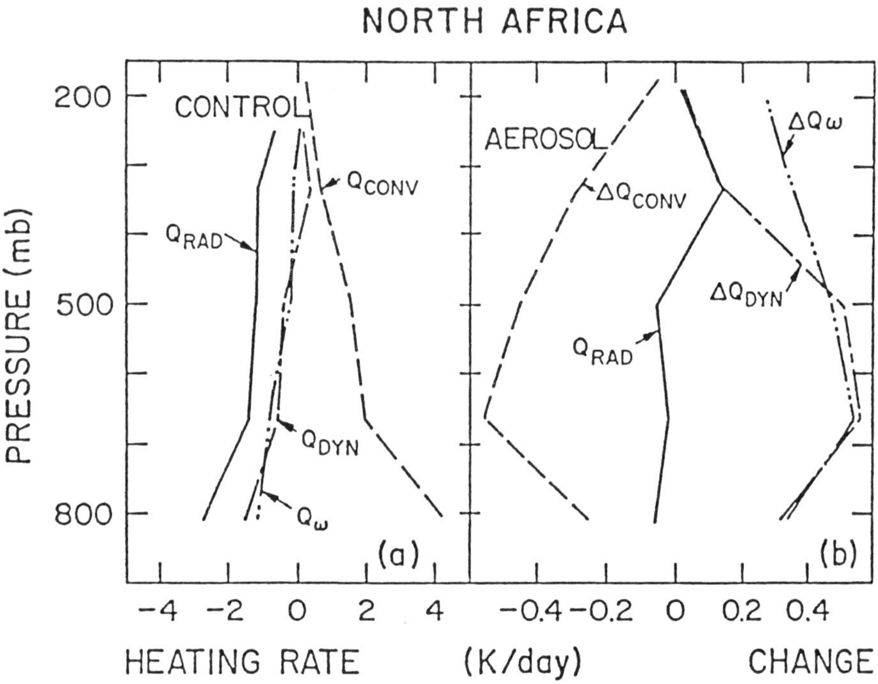


FIGURE 4. a) Heating rates and b) change in the heating rates for the North African region in the NCAR Community Climate Model (in degrees per day) due to the background tropospheric aerosol forcing. $Q_{rad}$ is the heating due to radiative processes, $Q_{conv}$ is the heating due to convective processes (including latent heat), $Q_{rad}$ is the heating due to dynamical processes. $Q_\omega$ is the heating rate due to the vertical wind field. (From Coakley and Cess, 1985.)

between the radiative cooling and the convective heating. The dynamical cooling term is small and nearly all of this is supplied by the uplift. There is a low-level convergence in this region and moisture is transported upwards by the vertical wind field. Condensation and latent heat release follow. In the perturbed situation (the changes in the heating rates with respect to the control case are shown in Fig. 4(b)), the reduction of the solar flux absorbed at the surface leads to a cooling and a decrease in the surface fluxes of sensible and latent heat. Convection is suppressed and $\Delta Q_{conv}$ is negative. Radiative cooling undergoes little change in the lower troposphere while there is a small decrease in the upper troposphere. Note that this is in contrast to the increase in the cooling of the column obtained for the radiative-convective response (scattering only case) in Section 3.2.

Since changes in $Q_{conv}$ are larger in magnitude relative to the changes in $Q_{rad}$ and because of the balance between the radiative and convective terms in the control (Fig. 4(a)), one may have expected the suppression of convective activity in the perturbed case to lead to a cooling throughout the troposphere. However, the decrease in tropospheric diabatic heating ($Q_{conv}$ + $Q_{rad}$) led to changes in the circulation. This resulted in a decrease in the dynamical cooling, thus offsetting the reduction in convective heating. Thus, the magnitude of the change in the temperature remained small everywhere (Fig. 4) and within the noise levels of the model. The value of $Q_{\omega}$ indicates that the dynamical changes were brought about primarily by a reduction in the vertical velocity (less adiabatic cooling). While the control case had an excess of precipitation over evaporation, the perturbed case had the two processes nearly in balance. There was also a decrease in cloudiness over this region. Other geographical regions also underwent cooling in the troposphere. For all the other regions, although it was possible to state in general terms that the reduction in the solar flux at the surface had reduced convection, the small changes in the terms did not permit an easy identification of the precise mechanisms.

Summarising, the GCM response, with fixed sea-surface temperatures, being controlled by moist convective processes in the Northern Hemisphere, led to small temperature changes throughout the column. The temperature response was primarily due to a suppression of the convective activity brought about by the reduction of solar flux at the continental surfaces. The dynamic response was limited to changes in vertical velocity. Thus, in assessing the response to an aerosol radiative forcing, the GCM results indicate that it is important to consider the effects of changes in the hydrologic cycle, convective heat transfer, and large-scale dynamics, although the magnitudes involved may be insignificant for "nominal" aerosol optical depths. Recalling the responses of the column models (Section 3), although the sign of the temperature changes obtained there does correspond to those obtained in the GCM, the magnitudes differ owing to the presence of nonradiative feedback mechanisms in the GCMs. It is, of course, yet to be verified whether the feedbacks occurring in the GCM simulations are similar in scale and magnitude to those occurring in the actual atmosphere. The GCM results also suggest that large local temperatures are not likely to occur for small perturbations involving predominantly scattering aerosols. Land surface temperature responses are likely to remain small if ocean surface temperatures are fixed, unlike the results from radiative-convective models with land surfaces of zero heat capacity. It may be noted again that this GCM study employed fixed sea-surface temperatures. The consequence of an explicit formulation of atmosphere-ocean interactions in GCMs on the aerosol-related impacts is yet to be investigated.

This section is concluded by discussing other experiments with GCMs that have employed different aerosol forcings. Geleyn and Tanre (1984), using the European Center (ECMWF) GCM, employed seven different aerosol models to estimate the short-term climatic response. They, too found that the compensation to the tropospheric aerosol forcing involved nonradiative effects. There was no detectable temperature response except above midtropospheric levels. An interesting result from their experiments was the large land surface coolings (3 K) in the Sahara due to a large desert aerosol optical depth and its vertical distribution.

Tanre and Geleyn (1984) also carried out a simulation of a stratospheric aerosol perturbation similar to that from El Chichon. For a globally-averaged optical depth of 0.15, a 1-3 K increase resulted in the stratospheric temperature at 30 mb (0-35 N). The temperature changes occur due to solar and longwave aerosol properties. Pitari et al. (1987) have also deduced temperature increases using a residual eddy circulation model of the stratosphere. It is important to point out that Pollack and Ackerman (1983) have estimated similar temperature changes using a radiative-convective model. Thus, the solutions from the radiative and the radiative-convective responses described in Section 3.2 are probably applicable in sign and, perhaps, even in magnitude, to aerosol-induced heating perturbations near the lower stratosphere. It may be noted from Tables 4 and 5 that in both the radiative and radiative-convective solutions, the layer situated near 20 km yields a substantially higher response than any of the other layers, for the same heating perturbations.

Randall et al. (1984) have also performed a GCM study of the response due to Saharan dust. In their experiment, the optical depth of the aerosol decreased from 1 near the West African Coast to 0 in the western Atlantic Ocean. Their study showed that a reduction occurred in the easterlies in the western Atlantic due to dust-induced heating--this is an interesting result since it is primarily the easterlies that transport dust in the first place.

Finally, with the introduction of the notion that a nuclear war or an asteroid colliding with the Earth may be accompanied by large aerosol loadings in the atmosphere (see NRC, 1985), several GCM studies have examined this climate problem. The basic climatic effects have been discussed in the "Nuclear Winter" session of the Symposium and in Ackerman (1987). Covey et al. (1984) found that the intense radiative heating due to smoke aerosols in the aftermath of a nuclear war caused a stabilization of the midtroposphere accompanied by substantial surface cooling. They also found that meridional circulation changes caused by aerosol heating could spread the aerosols well beyond the location of the original injections. Incidentally, it is worth pointing out that changes in stability were also found by Joseph (1984) in a simulation with hypothetical large dust aerosol optical depths.

## 5. AEROSOL MICROPHYSICS AND TRANSPORT

All the studies mentioned above assumed fixed aerosol physical properties both in the time and space domains. Thus, microphysical processes such as coagulation, aerosol-to-cloud transformation, washout, rainout, sedimentation and Brownian diffusion were ignored in these studies. Further, the transport of aerosol away from the source regions was also not possible in these models. This inhibits a comprehensive assessment of the radiative-microphysical-dynamical interactions. In recent years, some investigations have proceeded to rectify these deficiencies by incorporating aerosol microphysical processes and by treating the aerosols as a tracer. To do this accurately is a daunting task since the range involved in describing the physical processes extends from the micrometer scale (involving aerosols) to the kilometer scale (involving large-scale transport); the task can easily overwhelm available computational resources. A few of these efforts are mentioned below. The methodologies adopted in the attempts have necessarily involved varying degrees of parameterization of the aerosol transformation and transport processes.

Newiger et al. (1984) employed the large-scale zonally averaged dynamics and cloudiness to drive the equations for aerosol microphysical processes, aerosol radiative effects, and the meridional transport. Their two-dimensional model study indicated a transport of the aerosol radiative perturbations to lower latitudes for aerosol injected initially into the middle latitudes of the Northern Hemisphere.

Westphal et al. (1987) have also performed a two-dimensional simulation of the evolution of Saharan aerosols as they are transported over desert and the eastern Atlantic in a limited area dynamical model. By comparing with observed size distributions, they concluded that the dust sizes undergo modifications during the mesoscale transport. Such a transformation implies that radiative changes would also occur during the transport.

While two-dimensional models are useful in going beyond the scope of zero- and one-dimensional models, their usefulness is limited by the use of zonally-averaged circulation statistics of the atmosphere. One way to do the more complicated three-dimensional transport economically is to use the wind vectors from an actual GCM simulation to transport the aerosols; aerosol microphysical processes can also be treated simultaneously although feedbacks on circulation cannot be evaluated. Such modeling techniques have been pursued by Levy and Moxim (1987).

It is a natural step to make the transition to a three-dimensional GCM, which not only advects aerosols but also treats their microphysics and radiative effects explicitly and interactively. Such efforts, especially for long-term simulations, are computationally expensive so that some of the physics has to be approximated. The need for such models, however, is crucial in the investigation of the aerosol climatic effects.

One particular problem that has demanded attention to the details mentioned above is the climatic effects due to smoke aerosols in the aftermath of a nuclear war. Simulations with aerosol microphysical and interactive radiative processes have been performed in three different investigations. Malone et al. (1986) used a variation of the NCAR CCM along with aerosol absorption and bulk removal parameterizations to evaluate the smoke aerosol-induced climatic effects. The effect of introducing the aerosol removal by rain, although approximate, demonstrated that the lifetime of aerosols in the atmosphere has an important bearing on the radiative perturbations. Also, the aerosol lifetime is enhanced by reduction in convective activity, which, in turn, is initiated by the initial radiative effects. Thompson et al. (1987) have allowed for the temporal transformation of aerosol sizes, in addition to incorporating the washout and rainout processes. Their study indicates that the efficiency of removal by rain is the most important microphysical mechanism governing the radiative forcing in the middle and lower troposphere. Assumptions about aerosol optical properties were also found to be a significant factor in this study.

## 6. CONCLUDING COMMENTS

From the perspective of aerosol-climate interactions, the major uncertainties in the aerosol radiative forcing can be considered to arise from deficiencies in our knowledge on:

1) source specification and characterization of aerosol optical properties.
2) atmospheric effects on aerosols, and
3) aerosol effects on the atmosphere.

For the nominal stratospheric and tropospheric aerosols, the uncertainty in the spatial and temporal distribution of aerosol optical properties requires extensive observation networks, either ground-based or, more conveniently, through satellites. Since large perturbations in the aerosol radiative effects arise from events of an episodic nature, it will be difficult to characterize the forcing due to all such events. Examples of such events are the seasonal arctic haze and duststorms, and episodic volcanic injections. However, observations of the evolution of aerosol derived from these sources would provide a means to ascertain the temporal evolution of the radiative forcing and to improve our understanding of transport mechanisms.

The most significant atmospheric effect on aerosols is the dry and wet removal processes in the atmosphere. Efficiency of washout and rainout of aerosols impacts directly onto the radiative effects. Parameterizations used in models require stringent testing for their mesoscale and microscale treatments such as subgrid-scale processes, cloud and precipitation formation, nucleation scavenging, etc. Cloud formation, maintenance, and dissipation, as well as transport of water in all its phases are central problems in general circulation modeling and they affect the aerosol-related issues as well. It is pointed out that current GCMs, while able to reproduce several features of the real atmosphere, possess certain deficiencies (e.g., cloud physical processes, atmosphere-ocean interactions), aspects of which impinge directly on the assessment of aerosol effects.

Aerosol-induced effects on the atmosphere involve both microphysical and radiative processes. Only the solar aspects of the latter subject has been considered in the present paper. Among the uncertainties in microphysics are the external or internal mixtures of aerosols with water drops and the dependence of this process on their physicochemical properties. Uncertainties persist in the treatment of aerosol radiative effects even if the spatial and the vertical distribution is known accurately (which, for most of the globe, is not true currently). At present, practical solutions are possible only for homogeneous and spherical aerosols, although heterogeneous and nonspherical aerosols are the rule rather than the exception. Fundamental problems concerning the light scattering by arbitrarily shaped, inhomogeneous particles require attention. Perhaps, statistical and empirical formulations (Bohren, 1986) only can provide a practical solution to this dilemma. Cloud-aerosol radiative interactions are also, indirectly, a subset of this uncertainty.

The aerosol effects on the atmosphere concern more than merely the alteration of the thermal profile. As seen from the GCM simulations, for "nominal" tropospheric aerosol forcings, radiative perturbations in the troposphere are accompanied by changes in the convective heating and in the vertical velocity. The readjustment of the energy budget by the nonradiative components in the troposphere limits the magnitude of temperature change that is predicted by the radiative and radiative-convective response assumptions. In fact, the importance of nonradiative processes in the GCM tropospheric response to a "nominal" aerosol forcing may render the search for a temperature signal in the lower troposphere, and especially at the surface, a futile task.

The GCM solutions can be quite different from the response of the more simple models that perform averaging of sorts in either the vertical or horizontal dimensions. This does not imply that the one-dimensional radiative-convective models are suited only for estimating the initial radiative forcing. They remain useful for the analysis of the relative roles of radiative and nonradiative components in the GCM responses. The similarity between a GCM and a radiative or a radiative-convective model temperature response may exist only for the regions near or above the tropopause.

The qualitative nature of the aerosol radiative forcing on the atmosphere in terms of their absorbing and scattering properties are well understood at present. It is the quantitative determination of the quiescent and episodic aerosol radiative forcing (involving refractive indices, size distributions, vertical profiles of number densities, etc.), aspects of the microphysical forcing (e.g., changes in composition and size, with respect to space and time), and the radiative-convective-dynamical response of the atmosphere that constitute serious challenges for future investigations. For modeling efforts to be fruitful, improvements are required in the understanding and representation of physical processes and, whenever possible, in the comparison of simulations with observations.

## ACKNOWLEDGMENTS

The author wishes to thank Drs. J.A. Coakley, S.B. Fels, J.D. Mahlman, and an anonymous reviewer for their constructive comments and suggestions.

## REFERENCES

Ackerman, T.P., 1987: Aerosols in climate modeling. In P. Hobbs and M.P. McCormick, (Eds.), Aerosols and Climate, A. Deepak Publishing, Hampton, VA, in press.

Blanchet, J.P., and R. List, 1984: On Radiative Effects of the Anthropogenic Aerosol Components of the Artic Haze and Snow in a Surface-Atmosphere Coupled Model. In G. Fiocco (Ed.), International Radiation Symposium '84: Current Problems in Atmospheric Radiation, A. Deepak Publishing, Hampton, VA, 269-273.

Bohren, C., 1986: Review Paper on Optical Properties of Aerosols, Proceedings, Sixth Conference on Atmospheric Radiation, American Meteorological Society, Williamsburg, VA.

Carlson, T.N., and S.G. Benjamin, 1980: Radiative Heating Rates for Saharan Dust, J. Atmos. Sci., 37, 193-213.

Cess, R.D., G.L. Potter, S.J. Ghan and W.L. Gates, 1985: The Climatic Effects of Large Injections of Atmospheric Smoke and Dust: A Study of Climatic Feedback Mechanisms with One- and Three-Dimensional Models, J. Geophys. Res., 90, 12937-12950.

Charlson, R.J., J.E. Lovelock, M.O. Andreae and S.G. Warren, 1987: Oceanic Phytoplankton, Atmospheric Sulfur, Cloud Albedo and Climate, Nature, 326, 655-661.

Chylek, P., V. Ramaswamy and R. Cheng, 1984: Effect of Graphite Carbon on the Albedo of Clouds, J. Atmos. Sci., 41, 3076-3084.

Coakley, J.A., and R.D. Cess, 1985: Response of the NCAR Community Climate Model to the Radiative Forcing by the Naturally Occurring Tropospheric Aerosol, J. Atmos. Sci., 41, 1677-1692.

Coakley, J.A., R.D. Cess and F.B. Yurevich, 1983: The Effect of Tropospheric Aerosol on the Earth's Radiation Budget: A Parameterization for Climate Models, J. Atmos. Sci., 40, 116-138.

Coakley, J.A., R.L. Bernstein and P.A. Durkee, 1987: Effect of Ship-Stack Effluents on Cloud Reflectivity, Science, in press.

Covey, C., S.H. Schneider and S.L. Thompson, 1984: Global Atmospheric Effects of Massive Smoke Injections from a Nuclear War: Results from General Circulation Model Simulations, Nature, 308, 21-25.

Deepak, A., and H.E. Gerber (Eds.), 1983: Report on the Experts Meeting on Aerosols and Their Climatic Effects, WCP-55, World Meteorological Organization, Geneva, Switzerland.

Fels, S.B., and L.D. Kaplan, 1975: A Test of the Role of Longwave Radiative Transfer in a General Circulation Model, J. Atmos. Sci., 33, 779-789.

Fels, S.B., J.D. Mahlman, M.D. Schwarzkopf and R.W. Sinclair, 1980: Stratospheric Sensitivity to Perturbations in Ozone and Carbon Dioxide: Radiative and Dynamical Response, J. Atmos. Sci., 37, 2265-2297.

Fiocco, G. (Ed.), 1984: International Radiation Symposium '84: Current Problems in Atmospheric Radiation, A. Deepak Publishing, Hampton, VA, 438 pp.

Fouquart, Y., B. Bonnel, G. Brogniez, A. Cerf;, M. Chaoui, L. Smith and J.C. Vanhouette, 1984: Radiative Properties of Sahelian Aerosols. In G. Fiocco (Ed.), International Radiation Symposium '84: Current Problems in Atmospheric Radiation, A. Deepak Publishing, Hampton, VA, 71-73.

Geleyn, J.F., and D. Tanre, 1984: Short-Term Climatic Response to the Atmospheric Aerosol Radiative Forcing. In G. Fiocco (Ed.), International Radiation Symposium '84: Current Problems in Atmospheric Radiation, A. Deepak Publishing, Hampton, VA, 243-246.

Gerber, H.E., and A. Deepak (Eds.), 1984: Aerosols and Their Climatic Effects, A. Deepak Publishing, Hampton, VA, 297 pp.

Grassl, H., 1987: Review paper on Radiative Effects of Atmospheric Aerosols, Aerosols and Climate Symposium, IUGG, Vancouver, Canada.

Joseph, J.H., 1984: The Sensitivity of the Numerical Model of the Global Atmosphere to the Presence of Desert Aerosol. In H.E. Gerber and A. Deepak (Eds.), Aerosols and Their Climatic Effects, A. Deepak Publishing, Hampton, VA, 215-227.

Levy, H., and W.J. Moxim, 1987: Fate of U.S. and Canadian Combustion Nitrogen Emissions, Nature, 328, 414-416.

Malone, R.C., L.H. Auer, G.A. Glatzmaier, M.C. Wood and O.B. Toon, 1986: Three-Dimensional Simulations Including Interactive Transport, Scavenging and Solar Heating of Smoke, J. Geophys. Res., 91, 1039-1053.

McClatchey, R.A., R.W. Fenn, J.E.A. Selby, F.E. Volz and J.S. Garing, 1972: Optical Properties of the Atmosphere, third ed., Rep. Afcrl-72-0497, Bedford, MA.

Newiger, M., H. Grassl, P. Schussel and J. Rehkopf, 1984: Global Aerosol Transport and Consequences for the Radiation Budget. In G. Fiocco (Ed.), International Radiation Symposium '84: Current Problems in Atmospheric Radiation, A. Deepak Publishing, Hampton, VA, 247-251.

National Research Council, 1985: The Effects on the Atmosphere of a Major Nuclear Exchange, National Academy Press, Washington, D.C.

Pitari, G., M. Verdecchia and G. Visconti, 1987: A Transformed Eulerian Model to Study Possible Effects of the El Chichon Eruption on the Stratospheric Circulation, J. Geophys. Res., 92, 10961-10975.

Pollack, J.B., and T.P. Ackerman, 1983: Possible Effects of the El Chichon Volcanic Cloud on the Radiation Budget on the Northern Tropics, Geophys. Res. Letts., 10, 1057-1060.

Ramanathan, V., 1981: The Role of Ocean-Atmospheric Interactions in the $CO_2$ Climate Problem, J. Atmos. Sci., 38, 918-930.

Ramanathan, V., and R.E. Dickinson, 1979: The Role of Stratospheric Ozone in the Zonal and Season Radiative Energy Balance of the Earth-Troposphere System, J. Atmos. Sci., 36, 1084-1104.

Ramaswamy, V., and J.T. Kiehl, 1985: Sensitivities of the Radiative Forcing to Large Loadings of Smoke and Dust Aerosols, J. Geophys. Res., 90, 5597-5613.

Randall, D., T. Carlson and Y. Mintz, 1984: The Sensitivity of a General Circulation Model to Saharan Dust. In H.E. Gerber and A. Deepak (Eds.), Aerosols and Their Climatic Effects, A. Deepak Publishing, Hampton, VA, 123-133.

Shettle, E.P., and R.W. Fenn, 1979: Models for the Aerosols of the Lower Atmosphere and the Effects of Humidity Variations on Their Optical Properties, Air Force Geophysics Lab. Rep. AFGL-TR-79-0214, 94 pp.

Tanre, D., and J.F. Geleyn, 1984: Climatic Effects Induced by the El Chichon Cloud. In G. Fiocco (Ed.), International Radiation Symposium '84: Current Problems in Atmospheric Radiation, A. Deepak Publishing, Hampton, VA, 247-251.

Thompson, S.L., V. Ramaswamy and C. Covey, 1987: Atmospheric Effects of Nuclear War Aerosols in GCM Simulations: Influence of Smoke Optical Properties, J. Geophys. Res., 92, 10942-10960.

Toon, O.B., and J.P. Pollack, 1976: A Global Average Aerosol Model of Atmospheric Aerosols for Radiative Transfer Calculations, J. Appl. Meteor., 15, 225-246.

Twomey, S., 1976: Computations of the Solar Absorption of Solar Radiation in Clouds, J. Atmos. Sci., 33, 1087-1091.

Valero, F.P., T.P. Ackerman, and W.J.Y. Gore, 1984: The Absorption of Solar Radiation by the Arctic Haze During the Haze Season and Its Effects on the Radiation Budget, Geophys. Res. Letts., 11, 465-468.

Valero, F.P., T.P. Ackerman, and W.L. Starr, 1984: Radiative Effects of the El Chichon Volcanic Cloud. In G. Fiocco (Ed.), International Radiation Symposium '84: Current Problems in Atmospheric Radiation, A. Deepak Publishing, Hampton, VA, 144-147.

Westphal, D., O.B. Toon and T.N. Carlson, 1987: A Two-Dimensional Numerical Investigation of the Dynamics and Microphysics of Saharan Dust Storms, J. Geophys. Res., 92, 3027-3049.

# THE EFFECTS OF ARCTIC AEROSOLS ON A GENERAL CIRCULATION MODEL SIMULATION: A SENSITIVITY STUDY

*Jean-Pierre Blanchet*
Canadian Climate Centre, Atmospheric Environment Service,
North York, Ontario, Canada

## ABSTRACT

The effects of anthropogenic arctic aerosol on a general circulation model (GCM) spring simulation are determined by a progressively increasing aerosol amount. Aerosol properties are assumed to be a function of the wavelength and local relative humidity. Mean zonal and regional effects on temperature, moisture, pressure, circulation, and energy balance are examined. The main consequence of the diabatic warming induced by the aerosol is a weakening of the polar and Ferrel cells due to a reduction of the radiation deficit in the Arctic.

## 1. INTRODUCTION

During the last decade, arctic haze has become of interest because of the potential effects of tropospheric air pollution on regional climate. Particulate elemental carbon (EC) and sulphate of anthropogenic origin can interact with radiation to alter the earth-atmosphere energy budget. The complex interconnection between many atmospheric processes can be studied using a general circulation model (GCM). The objective of the current investigation is to examine the response of the Canadian Climate Centre (CCC) GCM to the presence of arctic aerosol enriched by anthropogenic constituents. At this exploratory stage of the investigation, only the direct solar and terrestrial infrared radiative effect of the aerosol is considered. Secondary effects related to transport, activation and wet processes, deposition, or alteration of surface albedo will not be examined here.

The aerosol composition, the regional distribution, and the size distribution have been the subject of several publications that were recently reviewed by Barrie (1986). In an attempt to bring many of these pieces of information into a coherent data set to be used by radiation and climate models, Blanchet and List (1983) have presented a moist aerosol model for the optical properties of arctic haze. In a second step, the radiative properties of EC and sulphate and their interactions with multiple snow layers have been investigated in a column model (Blanchet and List, 1987). In the next step, Blanchet, Heintzenberg and Winkler (1986) examined the radiative effects of extreme anthropogenic aerosol concentrations in urban environments during winter. Analysis of the specific case of the recordbreaking pollution episode around 15 January 1982, in northwestern Europe, with observed EC concentrations of 24 $\mu g\ m^{-3}$ and sulphates in excess of 400 $\mu g\ m^{-3}$, led to a calculated increase in diabatic heating of



ISBN 0-937194-11-5

2.2 $K.day^{-1}$. This is in agreement with the observed diurnal warming and the variations of relative humidity. The mass ratio of elemental carbon to sulphate and the midvisible single scattering albedo (0.92) found in that urban pollution are comparable to that of arctic aerosol.

## 2. MODEL AND METHOD

The GCM version used in this study is an extension of that presented by Boer et al. (1984a) for which a climatology has been published (Boer et al., 1984b). Cloud coverage and cloud optical properties are interactive with the GCM's prognostic variables. The radiation model is composed of two spectral intervals in the solar regime and five spectral intervals in the thermal regime. Both regimes include treatment of aerosols and clouds. Cloud albedo and transmittance are calculated using a delta-Eddington method and are parameterized from the evaluated liquid/ice water content. Aerosol radiative properties are calculated from a two-stream approximation. The solar radiation code is an updated version of the model described by Fouquart and Bonnel (1980). The essential improvements are the use of two spectral intervals, newer Air Force Geophysics Laboratory's transmittance data (1982 version), and the inclusion of aerosols. The thermal radiation code has been developed by Morcrette et al. (1986). Test calculations showed that the parameterized radiation model agrees with the detailed calculations for the effects of the arctic aerosols (Blanchet and List, 1987).

The main objective of the present work is to investigate the response of the atmospheric system to the radiative forcing from anthropogenic arctic aerosols. The model atmosphere is directly perturbed through the shortwave and longwave radiative processes. The reference GCM model is initialized from FGGE global data (Jan 3, 1979) and run at a resolution of 20 wavenumbers (triangular truncation) with 10 unequally spaced levels between the surface and 1 kPa. The model is integrated for a period of 17 months with diurnal and seasonal cycles. This procedure ensures that the model response is nearly independent of the initial conditions. The last 3 months of the simulation (second spring, March-April-May) constitute the period of investigation. The reference simulation has no aerosols. The second spring season is recomputed for various concentrations of aerosols. Four cases are examined by successively increasing the aerosol concentration from the base case (1x) to 5, 10, and 100 times the observed arctic aerosol concentration. The latter extreme case is comparable in concentration and composition to that found in extreme pollution events of some midlatitude urban air pollution episodes (Blanchet et al., 1986). However, it occupies a deeper layer (5 km) and could be produced only by extensive burning in midlatitude regions. Standard diagnostics are performed on each simulation, then differences with respect to the no aerosol case are compared. The results are examined to determine the sensitivity and the amplitude of the changes in the zonal mean atmospheric state and also in regional changes. A subjective criteria is used to determine whether a variable of the model is sensitive to the haze. The changes of a variable from the reference calculation should increase noticeably with increasing concentration of aerosols.

The aerosol model is based on Blanchet and List (1983). The optical parameters in each of the 7 spectral intervals are discretised into 5 relative humidity ranges bounded by 0.0, 0.5, 0.8, 0.93, 0.98, and 1.0. The optical depth in each layer is a function of the relative humidity and geometric depth of that layer. The aerosol concentration is constant in the lower 5 km and zero above that. This vertical profile of aerosol mass concentration is within the range of observed profiles. It corresponds to the mean observed concentration plus one standard deviation as reported by Barrie (1986). The accumulation mode of the reference aerosol is composed of 1.4 $\mu g\ m^{-3}$ of sulphuric acid and 3.2 $\mu g\ m^{-3}$ of other sulphate and soluble compounds, which compare to the range of 1.5 to 4 $\mu g\ m^{-3}$ of sulphate reported by Barrie (1986) for the 1980 and 1981 spring seasons. The EC concentration is fixed at 0.415 $\mu g\ m^{-3}$. Other bulk components in the aerosol including soil (0.8 $\mu g\ m^{-3}$) and sea salt (0.6 $\mu g\ m^{-3}$) are from natural origin. The aerosol is assumed to have a constant dry material concentration and composition at each of the 320 model grid points poleward of 60°N. At this stage, no attempt is made to include observed spatial variations in the aerosol concentration. The aerosol optical properties vary according to the local relative humidity and the effect of aerosol on a regional basis depends on cloud coverage and cloud optical properties. It will be seen that even in the absence of regional variation in the aerosol concentration, the simulated perturbation of the radiation terms vary spatially in the polar region; this is attributed to the fact that the aerosol radiative effects vary according to relative humidity, temperature, cloud coverage, and cloud optical properties.

At current arctic aerosol concentrations, the signal-to-noise ratio is too low to allow us to draw firm conclusions concerning the climatic effects from only one case study. The strategy adopted here consists of increasing the concentration to increase the signal-to-noise ratio. The trend of all diagnosed variables is examined. This work is in preparation for a more comprehensive climate study.

## 3. RESULTS AND DISCUSSION

### *3.1 ZONAL MEAN QUANTITIES*

In the base simulation (1x) the aerosol concentrations were set at the upper limit of actual observations. This produces a spacial and temporal mean optical depth of 0.1 in the visible wavelengths with local maxima in excess of 0.3 optical depth. This perturbation induces a zonal and seasonal mean solar radiative forcing of the order of 0.1 K $day^{-1}$ (Fig. 1a) and an increase of the zonal mean temperature of 1 to 2 K in the lower atmosphere poleward of 80°N (Fig. 2a). Large temperature changes above 40 kPa are due to a high degree of variability in that region and no direct link with the aerosol can be established here. The atmospheric warming is in agreement with the result of MacCracken et al. (1986), from a two-dimensional radiative convective model. As a consequence of this warming, the water vapour mixing ratio increases by approximately 0.1 g $kg^{-1}$. The thermal radiation response (for transient equilibrium) is a slight increase in the cooling rate (0 to 0.05 K $day^{-1}$) above 85 kPa. Below 85 kPa,

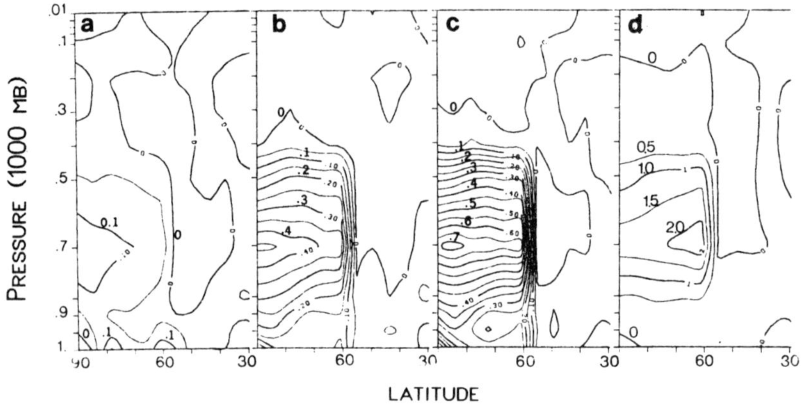


FIGURE 1. Changes of solar radiative heating rate (K day$^{-1}$) due to the introduction of arctic moist aerosol for the region poleward of 60°N and below 5 km (≈ 50 kPa = 500mb). Figures a, b, c, and d correspond to 1X, 5X, 10X, and 100X the observed aerosol concentration respectively. Note the change in contour interval in d.

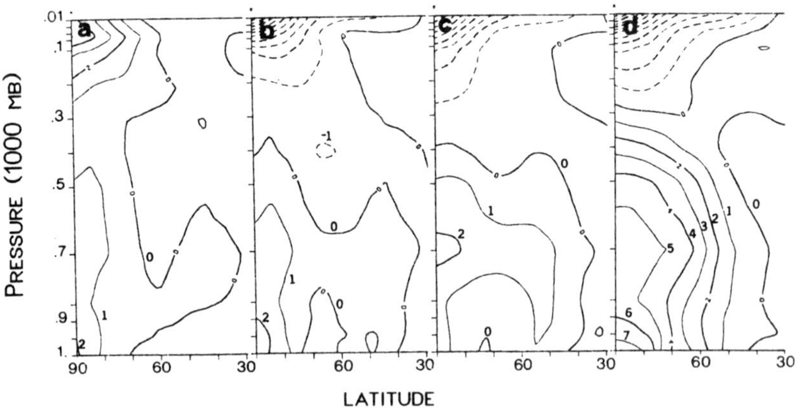


FIGURE 2. As Fig. 1 but for resulting temperature changes (K). Changes in the stratosphere are subject to greater variability, as shown by the change of sign between a and b, and do not appear significant at current aerosol concentrations.

particularly in the lowest model layer ($.97 \leq \sigma \leq 1$), large thermal heating or cooling rates of the order of 0.5 K day$^{-1}$ result from temperature changes of a few degrees in the lower atmosphere. This effect may be related to amplification due to positive feedback in the polar stable boundary layer and extremely cold surface temperatures.

The changes in mean zonal thermodynamic and dynamic variables are summarized in Table 1 for increasing aerosol concentrations. Here the solar radiative heating rate is the primary forcing term. At actual concentrations, the solar forcing is well in the linear range with respect to aerosol loading (Fig.1). The zonal mean radiative heating rate is about half that evaluated by Blanchet and List (1987) in a static column calculation. This is consistent with the overlying cloud coverage of about 40%, lower vertical resolution of the GCM, and the seasonal progression of the solar declination. At large concentrations (>10X) the attenuation of the incident radiation becomes a limiting factor for the forcing term. For terrestrial radiation, the changes below 1 km (Table 1) are up to 5 times larger than those obtained by Blanchet and List (-0.1 K day$^{-1}$) for static haze conditions. The terrestrial heating rate responds nonlinearly to increasing aerosol concentration. The change of terrestrial radiation may be divided into two components: 1) the change associated with the layer emissivity (direct effect), and 2) the change associated with temperature variation (indirect effect of solar heating). In the top portion of the aerosol region (2 to 5 km) the direct effect dominates with weak cooling (0 to -0.1 K day$^{-1}$), while in the lower region (0 to 1 km) the indirect effect dominates with large amplitude associated with temperature changes (±0.5 K day$^{-1}$), induced by the solar perturbation. The terrestrial radiation shows a net tendency to propagate the solar heating downward.

The zonal mean temperature change in the lowest 5 km (>50 kPa) is directly related to diabatic heating from the increase in aerosol amount. The temperature change (cases 1x to 10x in Fig.2) increases northward about 70°N at a rate of 0.1 K per degree of latitude. A tongue of warming centred around 70 kPa extends equatorward from the Arctic Circle to the northern limit of the Hadley cell (≈ 35°N). Consistent with the increase of temperature, the water vapour mixing ratio generally increases in corresponding regions. Changes in the mean meridional circulation are less obvious and certainly subject to caution in interpretation. No significant change is apparent in the mean zonal velocity except at extreme aerosol concentrations. An increase in low level westerly wind of the order up to 2 m/s stems from differential diabatic heating in the aerosol layer, increasing temperature, and decreasing geopotential height and lower surface pressure. Consequently, changes in mean meridional and vertical motion as well as the mass stream function show a broad tendency to weaken the polar and Ferrel cells. This is caused by the reduction of the radiation deficit in the polar region. There is a tendency for an increase in geopotential height above the aerosol layer that affects the upper troposphere and even the lower stratosphere.

TABLE 1. RANGE OF CHANGES (MINIMUM TO MAXIMUM WITHIN THE AEROSOL REGION) OF AVERAGED THERMODYNAMIC AND DYNAMIC QUANTITIES FOR PROGRESSIVE INCREASES OF THE AEROSOL CONCENTRATION

| Variables | Units | 1X | 5X | 10X | 100X | Range[1] |
|---|---|---|---|---|---|---|
| Heating rate | | | | | | |
| Solar | K $day^{-1}$ | (.05,.1) | (.1,.4) | (.3,.7) | (0,2.) | (.4,.5) |
| Thermal | | | | | | |
| 2 - 5 km | K $day^{-1}$ | (0,-.1) | (-.05,-.3) | (-.1,-.3) | (-.5,-1) | (-.5,-.8) |
| < 1 km | K $day^{-1}$ | ± 0.5 | (-.05,.5) | (-.1,1.2) | (.5,1.) | (-1,-2) |
| Temperature[2] | K | (0,2.) | (0,2.) | (0,2.) | (3.,7.) | (240,275) |
| Water Vapour[3] | g $kg^{-1}$ | (0,.1) | (0,.3) | (0,.1) | (.3,.8) | (1,3) |
| Geopotential | | | | | | |
| 100 kPa | m | (0,-10) | (0,-15) | (0,-30) | (0,-100) | 0 |
| 50 kPa | m | (0,20) | (-10,10) | (10,20) | (20,60) | 5000 |
| Mean circulation | | | | | | |
| Zonal U | m $s^{-1}$ | ±0.5 | ±1 | +1 | (+2,+3) | ±5 |
| Meridional V | m $s^{-1}$ | (.05,-.4) | (0,.8) | (0,.5) | (0,+1.4) | ±0.5 |
| Vertical $\omega$ | $10^{-3}$Pa $s^{-1}$ | ±2 | (0,-4) | (1,-6) | (0,-6) | ±5 |
| Stream function $\psi$ | $10^{10}$kg $s^{-1}$ | (0,-.2) | (-.2,-.4) | (-.1,-.3) | (-.5,-.8) | (0,1) |
| Meridional Momentum Transport[4] | | | | | | |
| $(U'V')$ | $m^{-2}$ $s^{-2}$ | 4 | 8 | 6 | 14 | -5 |
| $(U^*V^*)$ | $m^{-2}$ $s^{-2}$ | 2 | 6 | 6 | 12 | -5 |
| Meridional Heat Transport[5] | | | | | | |
| $(V'T')$ | m K $s^{-1}$ | (1,-2) | (-2,-4) | (0,-2) | (1,-6) | (3,15) |
| $(V^*T^*)$ | m K $s^{-1}$ | (0,+1) | (+2,+8) | (0,+2) | (-1,6) | (2,15) |
| Standard deviations | | | | | | |
| $\sqrt{(T'T')}$ | K | (-.5,+1) | (-.5,+1) | (.5,2.) | (1.5,4.5) | (2,8) |
| $\sqrt{(T^*T^*)}$ | K | (0,.2) | (0,1) | (0,1) | (0,-1) | (0,8) |
| $\sqrt{(U'U')}$ | m $s^{-1}$ | ±0.2 | ±0.5 | ±0.5 | (-.5,-1) | (5,7) |
| $\sqrt{(U^*U^*)}$ | m $s^{-1}$ | ±0.5 | ±0.5 | ±0.5 | ±0.5 | (1,4) |
| $\sqrt{(V'V')}$ | m $s^{-1}$ | ±0.1 | ±0.2 | ±0.5 | ±0.5 | (5,7) |
| $\sqrt{(V'V')}$ 3 kPa | m $s^{-1}$ | 0.5 | 1.0 | 1.2 | 1.7 | (10,22) |

[1] The last column (range) represents the range of each variable in the same region. All values are estimated with respect to the reference case without aerosols and in the aerosol region (latitude > 60°N, height <5 km) unless otherwise stated.

[2] Temperature changes extend equatorward to 35°N.

[3] Water vapour changes generally follow temperature changes.

[4] Transport of momentum by transient (') is maximum near (20 kPa, 70°N) and transport of momentum by standing (*) eddies is maximum near (20 kPa, 60°N).

[5] The meridional eddy heat transport changes are nearly opposed in phase.

Changes in the meridional eddy transport terms are related to changes in geopotential height and available potential energy for conversion to kinetic energy. The change in the transport of momentum exhibits a maximum in the upper troposphere. The maximum change in transient eddies occurs at 70°N and 30 kPa, while for the standing eddies this maximum is around 60°N and 20 kPa. All cases show similar positive anomalies that generally increase with aerosol concentration. The heat transport shows a trend that is less regular than momentum. The change of transient eddy term is generally negative (southward transport of heat) and modestly increases with aerosol concentration. However, the standing eddy term is positive and without apparent trend. These quantities are balance terms subject to large variations, consequently no definite conclusions can be made for those terms on the basis of this study.

The standard deviations of temperature and wind do not show any substantial variation except for the transient eddy components of temperature $\sqrt{(T'T')}$ and meridional wind $\sqrt{(V'V')}$. Both of these increase in the aerosol layer with increasing concentration. Furthermore $\sqrt{(V'V')}$ increases gradually in the lower polar stratosphere. This effect is remote from the region of primary forcing and the link between those changes, although apparently correlated, may be fortuitous and certainly needs to be confirmed.

### *3.2 REGIONAL RADIATION BUDGET*

The presence of EC in the aerosol increases the absorption of solar radiation by the atmosphere in all regions but with substantial regional variations in magnitude (Fig.3b). This modulation is essentially due to variations of local clouds and moisture. The base simulation indicates maxima around 6 to 8 W $m^{-2}$ in northern Alaska and over the Canadian arctic islands. These values agree well with column calculations for clear skies. In contrast, minima of 2 to 4 W $m^{-2}$ are obtained over Greenland, the European sector, and in the vicinity of the pole. The amplitude of the modulation due to clouds and moisture is at least as important as what is expected from spatial variations in aerosol concentration alone (viz., Fig. 9 of Barrie, 1986). At high (5X and 10X) aerosol concentrations, this geographical pattern is roughly maintained with a nearly constant amplitude of 10 W $m^{-2}$ but with a mean value over the Arctic increasing with aerosol concentration to about 20 and 30 W $m^{-2}$ respectively.

The change in the absorption of solar radiation at the surface (Fig.3a) shows variations with an amplitude (± 25 W $m^{-2}$) exceeding the effect of attenuation due to aerosol alone by one order of magnitude. The largest variations take place in regions of strong baroclinicity in the Barents and Greenland Seas, and north of the Bering Strait. In parallel to the variations of absorbed solar radiation at the surface, the change in absorbed atmospheric radiation at the surface is opposite in sign and lower in amplitude by about a factor of two. This behaviour is consistent with the variation in cloud cover and moisture amount.

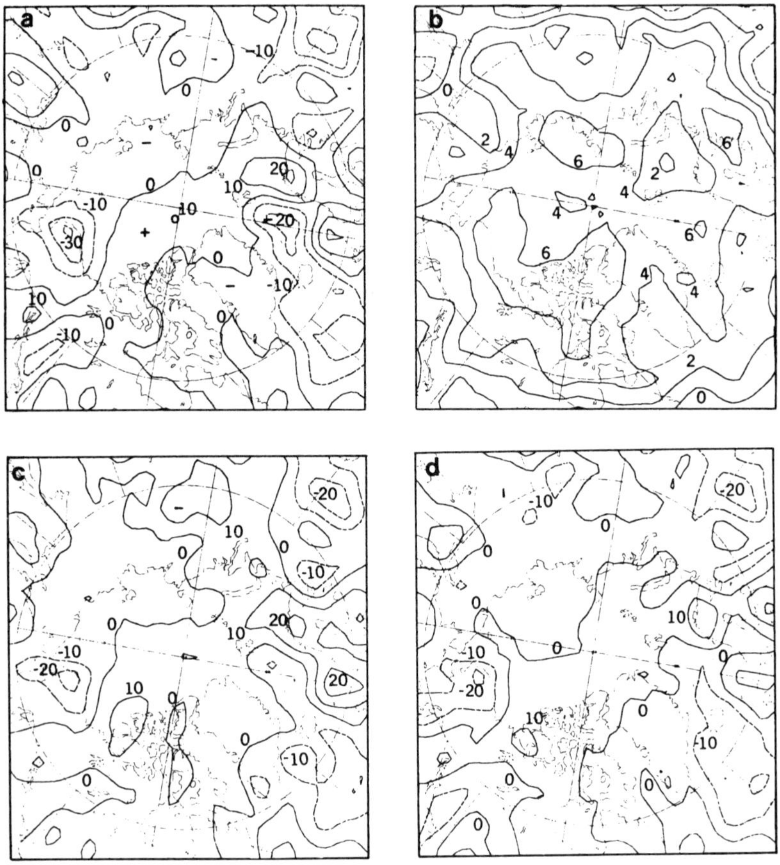


FIGURE 3. Changes of the radiation terms between the experiment (1X aerosol) and control runs. Figures are respectively the changes in: a) absorbed solar radiation at the surface, b) absorbed solar radiation by the atmosphere, c) the net radiation balance at the top (1 Kpa), and d) the net radiation balance at the surface.

The changes of the radiation balance at the top of the model and at the surface (Figs. 3c and 3d) are similar and in phase with a maximum increase (near 10 W $m^{-2}$) located on an axis from Finland to the North Pole and to Banks Island. Again this geographical pattern is maintained with increasing aerosol concentration. But an interesting result is that the amplitude of the change shows only a slight variation with the increase of aerosol loading. This is a result of the negative feedback processes incorporated in this model.

### *3.3 REGIONAL METEOROLOGY*

The seasonal mean surface pressure over the region shows a noticeable tendency to decrease with increasing aerosol concentration. However, the regional pattern may be quite variable. The present case study (Fig.4a) resulted in strengthening of the ridge over eastern Europe - Barents Sea on one side and Baffin Island - Davis Strait on the other side. Lower pressure developed principally over the Greenland, Laptev, and Beaufort Seas. A similar pattern is reproduced in all cases. The amplitude of the change increases from ±0.5 kPa to

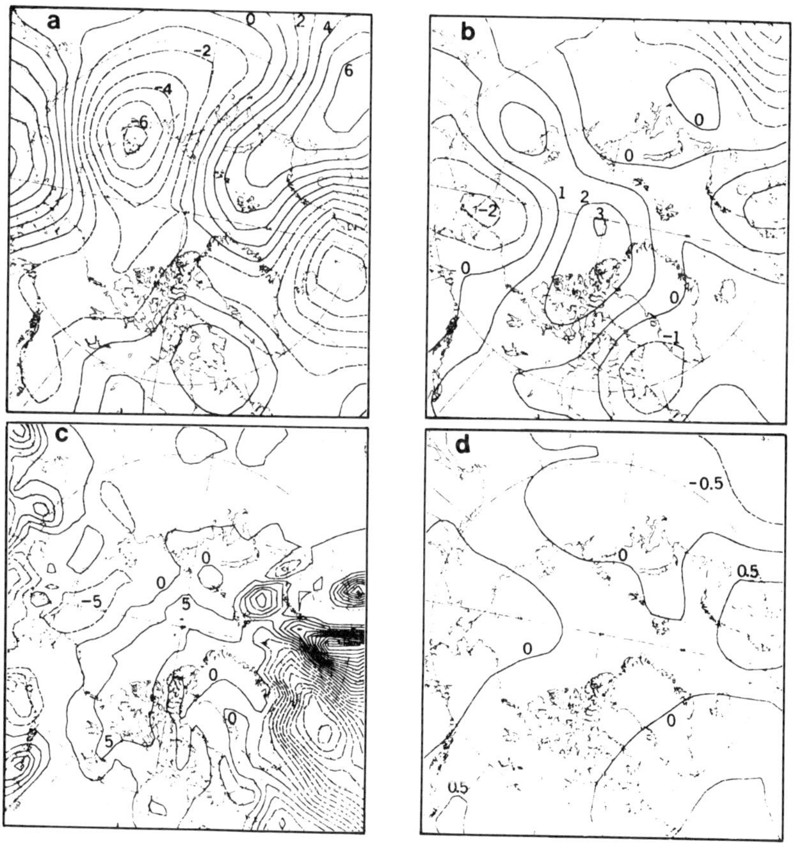


**FIGURE 4.** Changes in seasonal averaged surface fields between the experiment (1X aerosol) and control runs. Figures are respectively the changes in: a) surface pressure (mb = 0.1 kPa), b) lowest layer (98.5 kPa) air temperature (K), c) net surface energy balance (W $m^{-2}$, gain is >0), and d) lowest layer mixing ratio (g $kg^{-1}$). Dashed lines represent reduction, full lines increase. Similar anomaly patterns are found at 5X, 10X, and 100X aerosol concentrations, which suggest some significance with respect to the aerosol perturbation.

-1.6 kPa at extreme aerosol concentrations. The model response is such that the ridging is associated with increasing subsidence, lowered relative humidity in the midtroposphere, lower surface temperature, and lower specific humidity in the boundary layer, as well as lower cloud coverage and precipitation. The opposite is found in regions of decreasing surface pressure.

The surface energy balance over the northern Atlantic (Fig. 4c) shows large variations (± 60 W $m^{-2}$) over open sea area as expected from sea - air exchange in the storm track region. However, over thick ice, in the model the surface energy balance is mostly influenced by the radiation budget, which accounts for 50% to 80% of the change. The amplitude is near ±8 W $m^{-2}$ with a maximum near the pole and it is nearly constant with increasing aerosol concentration. At present aerosol concentration, the increase in solar radiation absorbed at the surface due to ridging and reduced cloud cover dominates over the increase in atmospheric thermal counter radiation associated with atmospheric warming, and over the direct attenuation of solar radiation by aerosols. With increasing concentration (5X to 100X), the longwave component progressively dominates as the solar component becomes negative in all regions (direct attenuation by the aerosol) and the air temperature increases above 1 km (from direct solar heating). In response to this surface energy budget, the surface and lowest model air temperatures increase by 1 to 3 K (Fig. 4b) in the vicinity of the pole and over the Canadian arctic islands. The specific humidity increases with temperature (Fig. 4d). The surface temperature generally increases with increasing aerosol amount to reach about 12 K at 100X concentration. The surface warming pattern is well correlated to the air temperature indicating strong coupling through longwave radiation.

## 4. CONCLUSIONS

The effect of arctic aerosols containing anthropogenic EC and sulphate on a spring simulation with the CCC/GCM is examined by gradually increasing the aerosol mass concentration. This procedure increases the signal to noise ratio and thereby enables one to identify the model variables responding with systematic tendencies. It is expected that in a longer climate simulation similar features will be identifiable in cases with realistic aerosol concentrations. In turn this method enables one to analyse and assess quantitatively the response of the atmospheric system to anthropogenic aerosols.

The results indicate for present aerosol concentrations a temperature increase of 1 to 2 K in the high Arctic and the development of an extensive and elevated tongue of warming from 70°N to 35°N. This structure results from the weakening of the polar and Ferrel cells as a consequence of a reduction in the radiation deficit. The increasing aerosol concentration tends to reduce the arctic mean surface pressure and to increase the overall humidity, cloud cover, cloud opacity, and precipitation. On a regional basis, the base simulation shows increased surface pressure along the main path of injection of aerosol into the Arctic, from eastern Scandinavia to the pole and to the Canadian arctic islands. The variations in clouds and atmospheric moisture are shown to be at least as important as the variation in aerosol concentration in influencing the regional heat balance.

## ACKNOWLEDGMENTS

Thanks are expressed to Drs. R. Laprise, L.A. Barrie, and S.M.Daggupaty for valuable suggestions on the manuscript.

## REFERENCES

Barrie, L. A., 1986: Arctic Air Pollution: An Overview of Current Knowledge, Atmospheric Environment, 20, 643-663.

Blanchet, J.-P., and R. List, 1983: Estimation of Optical Properties of Arctic Haze Using Numerical Model, Atmosphere-Ocean, 21, 444-464.

Blanchet, J.-P., and R. List, 1987: On Radiative Effects of Anthropogenic Aerosol Components Arctic Haze and Snow, Tellus, 39B, 293-317.

Blanchet, J.-P., J. Heintzenberg and P. Winkler, 1986: Radiative Heating During an Intense Pollution Episode in Hamburg, FRG, Beitr. Phys. Atmosph., 59, 359-374.

Boer, G.J., N.A. McFarlane, R. Laprise, J.D. Henderson and J.-P. Blanchet, 1984a: The Canadian Climate Centre Spectral Atmospheric General Circulation Model, Atmosphere-Ocean, 22, 397-429.

Boer, G.J., N.A. McFarlane and R. Laprise, 1984b: The Climatology of the Canadian Climate Centre General Circulation Model as Obtained From a Five-year Simulation, Atmosphere-Ocean, 22, 430-473.

Fouquart, Y., and B. Bonnel, 1980: Computation of Solar Heating of the Earth's Atmosphere: A New Parameterization, Beitr. zur Phys., 53, 35-62.

Morcrette, J.-J., L. Smith and Y. Fouquart, 1986: Pressure and Temperature Dependence of the Absorption in Longwave Radiation Parameterizations, Beitr. zur Phys, 59, 455-469.

MacCracken, M. C., R. D. Cess and G. L. Potter, 1986: Climatic Effects of Anthropogenic Arctic Aerosols: An Illustration of Climate Feedback Mechanisms with One- and Two-dimensional Climate Models, J. Geophys. Res., 91, No. D13, 14445 - 14450.

# ON THE POSSIBILITY OF CLIMATE INSTABILITIES AFTER VIOLENT VOLCANIC ERUPTIONS

*R. Reiter*
Garmisch–Partenkirchen, D-8100, Federal Republic of Germany

**ABSTRACT**

For more than a century the question whether volcanic eruptions can influence weather and climate has been discussed. After the eruption of Mt. Agung (1963) and Fuego (1974), scientific activity on this subject was stimulated. At the beginning of this decade, a new series of violent volcanic eruptions began; also, powerful new systems for monitoring stratospheric aerosols and computer-based theoretical models became available. In this paper results of stratospheric lidar measurements are compared with Northern Hemisphere temperature departures and some of their consequences to the biosphere are discussed. The facts clearly show that the El Chichon eruption (April 1982) was followed by an extraordinarily cold winter in 1984/1985 in many parts of the Northern Hemisphere.

## 1. INTRODUCTION

Lamb (1970) has introduced the dust veil index (DVI) in order to study correlations between volcanic activity and Northern Hemisphere surface temperatures (NHT) since A.D. 1500. Although he obtained reasonable results, some other indices have been developed recently that show better correlations (Smithsonian volcanism index (SVI), and the acidity index (AI) after Hammer, based on conductivity measurements in melted ice cores, see Schonwiese, 1986). The first efforts directed to better estimating and understanding possible reactions of the climate to volcanic aerosols in the stratosphere by Mitchell (1975), Oliver (1976), Pollack et al. (1976), and others have been of great value.

The detailed statistical analysis by Mass and Schneider (1977), using the DVI, showed a significant dip in the NHT some years after volcanic eruptions. Apart from the obvious climatic effects of Krakatoa, which erupted in 1883 (see Simkin and Fiske, 1983), the eruption of Mount Agung (1963) was followed by a drop in the NHT (Mitchell, 1975; Budyko, 1977; Hansen et al., 1978; and others).

Jones (1986) calculated the departure of the NHT for the period since 1851 from the mean of the reference period 1951-1970. Jones used a grid of representative stations to suppress the irregular local density of weather stations. By using recently available and very precise data for the NHT departures, annual mean values have been calculated for the years before, during, and after the eruptions of Krakatoa and Mt. Agung (Fig. 1). The surface cooling effect of the volcanic aerosol in the stratosphere seems apparent.



ISBN 0-937194-11-5

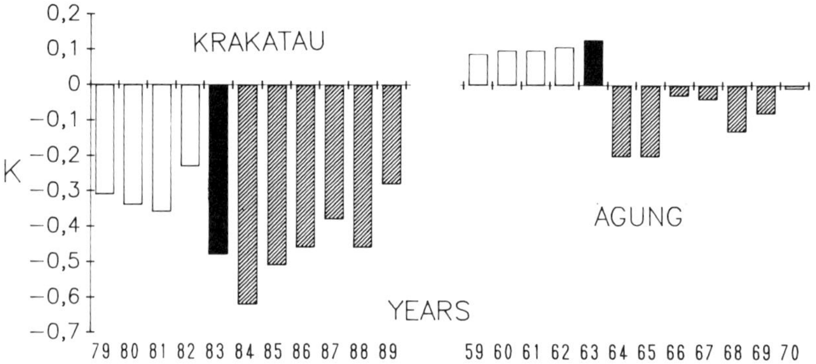


FIGURE 1. NHT annual mean departure in K during the year before (white columns), during (black column), and after (hatched columns) the volcanic eruptions of Krakatoa and Mt. Agung.

In the case of the tremendous Krakatoa eruption in 1883, the cooling set in during the year of the eruption itself. However, during the year of the Mt. Agung eruption (March 1963) there was no departure but an NHT jump from +0.12 (1963) to -0.2 K in 1964. These results generally confirm the earlier findings. In the case of Krakatoa we have to consider that the eruption occurred at the very end of the "small ice age," near the termination of the 19th century. This is the reason for the negative departures also in the years before the eruption. Nevertheless, the eruption enhanced the cold.

A recent work by Lough and Fritts (1987) shows that the mean surface temperature departure over the United States between 1602 and 1900 also reflects volcanic eruptions. The authors show that aside from coolings in some areas there are, in the same year, warmings in other areas. Thus, the effects of volcanic eruptions on surface temperatures appear on geographic and topographic conditions (e.g., strong cooling over the Rocky Mountains in the years after eruption).

Mt. St. Helens (18 May 1980) was the first strong volcanic eruption that caused a heavy transport of aerosols and sulfuric gases (becoming transformed in droplets of sulfuric acid) at a time when lidar systems were ready to monitor the stratospheric particulate load. However, the impact of this eruption on the surface temperatures was small compared to the natural noise (Newell and Deepak, 1982) and to other eruptions in earlier times (Toon and Pollack, 1980; Robock, 1981). The Mt. St. Helens eruption was followed by others of less importance (Ulawun, Alaid, Pagan, "mysterious cloud"), and then by the violent eruption of El Chichon on 3 and 4 April 1982.

## 2. SOME RESULTS FROM THE GARMISCH STRATOSPHERIC LIDAR

Some long-term observations from the lidar in Garmisch-Partenkirchen (Southern Bavaria), beginning with the period after the Fuego material had settled and background conditions again existed, are shown in Fig. 2. Since the El Chichon is situated at only 17 °N, the aerosol it injected into the stratosphere was, as Fig. 3 shows, first detected at Mt. Mauna Loa, then in Fukuoka, somewhat later at Langley, and at last in Garmisch-Partenkirchen (first peak on 3 May 1982). The data have converged since the winter of 1982/1983. The reason for this was the slow spread of the material in the stratosphere to the northern latitudes, finally

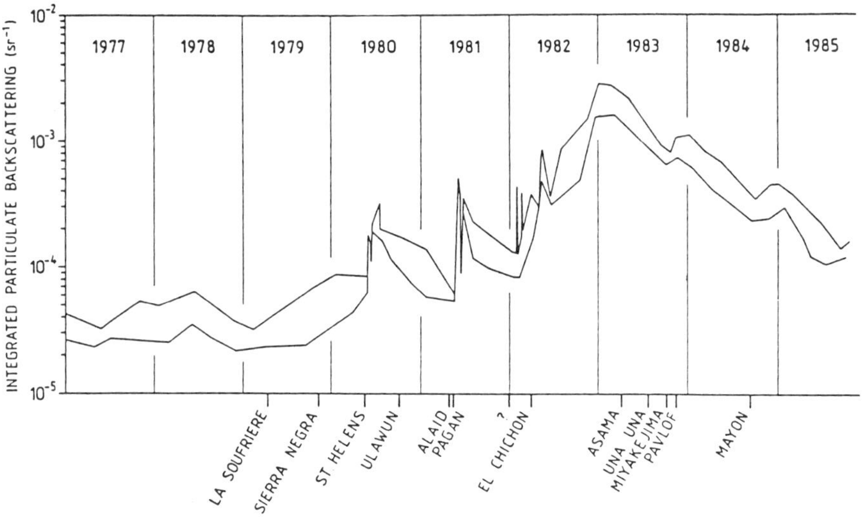


FIGURE 2. The maximum and minimum values of the integrated particulate backscatter from the lidar at Garmisch-Partenkirchen from 1977-1985.[1] Volcanic eruptions are noted. An increase of over two orders of magnitude was detected (the maximum value in the winter of 1982/1983) following the El Chichon eruption. (After Reiter, 1986.)

[1] Data available only while the author was Director of the Fraunhofer Institute of Atmospheric Environmental Research in Garmisch-Partenkirchen.

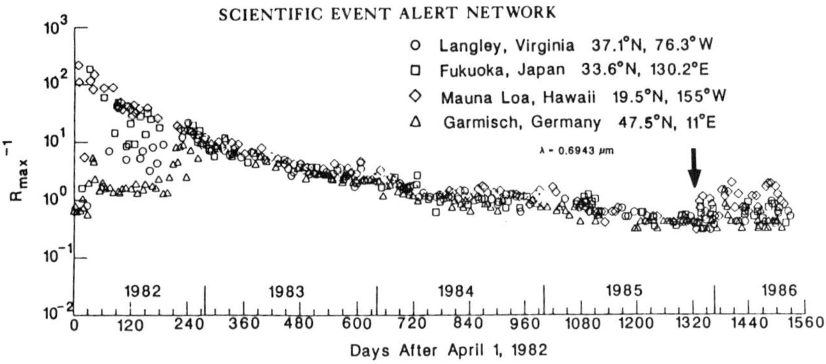


FIGURE 3. The maximum scattering ratio $R^{-1}$ simultaneously recorded at four stations around the Northern Hemisphere. (SEAN Bulletin, 1986).

forming a dense screen over the polar region (McCormick et al., 1982). For more details of the Garmisch results, see Reiter et al., 1980, 1982, and 1983, and Reiter, 1986.

The time lapse of the optical depth (under different assumptions) as calculated from the Garmisch-Partenkirchen backscatter results is shown in Fig. 4. It can be seen that a maximum increase in optical depth on the order of 0.1 - 0.15 occurred between December 1982 and February 1983.

## 3. EXPECTED INFLUENCE OF THE EL CHICHON ERUPTION

All speculations, models, and observations must respect the fact that the El Chichon material has spread more or less homogeneously in the Northern Hemisphere stratosphere (the altitude of the maximum load decreasing with time, see Reiter, 1986). Therefore, climatic effects can be expected in the Northern Hemisphere only. As a first consequence of the impact of El Chichon on the earth's climate, a warming in the stratosphere is to be expected as predicted by Hansen et al. (1978), and reported by Quiroz (1983) and Labitzke et al. (1983). They found an increase of approximately 5° in the tropical stratosphere (10 °N, July 1982, in the 30 hPa level), in agreement with model-based predictions by Hansen for Mt. Agung. See also the detailed studies by Turco et al. (1982). First findings, model calculations, and speculations (Mitchell, 1982a and b; American Geophysical Union, 1983) concerning the lower latitudes, weather, and climate have been somewhat premature for drawing consistent conclusions.

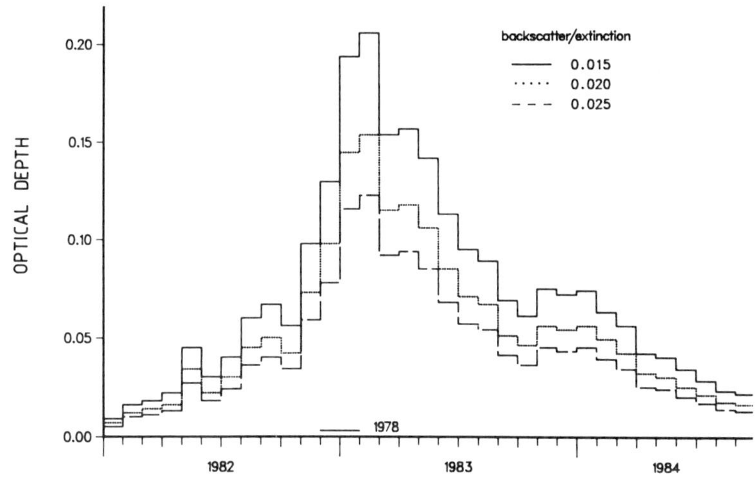


FIGURE 4. The optical depth, based on the Garmisch-Partenkirchen lidar measurements, for 3 assumptions of the ratio backscatter/extinction. (After Reiter, 1986).

## 4. CLIMATIC CONSEQUENCES RESULTING FROM THE EL CHICHON ERUPTION

Mitchell compares the results of his climatic model (Fig. 5)[2] with the volcanic index similar to the SVI (upper panel) and with the observed NHT departures (unbroken line, lower panel) reported by Borzenkowa et al. (1976). The model data (broken line) include aerosol and $CO_2$ effects. The numerical calculations end in 1980. However, Mitchell has predicted that El Chichon shall cause a cooling 2-3 years after its eruption on the order of 0.3 - 0.5 K.

Since Jones (1986) has made available the well defined, improved NHT departures, it is possible to compare these new data with the particulate backscatter measured in Garmisch-Partenkirchen (Reiter 1986). Figure 6 shows that the delay time between the maximum of the stratospheric particulate loading and the onset of surface cooling is on the order of 2 years. El Chichon caused a maximum in the NHT departure of -0.63 K in December 1985. The departure to negative values began in September 1984 and continued until February 1985. During this extremely cold winter

---

[2] Personal communication: the author received the data from J.M. Mitchell (NOAA, Rockville, MD) during a visit on 31 Aug 1982. Details of the model were made available by letter dated 27 April 1983.

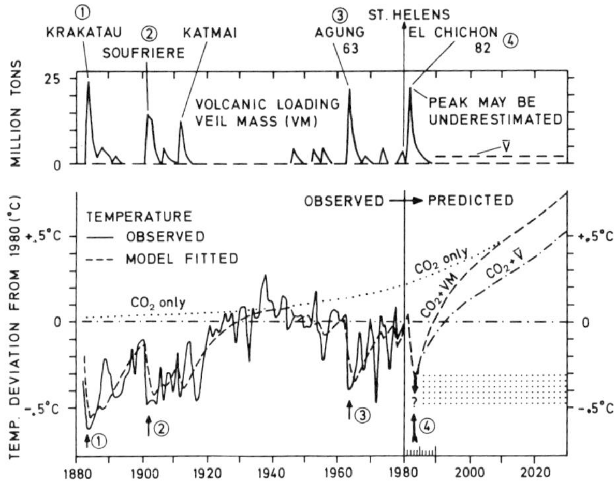


FIGURE 5. The result of a model calculation of the NHT departures from 1880-1980 by Mitchell (1982a) including volcanic mass erupted and $CO_2$ greenhouse effect with prediction of the El Chichon impact on the NHT departure.

sensitive plants were damaged. Unlike crops and fruit trees, grapevines are extremely susceptible to frost in winter. For this reason we used the annual vintage as an indicator of possible effects of a climatic instability. Figure 6 shows a weak reduction of the 1984 vintage in Germany caused by frozen grapes. However, in the winter of 1984/1985 many of the vines were destroyed and in the fall of 1985 the vintage was reduced drastically. Unfortunately, no numerical data on vintage yields were available from other European countries, although heavy damage to French and Italian vineyards has been reported. Meteorological data received from the Italian weather service are presented in Table 1. As a result of the extraordinarily cold winter of 1984/1985 along the coast of the Mediterranean Sea, the region suffered severe losses: many palms froze, the eucalyptus trees were destroyed totally and about 50% of the olive trees were lost. The World Meteorological Organization Bulletin (1986) reported extremely harsh winter conditions with unusual agricultural damages in France, Germany, Hungary, Spain, and Switzerland; other regions affected were China, Columbia, and also large parts of the USA and Canada.

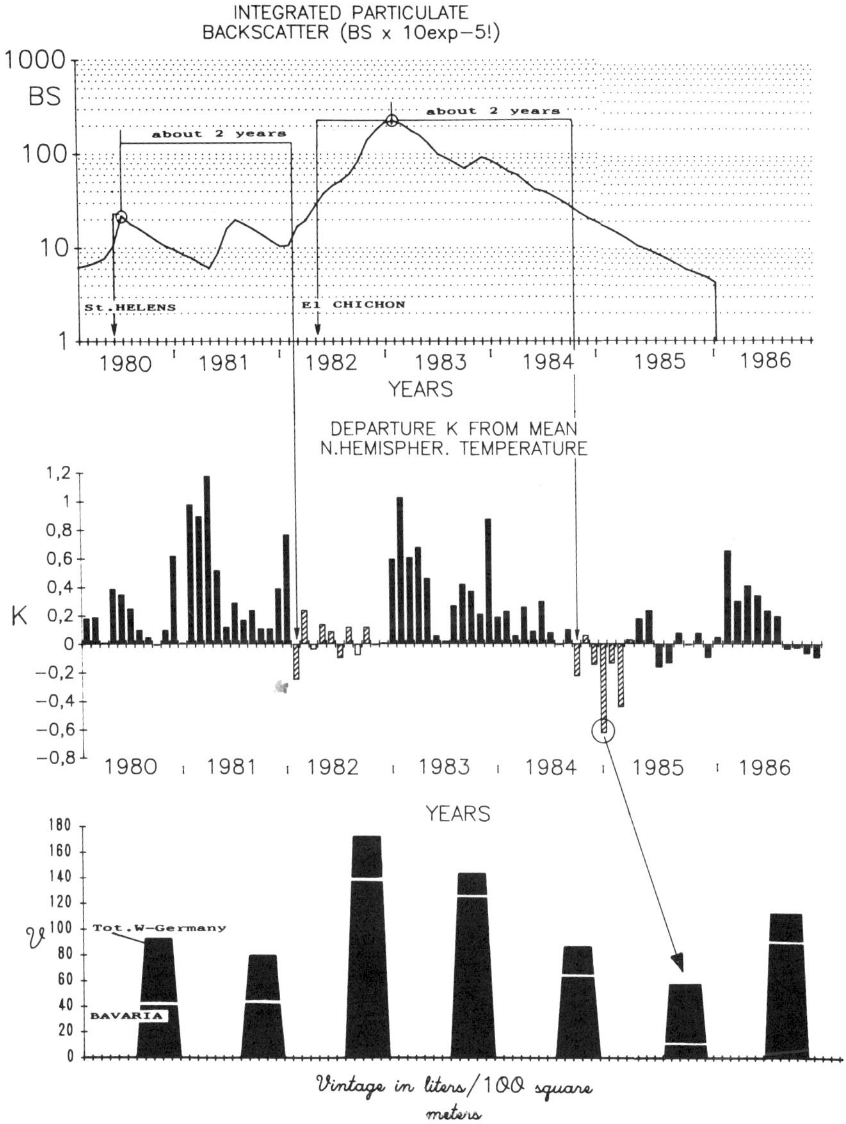


Figure 6. **1st panel:** Integrated particulate backscatter (measured in Garmisch-Partenkirchen) caused by stratospheric aerosols from the end of 1979 until the end of 1985.
**2nd panel:** Monthly mean value of the NHT departure in K. (After Jones, 1986.)
**3rd panel:** Vintage in Germany (total columns) and Bavaria (sections below the white lines) during fall of the respective years. (By courtesy of the Bayerisches Landesamt für Statistik, BLSt, 1980-1986.)

TABLE 1. METEOROLOGICAL DATA ON WINTER CONDITIONS 1984/1985, FROM THE LIGURIAN AREA OF THE ITALIAN MEDITERRANEAN SHORE[1,2]

| | 1984 | | 1985 | | |
|---|---|---|---|---|---|
| | November | October | January | February | March |
| Mean monthly temperature °C | 13.1 | 9.5 | 4.5 | 7.9 | 9.7 |
| Mean monthly minimum value | 8.4 | 5.0 | -0.1 | 3.1 | 5.5 |
| Deepest single value per month | 5.0 | -1.0 | -7.5[3] | -1.0 | 0.8 |

[1]Maximum snow cover: 60 cm; coldest winter in 100 years.
[2]During January 1985, the air temperature remained below 0°C over 14 days.
[3]During 14 days below zero.

## 5. CONCLUSIONS

Volcanic eruptions in low latitudes and of the El Chichon type (heavy stratospheric particle and droplet load) cause a maximum of stratospheric backscatter and optical depth (on the order of 0.15) about 10 months after the eruption (stratospheric delay time). The onset of a temporary hemispheric cooling of the Earth's surface can be expected about 2 years after the eruption (tropospheric delay time), and some additional months later (biological delay time) damage to sensitive plants may occur.

## REFERENCES

American Geophysical Union (Ed.), 1983: Climatic Effects of the Eruption of El Chichon, Special Issue of Geophys. Res. Letts., 10, No. 11, 989-1060.

BLSt, Bayerisches Landesamt für Statistik, Gesundheitswesen, Vol. 1980 - 1986.

Borzenkowa, I.I., K. Vinnikov, L.P. Spirina and D.I. Stekhovski, 1976: Izmenie temperatury vozducha severnogo polushariya za period 1881-1975, Met. Gidrol, 7, 27-35.

Budyko, M.I., 1977: Climatic Changes, Am. Geophys. Union, Washington, D.C., 261 pp.

Hansen, E.J., Wei-Chyung Wang and A.A. Lacis, 1978: Mt. Agung Eruption Provides Test of a Global Climatic Pertubation, Science, 199, 1065-1067.

Jones, P.D., 1986: Northern Hemisphere Temperatures 1951-1984, Climate Monitor, 14, No. 1, 14-21.

Labitzke, K., B. Naujokat, and M.P. McCormick, 1983: Temperature Effects on the Stratosphere of the El Chichon, Geophys. Res. Letts., 10, 24-26.

Lamb, H.H., 1970: Volcanic Dust in the Atmosphere, Philos. Trans. Roy. Soc., A 226, 425-533.

Lough, J.M., and H.C. Fritts, 1987: An Assessment of the Possible Effects of Volcanic Eruptions on the US Climate Using Tree Ring Data, 1602 to 1900 A.D., Clim. Change, 10, 219-239.

Mass, C., and St. H. Schneider, 1977: Statistical Evidence on the Influence of Sunspots and Volcanic Dust on Long-Term Temperature Records, J. Atm. Sci., 34, 1995-2004.

McCormick, M.P., H.M. Steele, P. Hamill, W.P. Chu and T.J. Swissler, 1982: Polar Stratospheric Cloud Sightings by SAM II, J. Atm. Sci., 39, 1387-1397.

Mitchell, J.M., 1975: Note on Solar Variability and Volcanic Activity, WMO GARP Publ. Ser. No. 16, 127-131.

Mitchell, J.M., 1982a: El Chichon, Weather Maker of the Century, Weatherwise, 35, 252-262.

Mitchell, J.M., 1982b: Empirical Modelling of Effects of Solar Variability, Volcanic Events, and $Co_2$, Proc. Conf. on Solar-Terr. Effects on Weather, Abstract, 15 p.

Mitchell, J.M., 1975: A Reassessment of Atmospheric Pollution as Cause of Long-Term Changes in Global Temperature. In S.F. Singer (Ed.), The Changing Global Environment, D. Reidel Publ. Co., Dordrecht.

Newell, J.M., and A. Deepak (Eds.), 1982: Mt. St. Helens Eruption 1980, NASA STIB, Washington, D.C., 117 pp.

Oliver, C.R., 1976: On the Response of Hemispheric Mean Temperature to Stratospheric Dust, J. Appl. Met., 15, 933-949.

Pollack, J.B., O.B. Toon, C. Sagan, A. Summers, B. Baldwin and W.V. Camp, 1976: Volcanic Explosions and Climatic Change: A Theoretical Assessment, J. Geophys. Res., 81, 1071-1083.

Quiroz, R.S., 1983: The Isolation of Stratospheric Temperature Change Due to El Chichon, J. Geophys. Res., 88, 6773-6780.

Reiter, R., H. Jäger, W. Carnuth and W. Funk, 1980: Lidar Observations of the St. Helens Clouds Over Mid-Europe, May to July 1980, Geophys. Res. Letts., 7, 1099-1101.

Reiter, R., H. Jäger, W. Carnuth and W. Funk, 1982: A Stratospheric Aerosol Increase During 1981, Observed by Lidar over Mid-Europe, Geophys. Res. Letts., 9, 469-472.

Reiter, R., H. Jäger, W. Carnuth and W. Funk, 1983: The El Chichon Cloud Over Central Europe, Observed by Lidar, Geophys. Res. Letts., 9, 1001-1004.

Reiter, R. 1986: Results of 8 Years Continuous Measurements of Aerosol Profiles in the Stratosphere, Meteorol. Atmos. Phys., 35, 19-48.

Robock, A., 1981: The St. Helens Volcanic Eruption: Minimal Climatic Effect, Science, 212, 1383-1384.

Schönwiese, C.D., 1986, Zur Parameterisierung der nordhemisphärischen Vulkantätigkeit, Meteorol. Rundsch., 39, 126-132.

SEAN BULLETIN, Vol. 12, 1986, No. 1, p. 15.

Simkin, T., and R.S. Fiske, 1983: Krakatau 1883: The Volcanic Eruption and Its Effects, Smithsonian Press, Washington, D.C.

Toon, O.B., and J.B. Pollack, 1980, Atmospheric Aerosols and Climate, Amer. Scientist, 68, 268-278.

Turco, R.P., R.C. Whitten and O.B. Toon, 1982: Stratospheric Aerosols, Observation and Theory, Rev. Geophys. and Space Phys., 20, 233-279.

WMO Bulletin, 35, July 1986.

# CLIMATIC EFFECTS OF SMOKE AND DUST PRODUCED FROM NUCLEAR CONFLAGRATIONS[1]

*A.B. Pittock, J.S. Frederiksen, J.R. Garratt and K. Walsh*
CSIRO Division of Atmospheric Research
Mordialloc, Victoria 3195, Australia

ABSTRACT

Despite many acknowledged uncertainties, a large body of research now supports the theory that a major nuclear war would generate climatically significant quantities of airborne smoke and dust. Current estimates put the absorption optical depth of smoke and dust in the first few weeks over the Northern Hemisphere, following a summer war in that hemisphere, at about 1.0 to 3.0 over large areas. Once the smoke has stabilised in the upper atmosphere it could have a 1/e residence time of the order of 180 days. A short summary of the estimated surface climatic effects is given, including new results for absorption optical depths around 0.2 that could prevail in the Southern Hemisphere, and perhaps in the Northern Hemisphere in the second growing season after a nuclear war.

## 1. INTRODUCTION

The idea that serious climatic effects might follow from a major nuclear war was first supported scientifically by Crutzen and Birks (1982) and Turco et al. (1983). Since then the level of sophistication of the models applied to the problem has increased enormously (Pittock et al., 1986). Figure 1 is a schematic diagram of the processes involved, in which it is possible to identify three levels of sophistication.

The early calculations, typified by Turco et al. (1983), were confined to the area labelled "1", and considered only what might happen in a vertical column in which a vertically distributed amount of smoke was assumed to be present. The quantity of smoke was calculated from estimates of fuel density, targeting scenarios, and estimates of the amount of smoke emitted per unit fuel burned, and the fraction lost by prompt washout. The latter was taken as 50% by Turco et al. (1983) and 30% by Crutzen et al. (1984). Smoke coagulation was calculated and removal rates were generally taken as those applicable to particular matter in the unperturbed lower troposphere, i.e., with a 1/e residence time of a few days to a week. Residence times of 10-14 days were assumed for injections into the upper troposphere.

---

[1]This work was supported by a grant from the Australian Department of Foreign Affairs.




ISBN 0-937194-11-5

The results from these first generation models suggested, for annual average conditions, cooling of surface air over land within the first week or two of the order of 30-40°C, but negligible cooling over the oceans. A common sense allowance for the advection of heat from the oceans over land thus suggested that land average surface coolings might be in the range of 10-20°C, although continental interiors might approach the larger land-only figure. The assumed removal rates for the smoke led to an estimated duration of a few weeks for such coolings.

The next generation of models, run for specific fixed seasons, utilised 2- and 3-dimensional calculations with prescribed smoke and dust distributions, in some cases decreasing in amount with time, to calculate the changes in temperature and circulation in the atmosphere, as indicated in the second area of Fig.1 (see Pittock et al., 1986 for references). These models showed that the presence of smoke and dust would substantially change the temperature structure and circulation of the atmosphere in such a way as to tend to stabilise the upper troposphere and move smoke upwards and southwards in the case of a war in the northern summer. However, lack of feedback between the changed temperature structure and circulation on the one hand, and the smoke and dust on the other, rendered these results suggestive rather than conclusive.

The third generation of model simulations has incorporated fully interactive smoke as indicated in the third area of Fig.1. Thus the heating of the smoke by absorption of solar radiation

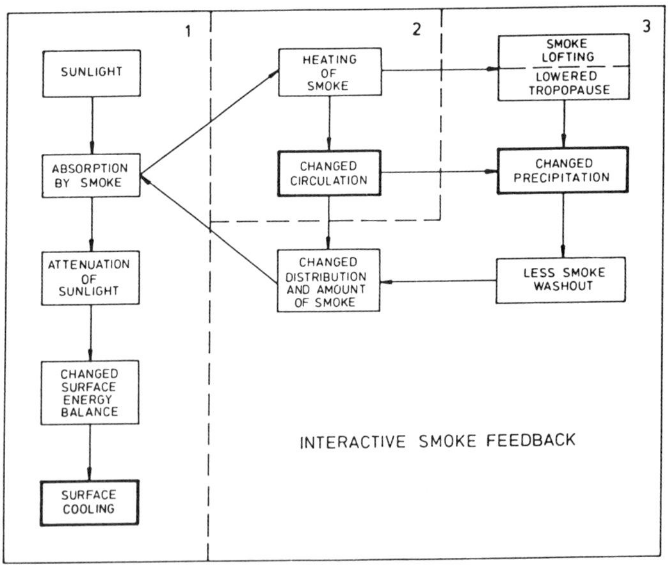


FIGURE 1. Schematic diagram of smoke-weather/climate effects. The area labelled 1 shows interactions in the first generation models. Second generation models added the effects in area 2. Third generation models completed the feedback loops shown in area 3. Adapted from Malone et al., 1986.

causes the smoke to rise, while the changed temperature lowers the tropopause, reducing the rainfall and washout. This is illustrated in Fig.2, from Malone et al. (1986), where we see almost complete separation of the lofted smoke in the third week from the precipitation, which is confined below the lowered tropopause. Thus the changed circulation and temperatures lead to slower smoke removal and a changed distribution of the smoke, notably with more reaching high altitudes and moving across the equator to affect the Southern Hemisphere.

Most of these models still neglect one or more at least marginally important processes such as infrared absorption and emission by the smoke and stability-dependent boundary layer processes, or are otherwise crude in their representation of the soil heat content or vertical structure of the atmosphere. Nevertheless, for initial summertime smoke injections in the plausible range of 100-200 million tonnes, all of these 3-dimensional models tended to obtain surface air coolings in the first few weeks over the northern continents in the range 10-20°C. More recently Thompson and Schneider (1986) have reported a result in the 10-15°C cooling range after allowance for smoke infrared. Other as yet unpublished results from several laboratories give land-averaged surface coolings in the range of 5-15°C, although all show coolings of 15-20°C over large areas in the interiors of the northern continents, and short-lived extremes under particularly thick smoke clouds may be even greater.

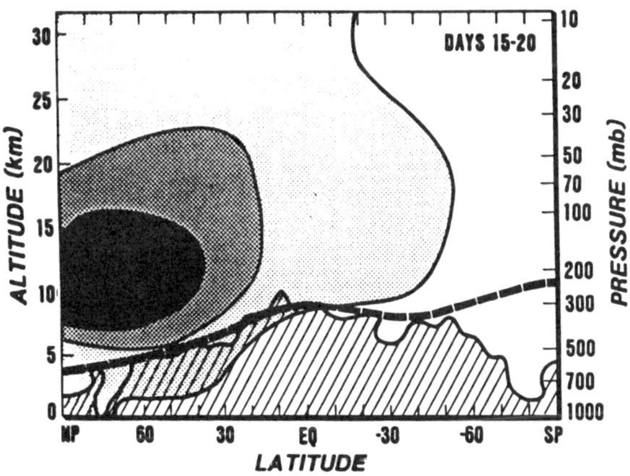


FIGURE 2. Meridional cross-section of the atmosphere, showing the position of the perturbed tropopause (heavy dashed line) and the precipitation distribution (cross-hatched area below the tropopause), both averaged over days 15-20, and the smoke distribution at day 20 (stippled area mainly above the tropopause). Results are for the July interactive-smoke case with 170 million tonnes of smoke injected between 0 and 9 km altitude. (From Malone et al., 1986; copyright by American Geographical Union.)

## 2. THE CURRENT CONSENSUS

In 1982 the Scientific Committee on Problems of the Environment (SCOPE) undertook a major project on the environmental consequences of nuclear war (ENUWAR), which led by 1986 to a consensus report involving scientists from some 30 countries (Pittock et al., 1986; Harwell and Hutchinson, 1985). Since then SCOPE has managed to keep some degree of international coordination going between researchers in the field, and in February 1987 convened another international workshop in Bangkok. This led to another consensus statement (Warner et al., 1987), which incorporated an evaluation of recent published and unpublished work.

Participants at the Bangkok workshop considered the controversy that had arisen involving the terms *nuclear winter* (first used by Turco et al., 1983) and *nuclear autumn* (first used by Thompson and Schneider, 1986). The participants comment that:

> *The authors of the SCOPE ENUWAR report had carefully chosen to avoid using the term nuclear winter because its meteorological implications do not adequately portray the full complexity and diversity of the many possible environmental stresses that could result; this criticism applies to the term nuclear autumn as well.*

The statement goes on to say that:

> *Taking into account the various uncertainties remaining in the models, the calculations suggest that, for the baseline smoke injections, summertime land temperatures in the Northern Hemisphere midlatitudes could decrease rapidly to conditions superficially similar to conditions in fall or early winter, including the possibility of widespread frost events and discontinuance or interruption of the summer monsoon.*

The Bangkok participants noted that recent controversy had tended to focus on claims that there could be a global "deep freeze", which were based on an extreme worst-case interpretation of temperature estimates from Turco et al. (1983). The participants noted that:

> *Focussing only on the changing amplitude of one factor -- namely the large-scale land average temperature depression . . . without considering the increasing amplitude of other factors (for example, the soot fraction of the smoke, the increased smoke residence time from induced lofting of smoke into the stratosphere, the possibility of monsoon disruption for even low smoke amounts) fails to recognize the complexity and interconnectedness of the SCOPE-ENUWAR analysis.*

The Bangkok statement went on to note the general agreement in the various models that there would be a rapid onset of substantial temperature decreases and reduction of precipitation over wide areas of the Northern Hemisphere midlatitudes. These are well illustrated in Figs.3 and 4.

Figure 3, from MacCracken at the Lawrence Livermore Laboratory, shows surface air temperatures calculated for a point near the coast

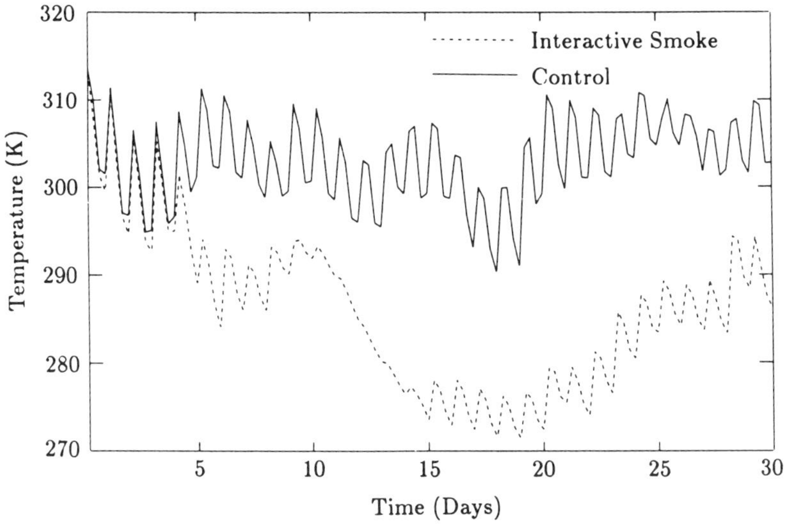


FIGURE 3. Time series of surface air temperatures for control (no smoke) and interactive smoke case, for a point on the coast of China. The curves show diurnal cycles and fluctuations due to passing weather systems and moving smoke clouds. Actual variations at a particular place would depend on the weather situation at the time. (Model results from M. MacCracken at Lawrence Livermore National Laboratory; work performed under the auspices of the Department of Energy.)

of China north of Shanghai during the first 30 days after smoke injection in target areas (not including China). The dotted curve shows the temperatures varying with a reduced diurnal cycle as the smoke comes overhead, with coolings relative to the control run varying between 10 and 30°C depending on smoke thickness and wind direction. One significant point arising out of this figure is that even coastal areas may experience substantial coolings under offshore wind conditions. The detailed variations are, of course, a function of the varying weather patterns at this point.

Figure 4a shows the zonal mean land area precipitation rate for successive 10-day intervals after injection of 150 million tonnes of smoke. Figure 4b shows the same quantity for the third 10-day period for varying quantities of smoke injected. It is apparent that precipitation rates, at least at some latitudes, are very sensitive to injections of as little as 15 million tonnes of smoke, which is one tenth of the baseline case considered by Ghan et al. (1987).

(a)

(b)

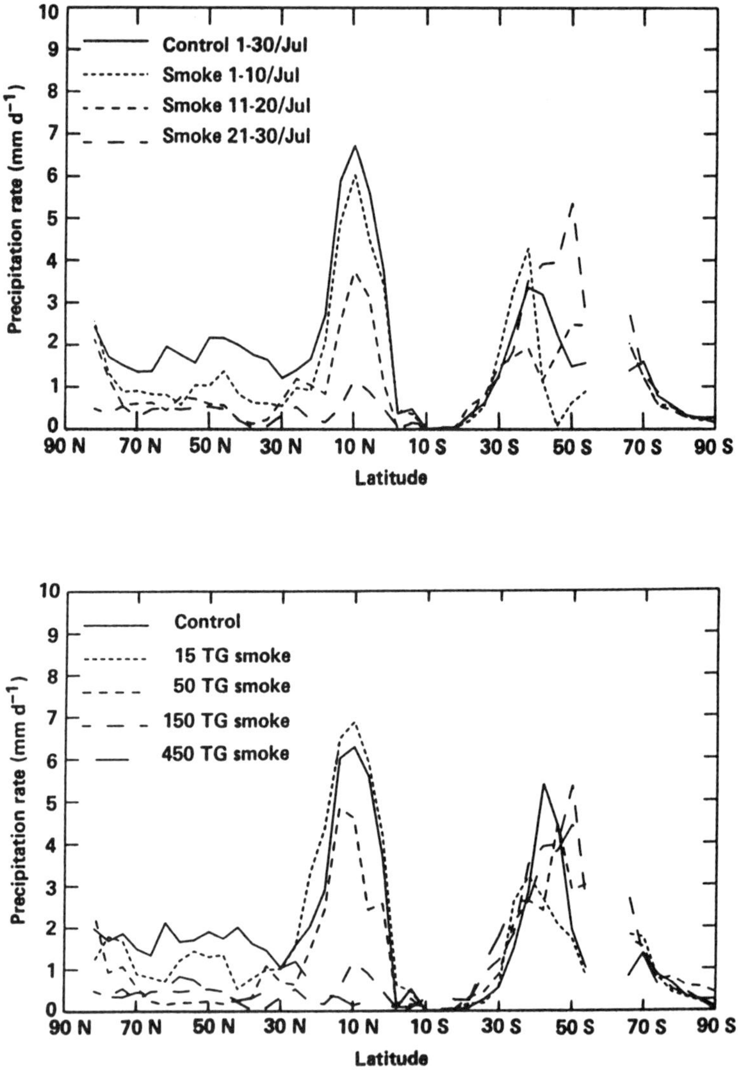


FIGURE 4. Zonal mean land precipitation rate as a function of latitude in July. The top panel (a) shows results for the control and for the perturbed run for days 1-10, 11-20 and 21-30 following an injection of 150 million tonnes of smoke. The bottom panel (b) shows results for days 21-30 for various amounts of smoke injected into the atmosphere. (From Ghan et al., 1987; work performed under the auspices of the Department of Energy.)

# 3. SOME CURRENT ISSUES

## 3.1 SMOKE AMOUNTS AND PROPERTIES

Penner (1986) has reviewed the published estimates of the quantity and optical properties of the smoke that might be expected after a nuclear war. She finds a wide range of estimates of resulting smoke optical depth. If the primary stocks of petroleum are not included, the absorption optical depth varies from 0.2 to 4.2. Including these stocks raises the range of absorption optical depth to 0.4 to 6.1. In the first case, the low estimate is equivalent to 50 million tonnes of smoke with the optical properties assumed in NAS (1985), i.e., with a specific absorption of 2 $m^2 g^{-1}$, while the absorption optical depth of 6.1 is equivalent to about 750 million tonnes of smoke.

These values assume that coagulation of smoke leads to a decrease in absorption optical depth, but there is strong observational evidence that the primary submicron spherules of soot agglomerate into sparse random clusters or chains rather than into solid spheres, unless they have gone through an intermediate water droplet stage or are oily. Berry and Percival (1986) have used fractal theory to show that the scattering and absorption cross-sections of such particles increase in proportion to the number of primary spherules, so that the specific absorptivity does not decrease with coagulation. The Mie theory is clearly not applicable. Under these circumstances the absorption optical depths range from minimum estimates of 0.4 without petroleum stocks, or 1.0 including these stocks, to much higher values. Extinction optical depths are greater again, by about a factor of two. These include the effect of scattering, which is relatively unimportant in terms of the solar radiation reaching the surface, since most scattering is in the forward direction, except at large optical depths when scattering may significantly increase the absorption pathlength.

Further refinements in these estimates are dependent on more representative surveys of fuel densities in likely target areas, a better understanding of the combustion processes in mass fires, and a better understanding of the prompt removal mechanisms in fire plumes. The last is related to the hygroscopic properties and coagulation rates of smoke, and the dynamics of fire plumes of varying intensities under representative meteorological conditions. As most smoke in a nuclear war will come from urban fires, information from experimental forest fires may be of very limited value.

## 3.2 EFFECTS OF SMOKE INFRARED AND SOIL HEAT CAPACITY

Most of the models applied to the nuclear winter problem have had only a crude representation of the soil heat capacity, a crude representation of planetary boundary layer processes with no diurnal cycle and transfer coefficients independent of stability, and have neglected the infrared absorption and emission by the smoke. We have used a modified version of the Colorado State University (CSU) mesoscale model (Mahrer and Pielke, 1977; McNider and Pielke, 1981), run in a quasi-one-dimensional mode in which calculations are made at five equidistant horizontal grid points. We had 18 levels in the atmosphere from 1.5 m to 13 km for wind and pressure (with temperatures calculated at intermediate levels starting at 1 m), and 20 levels in the soil from 0 to 1 m depth. Most runs were done with a soil density of 1.6 kg $m^{-3}$, a soil specific heat of 836 J $kg^{-1}$ $K^{-1}$, a soil heat diffusivity of 6 x $10^{-7}$ $m^2$ $s^{-1}$, and a time step of 30 sec. Starting conditions were taken from climatological means, with a geostrophic wind speed of 10 m $s^{-1}$. A surface albedo of 0.2 and a soil wetness factor of 0.1 were assumed unless otherwise indicated.

This model was run for 20 days to reach quasi-equilibrium before the introduction of smoke above the top of the model at 13 km. The smoke was assigned an absorption optical depth, and the solar flux modified according to the appropriate slant path. In general the control runs did not reach a true equilibrium, presumably due to limitations in the radiation code, which does not achieve an exact energy balance, and more importantly because of a lack of advective heat transport when operating the model in a one-dimensional mode. Instead, there is a slow drift in surface temperatures, which must be allowed for in interpreting the effect of smoke. This drift would not be a problem in short runs for studying mesoscale processes, for which purpose the model was originally designed. In the present application the drift will affect the absolute temperatures reached, but should not seriously affect the relative coolings in the first few days after smoke is inserted in the model.

Here we will not discuss all of the many runs that we have done. However, it is instructive to look at results from runs with and without smoke infrared emissions, to illustrate the effect of smoke infrared. In Fig.5, the upper pair of curves are the daily maximum soil surface temperatures for a run at 30 S in July with an absorption optical depth of 0.2. We see the end of the control run from days 15 to 20. Then smoke was introduced, and the lower curve shows how the temperature evolved when no smoke infrared was included. After an initial cooling of some 4°C, the soil surface continues to cool slowly relative to the extrapolation of the control run. The upper of the pair of curves shows the response when an additional downwards infrared flux was added at the top of the model, equal to 50% of the 24-hr average of the solar energy absorbed by the smoke during daylight. In this case we see that the added infrared flux has largely stopped the cooling after the first day of smoke. This is not always the case, but it does illustrate that the additional downwards infrared flux has an appreciable but not dominant effect on the daytime surface temperature changes.

The lower pair of curves is similar except that they are for the overnight minimum soil surface temperatures. Here we see that, without the smoke infrared, slow cooling sets in on the first night and continues throughout the run. However, when smoke infrared is added, we see an initial warming relative to the control over the first few nights, after which the temperature follows the control curve fairly closely. This is because at night the smoke does not remove any solar flux, but its warmth imposes an extra downwards

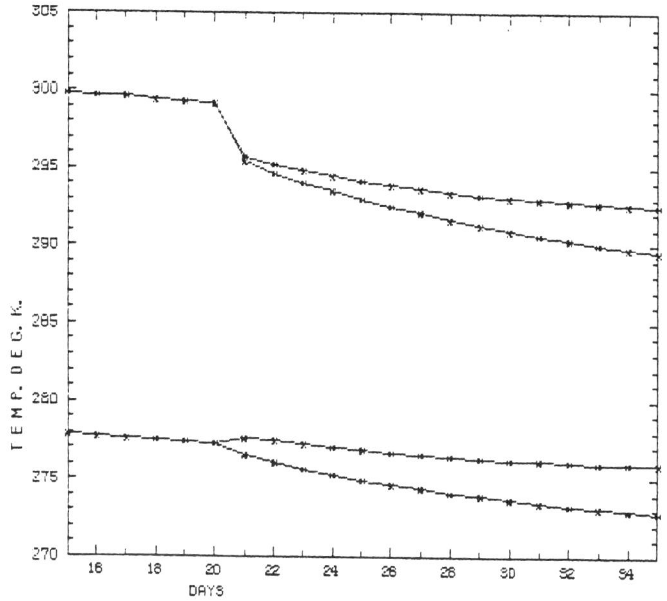


FIGURE 5. Soil surface temperatures calculated with the CSU meso-scale model, showing days 15-20 of the control (no smoke) run, and the following 15 days with smoke of absorption optical depth 0.2 added at the top of the model at 13 km altitude. Results are for 30 S in July, with soil wetness factor 0.1. The upper pair of curves are for the daily maximum temperature, while the lower pair are for the overnight minimum. In each pair the lower perturbed curve is without allowance for the downwards infrared flux from the smoke, while the upper curve allows for the infrared.

infrared flux. This is a significant part of the nocturnal energy balance, making up for the reduction in nighttime heat flux from the soil to the surface (which is less due to reduced heating of the soil during the day). The result in this case is that the overall effect of the smoke is to cause significant cooling of the soil surface during the day, but no reduction in the overnight minima. Such an effect could be very important in relation to agricultural impact, since morning frost frequencies will be less affected than daytime temperatures associated with the thermal time required for crop growth.

The present treatment is not a complete allowance for the smoke infrared properties, since the smoke would also absorb some of the outgoing infrared from the soil and lower atmosphere. However, if the smoke has an infrared absorptivity only about one tenth that in the visible, we could expect it to absorb only one tenth the fraction of outgoing infrared as of incoming solar radiation. As these are of the same order of magnitude when diurnally averaged at the altitude of the smoke, this suggests that the effect on the surface energy balance of the smoke absorption of outgoing infrared will be only about a tenth of that due to the smoke infrared originating from absorption of the solar radiation. The smoke infrared absorption effect is thus negligible, at least for relatively thin smoke.

Figure 6 shows results at 45 S in July, with an absorption optical depth of 1.0, and including the downwards infrared flux from the smoke. Here the control runs for days 15-20 are for the soil surface temperature at daily maximum (top curve), the boundary-layer average air temperature at the time of daily maximum soil temperature (middle curve), and the soil surface daily minimum temperature (bottom curve). Here we see that after smoke is introduced the daily maximum soil temperature drops by some 9°C in the first day, but then the rate of cooling slows down. In contrast, the overnight minimum soil surface temperature initially warms slightly, but then starts to cool at a rate in excess of 1°C per day after about three days. The daytime boundary layer average air temperature drops more slowly than the soil surface temperature on the first day, allowing the vertical stability to change from unstable in the control to stable. The air then starts to cool more rapidly, and eventually reaches about the same temperature as the soil. By day 35, i.e., after 15 days of smoke overhead, the daily maximum soil surface has cooled by some 16°C, while the air

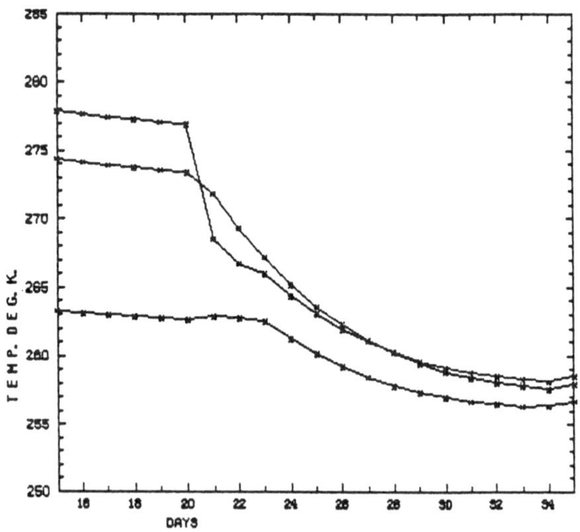


FIGURE 6. Soil surface daily maximum temperatures (top curve at left), boundary layer average air temperatures at the same time (middle curve at left), and soil surface daily minimum temperatures, calculated with the CSU mesoscale model for 45 S in July. Days 15-20 are for the control (no smoke) case, while days 21-35 are for a smoke absorption optical depth of 1.0.

temperature has dropped by some 14°C. In contrast, the overnight minimum soil surface temperature has dropped by only some 5°C. These variations in soil surface temperature are explicable in part in terms of the fast response of the surface layer of soil being moderated on a longer time scale by the larger heat capacity of the underlying soil.

These results, and others from the CSU mesoscale model for different conditions, can usefully be compared with results from the general circulation models, after due allowance for the lack of advection from the oceanic heat reservoir. It is particularly instructive to note the rapid cooling of the daytime soil surface temperatures in the first day or two, despite the finite soil heat capacity in the model. This has implications for the rapidity of onset of cooling with distance inland from coasts, especially under light or offshore wind conditions.

### 3.3 DURATION OF EFFECTS

Malone et al. (1986) demonstrated the importance of lofting of smoke in quickly separating it from the region of fast washout due to precipitation (see Fig.2). For a war in July they showed that lofting rendered the initial height of injection of the smoke almost irrelevant, and that some 40% or so of smoke would remain after 40 days, beyond which it would have a 1/e residence time of 5 or 6 months. In the January case, in which much less lofting occurs, only some 5 to 15% of the initial smoke would remain after 40 days, depending on the height of injection, and the 1/e residence time after 40 days would range from only 2 weeks for low injection heights, to about 1.5 months for a higher initial injection profile. However, in some model treatments (e.g., Thompson and Schneider, 1986), assumed relatively low heights of injection still lead, despite lofting, to substantially greater smoke deposition rates.

More recently Malone and co-workers have run their model out for more than 40 days, and find absorption optical depths in the range 0.3 to more than 1.0 persisting in the Northern Hemisphere, and 0.1 to 0.3 in the Southern Hemisphere after several months. Most of this smoke is contained in the modified stratosphere at altitudes between 10 and 20 km where the only removal mechanisms are expected to be gravitational settling, chemical destruction, or exchange of air with the troposphere.

Gravitational settling may be expected to be very slow, especially as the coagulation of smoke into fluffy particles or long chains is not likely to increase their fall velocity appreciably. Preliminary results of laboratory measurements of the oxidation of smoke particles by ozone (de Haas et al., 1986) suggest a half-life in the stratosphere of the order of several weeks to several months. Apart from gravitational settling of larger particles, the main sink of volcanic particles in the stratosphere is thought to be passage of air into the troposphere. This occurs mainly in the vicinity of the jet streams and at the winter pole. A recent estimate by Post (1986) suggests a stratospheric residence time of order of 200 days.

Such long lifetimes for the smoke in the stratosphere suggest that for the baseline case of some 150 million tonnes of smoke in the initial injection, smoke absorption optical depths of the order 0.1-0.3 could remain in the Northern Hemisphere up to a full year after a spring or summer war, and of the order 0.1-0.2 in the Southern Hemisphere during the first post-war southern summer. It is therefore of considerable interest to calculate the surface climatic effects of such moderate smoke loadings.

3.4 EFFECTS OF MODERATE SMOKE AMOUNTS

Besides simulations with the CSU mesoscale model with smoke absorption optical depth 0.2, we have performed numerical simulations using the Australian Numerical Meteorology Research Centre (ANMRC) general circulation model (GCM) (McAvaney, Bourke and Puri, 1978) assuming a uniform layer of smoke of the same optical depth situated above the top of the model atmosphere. In this case the additional downwards infrared flux from the smoke has been neglected.

The version of the ANMRC model used here has 9 layers in the atmosphere, three sub-soil layers, a stability dependent Monin-Obukhov boundary layer, convective adjustment, three layers of zonally averaged climatological cloud, prognostic precipitation and soil hydrology, and prescribed climatological mean sea surface temperatures. It is a spectral model, and was run at rhomboidal wave number 21 resolution. It was run for constant January and July conditions in a diurnally averaged mode from real (observed) initial conditions for 95 days to reach equilibrium. It was then run for 105 days of control conditions, and, from the same model initial conditions at day 95, for 105 days of perturbed conditions (i.e., with smoke).

The resulting zonally averaged soil surface temperatures at 27 S for the control and perturbed runs are shown in Fig.7 for July. Coolings due to smoke vary between about 2 and 6$^{\circ}$C. Figures 8a-d show resulting frequency distributions of the zonally averaged soil surface temperatures for: (a) the July control and (b) perturbed runs, and (c) the January control and (d) perturbed runs. Grid point data show similar coolings, although the spread and noise is rather greater. Inland grid points, such as one near Alice Springs, show slightly greater coolings. Based on the CSU mesoscale model results, the estimated coolings from the GCM runs should be reduced by perhaps a degree or so to allow for the infrared emission from the smoke. We would also expect that daytime coolings would be greater than the diurnally averaged coolings, and the nocturnal coolings less.

Zonally averaged precipitation rates over land in the control and perturbed cases have also been calculated over successive 10- and 30-day periods. Means are shown for days 31-60 of the January case in Fig.9. Despite considerable fluctuation in the 10-day means, the 30-day means clearly show significant reductions in the precipitation rates over land in the tropics and in the Southern Hemisphere subtropics. A July simulation shows similar reductions

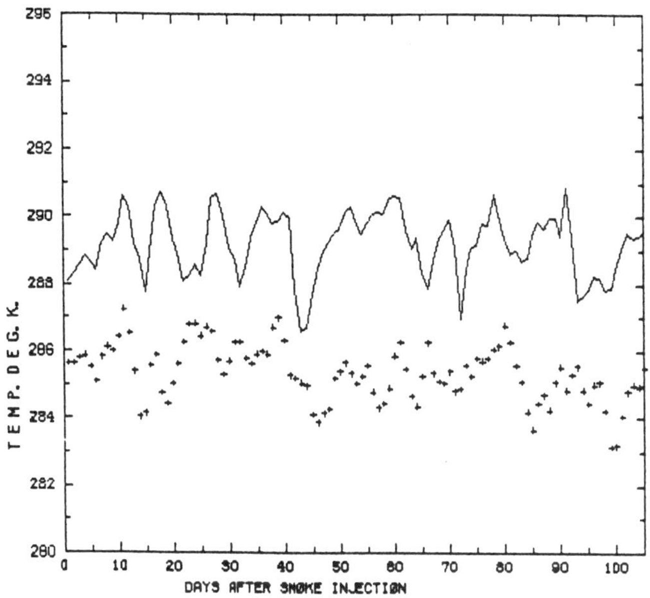


FIGURE 7. For the July case, 105 days of control (no smoke) and perturbed (uniform smoke absorption optical depth 0.2) zonally averaged land surface temperatures at 27 S, obtained with a version of the ANMRC general circulation model of McAvaney et al., 1978.

in precipitation in the tropics and the Northern Hemisphere monsoon regions. These results are consistent with those shown in Fig.4b in regard to the sensitivity of precipitation to moderate smoke loadings. The suppression of precipitation in our simulations is, however, confined to regions of monsoonal or convective rainfall, and is not evident in regions dominated by synoptically generated precipitation. This is not unreasonable, given the probable maintenance of synoptic activity by the heat storage in the oceans. We will be investigating this aspect further.

## 4. CONCLUSIONS

The large body of recent research on the possible climatic effects of a major nuclear war, including new results presented here, tends to support the hypothesis that the smoke and dust generated could have climatic consequences of a magnitude likely to be highly detrimental to agriculture and human survival.

While current estimates of the surface coolings over large continental areas in the early acute stages are of the order of "only" some 10-20°C in the Northern Hemisphere after a spring or summer war, the duration of lesser, but still substantial, coolings must now be measured in terms of several months to a year or more.

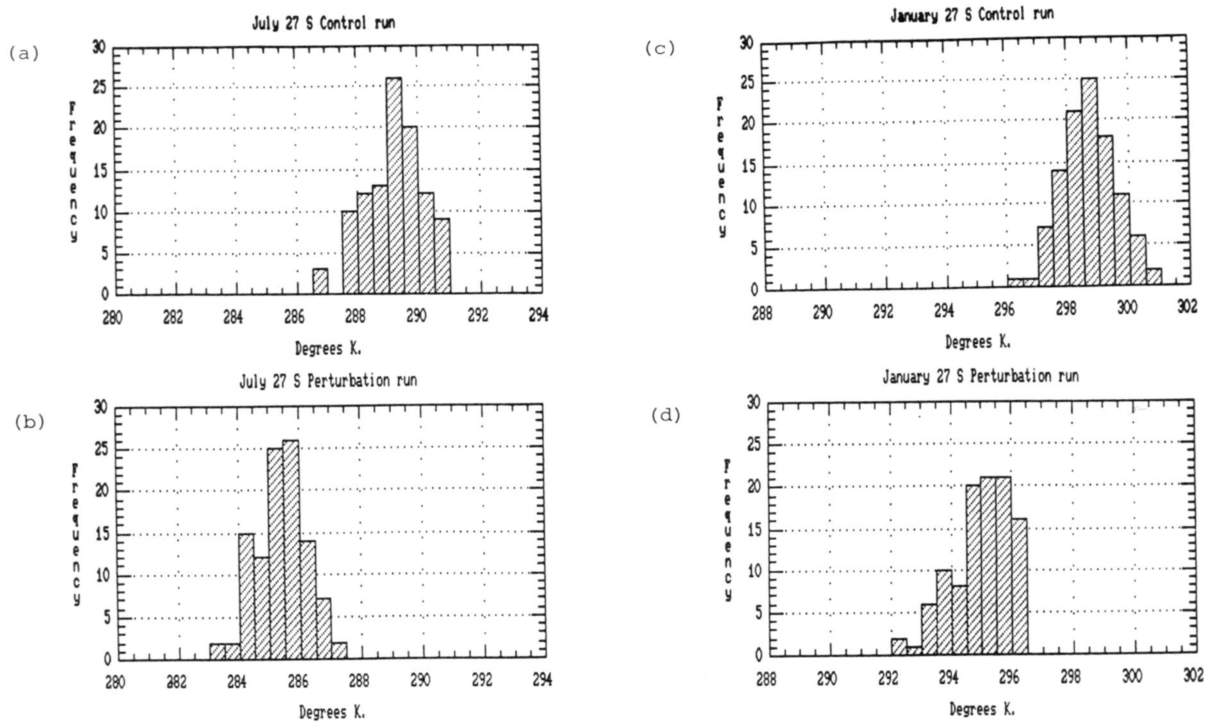


FIGURE 8: Frequency of occurrence of given soil surface temperatures at 27 S in 105 days of: (a) July conditions with no smoke, (b) July conditions with smoke of absorption optical depth 0.2, (c) January with no smoke, and (d) January with smoke.

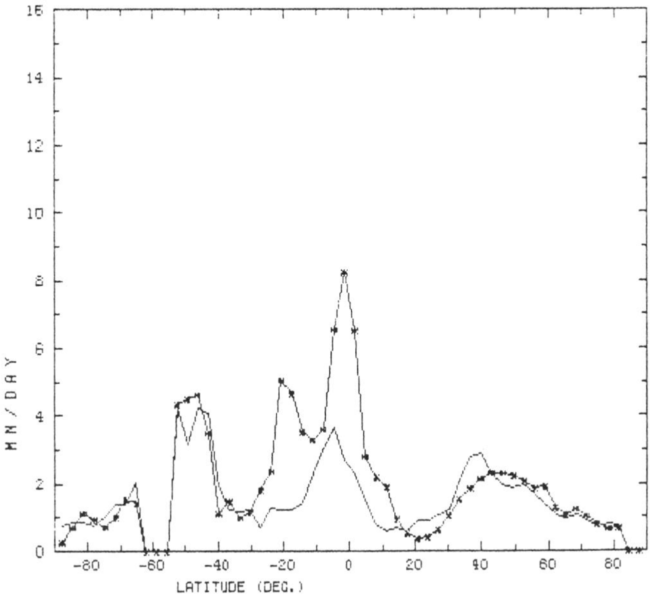


FIGURE 9: Zonal mean precipitation rates over land, as a function of latitude, averaged over days 31-60 in the control (curve with asterisks) and perturbed (plain curve) cases in January, for a uniform smoke layer of absorption optical depth 0.2. Results obtained with a version of the ANMRC general circulation model of McAvaney et al. (1978).

Agriculturally significant coolings may well occur in the Southern Hemisphere in the first growing season, and in the Northern Hemisphere in the second growing season, after such a nuclear war.

Tropical rainfall, and summer rains in the subtropics, may be reduced by 50% or more by quite moderate smoke amounts such as may reach the Southern Hemisphere, or persist in the Northern Hemisphere for a year or more. Even quite moderate initial injections of smoke, of the order of 50 million tonnes or less, could cause serious climatic effects in the Northern Hemisphere.

## ACKNOWLEDGMENTS

Special thanks are due to Dr. W. Bourke and colleagues at the Bureau of Meteorology Research Centre in Melbourne for help in getting the ANMRC model running, and to Dr. R.C. Malone of the Los Alamos National Laboratory for extensive discussions and collaboration. Eva Kowalczyk from CSIRO helped in the early stages with the CSU mesoscale model.

## REFERENCES


Berry, M.V., and I.C. Percival, 1986: Optics of Fractal Clusters Such as Smoke, Optica Acta, 33, 577-591.

Crutzen, P.J., and J.W. Birks, 1982: The Atmosphere After a Nuclear War, Ambio, 11, 114-125.

Crutzen, P.J., I.E. Galbally, and C. Bruhl, 1984: Atmospheric Effects From Post-Nuclear Fires, Climatic Change, 6, 323-364.

de Haas, N., R.M. Fristrom, M.J. Linevsky and D.M. Silver, 1986: Smoke Scavenging by Atmospheric Ozone as a Possible Factor in the Nuclear Winter Problem, Johns Hopkins APL Technical Digest, 7, 181-186.

Ghan, S.J., M.C. MacCracken and J.J. Walton, 1987: The Climatic Response to Large Atmospheric Smoke Injections: Sensitivity Studies With a Tropospheric General Circulation Model, UCRL-95984 Preprint (available from Lawrence Livermore National Laboratory, Livermore, CA), 80 pp.

Harwell, M., and T.C. Hutchinson, 1985: Environmental Consequences of Nuclear War: Vol. II. Ecological and Agricultural Effects, SCOPE 28, John Wiley and Sons, Chichester, 554 pp.

Mahrer, Y., and R.A. Pielke, 1977: A Numerical Study of the Airflow Over Irregular Terrain, Beitr. z. Physik d. Atmos., 50, 98-113.

Malone, R.C., L.H. Auer, G.A. Glatzmaier and M. Wood, 1986: Nuclear Winter: Three-Dimensional Simulations Including Interactive Transport, Scavenging, and Solar Heating of Smoke, J. Geophys. Res., 91 D1, 1039-1053.

McAvaney, B.J., W. Bourke and K. Puri, 1978: A Global Spectral Model for Simulation of the General Circulation, J. Atmos. Sci., 35, 1557-1583.

McNider, R.T., and R.A. Pielke, 1981: Diurnal Boundary-Layer Development Over Sloping Terrain, J. Atmos. Sci., 38, 2198-2212.

National Academy of Sciences, 1985: The Effects on the Atmosphere of a Major Nuclear Exchange, National Academy Press, Washington, D.C., 193 pp.

Penner, J.E., 1986: Uncertainties in the Smoke Source Term for 'Nuclear Winter' Studies, Nature, 324, 222-226.

Pittock, A.B., T.P. Ackerman, P.J. Crutzen, M.C. MacCracken, C.S. Shapiro and R.P. Turco, 1986: Environmental Consequences of Nuclear War: Vol. I., Physical and Atmospheric Effects, SCOPE 28, John Wiley and Sons, Chichester, 374 pp.

Post, M.J., 1986: Atmospheric Purging of El Chichon Debris, J. Geophys. Res., 91 D4, 5222-5228.

Thompson, S.L., and S.H. Schneider, 1986: Nuclear Winter Reappraised, Foreign Affairs, 65, 981-1005.

Turco, R.P., O.B. Toon, T.P. Ackerman, J.B. Pollack and C. Sagan, 1983: Nuclear Winter: Global Consequences of Multiple Nuclear Explosions, Science, 222, 1283-1292.

Warner, F., and 11 others, 1987: Severe Global-Scale Effects of Nuclear War Reaffirmed, Environment (Washington, DC), 29, 4-5 and 45.

# AIRBORNE MEASUREMENTS ON SMOKES FROM BIOMASS BURNING[1]

*L.F. Radke, D.A. Hegg, J.H. Lyons, C.A. Brock and P.V. Hobbs*
University of Washington
Seattle, Washington 98195, USA

*R. Weiss*
Radiance Research
Seattle, Washington 98177, USA

and

*R. Rasmussen*
Oregon Graduate Center
Beaverton, Oregon 97005, USA

## ABSTRACT

Airborne measurements have been made in the smokes from large fires of standing coniferous trees and logging debris, standing chaparral, fallen jack pine, and wheat stubble. Particle emission factors, particle size distributions, optical properties of the smokes, and trace gas concentrations in the emissions are derived from the airborne measurements. Major differences exist in these factors during any one fire, as well as from one fire to another. Measurements of optical scattering and extinction coefficients, with a time resolution of a few seconds, reveal significant spatial inhomogeneities in optical absorption within the smoke plumes.

## 1. INTRODUCTION

Wild forest fires, prescribed burning of forest logging debris, and silvacultural and agricultural burning combined constitute one of the largest sources of particles injected into the atmosphere and also a large source of trace gases (Wilson, 1970). However, despite the fact that 2 - 5% of the global land area is burned annually (National Academy of Sciences, 1984), there are still "...no satisfactory published estimates of amounts of biomass converted to gaseous and particulate matter..." (National Academy of Sciences, 1986). Uncertainties in the efficiencies with which biomass is converted into particles and trace gases in large fires is also a major uncertainty in quantifying the nuclear winter hypothesis (National Academy of Sciences, 1985).

In an effort to improve this situation we have made airborne measurements in the smokes from relatively large fires of standing coniferous trees, standing chaparral, fallen jack pine, coniferous logging debris, and wheat stubble. These fires, whether uncontrolled or prescribed for agricultural or forestry management, were intense, producing vigorous ascending columns of smoke often capped by a convective cloud. The sizes of the fires ranged from a few tens to many hundreds of hectares in area.

---

[1] This research was supported by the Defense Nuclear Agency, through funds made available under NRL Contract N00014-86-C-2246, and Sandia National Laboratory under Document 57-0343, and by the USDA Forest Service under Agreement PSW-87-0020CA.



ISBN 0-937194-11-5

Our measurements to date have emphasized determination of particle emission factors (mass of particles emitted into the air per unit mass of fuel burned), particle size distributions and trace gas concentrations in the smokes, and the optical properties of the smokes. A summary of our results on particle emissions from these fires is presented here. Results on trace gas emissions may be found in Hegg et al. (1987; 1988).

## 2. AIRBORNE INSTRUMENTATION AND PROCEDURES

The measurements were made aboard the University of Washington's C-131A research aircraft. This twin-engined, propeller-driven airplane carries instrumentation for measuring the size and nature of aerosol and cloud particles, trace gas concentrations, and meteorological parameters. Major portions of the aerosol system have been described by Radke (1983) and the trace gas instrumentation has been described by Hegg et al. (1987). The primary instruments used in the present study are listed in Table 1.

The aircraft made horizontal passes either through the ascending smoke column or perpendicular to the axis of a stabilized smoke plume from the fires. "Grab" samples of air for determination of aerosol size distributions were averaged over ~0.5 km of flight path, and those for carbonaceous trace gases and aerosol total mass over ~1 km of flight path. Plume structures were resolved with continuously measured variables (e.g., the light-scattering coefficient).

Particle emission factors were determined using the carbon-balance method (Ward et al., 1982). This requires measurements of the concentrations of carbonaceous gases and particles across a section of the plume and simultaneous measurements of the horizontal wind through this cross section. The emission factor (EF) for particles in a specified size range is given by:

$$EF = \frac{\text{mass flux of particles along the plume}}{\text{fuel consumption rate}} \tag{1}$$

Therefore,

$$EF = \frac{\overline{P}\,A\,V}{F} \tag{2}$$

where $\overline{P}$ is the average mass concentration of the particles across a section perpendicular to the axis of the plume, A the cross-sectional area of the plume, V the horizontal wind speed through the section, and F the fuel consumption rate. Since the carbon burned per unit time ($C_B$) is given by:

$$C_B = F\,C_f \tag{3}$$

where $C_f$ is the fractional mass of carbon in the fuel, we have from Eqs. (2) and (3):

$$EF = \frac{\overline{P}\,A\,V\,C_f}{C_B} \tag{4}$$

If deposition of the particles is negligible and the plume is in steady state:

TABLE 1. RELEVANT INSTRUMENTATION ABOARD THE UNIVERSITY OF WASHINGTON'S C-131A RESEARCH AIRCRAFT

(a) Particle Sizes and Concentrations

| Instrument | Equivalent Particle Diameter (μm) |
|---|---|
| Condensation nucleus counter* | > 0.005 |
| Diffusion battery with condensation nucleus counter† | ~ 0.01 - 0.1 |
| Electric aerosol analyzer† | ~ 0.01 - 1.0 |
| Active scattering aerosol spectrometer† | 0.09 - 3 |
| Laser aerosol spectrometer† | 0.5 - 11 |
| Optical particle counter† | 2.6 - 48 |
| Forward-scattering spectrometer probe* | 3 - 45 |

(b) Trace Gas Chemistry

| Gas | Technique | Detection Limit (ppbv) |
|---|---|---|
| $SO_2$* | pulsed fluorescence | > 1 |
| $O_3$* | chemiluminescence ($C_2H_4$) | > 5 |
| NO, $NO_2$* | chemiluminescence ($O_3$) | > 1 |
| CO† | correlation IR spectrometer | > 100 |
| $CO_2$† | IR spectrometer | ± 2 ppm |
| $NH_3$† | impregnated filter | % variable |
| Whole air grab sample | GC/mass spectrometer (post flight) | variable |

(c) Aerosol Characteristics

| Parameter | Instrument | Range |
|---|---|---|
| Light-scattering coefficient | integrating nephelometer | $1 \times 10^{-6}$ - $2.5 \times 10^{-3}$ $m^{-1}$ |
| Light-extinction coefficient | optical extinction cell | $5 \times 10^{-5}$ - $10^{-2}$ $m^{-1}$ |
| Particle mass | quartz microbalance impactor | < 2 μm diameter |
| Bulk aerosol chemistry (soluble anions concentration) | Teflon filters, ion chromatography | $Cl^-$ ± 14%; $SO_4^=$ ± 4%; $NO_3^-$ ± 11% |
| Particle morphology & chemistry | impaction, electron microscopy | > 0.2 μm diameter |
| Carbon soot§ | filtration, light absorption, pyrolysis | - |

* Continuous measurements.
† Grab sample measurements.
§ Analysis by Radiance Research and Sunset Laboratories.

$$C_B = [(\overline{P}\,C_p) + (\overline{CO_2}\,C_{CO_2}) + (\overline{THC}\,\,C_{THC}) + (\overline{CO}\,C_{CO})]AV \qquad (5)$$

where $C_p$, $C_{CO_2}$, $C_{THC}$, and $C_{CO}$ are the fractional masses in the smoke plume of carbon in the particles, $CO_2$, total hydrocarbons, and CO, respectively, and $\overline{CO_2}$, $\overline{THC}$, and $\overline{CO}$ are, respectively, the average mass concentrations of $CO_2$, total hydrocarbons, and CO in the cross section of the plume. From Eqs. (4) and (5):

$$EF = \frac{\overline{P}\,C_f}{(\overline{P}\,C_p) + (\overline{CO_2}\,C_{CO_2}) + (\overline{THC}\,\,C_{THC}) + (\overline{CO}\,C_{CO})} \qquad (6)$$

If it is assumed that the particles move with the gases, so that they bear the same ratio to each other from one point in the plume to another, an expression similar to Eq. (6) will relate instantaneous measurements made at any point in the plume, where the average concentrations in Eq. (6) would then be replaced by point concentrations. It is this latter modification of Eq. (6) that we use to determine particle emission factors. Particle mass concentrations, over various size ranges, can be derived from the "grab" samples by measuring the particle size distribution, from weighing of filters through which the "grab" samples were passed, and from microbalance impactor measurements. The mass concentrations of the various carbon gases are obtained from GC-mass spectrometer analysis (Westberg et al., 1974; Rasmussen et al., 1974). $C_p$ is measured through elemental carbon analysis of sequenced pyrolysis of quartz filters exposed to the smokes (Johnson et al., 1981). The value of $C_f$ is taken as 0.5 (Bryam and Davis, 1959).

Measurement of the light-extinction, scattering, and absorption coefficients were obtained from a combination of continuous measurements and grab-bag sampling. Continuous measurements of the extinction and scattering coefficients were obtained with a highly stable 7 m long extinction cell and an integrating nephelometer (Weiss and Radke, 1987). This technique produces excellent agreement with filter samples analyzed by the integrating plate method (Weiss et al., 1979) and with the elemental carbon analysis of sequenced pyrolysis of the quartz filters.

The derived emission factor is strictly valid only at the point of measurement. Any process that adds or removes a pollutant between the source and the location of the measurement will change the derived emission factor for that pollutant. Such changes are often small if the measurements are made close to the fire, but cloud scavenging and gas and particle chemistry have the potential to significantly change the measured values of emission factors. The emission factors for particle mass given below were derived from measurements in smoke columns above the fire or at short distances downwind in stabilized smoke plumes. In some cases, however, the plume had been scavenged by capping cumulus clouds.

## 3. RESULTS

Using the techniques described in the previous section, we have examined emissions from nine major fires and one minor fire (of wheat stubble). Information on these fires is summarized in Table 2. Eight of the fires were prescribed burns, and two were portions of major wild fires on the Oregon-California border that burned largely uncontrolled for over a month in the late summer of 1987.

TABLE 2. FIRES EXAMINED IN PRESENT STUDY

| Fire | Date | Location | Size (hectare)* | Type of Fire | Fuel |
|---|---|---|---|---|---|
| Abee† | 22 September 1986 | Montesano, Washington | 40 | Prescribed | Debris from Douglas Fir, & Hemlock |
| Eagle | 3 December 1986 | Ramona, California | 30 | Prescribed | Standing Black Sage, Sumac & Chamise |
| Lodi 1 | 12 December 1986 | Los Angeles, California | 40 | Prescribed | Standing Chaparral, Chamise |
| Lodi 2 | 22 June 1987 | Los Angeles, California | 150 | Prescribed | Standing Chaparral, Chamise |
| Hardiman† | 28 August 1987 | Chapleau, Ontario | 325 | Prescribed | Debris from Jack Pine, Standing Aspen & Paper Birch |
| Wheat | 31 August 1987 | Rosalia, Washington | ~10 | Prescribed | Wheat stubble |
| Myrtle/Fall Creek† | 2 September 1987 | Roseburg, Oregon | 2,000 | Wild Fire | Standing Pine, brush & Douglas Fir |
| Silver | 17-19 September 1987 | Grants Pass, Oregon | 20,000 | Wild Fire | Douglas Fir, True fir, Hemlock |
| Satsop† | 19 September 1987 | Satsop, Washington | 40 | Prescribed | Debris from Douglas Fir & Hemlock |
| Troy† | 8 October 1987 | Troy, Montana | 70 | Prescribed | Debris from Pine, Douglas Fir & True fir |

* Area estimates for the wild fires are very approximate; actual active flaming zones must have been much smaller.
† These fires produced a capping cumulus cloud at some time during the period of the measurements.

## 3.1 PARTICLE SIZE DISTRIBUTIONS

The particle size distributions, particularly the volume distributions, in the smokes showed comparatively little variation in shape from one fire to another or during any one fire while near the source, although particle concentrations varied widely. Shown in Fig. 1 are average particle size distributions for three of the fires. The nucleation mode (<0.1 μm diameter) was the most variable: frequently it was nearly completely suppressed but occasionally it was prominent. The accumulation mode (0.1 - 2.0 μm) often dominated both the number and volume size distributions and it always contributed overwhelmingly to the light-scattering coefficient. The peak in the accumulation mode tended to move to slightly larger particle sizes as concentrations increased and as the age of the smoke increased. Particles in the accumulation mode consisted primarily of tarry, condensed hydrocarbons, typically spherical in shape, but they also contained some water-soluble inorganics (primarily $SO_4^=$ and $NO_3^-$) (Hegg et al., 1987); this accounts, in part, for their activity as very efficient cloud condensation nuclei (Radke et al., 1978).

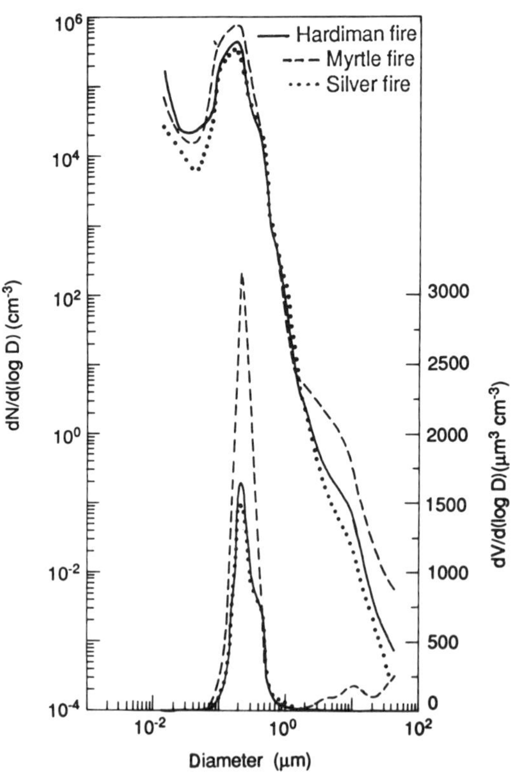


FIGURE 1. Average number and volume distributions of particles measured in the smoke plumes near the sources of three fires.

The coarse particles (>2 μm) in the smokes showed considerable variations. In the Lodi 1 fire there was a mode near 10 μm that consisted largely of soil and condensed hydrocarbons (personal communication, 20 June 1987, W. R. Coffer, NASA Langley Research Center, Hampton, Virginia). Airborne measurements (not shown here) indicated that particles up to at least 4 mm in maximum dimensions, in concentrations of $\sim 10^{-6}$ cm$^{-3}$, were present in the smoke from the Lodi 1 fire. Such particles would produce another peak in the volume distribution at a size beyond that plotted in Fig. 1. A hint, at smaller sizes, of this ultra-giant mode can be seen on the right-hand side of the volume distribution plots shown in Fig. 1.

## 3.2 PARTICLE EMISSION FACTORS

The accumulation mode plays an important role in the particle emission factor, both because most of the mass resides in this mode and also because atmospheric residence times for larger and smaller particles are comparatively short.

Shown in Fig. 2 are the particle emission factors as a function of time for the Lodi 1 fire. Three independent measurements of particle mass are shown in this figure. The agreement between these measurements is excellent; the occasional discrepancies demonstrate both experimental errors and the danger of depending on a single technique for measuring particle emission factors. The large variations with time in the particle emission factors were probably caused by various attempts to reignite the fire. The peaks in the emission factor appeared to correspond to periods when the fire was spreading rapidly. This was contrary to expectation, since Patterson and McMahon (1984) and Hardy and Ward (1986) reported that increases in flaming combustion reduce the particle emission factor.

The time dependence of the emission factor for the Abee fire (Fig. 3) was quite different from the Lodi 1 fire. All of our measurements in the Abee fire were taken in

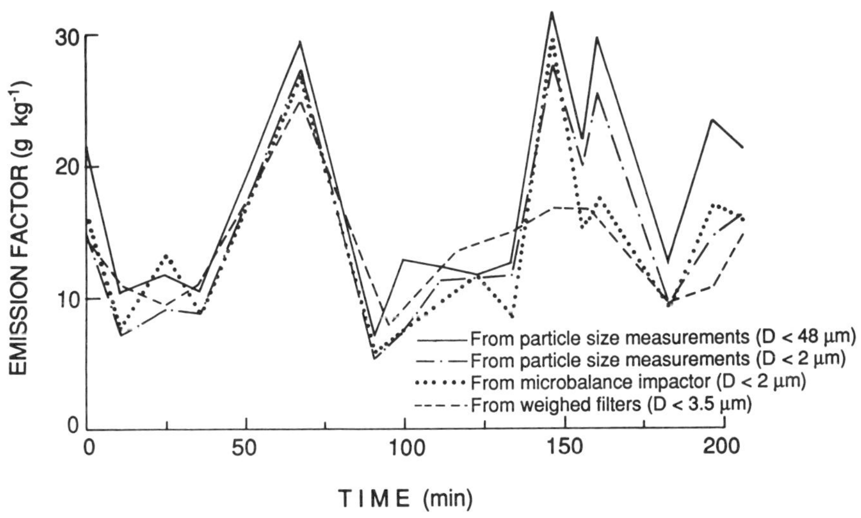


FIGURE 2. Emission factors for particle mass as a function of time after ignition for the Lodi 1 fire.

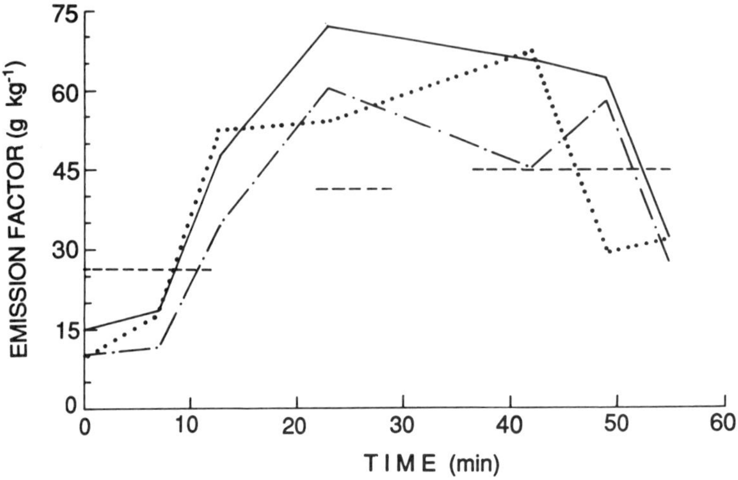


FIGURE 3. Particle emission factors as a function of time from the beginning of observations on the Abee fire. A vigorous cumulus cloud capped the smoke column for the first 10 - 15 min. The measurements were made downwind of this capping cumulus in the stabilized plume. (Data key as in Fig. 2.)

the stabilized (bent over) portion of the plume. During the first 10 - 15 min of observations, the particle emission factors were quite low. This could have been due to the scavenging of particles by a vigorous cumulus cloud that capped the smoke column during this period, or to more efficient combustion of fuel during this period. This fire yielded the highest sustained particle emission factor (37 g kg$^{-1}$ for particles <3.5 μm diameter) measured in these studies. It should be noted that the Abee fire had a higher fuel loading and was smaller in area than the Lodi 1 fire.

Listed in Table 3 are the average particle emission factors for the ten fires. Note that the two wild fires (Myrtle/Fall Creek and Silver) produced some of the highest measured particle emission factors. The average particle emission factor (for particles <2 μm diameter) for all our measurements in all fires (71 data points) is ~18 g kg$^{-1}$ with a standard deviation of 12.8 g kg$^{-1}$. For particles <48 μm, the average particle emission factor is 24 ± 16 g kg$^{-1}$.

3.3 OPTICAL PROPERTIES OF SMOKES

Our real-time measurements of the light-extinction coefficient ($\sigma_E$) and light-scattering coefficient ($\sigma_s$) allow us to examine the single-scattering albedo ($\tilde{\omega} \equiv \sigma_s/\sigma_E$), the light-absorption coefficient ($\sigma_A = \sigma_E - \sigma_s$), and the specific absorption $B_A = \sigma_A$ x (particle mass per unit volume of air)$^{-1}$. Values of $\sigma_E$ and $\sigma_s$ measured across the smoke column from the Lodi 2 fire are shown in Fig. 4a. The center of the smoke column was markedly darker than its edges. Well downwind in the stabilized smoke plume not only was the center of the plume less dark but the entire plume was whiter (Fig. 4b). This may have been due to entrainment into the plume of cooler, whiter puffs from smoldering combustion and to "whiter" particle mass added to the plume by gas-to-particle conversion.

TABLE 3. PARTICLE EMISSION FACTORS (IN GRAMS OF SMOKE PER KILOGRAM OF FUEL BURNED) AND SINGLE-SCATTERING ALBEDO AND SPECIFIC ABSORPTION FOR ALL FIRES STUDIED

| | Emission Factor (derived from size distribution of particles < 2 μm diameter) | | | Emission Factor (derived from weighed filters for particles < 3.5 μm diameter) | | | Emission Factor (derived from size distribution of particles < 48 μm diameter) | | | Mean Value of Single-Scattering Albedo $\tilde{\omega}$ (= $\sigma_S/\sigma_E$) | Mean Values of Specific Absorption $B_A$ ($m^2\ g^{-1}$) |
|---|---|---|---|---|---|---|---|---|---|---|---|
| | Mean | Standard Deviation | Number of Samples (N) | Mean | Standard Deviation | Number of Samples (N) | Mean | Standard Deviation | Number of Samples (N) | | |
| Abee | 35.1 | 20.3 | 7 | 37.4 | 9.8 | 3 | 44.5 | 23.1 | 7 | ND† | ND |
| Eagle | 7.9 | 5.7 | 8 | 11.3 | 4.8 | 2 | 10.8 | 6.0 | 8 | 0.85 (N = 2) | 0.67 (N = 2) |
| Lodi 1 | 14.3 | 7.3 | 16 | 13.5 | 4.4 | 13 | 17.6 | 7.9 | 16 | 0.89 (N = 13) | 0.74 (N = 13) |
| Lodi 2 | 15.5 | 6.5 | 9 | 23.0 | 19.6 | 8 | 17.7 | 7.0 | 9 | 0.80 (N = 7) | 0.68 (N = 7) |
| Hardiman | 16.2 | 12.4 | 11 | 10.5 | 3.0 | 9 | 21.7 | 14.6 | 11 | 0.87 (N = 9) | 0.79 (N = 5) |
| Wheat | 43.8 | - | 1 | ND | - | - | 47.9 | - | 1 | NA§ | ND |
| Myrtle/ Fall Creek* | 19.5 | 12.1 | 10 | NA | - | - | 29.3 | 16.5 | 10 | NA | 0.73 (N = 16) |
| Silver* | 26.4 | 13.6 | 2 | 20.2 | 12.7 | 3 | 32.5 | 17.1 | 2 | 0.88 (N = 3) | 0.36 (N = 3) |
| Satsop | 24.6 | 7.1 | 2 | 12.0 | - | 1 | 34.3 | 10.8 | 2 | NA | 0.37 (N = 1) |
| Troy | 17.1 | 12.4 | 5 | 9.7 | 3.0 | 5 | 30.2 | 19.5 | 5 | 0.82 (N = 3) | 0.82 (N = 3) |
| Averages | 18 ± 13 (N = 71) | | | 16 ± 12 (N = 44) | | | 24 ± 16 (N = 71) | | | 0.86 ± 0.07 (N = 37) | 0.72 ± 0.18 (N = 44) |

* Wild fires.
† ND = No data.
§ NA = Not analyzed.

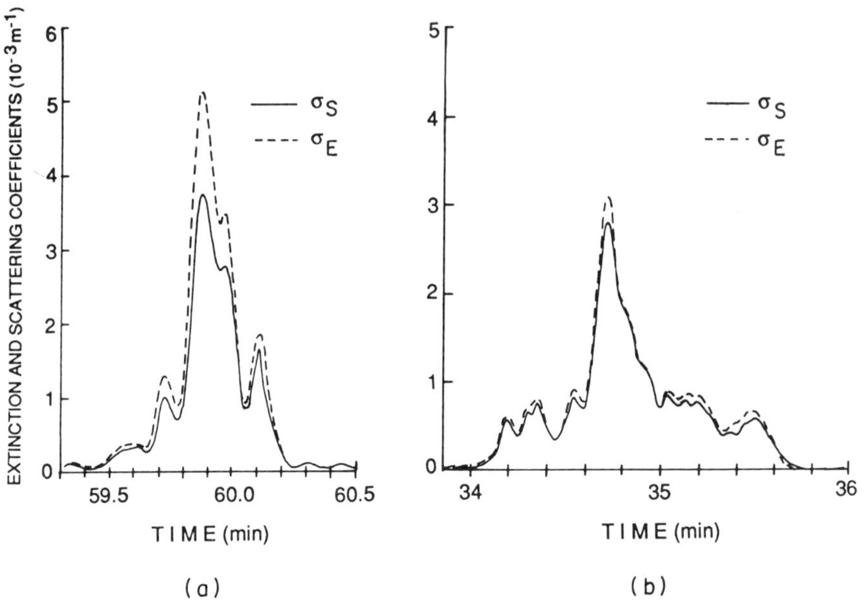


FIGURE 4. Light-extinction ($\sigma_E$) and light-scattering ($\sigma_S$) coefficients in the smoke from the Lodi 2 fire: (a) in the vertical column above the fire ($\tilde{\omega} = 0.78$), and (b) in the stabilized plume well downwind of the fire ($\tilde{\omega} = 0.93$).

Table 3 lists average values of $\tilde{\omega}$ and $B_A$ for the fires studied. The average values of $\tilde{\omega}$ and $B_A$ for all ten fires were $0.86 \pm 0.07$ and $0.72 \pm 0.18$ $m^2$ $g^{-1}$, respectively. Interestingly, the whitest smoke came from the Silver wild fire ($\tilde{\omega} = 0.88$) and the rather wet (late season) Lodi 1 fire ($\tilde{\omega} = 0.89$). Three of the prescribed fires (Lodi 2, Hardiman and Troy) had significant periods when $\tilde{\omega}$ was less than 0.75. During these periods the smokes were much darker (viewed from the aircraft) and, in the case of the Hardiman fire, these periods were clearly related to intense flaming combustion of the lighter fuels.

## 4. DISCUSSION

These results significantly increase the available data base on particle emission factors and the optical properties of smokes from forest fires. The average particle emission factor for the ten fires discussed here ($18 \pm 13$ g $kg^{-1}$ for particles <2 μm diameter) is significantly larger than the average value ($4.1 \pm 1.5$ g $kg^{-1}$ for particles <2 μm diameter) measured in a series of smaller, but rather intense, fires of logging debris by Radke et al. (1983). These results suggest some scaling of the particle emission factor with the intensity of the fire (lower EF with increased intensity) and/or the size of the fire (higher EF with larger fires). The particle emission factors for the two wild fires (19.5 and 26.4 g $kg^{-1}$ for particles <2 μm diameter), while greater than the average for the smaller prescribed fires, were not the largest emission factors measured. The highest particle emission factor for particles <2 μm diameter for all of the fires (with the exception of the wheat fire) was for the Abee fire, which, if the period of apparent nucleation scavenging by cloud droplets is removed, averaged

45 g kg$^{-1}$. Our single observation of a very intense fire in wheat stubble gave a particle emission factor (for particles <2 μm diameter) of 44 g kg$^{-1}$; this suggests that further attention should be paid to this source of smoke. These were the only fires where the measured particle emission factors sustained values in excess of 30 g kg$^{-1}$, which is the value used in the National Academy of Sciences (1985) nuclear winter calculations for wild fires. Thus, our measurements suggest that the values for particle emission factors used in the National Academy Report are reasonable, but possibly on the high side.

In the National Academy of Sciences (1985) report it was concluded that, following a nuclear war, smokes from the burning of biomass would contribute about 20% to the smoke in the atmosphere. It was assumed that the smoke from biomass burning would have a significantly higher single-scattering albedo than the (blacker) smokes from urban fires. It was estimated that the globally-averaged, single-scattering albedo of the mixed smokes produced by urban and biomass burning would be ~0.6. In a recent modeling study of the nuclear winter hypothesis, Thompson et al. (1987) suggested that the globally-averaged, single-scattering albedo of the mixed smokes might be as high as 0.75.

If, as seems likely, smokes from urban fires are, on average, blacker than those from biomass burning, globally-averaged values of 0.6 and 0.75 for the single-scattering albedo are not inconsistent with the average value given in this paper of $0.86 \pm 0.07$ for the single-scattering albedo of smokes from biomass burning. However, our measurements also show that for significant periods of time the smokes from biomass fires can be quite dark. Thus, for three of the fires studied (Lodi 2, Hardiman, and Troy), single-scattering albedos $\leq 0.75$ were measured for extended time periods, and the darkest smokes in the Lodi 2 and Troy fires had average single-scattering albedos of 0.65. The darkest smokes appeared to be associated with the most intense burning and long flame lengths (and, perhaps, associated lack of oxygen). Since these conditions are likely to be those that would occur in large fires produced by a nuclear war, and assuming smokes from urban fires are generally darker than those from biomass burning, our measurements suggest that the values for the globally-averaged, single-scattering albedo for the mixed smokes that have been used in modeling studies of the nuclear-winter hypothesis may be high (i.e., the mixed smoke could be darker than assumed in the modeling studies).

## ACKNOWLEDGMENTS

Special thanks are due our Chief Pilot, Mr. Donald Veach, our research flight crew, and the many foresters and firemen who worked to ignite and/or extinguish the fires studied. We are particularly grateful to Drs. Philip Riggan, Darold Ward, Brian Stocks and Robert Cary for their help in these studies.

## REFERENCES

Bryam, G. M., and K. P. Davis (Eds.), 1959: Forest Fire Control and Use, McGraw-Hill, New York, NY, 321 pp.

Hardy, C. C., and D. E. Ward, 1986: Emission Factors for Particulate Matter by Phase of Combustion From Prescribed Burning. In Preprints, 1986 PNWIS-APCA Annual Meeting, Air Pollution Control Association, Pittsburgh, PA, 151-157.

Hegg, D. A., L. F. Radke, P. V. Hobbs, C. A. Brock and P. J. Riggan, 1987: Nitrogen and Sulfur Emissions From the Burning of Forest Products Near Large Urban Areas, J. Geophys. Res., 92, 14701-14709.

Hegg, D. A., L. F. Radke, P. V. Hobbs and P. J. Riggan, 1988: Ammonia Emissions From Biomass Burning, Geophys. Res. Letts., (in press).

Johnson, R. L., J. J. Shah, R. A. Cary and J. J. Huntziker, 1981: An Automated Thermal-Optical Method for the Analysis of Carbonaceous Aerosol. In E. S. Marcias and P. K. Hopke (Eds.), Atmospheric Aerosol: Source/Air Quality Measurements, ACS Symposium Series, No. 167, American Chemical Society, Washington, DC, 223-233.

National Academy of Sciences, 1984: Global Tropospheric Chemistry: A Plan for Action, National Academy Press, Washington, DC, 194 pp.

National Academy of Sciences, 1985: The Effects on the Atmosphere of a Major Nuclear Exchange, National Academy Press, Washington, DC, 193 pp.

National Academy of Sciences, 1986: Global Tropospheric Chemistry, National Academy Press, Washington, DC, 110 pp.

Patterson, E. M., and C. K. McMahon, 1984: Absorption Characteristics of Forest Fire Particulate Matter, Atmos. Environ., 18, 11, 254-255.

Radke, L. F., J. L. Stith, D. A. Hegg and P. V. Hobbs, 1978: Airborne Studies of Particles and Gases from Forest Fires. J. Air Pollut. Control Assoc., 28, 30-34.

Radke, L. F., 1983: Preliminary Measurements of the Size Distribution of Cloud Interstitial Aerosol. In H. R. Pruppacher, R. G. Semonin and W. G. N. Slinn (Eds.), Precipitation Scavenging, Dry Deposition and Resuspension, Elsevier, New York, NY, 71-78.

Radke, L. F., J. H. Lyons, D. A. Hegg and P. V. Hobbs, 1983: Airborne Monitoring and Smoke Characterization of Prescribed Fires on Forest Lands in Western Washington and Oregon, EPA Report 600X-83-047, available from Environmental Monitoring Systems Laboratory, Office of Research and Development, U.S. Environmental Protection Agency, Las Vegas, NV, 120 pp.

Rasmussen, R. A., H. H. Westberg and M. W. Holdren, 1974: Need for Standard Referee GC Methods in Atmospheric Hydrocarbon Analyses, J. Chromatographic Science, 12, 80-84.

Thompson, S. L., V. Ramaswamy and C. Covey, 1987: Atmospheric Effects of Nuclear War Aerosols in General Circulation Model Simulations: Influence of Smoke Optical Properties, J. Geophys. Res., 92, 10,942-10,960.

Ward, D. E., D. V. Sandberg, R. D. Ottman, J. A. Anderson, G. G. Hofner and C. K. Fitzsimmons, 1982: Measurements of Smoke From Two Prescribed Fires in the Pacific Northwest. Seventy-Fifth Annual Meeting of the Air Pollution Control Association, available from D. E. Ward, U. S. D. A. Forest Service, Seattle, WA 98105.

Weiss, R. E., A. P. Waggoner, R. J. Charlson, D. A. Thorsell, J. S. Hall and L. A. Richy, 1979: Studies of the Optical, Chemical and Physical Properties of Light Absorpting Aerosols. In T. Novakov (Ed.), Proceedings of the Conference on Carbonaceous Particles in the Atmosphere, LBL-9037, Conf. #7803101, Berkeley, CA, 257-262.

Weiss, R. E., and L. F. Radke, 1987: Optical Properties of Lodi Canyon Forest Smoke: Airborne Plume Measurements Using "In Situ" Optical Extinction. In M. J. Pilat and E. J. Davis (Eds.), Aerosols '87, Elsevier, New York, NY, 261 pp.

Westberg, H. H., R. A. Rasmussen and M. W. Holdren, 1974: Gas Chromatographic Analyses of Ambient Air for Light Hydrocarbons Using a Chemically Bonded Stationary Phase, Anal. Chem., 46, 1852-1854.

Wilson, C. L. (Ed.), 1970: Study of Critical Environmental Problems (SCEP), MIT Press, Cambridge, MA, 318 pp.

# POTENTIAL NUCLEATION SCAVENGING OF SMOKE PARTICLES OVER LARGE FIRES: A PARAMETRIC STUDY

*Leslie L. Edwards and Joyce E. Penner*
University of California, Lawrence Livermore National Laboratory
Livermore, California 94550, USA

## ABSTRACT

During hypothesized nuclear exchanges, massive fires may be ignited and inject large amounts of smoke into the atmosphere. Considerable evidence has been accumulated to suggest that nucleation scavenging where smoke particles serve as cloud condensation nuclei (CCN) may be an important mechanism for incorporating these particles into cloud water. The fraction of smoke particles that acts as CCN depends on the cloud environment as well as the affinity of the smoke particles to water.

A numerical model of the detailed microphysics of condensation growth on aerosol and cloud drop distributions is employed to produce a parametric study of the dependence of nucleation to a range of conditions. We consider aerosol number concentrations of $10^9$ to $10^{13}$ $m^{-3}$, updraft speeds from 1 to 100 m $s^{-1}$, and aerosol particles from fully water soluble to insoluble but completely wettable. The study provides insight into how well we must characterize smoke particles in order to predict the fraction that acts as CCN given the dynamical environment.

## 1. INTRODUCTION

Crutzen and Birks (1982) first pointed out the role of smoke and dust in the atmosphere and the potential global effects of a hypothetical large-scale nuclear war. Turco et al. (1983) further alerted the world to the potential for severe climatic effects. Subsequent works (National Research Council, 1985; Pittock et al., 1986; Penner, 1986) have discussed the issues and uncertainties involved. The potential climatic effects are due to solar attenuation (thought to be dominated by absorption due to smoke rather than reflection due to dust) brought about by the large amounts of smoke and dust particles injected into the upper atmosphere and remaining there for days to even years. These smoke and dust particles are lofted into the atmosphere not only directly by blast, but subsequently by the large-scale fires ignited by nuclear detonations.

The estimated magnitude of the climatic effects due to a nuclear war are very uncertain primarily because of the uncertainty in the amount of smoke that might be injected. The course of events in a nuclear war depends not only on the classified war plans of the combatants, but also on the human decision-makers involved in the battle. That is, the number and sizes of nuclear weapons exploded in an exchange and the targets selected can only be hypothesized and consequently will remain uncertain. Further, the dust and smoke raised by blast and fires depend on the target, weapon size, height of burst, type and amount of fuels burned, fire intensity, ambient atmospheric conditions, amount of smoke affected by cloud processes, etc. (Penner, 1986; Penner et al., 1986), all leading to major uncertainties.


A. DEEPAK Publishing
ISBN 0-937194-11-5

Considerable evidence has been accumulated to suggest that nucleation scavenging[1] in which smoke particles serve as cloud condensation nuclei (CCN) may be an important mechanism for incorporating smoke particles into cloud water (Hobbs et al., 1984; Pruppacher, 1985; Penner and Edwards, 1986). Data indicate that some highly carbonaceous smoke particles can act as CCN (Hallett, et al., 1986) although smoke from an oil fire was apparently not very hygroscopic (Radke, et al., 1980). Even if the smoke particles are not removed from the atmosphere, incorporation into cloud water might significantly alter the optical properties of the particles. This would occur if, for example, chained agglomerate smoke particles collapse to a more compact morphology.

The characterization of smoke particles, their shapes, affinity to water, size distributions, etc., are the subject of current research (Hallett et al., 1986). This paper describes results from a mathematical model that computes the condensation and evaporation of water on a distribution of spherical particles (nuclei) surrounded by a spherical shell of water. The nucleus is assumed to have a specified composition of soluble and insoluble materials. "Nucleation to drops" is computed. Aerosol coagulation due to Brownian diffusion, turbulent shear and impaction, and gravitational sedimentation is also computed. Aerosol coagulation due to electrical effects will be the subject of future work and is not included. To provide insight as to how well we must characterize the smoke particles for use in numerical model calculations, we present a series of calculations over a wide range of parameters.

## 2. THE MATHEMATICAL MODEL

The mathematical model is based on equations taken from Pruppacher and Klett (1978) and formulated with methodology similar to that of Flossmann et al. (1985). It is assumed that the overall physical processes can be evaluated by the separate processes outlined in Eq. (1). Accordingly, we introduce the number density functions $n_{AP}(r, e_m, t)$ for interstitial aerosols and $n_D(a, e_m, t)$ for the drops, where $r$ is the aerosol nucleus (dry) radius, $a$ is the drop (wet) radius, $e_m$ is the mass fraction of soluble material in the aerosol nucleus or solubility ratio, and $t$ is the time. Aerosols that contain a soluble material may grow due to water condensation, but they will be termed "interstitial" as long as their wet radius is less than the critical radius defined by Eq. (4). For the drops we also define the aerosol mass density function $m_{AP,D}(a, e_m, t)$. This function is used to compute an "average" nucleus radius for drops of a given size. Since the interstitial aerosols rapidly grow to an equilibrium size and are defined as a function of dry radius, an average wet radius is computed at each size and time step using Eq. (2).

For this study we compute the time evolution of the distributions considering the processes:

$$\frac{d}{dt}n_{AP}(r, e_m, t) = \frac{d}{dt}n_{AP}\mid_{dyn} + \frac{d}{dt}n_{AP}\mid_{c/e} + \frac{d}{dt}n_{AP}\mid_{act} + \frac{d}{dt}n_{AP}\mid_{coag} \tag{1a}$$

$$\frac{d}{dt}n_D(a, e_m, t) = \frac{d}{dt}n_D\mid_{dyn} + \frac{d}{dt}n_D\mid_{c/e} + \frac{d}{dt}n_D\mid_{act} \tag{1b}$$

$$\frac{d}{dt}m_{D,AP}(a, e_m,) = \frac{d}{dt}m_{D,AP}\mid_{dyn} + \frac{d}{dt}m_{D,AP}\mid_{c/e} + \frac{d}{dt}m_{D,AP}\mid_{act} \tag{1c}$$

[1] In this work scavenging does not mean removal from the atmosphere. Nucleation scavenging merely means incorporation of interstitial aerosol into water drops that are larger than the critical radius as defined by Eq. (4).

where *dyn* refers to terms determined by the dynamical environment; *c/e* refers to condensation/evaporation of water vapor; *act* refers to the activation of aerosols to drops (nucleation) and deactivation of drops to interstitial aerosols; and *coag* refers to aerosol coagulation. In this work we have neglected the time rate of change in number and mass density due to:

1) collection of interstitial aerosol by drops, expected to be a major factor particularly when electrical effects are important;
2) drop coalescence and breakup, which are important in the formation of precipitation size drops and, consequently, important to removal; and
3) ice formation and collection processes, which may determine the primary precipitation pathway especially for high intensity fires that lift smoke to high altitude.

Models for the description of these processes are currently under development.

For this model the dynamics are those of a closed parcel of well-mixed air, water vapor, condensed water, and aerosol. The parcel is assumed to rise with a prescribed updraft velocity. Its pressure is equal to that of a standard atmosphere at the current altitude of the parcel. The temperature, density, and mixing ratios are computed to be consistent with the current rates of condensation/evaporation occurring on both the aerosol and drop distributions.

The condensation/evaporation growth rate equation for both aerosols and drops is computed according to Eqs. 13–26 of Pruppacher and Klett (1978). Using their notation:

$$a\frac{da}{dt} = \frac{F_v D'_v e_{sat,w}}{\rho_w R_v T}\{S - \frac{1}{(1+\delta)}\, exp\,[\frac{L_e}{R_v T}(\frac{\delta}{(1+\delta)}) + \frac{2\sigma_{s/a}}{\rho_w R_v T(1+\delta)a} - \frac{\nu\phi_s \rho_N M_s e_m}{\rho_w M_w}(\frac{r^3}{a^3 - r^3})]\} \tag{2}$$

where, $S$ is the saturation ratio and $F_v$ is the ventilation coefficient.

A discussion of this equation and the coefficients used are beyond the scope of this report. The interested reader is referred to, for example, Pruppacher and Klett (1978), Fukuta and Walter (1970) and Mozurkewich (1986). For clarification, in the results reported here, we have used $\sigma_{s/a} = \sigma_{w/a}$ and $\phi_s = 1$, even though both of these quantities are functions of solution molality, surface impurities, etc.

The "modified" diffusivity, $D'_v$, of water vapor has been computed with an "average" condensation coefficient, $\alpha_c = 0.036$, and "vapor jump" distance $\Delta_v = 1.3\lambda_a$. The "modified" thermal conductivity, $k'_a$, implicit in $S$, has been computed with $\alpha_T = 0.7$ and $\Lambda_T = 2.7\lambda_a$. It is possible that "dirty droplets" nucleated on the smoke particles are likely to carry some molecules at their surfaces and thereby reduce the values of the condensation and thermal accommodation coefficients.(Personal communication 1 Oct. 1987, N. Fukuta, Meteorology Dept., U. of Utah, Salt Lake City, Utah.) This would reduce the growth rates, increase the maximum saturation achieved, and potentially affect the results of fraction nucleated.

We have chosen to use "average values," however, additional parametric studies and experimental data will guide further work in this area.

The term $S$ is a function of $\frac{da}{dt}$ so that Eq. (2) is an implicit function of the growth rate. For usual conditions of drop growth $\delta < 10^{-5}$, and is usually neglected. However, for small aerosols in the early stages of growth we observe $\delta$ to be more than two orders of magnitude larger so we retain the term and numerically solve Eq. (2) in its implicit form.

If the interstitial aerosol is at its equilibrium size with the current saturation ratio, temperature, etc., $da/dt = 0$ and Eq. (2) becomes:

$$s \equiv lnS = \frac{A}{a} - Be_m \frac{r^3}{a^3 - r^3} \tag{3}$$

where A and B are combinations of coefficients. For nonzero $e_m$, the solution of these Köhler curves at $ds/da = 0$ and supersaturation, $s > 0$, gives the critical radius, $a_c$:

$$a_c = -3A/(2Be_m - 6s) \text{ if } s = Be_m \tag{4a}$$

$$a_c = -D/2 + sqrt(D^2/4 - E) \text{ if } s \neq Be_m \tag{4b}$$

where $D = 2A(Be_m - 3s)/3s(s - Be_m)$ and $E = A^2/s(s - Be_m)$.

If the aerosol is insoluble ($e_m = 0$) the Köhler curves have no maximum and we set the critical radius to

$$a_c = s/A \text{ for } s > 0. \tag{4c}$$

Whenever the wet radius is greater or equal to the critical radius, the aerosol is nucleated (activated) to a drop. If the drop wet radius becomes less than $a_c$, the drop is deactivated, that is, it is moved from the drop distribution to the interstitial aerosol.

Aerosol coagulation is computed in the usual manner with the collection kernels evaluated,using the current values of temperature, pressure, etc., and the wet radii of the colliding particles.

The numerical algorithms of the mathematical model are programmed using finite difference techniques. In particular, the distribution arrays describe quantities whose values represent an average over a bin with center defined by:

$$r_j = r_1 2^{(j-1)/3J_d} \text{ and } a_i = a_1 2^{(i-1)/3J_d} \tag{5}$$

for $j = 1, 2, \ldots, J$; $i = 1, 2, \ldots, I$; and a mass doubling parameter $J_d$ taken to be 2 for the calculations reported in this paper. All numerical algorithms are derived and implemented such that aerosol mass, water vapor and liquid mass, and number density are conserved exactly. Of course, in coagulation the number density changes whenever two particles collide and form a single particle.

## 3. SIMULATIONS AND RESULTS

The simulations reported here are for a non-entraining parcel with initial saturation ratio, temperature, pressure, and height of 0.75, 290 K, 900 kPa, and 1000 m, respectively. The temperature and pressure are adjusted self-consistently with adiabatic expansion and condensation as the parcel rises with a prescribed constant updraft in a standard atmosphere. The initial aerosol number density distribution is lognormal with a geometric mean radius of 0.05 $\mu$m and a standard deviation of 2. Both the soluble and insoluble portions of the aerosol mass have densities of 1 g cm$^{-3}$ and the soluble fraction is characterized using $\nu = 3$, $M_s = 132$ and $\phi_s = 1$ (see Pruppacher and Klett, 1978). The aerosol coagulation due to turbulence is computed with a constant turbulent energy dissipation rate of 1000 $cm^2\ s^{-3}$. This value is approximately equal to the average value computed below cloud base over a high intensity fire (Penner et al., 1986).

The parameter space explored considers a range of solubility ratios, $e_m$, from 0 to 1, a range of initial number densities from $10^9$ to $10^{13}$ particles $m^{-3}$ of air, and a range of updraft speeds from 1 to 100 m $s^{-1}$.

Figures 1a through 3b show the saturation ratio as a function of height over these parameter ranges. As shown there the peak saturation ratio increases for decreasing soluble mass fraction, increasing updraft velocity, and decreasing number density. With the forced updraft the parcel remains slightly supersaturated throughout the time of interest and the curves essentially overlay. For example, in Fig. 1a, at 5000 m, the saturation ratio is about 1.0007, independent of $e_m$. Although not shown explicitly in these plots, for a constant updraft and high solubility ratio, enough particles may be available to take up the condensed water at a rate fast enough so that supersaturation

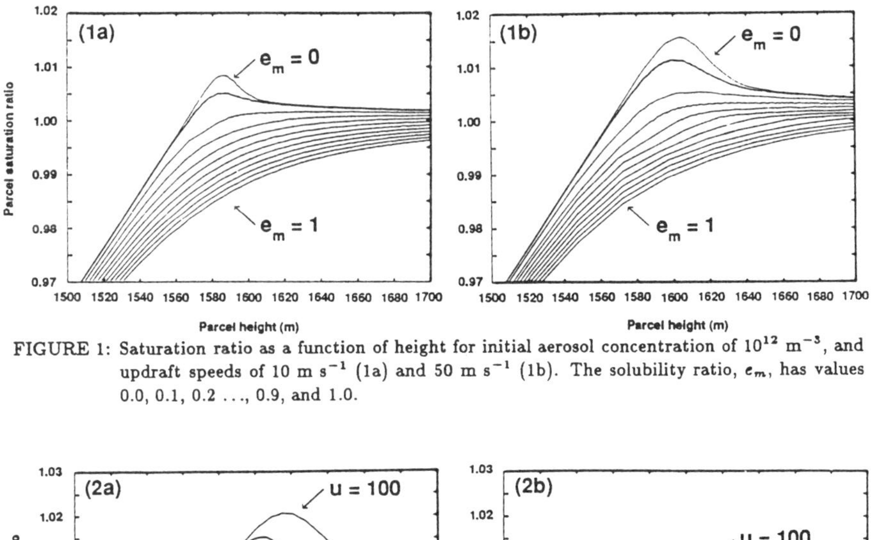


FIGURE 1: Saturation ratio as a function of height for initial aerosol concentration of $10^{12}$ $m^{-3}$, and updraft speeds of 10 m $s^{-1}$ (1a) and 50 m $s^{-1}$ (1b). The solubility ratio, $e_m$, has values 0.0, 0.1, 0.2 ..., 0.9, and 1.0.

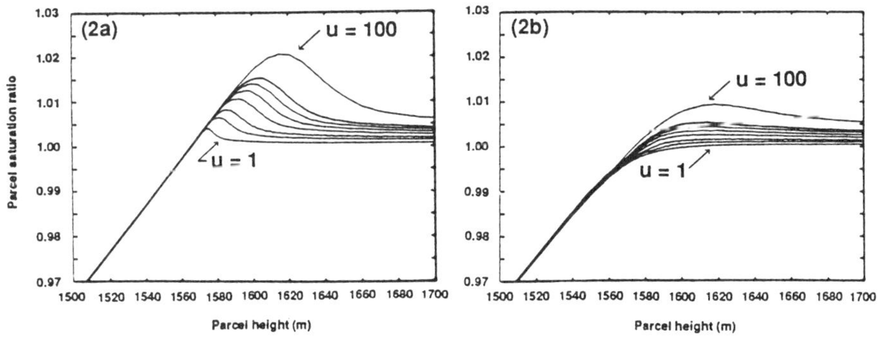


FIGURE 2: Saturation ratio as a function of height for initial aerosol concentration of $10^{12} m^{-3}$ and solubility ratios, $e_m$, of 0 (2a) and 0.01 (2b). The updraft speed, $u$, has values 1, 5, 10, 20, 30, 40, 50, and 100 m $s^{-1}$.

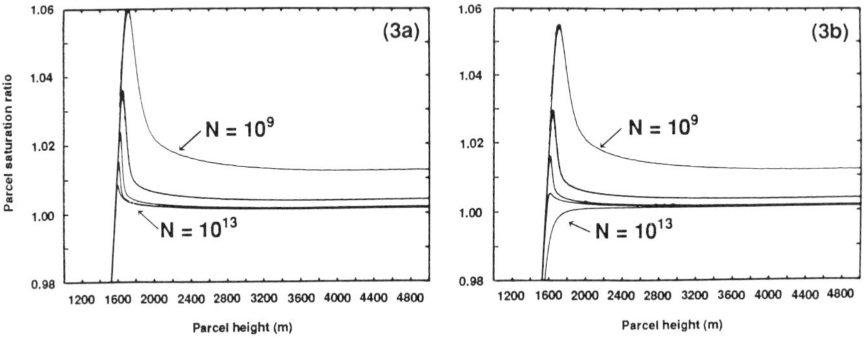


FIGURE 3: Saturation ratio as a function of height for updraft speed of 50 m $s^{-1}$ and solubility ratio, $e_m$, of 0. (3a) and 0.1 (3b). The initial aerosol concentration has values of $10^9$, $10^{10}$, $10^{11}$, $10^{12}$, and $10^{13} m^{-3}$.

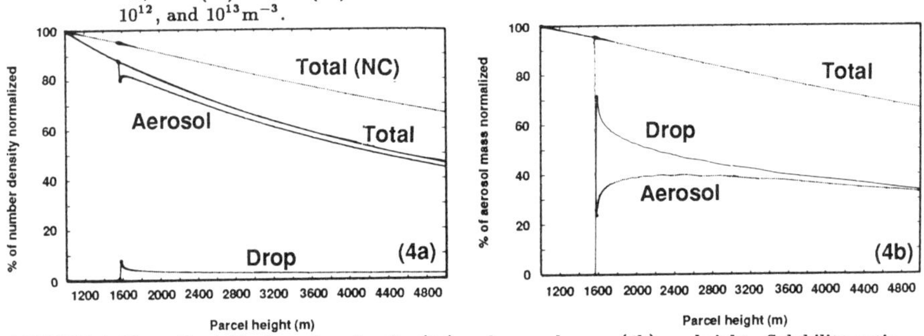


FIGURE 4: Normalized % of number density (4a) and aerosol mass (4b) vs. height. Solubility ratio $e_m = 0.0$. Initial $10^{12}$ particles $m^{-3}$ air. Updraft = 10 m $s^{-1}$.

is not reached. In such a case there would be no nucleation in the sense of Eq. (4) even though the wet radius size of aerosols may be of the order of 10 $\mu$m or larger.

Figures 4a, 5a, 6a, and 7a show curves of number density percentages normalized to the initial number density. As the parcel moves up with height, the effect of parcel expansion causes a decrease in the total number density. The curve labeled TOTAL(NC) is where the total would be if there were no coagulation. The difference between this curve and the one labeled TOTAL indicates the change in number density due to coagulation. The curves labeled AEROSOL and DROPS show the split of those particles that remain as interstitial aerosol and those incorporated in drops. Figures 4b, 5b, 6b, and 7b are similar curves representing the aerosol mass percentages. Since mass is conserved during coagulation, the only change in the total mass is that due to parcel expansion.

Once supersaturation is reached aerosol particles activate to drops until the peak saturation ratio is reached. As the saturation ratio falls, those drops that do not grow at a rate faster than the rate of increase of critical radius are deactivated, that is, returned to the interstitial aerosol distribution. This effect is shown in the curves of Figs. 4 through 7.

Figures 8a through 11b are representative plots of number density and aerosol mass distribution functions as a function of the wet radius. These plots are for 5000 m, that is, after 4000 m of parcel rise. The "separation" of the two distributions for

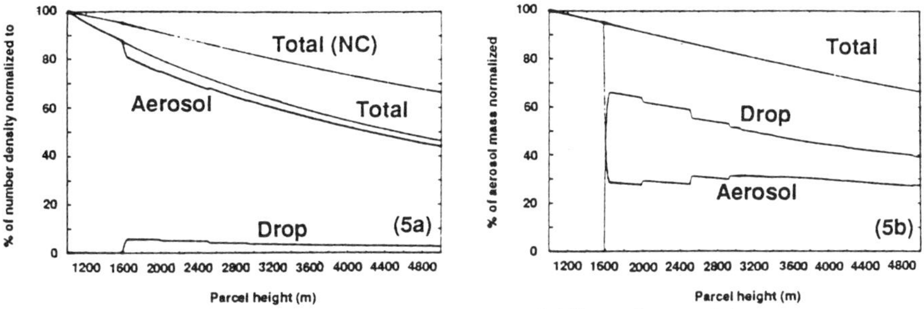


FIGURE 5: Same as Figure 4 but for solubility ratio $\epsilon_m = 0.1$.

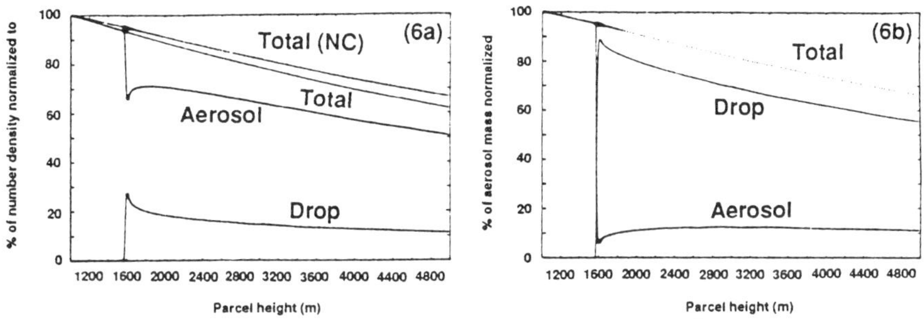


FIGURE 6: Same as Figure 4 but updraft $= 50 \text{ m s}^{-1}$.

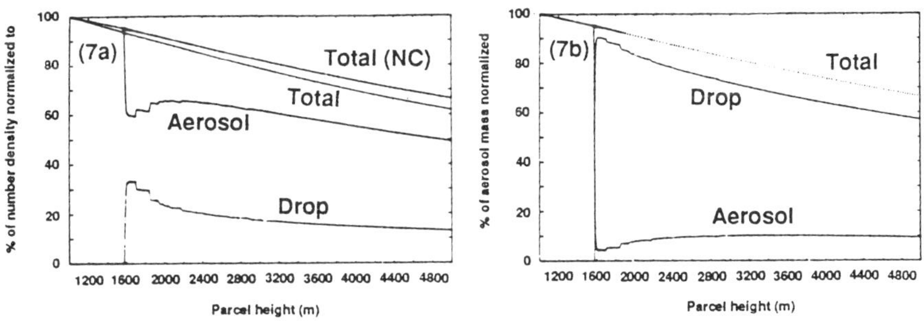


FIGURE 7: Same as Figure 6 except for solubility ratio $\epsilon_m = 0.1$.

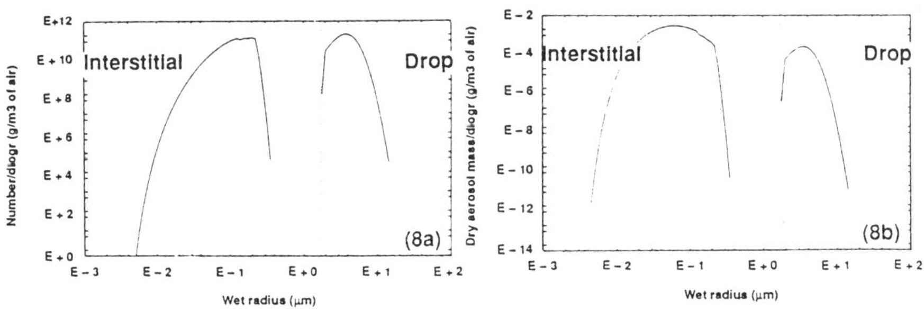


FIGURE 8: Interstitial aerosol and drop distributions vs. wet radius. Number density (8a) and aerosol mass density (8b) for parcel at 5000 m. Solubility ratio $\epsilon_m = 0.0$. The initial aerosol concentration was $10^{12}$ particles $m^{-3}$.

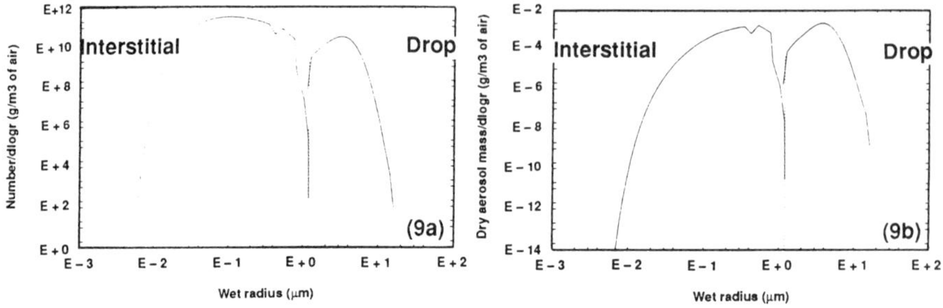


FIGURE 9: Same as Figure 8 except solubility ratio $e_m = 0.1$.

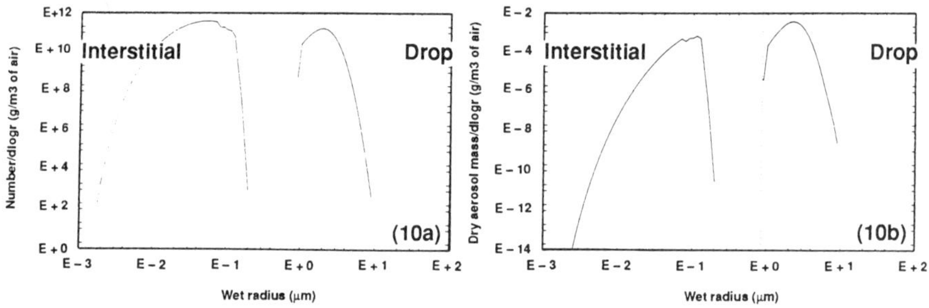


FIGURE 10: Same as Figure 8 except updraft $= 50$ m s$^{-1}$.

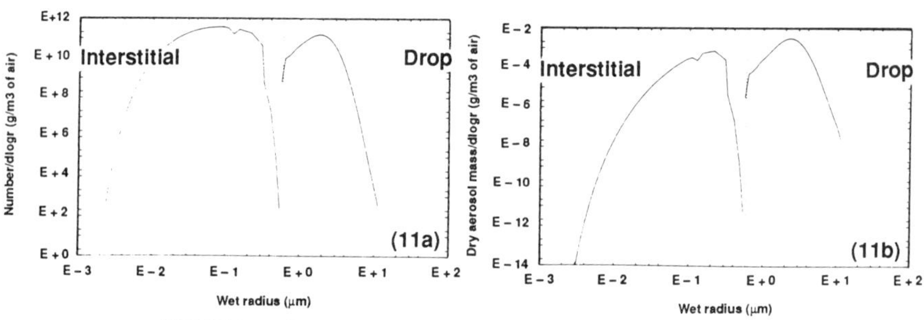


FIGURE 11: Same as Figure 10 except solubility ratio $e_m = 0.1$.

insoluble aerosols is distinct. In this case the interstitial aerosol does not grow due to condensation, whereas, once activated, the drops grow at a substantial rate. This separation is not very pronounced even at a soluble mass fraction of 0.1 since the aerosol does contain condensed water. Given enough time the drops continue to grow and produce a separation because the aerosols will have achieved and maintain an equilibrium size.

Due to the finite bin structure, as drops grow in size the resolution is decreased. Further, the condensation growth rate for small drops is much faster than that for larger drops, resulting in the combination of several average nucleus sizes into a single average for a given drop wet radius category. Evaporation and/or increase in critical radius then causes the deactivation of smaller drops. These effects combine to result in the perturbations visible on the interstitial aerosol distributions shown in Figs. 8 through 11. The perturbations indicate those bins where the aerosol has been activated resulting in "average" nucleus sizes, then deactivated to the interstitial aerosol array. Note that these aerosols tend to be in the 0.1 to 1.0 $\mu$m range (wet radius).

Tables 1 through 6[2] summarize the percentages of mass and number of interstitial aerosol in the parcel at 5000 m. These percentages are not normalized to initial values as in Figs. 4 through 7. Statistical characteristics of the interstitial aerosol number distribution as a function of wet radius are also listed.

TABLE 1: INTERSTITIAL AEROSOL PROPERTIES AS A FUNCTION OF $e_m$; $u = 10$ m s$^{-1}$, $N_i = 10^{12}$

| $e_m$ | % mass | % nr | $d\%_{coag}$ | $r_{max}$ | $a_{max}$ | Mode | Median | Mean | Std. Dev. |
|---|---|---|---|---|---|---|---|---|---|
| 0.0 | 49 | 96 | −30 | 0.35 | 0.35 | 0.063 | 0.063 | 0.062 | 1.75 |
| 0.01 | 48 | 96 | −30 | 0.35 | 0.53 | 0.067 | 0.067 | 0.067 | 1.81 |
| 0.1 | 41 | 95 | −30 | 0.31 | 1.19 | 0.104 | 0.104 | 0.104 | 2.07 |
| 0.5 | 30 | 92 | −27 | 0.22 | 1.48 | 0.214 | 0.179 | 0.193 | 2.20 |
| 0.8 | 29 | 92 | −26 | 0.22 | 1.72 | 0.225 | 0.225 | 0.240 | 2.25 |
| 1.0 | 29 | 92 | −26 | 0.22 | 1.86 | 0.252 | 0.252 | 0.265 | 2.26 |

TABLE 2: INTERSTITIAL AEROSOL PROPERTIES AS A FUNCTION OF $e_m$; $u = 50$m s$^{-1}$, $N_i = 10^{12}$

| $e_m$ | % mass | % nr | $d\%_{coag}$ | $r_{max}$ | $a_{max}$ | Mode | Median | Mean | Std. Dev. |
|---|---|---|---|---|---|---|---|---|---|
| 0.0 | 16 | 82 | -6.9 | 0.20 | 0.20 | 0.056 | 0.044 | 0.043 | 1.72 |
| 0.01 | 16 | 81 | -6.8 | 0.20 | 0.25 | 0.059 | 0.046 | 0.046 | 1.77 |
| 0.1 | 14 | 79 | -6.8 | 0.18 | 0.54 | 0.077 | 0.066 | 0.065 | 2.00 |
| 0.5 | 9 | 76 | -5.9 | 0.12 | 0.64 | 0.157 | 0.130 | 0.116 | 2.10 |
| 0.8 | 9 | 76 | -5.8 | 0.12 | 0.75 | 0.201 | 0.166 | 0.149 | 2.19 |
| 1.0 | 9 | 76 | -5.7 | 0.11 | 0.80 | 0.227 | 0.187 | 0.167 | 2.22 |

TABLE 3: INTERSTITIAL AEROSOL PROPERTIES AS A FUNCTION OF $u_i$; $e_m = 0$, $N_i = 10^{12}$

| $u$ | % mass | % nr | $d\%_{coag}$ | $r_{max}$ | $a_{max}$ | Mode | Median | Mean | Std. Dev. |
|---|---|---|---|---|---|---|---|---|---|
| 1 | 81 | 99 | -76 | 0.71 | 0.71 | 0.111 | 0.111 | 0.112 | 1.60 |
| 10 | 51 | 96 | -30 | 0.35 | 0.35 | 0.063 | 0.063 | 0.062 | 1.75 |
| 20 | 34 | 92 | -18 | 0.28 | 0.28 | 0.056 | 0.056 | 0.053 | 1.75 |
| 30 | 26 | 89 | -12 | 0.25 | 0.25 | 0.056 | 0.050 | 0.049 | 1.74 |
| 40 | 20 | 85 | -9 | 0.22 | 0.22 | 0.056 | 0.050 | 0.046 | 1.73 |
| 50 | 16 | 82 | -7 | 0.20 | 0.20 | 0.056 | 0.044 | 0.043 | 1.71 |
| 100 | 5 | 63 | -3 | 0.12 | 0.13 | 0.050 | 0.039 | 0.036 | 1.63 |

[2] In the tables, the % mass column is the percentage of aerosol mass remaining in the interstitial aerosol, % nr is that of number density. The $d\%_{coag}$ reflects the change from initial number density due to coagulation. The $r_{max}$ and $a_{max}$ columns are the maximum dry and wet radii ($\mu$m) of the interstitial aerosol number distribution. Mode and Median are the computed mode and median wet radius ($\mu$m) of the interstitial aerosol number distribution, and Mean and Std. Dev. are the computed geometric mean radius ($\mu$m) and geometric standard deviation. The labels $e_m$, $u$, $N_i$ are for soluble aerosol mass fraction, updraft velocity (m s$^{-1}$), and initial aerosol concentration per m$^3$, respectively.

TABLE 4: INTERSTITIAL AEROSOL PROPERTIES AS A FUNCTION OF $u$; $e_m = 0.1$, $N_i = 10^{12}$

| $u$ | % mass | % nr | $d\%_{coag}$ | $r_{max}$ | $a_{max}$ | Mode | Median | Mean | Std. Dev. |
|---|---|---|---|---|---|---|---|---|---|
| 1 | 74 | 99 | −74 | 0.56 | 3.07 | 0.231 | 0.231 | 0.229 | 1.93 |
| 10 | 42 | 95 | −30 | 0.31 | 1.18 | 0.104 | 0.104 | 0.105 | 2.07 |
| 20 | 29 | 91 | −18 | 0.25 | 0.87 | 0.089 | 0.089 | 0.086 | 2.08 |
| 30 | 22 | 87 | −12 | 0.20 | 0.69 | 0.090 | 0.077 | 0.077 | 2.06 |
| 40 | 17 | 83 | −9 | 0.18 | 0.58 | 0.090 | 0.066 | 0.070 | 2.03 |
| 50 | 14 | 79 | −7 | 0.18 | 0.54 | 0.077 | 0.066 | 0.065 | 2.00 |
| 100 | 5 | 62 | −3 | 0.11 | 0.32 | 0.057 | 0.049 | 0.050 | 1.88 |

TABLE 5: INTERSTITIAL AEROSOL PROPERTIES AS A FUNCTION OF $n_i$; $e_m = 0$, $U = 50$ m s$^{-1}$

| $Ni$ | % mass | % nr | $d\%_{coag}$ | $r_{max}$ | $a_{max}$ | Mode | Median | Mean | Std. Dev. |
|---|---|---|---|---|---|---|---|---|---|
| $10^9$ | 0.02 | 7.6 | −0.002 | 0.02 | 0.02 | 0.018 | 0.014 | 0.014 | 1.31 |
| $10^{10}$ | 0.22 | 22.3 | −0.023 | 0.04 | 0.04 | 0.028 | 0.022 | 0.020 | 1.39 |
| $10^{11}$ | 1.49 | 46.2 | −0.37 | 0.07 | 0.07 | 0.044 | 0.031 | 0.028 | 1.50 |
| $10^{12}$ | 16.4 | 81.9 | −6.9 | 0.20 | 0.20 | 0.057 | 0.044 | 0.044 | 1.72 |
| $10^{13}$ | 53.3 | 96.3 | −44.6 | 0.40 | 0.40 | 0.070 | 0.070 | 0.071 | 1.69 |

TABLE 6: INTERSTITIAL AEROSOL PROPERTIES AS A FUNCTION OF $N_i$; $e_m = 0.1$, $U = 50$ m s$^{-1}$

| $Ni$ | % mass | % nr | $d\%_{coag}$ | $r_{max}$ | $a_{max}$ | Mode | Median | Mean | Std. Dev. |
|---|---|---|---|---|---|---|---|---|---|
| $10^9$ | 0.001 | 1.2 | −0.002 | 0.01 | 0.01 | 0.011 | 0.011 | 0.010 | 1.26 |
| $10^{10}$ | 0.012 | 5.5 | −0.02 | 0.02 | 0.02 | 0.018 | 0.016 | 0.014 | 1.33 |
| $10^{11}$ | 0.26 | 22.6 | −0.22 | 0.06 | 0.31 | 0.032 | 0.028 | 0.025 | 1.50 |
| $10^{12}$ | 13.8 | 79.2 | −6.8 | 0.18 | 0.54 | 0.077 | 0.066 | 0.065 | 2.00 |
| $10^{13}$ | 35.1 | 92.5 | −43.4 | 0.25 | 0.78 | 0.125 | 0.125 | 0.126 | 1.99 |

As shown in the tables, the aerosol mass fraction remaining in the interstitial aerosol ranges from near 0 to 81%. The aerosol number fraction remaining in the interstitial aerosol ranges from 1 to 99%. These ranges are very sensitive to initial number density and updraft velocity. They are mildly sensitive to fraction of soluble material in the aerosol. The size distribution of the wet interstitial aerosol is very sensitive to aerosol soluble mass fraction. This is primarily due to the amount of water on the aerosol particles, which varies with $e_m$. Conversely, although not included in the tables, the aerosol number distribution statistics computed on the dry radius are not very sensitive to the solubility ratio. For example, with an initial concentration of $10^{12}$ particles per m$^3$ and updraft of 50 m s$^{-1}$, for $e_m = 0$, 0.1, and 1.0, the mode radius is 0.056, 0.050, and 0.050 $\mu$m; the median radius is 0.044 in all cases; the geometric mean radius is 0.044, 0.043, and 0.040 $\mu$m; and the standard deviation is 1.72, 1.68, and 1.60, respectively.

## 4. CONCLUSIONS

We have presented a parameter study of aerosol nucleation to drops over a range of conditions that include those expected over fires initiated by nuclear blast. We find that the saturation ratio reaches a steady value for forced updraft velocity. At this saturation, the fractions of aerosol mass and number nucleated are mildly sensitive to the fraction of soluble material in the aerosol and highly sensitive to the initial particle loading and updraft speed. The number of particles that nucleate at peak supersaturation and do not grow at a rate exceeding the rate of increase of the critical radius is quite sensitive to solubility. Thus, we conclude that the amount of aerosol that is affected by condensing water is very sensitive to the affinity of the aerosol to water.

For updraft velocities near those expected above high intensity fires, i.e., 50 m $s^{-1}$, in all cases studied, less than 55% of the aerosol mass remained as interstitial aerosol independent of the aerosol characteristics. Thus, we predict that a substantial fraction of smoke mass could be incorporated into drops. Computations (Malone et al., 1986) estimate a surface temperature decrease of about 15 K over large regions for 170 Tg of smoke injected, 5–10 K for 60 Tg and 4–6 K for 20 Tg. If the indicated amounts of smoke incorporated into drops by nucleation were removed by precipitation or the optical properties significantly altered, one might hypothesize a 4–5 K change in surface temperature due to these processes. More realistic conditions will be explored with our trajectory model (Penner and Edwards, 1986) and as experimental results become available.

This model considers only spherical particles and drops, so that a droplet consists of a spherical nucleus surrounded by a spherical shell of water. Also, we have only considered a single size distribution defined as lognormal with geometric mean dry radius of 0.05 μm and a standard deviation of 2. Further, we have neglected drop coalescence, aerosol impaction scavenging, entrainment, rain-out, ice processes, and electrical effects, as well as differing values of coefficients in the growth equation, any or all of which could produce large effects on the aerosol mass and number that might survive cloud processes and remain in the upper atmosphere for substantial time periods. These topics will be the subject of future research.

## ACKNOWLEDGMENTS

This work was performed under the auspices of the U.S. Department of Energy by the Lawrence Livermore National Laboratory under contract No. W-7405-Eng-48. Partial support for this research was provided by the Defense Nuclear Agency.

## REFERENCES

Crutzen, P.J., and J.W. Birks, 1982: The Atmosphere After a Nuclear War: Twilight at Noon, *Ambio 11,* 114–125.

Flossmann, A.I., W.D. Hall and H.R. Pruppacher, 1985: A Theoretical Study of the Wet Removal of Atmospheric Pollutants. Part I: The Redistribution of Aerosol Particles Captured Through Nucleation and Impaction Scavenging by Growing Cloud Drops, *J. Atmos. Sci., 42,* 583–606.

Fukuta, N., and L.A. Walter, 1970: Kinetics of Hydrometer Growth from a Vapor-Spherical Model, *J. Atmos. Sci. 27,* 1160–1172.

Hallett, J., B. Gardiner, J. Hudson and F. Rogers, 1986: Cloud Condensation and Ice Nucleation of a Range of Carbonaceous Aerosols, *Proc. 23rd Conf. on Radar Meteor. and Conf. on Cloud Phys.,* Am. Meteor. Soc., Snowmass, Colo., c9–c12.

Hobbs, P.V., L.F. Radke and D.A. Hegg, 1984: Some Aerosol and Cloud Physics Aspects of the "Nuclear Winter" Scenario, *Proc. ICSU-SCOPE Workshop, Supplemental Vol. to Proc. 9th Int'l Cloud Phys. Conf.,* Tallinin, Estonia, USSR.

Malone, R.L., L.H. Auer, G.A. Glatzmaier, M.C. Wood and O.B. Toon, 1986: Nuclear Winter: Three-Dimensional Simulations including Interactive Transport, Scavenging and Solar Heating of Smoke, *J. Geophys. Res. 91, 1039–1053.*

Mozurkewich, M., 1986: Aerosol Growth and the Condensation Coefficient for Water: A Review, *Aerosol Sci. Technol. S.* 233–236.

National Research Council (NRC), 1985: *The Effects on the Atmosphere of a Major Nuclear Exchange,* National Academy Press, Washington, D.C., 193 pp.

Penner, J.E., 1986: Uncertainties in the Smoke Source Term for "Nuclear Winter" Studies, ***Nature 324***, 222–226.

Penner, J.E., and L.L. Edwards, 1986: Nucleation Scavenging of Smoke and Aerosol Particles in Convective Updrafts, ***Proc. 23rd Conf. on Radar Meteor. and Conf. on Cloud Phys.***, Am. Meteor. Soc., Snowmass, Colo., c83–c86.

Penner, J.E., L.C. Haselman, Jr., and L.L. Edwards, 1986: Smoke Plume Distribution Over Large Scale Fires: Implications for Simulations of "Nuclear Winter," ***J. Climate and Appl. Meteor., 25***, 1434–1444.

Pittock, A.B., T.P. Ackerman, P.J. Crutzen, M.C. MacCracken, C.S. Shapiro and R.P. Turco, 1986: ***Environmental Consequences of Nuclear War***, Vol. 1, John Wiley and Sons, Chichester.

Pruppacher, H.R., and J.D. Klett, 1978: ***Microphysics of Clouds and Precipitation***, D. Reidel Pub. Co., Boston, 714pp.

Pruppacher, H.R., 1985: On the Scavenging of Aerosol Particles by Atmospheric Clouds and Precipitation. ***International Seminar on Nuclear War, 5th Session: SDI, Computer Simulations, New Proposals to Stop the Arms Race***, held at the "E Majorana" Centre for Scientific Culture, Italy, August 19–24, 1985., 163–174.

Radke, L.F., B. Benech, J. Dessens, M.W. Eltgrowth, V. Henvuon, P.V. Hobbs and M. Ribon, 1980: Modifications of Cloud Microphysics by a 1000 MW Source of Heat and Aerosol (the Meteotron Project), ***Proc. Third WMO Conf. on Weather Modification***, World Meteor. Org., Geneva, Switz., 119–126.

Turco, R.P., O.B. Toon, T.P. Ackerman, J.B. Pollack and C. Sagan, 1983: Nuclear Winter: Global Consequences of Multiple Nuclear Explosions, ***Science, 222***, 1283–1292.

# SURFACE TEMPERATURE EFFECTS OF FOREST FIRE SMOKE PLUMES[1]

*Alan Robock*
University of Maryland
College Park, Maryland 20742, USA

ABSTRACT

The location of smoke clouds from extensive Canadian forest fires was determined from satellite imagery for cases in the summers of 1981 and 1982. As these smoke clouds passed over the midwestern United States, surface air temperature effects were determined by comparing actual surface air temperatures with those forecast using Model Output Statistics (MOS) by the United States National Weather Service. A cooling of 1.5 °C to 4 °C was found in the daytime under the smoke plumes. No effect was found at night. This corresponds to theoretical estimates of the effects of an elevated smoke plume. It serves as observational confirmation of a portion of the nuclear winter theory.

## 1. INTRODUCTION

Crutzen and Birks (1982) suggested that smoke from forest, urban, and industrial fires ignited by nuclear weapons would be extensive enough to block out significant amounts of sunlight. Subsequent works, such as Turco *et al.* (1983), National Academy of Sciences (1985), and Pittock *et al.* (1986), have pointed out that the smoke from urban and industrial fires (especially oil refineries) would probably be much more effective at preventing solar radiation from reaching the earth's surface than forest fire smoke after a large-scale nuclear war. With both urban and rural targets, not only would more urban smoke be generated, but its optical properties would make it more effective at blocking sunlight.

The effects of forest fire smoke are still important, however, for two reasons. First, a lot of forest fire smoke would still be generated in many nuclear war scenarios, especially those that include only nonurban military targets. The optical properties and surface temperature effects of this smoke are important parts of the study of nuclear winter. Second, it is useful to have some actual observations of the effects of smoke to compare to theoretical models of nuclear winter. Extensive urban and industrial smoke plumes are not readily available for study. Each year, however, forest fires are generated by lightning, and in some cases produce extensive smoke plumes. Ginsburg and Golitsyn (1986) presented observations of daytime surface cooling from extensive Siberian forest fires in 1915. Wexler (1950) presented anecdotal evidence of daytime cooling at the surface of 2 °C to 5 °C from "The Great Smoke Pall" caused by extensive Canadian forest fires in 1950. In this paper, two such cases in which smoke from western Canadian forest fires drifted across the midwestern United States are investigated to attempt to measure their surface temperature effects. It is believed that The Great Smoke Pall of 1950 and the Siberian fires of 1915 were more optically thick clouds than the ones investigated here.

## 2. DATA ANALYSIS

The location of smoke clouds from extensive Canadian forest fires was determined from satellite imagery for cases in the summers of 1981 and 1982. The smoke clouds were easy to detect with visible wavelength imagery (Figs. 1a, 2a and 3a), but were invisible in infrared

[1] This work was supported by NOAA grant NA87AA-D-CP003 from the National Climate Program Office, which was funded by the Defense Nuclear Agency.




ISBN 0-937194-11-5

(a)

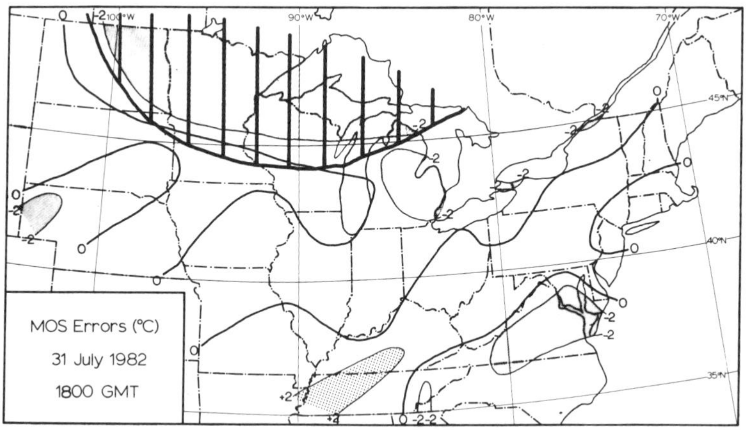


(b)

Figure 1. (a) GOES satellite image in visible wavelengths, 1800 GMT, 31 July 1982. (b) MOS surface air temperature forecast errors for same time. The smoke cloud location is indicated with heavy vertical lines.

(a)

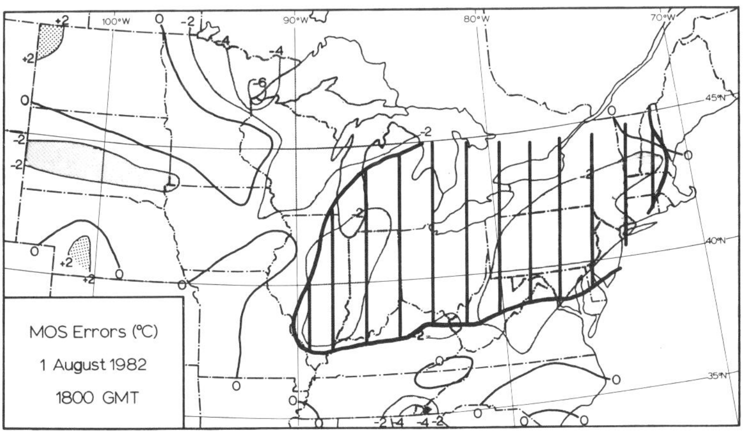


(b)

Figure 2. (a) GOES satellite image in visible wavelengths, 1800 GMT, 1 August 1982. (b) MOS surface air temperature forecast errors for same time. The smoke cloud location is indicated with heavy vertical lines.

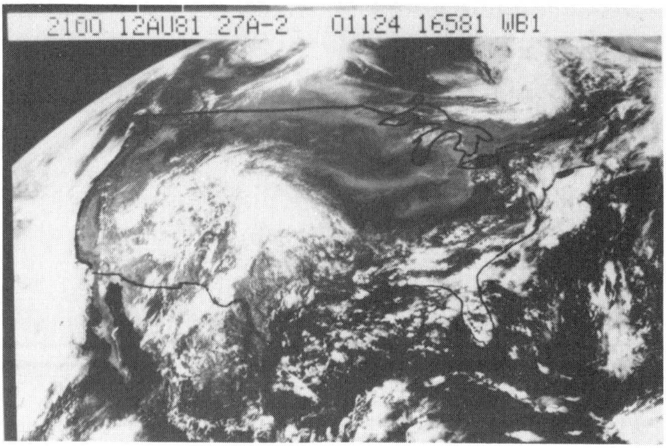


(a)

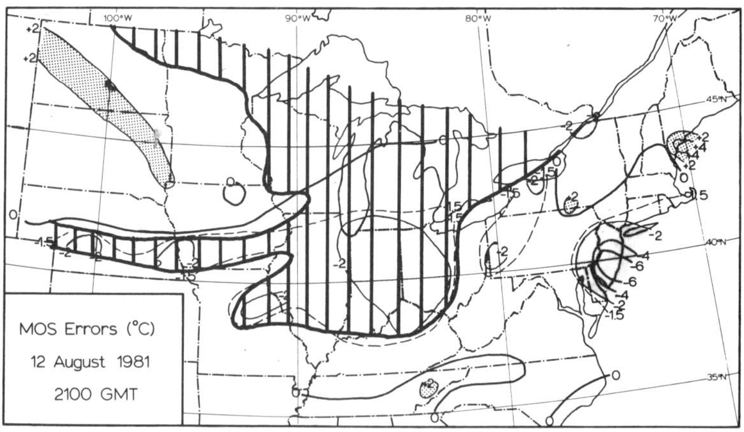


(b)

Figure 3. (a) GOES satellite image in visible wavelengths, 2100 GMT, 12 August 1981. (b) MOS surface air temperature forecast errors for same time. The smoke cloud location is indicated with heavy vertical lines.

imagery. Within 3 days of being produced, the smoke plumes in the 1982 case grew into an extensive shield covering about $10^6$ $km^2$. There was some patchiness evident (Figs. 1a and 2a), but the smoke veil maintained a coherent structure for several days. In the 1981 case (Fig. 3a), the smoke appeared to be in streaks, with patches of less dense smoke in between. The total area covered was about the same as the 1982 case.

As these smoke clouds passed over the midwestern United States, surface air temperature effects were determined by comparing actual temperatures with those forecast using Model Output Statistics (MOS) by the United States National Weather Service (Glahn and Lowry, 1972). This MOS error technique had been used successfully before, to determine surface air temperature effects of the Mount St. Helens volcanic eruption of 1980 (Robock and Mass, 1982). The analysis was done in regions of high pressure where synoptic disturbances were not affecting the temperature. The errors are attributed to the presence of aerosols in the atmosphere, since the aerosol content is not a MOS predictor. The locations of all the MOS stations used in this analysis are shown in Fig. 4.

Intense forest fires burned in British Columbia, Canada, on 29 July 1982 (Cowell, 1983). Smoke from these fires was transported by the prevailing winds, and crossed the U.S. border in North Dakota on 31 July. It proceeded across the Midwest and then east over the Mid-Atlantic states, and was reported over Western Europe on 5 and 6 August. Such was also the case for The Great Smoke Pall (Wexler, 1950) and demonstrates the ability of the atmosphere to transport smoke over long distances before it is removed.

Figures 1 and 2 show both visible wavelength GOES satellite images and MOS surface temperature forecast errors for 31 July and 1 August 1982 at 1800 GMT (1 p.m. CDT - local time in the Midwest) each day. Smoke plumes appear gray in these images, and can easily be detected over the Midwest. On the original images, the smoke plume that was headed for Europe can be seen on 1 August over the Atlantic Ocean. Figure 5 shows MOS forecast errors for 1 August 1982 at 0600 GMT (1 a.m. CDT).

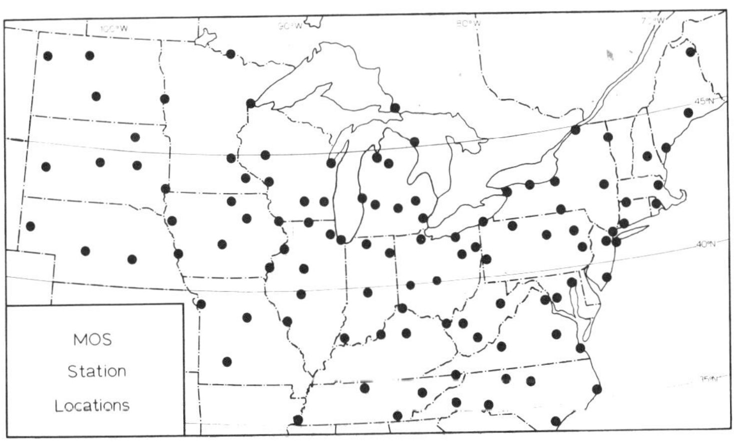


Figure 4. MOS station locations.

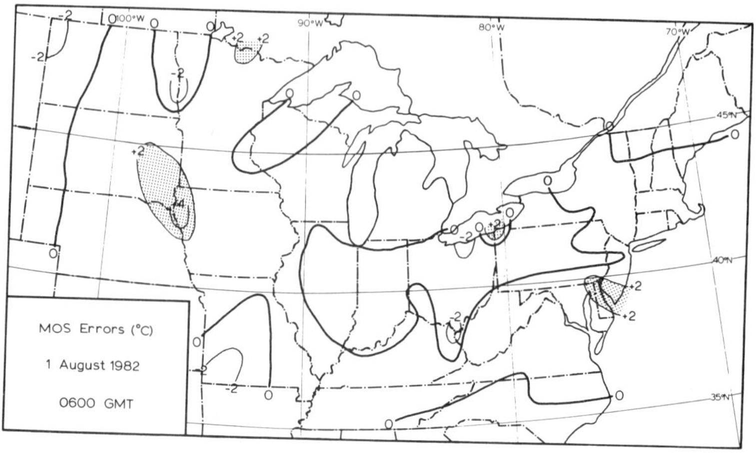


Figure 5. MOS surface air temperature errors, 0600 GMT, 1 August 1982. The smoke cloud location cannot be determined from satellite images. It is intermediate between those in Fig. 1 and Fig. 2.

Forecast errors of -2 °C to -4 °C are evident under the smoke plumes in Figs. 1b and 2b, and at 1500 and 2100 GMT (not shown here due to lack of space). Forecast errors are not evident under the smoke plumes at other times of the day. Since the errors are only evident at times of maximum daily insolation, their predominant effect on shortwave radiation is demonstrated.

Forecast errors of the same or greater amplitude are also evident in other locations in the figures. Upon close inspection, each can be attributed to the presence or absence of water clouds or mesoscale features that were not well forecast by the Limited-Area Fine Mesh (LFM) numerical forecasting model, which provides the predictors for the MOS forecasts. In clear areas, however, the only large errors are under the smoke cloud. For example, at 1500 GMT on 31 July (not shown), there is a negative area in eastern Michigan that can be attributed to an unforecast mesoscale region of cloudiness. A positive error region in western Tennessee and Kentucky is associated with a mesoscale clear area in an otherwise overcast region.

In Fig. 1 for 1800 GMT, 31 July, the same error regions are present as at 1500 GMT, except that an additional negative area appears in Virginia under a heavy water cloud bank, and small negative areas appear east and west of Lake Ontario. These latter areas become more extensive and of higher amplitude three hours later (not shown). They are associated with strong cold advection behind a rapidly developing mesoscale low, the center of which is located on the Vermont-Canadian border. This is easily seen on the synoptic weather maps (not presented here) and presumably was not well forecast by the LFM, which does not handle small-scale weather systems well. In Fig. 1 and three hours later the moderating effects of Lakes Erie and Ontario are evident, as the maximum cooling associated with this cold advection is over the land areas between the lakes. The only other negative error area, and by far the largest one, is under the forest fire smoke veil. Areas of fair weather cumulus over Illinois, Indiana, and Ohio do not produce large MOS errors.

In Fig. 2, on 1 August during the daytime, the only large error regions (both negative) shown are associated with a small region of water clouds over Wisconsin and the forest fire smoke veil. The region around the Great Lakes under the smoke veil has errors between -1 °C and -2 °C, evidence again of a moderating effect of the lakes. The thickest patch of smoke appears over Baltimore, Maryland, at 2100 GMT, and has an error of -4.1 °C associated with it.

The results presented here are from forecasts made from LFM runs 12 to 24 hours before the forecast time. Comparisons were made with forecasts made with earlier and later model runs, and the results were virtually the same. The errors (cooling) found were much larger than the run-to-run differences.

Attempts were made to estimate the height of the smoke as it passed over the Midwest. Winds were examined at different levels during this time and compared to the motion of the smoke. The winds corresponding most closely to the motion were at 700 mb (3.2 km). The jet of smoke headed to Europe on 1 August corresponded to winds between the 700 mb and 500 mb levels (3 km to 5.5 km).

A vertical profile study of lower level haze was conducted by chance on the next day over the eastern shore of Chesapeake Bay, due east of Baltimore at 1700 GMT (personal communication, 1 April 1987, Edward Eloranta, University of Wisconsin, Madison, Wisconsin). The portion of the smoke veil that was over Kentucky on 1 August appears from satellite images to have been over Maryland at that time. Eloranta flew up into the base of the forest fire smoke plume at an altitude of approximately 3.5 km and at 4.5 km (the highest altitude reached) he was still in the smoke.

A preliminary analysis of the 12 August 1981 case (Fig. 3) also shows daytime cooling under the smoke. In this case, the smoke is not as uniform, but appears in streaks. The cooling in this case is -1.5 °C to -3 °C. An examination of the satellite images from this case shows shorter shadows at sunrise and cumuliform clouds casting shadows on the smoke later in the day. These both suggest that the smoke was at a lower elevation on these days.

## 3. DISCUSSION

A cooling of 1.5° C to 4° C was found in the daytime under forest fire smoke plumes in two cases in 1981 and 1982. No effect was found at night. This corresponds to theoretical estimates of the effects of an elevated smoke plume (Ginsburg and Golitsyn, 1986) with optical depth of approximately 2. The optical depth of this smoke cloud, based on multispectral satellite measurements, has been reported to be 2 to 3 (personal communication, 30 March 1987, Richard Ferrare, NASA Goddard Space Flight Center, Greenbelt, Maryland).

This result serves as observational confirmation of a portion of the nuclear winter theory. Smoke from burning cities and industries, which could result from a large-scale nuclear war, would be a better absorber of shortwave radiation than forest fire smoke, and should therefore produce even larger surface temperature drops. The relative amounts of forest fire and city fire smoke resulting from a nuclear war would be scenario-dependent.

## ACKNOWLEDGMENTS

I thank Rich Ferrare for valuable discussions of the 1982 case and for providing preliminary satellite photos, Ed Eloranta for providing in situ smoke cloud data taken 2 August 1982 over Maryland, Mike Matson and Will Gould for information on availability of satellite images of smoke, Paul Dallavalle for MOS forecasts and surface temperature observations, Sunny Bae for plotting the maps, Marco Rodriguez for drafting the figures, and Sherri West for proofreading the manuscript.

## REFERENCES

Cowell, D., 1983: The Roar of the Fire, Forestalk, 7, No. 2, 14-20.

Crutzen, P. J., and J. W. Birks, 1982: The Atmosphere After a Nuclear War: Twilight at Noon, Ambio, 11, 115-125.

Ginsburg, A. S., and G. S. Golitsyn, 1986: Comparative Analysis of Mass Forest and "Nuclear" Fires. Paper presented at the Second All-Union Conference of Soviet Scientists for Peace and Against Nuclear Threat, Moscow, 27-29 May 1986.

Glahn, H. R., and D. A. Lowry, 1972: The Use of Model Output Statistics (MOS) in Objective Weather Forecasting, J. Appl. Meteor., 11, 1203-1211.

National Academy of Sciences, 1985: The Effects on the Atmosphere of a Major Nuclear Exchange, NAS Press, Washington, D.C., 193 pp.

Pittock, A. B., T. P., Ackerman, P. J. Crutzen, M. C. MacCracken, C. S. Shapiro and R. P. Turco, 1986: Scientific Committee on Problems of the Environment 28, Environmental Consequences of Nuclear War, Vol. I: Physical and Atmospheric Effects, John Wiley & Sons, New York, 359 pp.

Robock, A., and C. Mass, 1982: The Mount St. Helens Volcanic Eruption of 18 May 1980: Large Short-Term Surface Temperature Effects, Science, 216, 628-630.

Turco, R. P., O. B. Toon, T. P. Ackerman, J. B. Pollack and C. Sagan, 1983: Nuclear Winter: Global Consequences of Multiple Nuclear Explosions, Science, 222, 1283-1292.

Wexler, H., 1950: The Great Smoke Pall — September 24-30, 1950, Weatherwise, December, 129-142.

# PERSISTENT EFFECTS OF RESIDUAL SMOKE LAYERS

*T.P. Ackerman,*[1] *R.P. Turco*[2] *and O.B. Toon*
NASA Ames Research Center
Moffett Field, California 94035, USA

## ABSTRACT

In this paper, we consider the climatic effects of residual smoke layers following a large-scale nuclear exchange. We estimate that the most plausible e-folding lifetime of stratospheric smoke following a nuclear exchange is about 6 months to 1 year. The climatic impact of this smoke is examined using a one-dimensional radiative-convective model. Our results suggest that surface temperature perturbations are likely to persist for at least as long as the smoke e-folding time. The sensitivity of these perturbations to various values of the smoke decay time and to infrared properties of the smoke layer is investigated.

## 1. INTRODUCTION

In our original paper on the climatic impact of a large-scale nuclear exchange (Turco et al., 1983), we assumed that the injected particulate matter would be removed from the atmosphere at a rate consistent with observed particle lifetimes. We also speculated that the residence time of the particles could be substantially increased by: 1) a reduction in precipitation scavenging and deposition in the stably-stratified post-exchange atmosphere, and 2) self-induced lofting due to solar absorption and heating. These speculations were confirmed in the three-dimensional general circulation model (GCM) simulations carried out by Malone et al. (1985, 1986). Using a 20-level version of the National Center for Atmospheric Research Community Climate Model coupled with an aerosol transport model, they showed that, in the case of a summer injection, by about 2 weeks after the smoke injection the region of significant precipitation was confined to below approximately 5 km. The estimated lifetime of the particles remaining after several months of simulated time was on the order of 6 months to 1 year. Furthermore, the smoke remaining after 40 days was sufficient to produce an absorption optical depth on the order of 0.2 if spread uniformly around the globe.

---

[1]Now affiliated with Department of Meteorology, Pennsylvania State University, University Park, PA 16802.

[2]R&D Associates, P.O. Box 9695, Marina del Rey, CA 90295.


A. DEEPAK Publishing
ISBN 0-937194-11-5

In this paper, we are concerned with the ultimate fate of that remaining aerosol. We first briefly review the possible removal mechanisms and timescales for the residual smoke particles. This is followed by an estimate of the climatic impact and the climate sensitivity based on radiative-convective model simulations.

## 2. REMOVAL MECHANISMS AND PARTICLE LIFETIMES

Smoke and dust particles injected or transported into an elevated, stable layer (which we will call the stratosphere, although it could be somewhat different in altitude and thermal structure than the current stratosphere) could remain in that layer for a long time. Removal would occur via one of three possible mechanisms: 1) coagulation followed by gravity-induced sedimentation into the troposphere, 2) large-scale downward transport into the troposphere, or 3) photochemical destruction of the smoke particles in place.

The fall velocity of spherical particles in air can be calculated in a straightforward manner. It is readily estimated that a sphere with a density of 2 $g\,cm^{-3}$ and a radius of 0.3 $\mu$m would take about 6 months to fall from 20 to 10 km. Since nuclear winter simulation studies suggest that the radius of a solid sphere containing a mass equivalent to that of a typical coagulated residual smoke particle would be less than 0.3 $\mu$m (Turco et al., 1983), only the largest smoke particles could be removed from an elevated debris layer by sedimentation on this timescale. On the one hand, in the size and altitude regime of interest, the particle fallspeed for a *fixed particle density* (but variable mass) varies in proportion to the particle radius, $r$ (Pruppacher and Klett, 1978). Thus the fallout time varies as $r^{-1}$. If the particle radius is doubled, the fallspeed will be doubled and the residence time will be halved. On the other hand, in this regime for spherically shaped particles with a *fixed particle mass* (but variable density), the fallspeed varies as $r^{-2}$ and the fallout time as $r^2$.

Several factors may act to increase nuclear-exchange generated smoke particle lifetimes against sedimentation. The sooty particles generated in intense urban fires would be nonspherical in shape and consist of agglomerated chains and fluffy balls of tiny spherules[2] (Pittock et al., 1986). Such particles fall more slowly than solid spheres with an equivalent mass due to their larger physical

[3]We differentiate between "smoke" and "soot" as follows. Smoke is all of the particulate matter emitted by combustion. It consists of basically transparent, refractory organic compounds mixed with highly absorbing elemental carbon particulates (and may also contain some liquid water). Soot is just the elemental carbon component of the smoke, consisting of complex, irregular aggregates of very small (about 50 to 100 nm radius) solid spherules. The relative fraction of smoke that is soot is much greater in the case of flaming combustion, particularly for fuel consisting of petroleum, plastics, and related compounds (Pittock et al., 1986; Turco et al., 1987).

size and lower effective density. (These combine to increase the drag force, which is responsible for the $r^{-2}$ dependence of the fallspeed mentioned above.) Soot aggregates are observed to have a physical size of 2 to 10 times the solid-carbon spherical equivalent size (*e. g.*, Medalia and Richards, 1972) and thus would fall much more slowly than solid spheres of the same mass. In addition, a spherical shape (which was used to arrive at the $r^{-2}$ dependence) minimizes drag, while irregularly shaped soot particles would experience more drag than an equivalent diameter sphere.

These factors, as well as uncertainties in particle size, introduce uncertainty in the sedimentation timescale. Considering all the information at hand, we conclude that 6 months is probably a minimal fallout timescale for residual soot particles; the timescale could easily approach a year or more for these particles in the lower stratosphere.

The lifetime of stratospheric smoke and dust particles due to large-scale transport processes is also uncertain. In the current atmosphere, there is little downward transport of stratospheric particles by small-scale mixing and convection, because of the high stability of the lower stratosphere. The major mechanisms for downward transport seem to be tropopause folding events (see, for example, Raatz et al., 1985) and subsidence in the polar vortex during the winter season (Kent et al., 1985). These processes and the transport associated with them are not completely understood, nor are they well-resolved in GCMs. Malone and Glatzmaier (1986) have compared model simulated particle lifetimes against observed lifetimes of radioactive tracers from nuclear tests and of volcanic aerosols, and found good agreement with observed lifetimes. Using this model to study smoke generated by a nuclear-exchange, Malone et al. (1986) reported lifetimes on the order of a year for the particles remaining in the atmosphere a month or two after injection. However, because stratosphere-troposphere exchange is highly parameterized in GCMs and tuned to current climatic conditions, this lifetime is uncertain and could be impacted by changes in the thermal structure of the atmosphere following a nuclear exchange. Model simulations show significant heating of the upper troposphere and lower stratosphere by the soot, which results in an enhanced atmospheric stability. This, in turn, could substantially alter exchange mechanisms. It seems plausible to expect an increase in particle lifetimes associated with increased stability, but further model simulations are warranted. As an added complexity, heating of the air by the particles and the probable subsequent lofting of the particles could lead to a significant increase in particle lifetimes.

Additional insight into the typical lifetimes of stratospheric particles can be obtained from the fate of particles injected by volcanic eruptions. Since the recent eruption of El Chichon in March/April of 1982 was fairly large and well documented, it provides a useful example. Following the fallout of large ash particles and the conversion of emitted $SO_2$ gas into $H_2SO_4$ particles, which occur during the first few months, the volcanic aerosol consists of essentially nonvolatile sulfuric acid droplets of about 0.3 $\mu$m radius, which are excellent tracers of stratospheric lifetimes. The best time-series of volcanic particle loading is provided by lidar measurements of particle backscatter. Following the eruption of El Chichon, measurements at a variety of sites in the tropics and midlatitudes (Hofmann, 1987) show that it took on the order of 18 months to 2 years for the integrated backscatter to decrease by an order of magnitude from its peak value. Three to 4 years after the peak value, the backscatter was still above the pre-eruption background. While backscatter is not strictly proportional to particle concentration, for our purposes it can be used as a proxy indicator. This slow decrease in particle concentration is confirmed by direct measurements made periodically over Wyoming and Texas with balloonborne particle counters (Hofmann and Rosen, 1987). As indicated above, the relevance of volcanic aerosol lifetimes to smoke particle lifetimes is debatable due to potential changes in circulation patterns and particle transport following large smoke injections. However, the volcanic analog does suggest that smoke particle lifetimes determined by large-scale transport processes in the stabilized upper atmosphere (the stratosphere, but not necessarily with the same altitude and thermal structure as the current stratosphere) may be on the order of a year or longer. This is consistent with the results reported by Malone et al. (1985, 1986).

In situ destruction of soot particles[3] by ozone has been examined by Stephens et al. (1986). One possible reaction mechanism is

$$\begin{aligned} C(s) + O_3(g) &\longrightarrow C \cdot O_3(s) \longrightarrow C \cdot O(s) + O_2(g) \\ C \cdot O(s) &\longrightarrow CO(g) \\ C \cdot O(s) + O_3(g) &\longrightarrow CO_2(g) + O_2(g) \end{aligned}$$

where *(s)* and *(g)* indicate surface (adsorbed) and gaseous components, respectively. The mechanism described by Stephens et al. (1986) has two distinct phases: 1) an initial phase in which ozone molecules react rapidly with the available active surface sites to form an adduct, $C \cdot O$, and 2) a second slower phase in which the reaction rate is limited principally by desorption or reaction of the adduct (the second and third steps in the mechanism above) to re-expose the active sites. The overall rate of reaction may be characterized by a parameter $\gamma$, which is the fraction of the total number of collisions of ozone molecules with the carbon surface that leads to a net reaction (steps 1 and 3 above). The rate of reaction is then calculated as the product of $\gamma$ and the ozone collision rate. The latter may be estimated using gas kinetic theory. (Although this definition of $\gamma$ greatly oversimplifies the surface physical and chemical processes

[4] The other transparent components of the smoke can also be chemically decomposed, probably at a faster rate than soot (Turco et al., 1987). Therefore, we will simply consider soot in the discussion on chemical lifetime.

at work, it provides a useful means to arrive at an order of magnitude estimate for the chemical rates under atmospheric conditions.)

The data of Stephens et al. (1986) indicate that at room temperature the initial value of $\gamma$ may be as large as $\sim 10^{-3}$, but that it rapidly decreases to a value of $\sim 10^{-5}$ as surface adduct accumulates. Hence, $\gamma \sim 10^{-5}$ is the appropriate value for atmospheric calculations of long-term soot removal at room temperature. The lifetime of a soot particle against ozone decomposition under such conditions can then be roughly estimated using the following relationship:

$$\tau_S \approx \frac{r\,\rho_C}{[O_3]\,v\,\gamma\,m_C} \quad (sec)$$

where $r$ is the equivalent particle spherical radius in cm, $\rho_C = 2\ g/cm^3$, $m_C$ is the mass of a carbon atom in grams, $[O_3]$ is the ozone concentration, and $v$ is the thermal speed of an ozone molecule (cm/sec). For a ratio of $r/\gamma \sim 1\ cm$ and an ozone concentration of $10^{12}/\ cm^3$, $\tau_S$ would be roughly 40 days. This time constant is consistent with observations of the changes in smoke optical extinction upon exposure to high ozone concentrations at room temperature (Stephens et al., 1986).

In a nuclear winter stratospheric smoke cloud, the soot lifetime against chemical decomposition is likely to be much longer for several reasons:

1. The soot lifetime estimated above applies to ambient temperatures of 300 – 330 K (Stephens et al., 1986). However, more recent laboratory measurements show that the soot/ozone reaction has a substantial activation energy (Stephens and Birks, 1987); at temperatures of 270 – 330 K (which have been projected for lofted smoke clouds – see Section 3.2), the rate of reaction would be more than an order of magnitude slower.

2. Unlike relatively fresh soot particles studied in the laboratory, nuclear winter soot that has been "aged" for several days in an atmospheric environment prior to exposure to stratospheric levels of ozone would have surfaces highly contaminated by a variety of chemically "inert" compounds such as sulfates, nitrates, chlorides, and organics present in the smoke. These compounds would compete with ozone to occupy active surface sites and thus would inhibit ozone reactions. The mass of material needed to "poison" all of the reactive sites on a soot particle is a relatively small fraction of the carbon mass of the particle. (Note that the collection of a substantial "foreign" mass on a soot particle by condensation or coagulation could increase its fallspeed and reduce its stratospheric residence time.)

3. In a perturbed environment, the concentration of stratospheric ozone might be considerably lower than the background value assumed above

due to ozone depletion by chemical reactions (Crutzen and Birks, 1982), stratospheric heating (Vupputuri, 1986), and large-scale transport (Pittock et al., 1986).

Accordingly, it is reasonable to project a soot chemical lifetime of a year or more in a nuclear perturbed stratosphere.

We have also considered the decomposition of soot through reactions with OH and O. Based on existing laboratory data, these processes are expected to be at least an order of magnitude less efficient than the ozone process.

From the above analysis, it appears that the lifetime of heated, self-lofted smoke produced by a large-scale nuclear exchange is uncertain, but may fall in the range of 6 to 12 months. Currently available data on chemical destruction suggest a minimum lifetime of several months to a year. However, there are a number of mitigating circumstances that probably act to increase the chemical lifetime appreciably. It is also possible, but we think unlikely, that some other, as yet unidentified, chemical mechanism exists that would destroy soot particles on a much shorter timescale. Gravitational sedimentation and large-scale transport are unlikely to remove aggregated smoke particles on timescales shorter than 6 months and could take much longer depending on the stability of the post-exchange atmosphere. Thus, we conclude that the most plausible value for the smoke lifetime is roughly 6 months to a year and that all three of the mechanisms discussed above probably contribute to particle removal on this timescale. Since the particle lifetimes are e-folding lifetimes, the effect of the smoke on solar energy deposition could remain significant for a period of about twice the lifetime.

## 3. RADIATIVE-CONVECTIVE MODEL SIMULATIONS

### 3.1 MODEL DESCRIPTION

Assessing the long-term climatic impact of a residual smoke layer is difficult. Obviously, because of the long timescales and possible surface feedback processes, one would like to use a general circulation model, ultimately one with a coupled ocean. While progress is being made in this direction (*e. g.*, Pittock et al., 1987; Ghan et al., 1987), only a limited number of simulations can be carried out due to the expense involved.

We have used a much simpler model, a 1-D radiative-convective model (RCM) with a land surface, to look at possible surface temperature responses and examine the sensitivities to some possible variations in input parameters. Clearly, the predicted changes in surface temperature in the RCM cannot be equated with expected actual changes in global land surface temperatures.

However, the predicted changes can be used to estimate the potential impact of a long-lasting, elevated smoke layer and the relative importance of various processes in determining that impact. We hope that such studies can provide useful guidelines for higher resolution models.

The model used here is basically the same as that employed in previous work (Turco et al., 1983; Pollack and Ackerman, 1983). It consists of an atmospheric column extending from the surface to 40 km divided into 2 km thick layers. The underlying land surface has essentially no heat capacity but is tightly coupled to the lowest atmospheric layer, which means that the thermal inertia of the surface is the thermal inertia of the lowest 2 km of the atmosphere, and that the land surface temperature is identical to the atmospheric temperature at the bottom of the lowest layer. The model assumes annual-average solar insolation and 50% cloud cover with fixed cloud height and cloud properties. Water vapor is distributed with a fixed relative humidity according to the Manabe and Wetherald (1967) prescription.

Solar radiative transfer is computed with a multilayer delta-four-stream algorithm (Cuzzi et al., 1982) that explicitly accounts for multiple scattering by particles and clouds and for Rayleigh scatter. Gaseous absorption is treated using exponential sum fits to laboratory data with an assumed pressure scaling factor. A total of 77 intervals are used to span the solar spectrum from 0.25 to 4.5 $\mu$m. Infrared radiation is computed using a band-model formulation for 20 spectral intervals based on the Rodgers and Walshaw (1966) grid and using their water vapor parameters. A modified $CO_2$ band formulation, an $O_3$ band model, and the Roberts et al. (1976) continuum have also been incorporated into the model. Particle absorption is included in the infrared by treating the particles as gray absorbers within each spectral interval but allowing the absorption coefficient to vary from interval to interval. Neglecting particle scatter in the infrared is a good approximation for typical atmospheric particles (radius $\leq$ 1.0 $\mu$m) but may not be adequate if the aerosols grow to supermicron sizes.

In order to simplify the model computations for this study, we approximated the smoke spectral extinction and absorption by scaling them with wavelength as $\lambda^{-n}$. The particle extinction and absorption optical depths are specified at 0.55 $\mu$m. Then, for both quantities, the ratio of the value at 0.55 $\mu$m to the value at 10.0 $\mu$m is specified, from which values of $n$ are derived separately for both extinction and absorption. These values of $n$ are then used to scale the optical depths to the necessary wavelength intervals for the model calculations. This particular approximation works well for carbon particles because carbon lacks significant spectral features. However, this scaling would be inappropriate for a material that exhibits strong absorption bands in either the solar or thermal infrared spectrum.

For our baseline smoke case, we adapted parameter values from Malone et al. (1986) at the end of their 40-day simulation. We distributed the smoke between 10 and 20 km with a uniform mixing ratio. The extinction and absorption optical depths at 0.55 $\mu$m are 0.40 and 0.14, respectively. The ratio of extinction optical depths at 0.55 and 10 $\mu$m is 10:1 and the absorption ratio is 5:1.

The water vapor algorithm in the model was modified to fix the absolute, rather than the relative humidity above 10 km. This was done to prevent large quantities of water vapor from appearing artificially in stratospheric regions that are warmed due to the absorption of solar radiation by smoke. Since the lower atmosphere cools in the model and a cold trap is maintained in the upper troposphere, there is no source for this increased water vapor. Hence, it is inconsistent to maintain a fixed relative humidity in the heated, elevated layer.

### 3.2 SIMULATIONS

Using the model described above, we carried out a number of simulations. In the first series, we assumed that the smoke optical depth decayed exponentially and uniformly at all altitudes. The decay constant (smoke lifetime) was set to 60 days, to 180 days, and to infinity (this last case corresponds to a fixed smoke optical depth). The results of these calculations are shown in Fig. 1 for days 0 to 300. For the fixed smoke case, the model equilibrates after about 200 days at a surface temperature of 281 K, which is about 5 K less than the unperturbed surface temperature. The long equilibration time and the partial recovery from the minimum temperature is due primarily to the long radiative timescales of the lower stratosphere. Initially the surface and lower troposphere cool quickly responding to the loss of solar radiation. The absolute humidity of the troposphere falls due to the cooler temperatures and a minimum surface temperature of about 277 K is reached. As the stratosphere warms, additional thermal radiation is emitted downward and heats the surface and troposphere. This increases the absolute humidity somewhat, further increasing the surface temperature. The final equilibrium temperature is less than in the unperturbed case because the additional downward thermal radiation cannot compensate for the loss of solar radiation. The slight overcompensation of the surface temperature in the case of a 60-day lifetime is similarly related to the radiative timescale of the stratosphere. While much of the smoke has been removed by day 100, the stratosphere is still warmer than normal and requires another 100 days or so to reestablish radiative equilibrium.

The curves in Fig. 1 suggest several plausible hypotheses about the long-term effects of smoke injections. All other things (*e. g.*, season, proximity to the ocean, surface heat capacity) being equal, the initial temperature decrease is most strongly controlled by the average optical depth during the first week to month. The length of time for return to normal is somewhat longer than the e-folding decay time of the smoke optical depth. This reinforces the conclusion

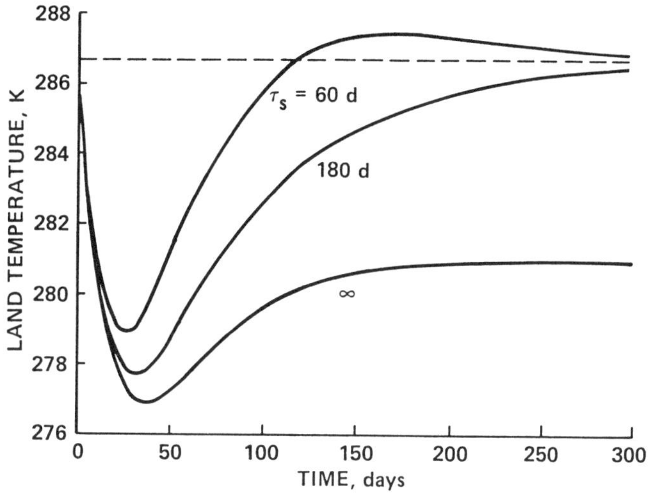


FIGURE 1. Surface land temperatures as a function of model time in days following the smoke injection. The three curves are for e-folding decay times of 60 days, 180 days, and ∞.

that determining the lifetime of particles in this elevated layer is crucial to understanding the long-term consequences.

A comparison of the equilibrium energy balance for the unperturbed case and for the fixed smoke case provides some additional insight into effects on surface temperature. When smoke is added to the model stratosphere, the solar flux absorbed at the surface decreases from 172.5 to 129.2 W m$^{-2}$. At the same time the downward infrared (IR) flux absorbed by the surface *decreases* from 285 to 261 W m$^{-2}$. Since the surface temperature drops, the surface emission decreases and the net infrared flux lost from the surface (emitted − absorbed) decreases slightly from 97 to 91 W m$^{-2}$. One might have expected the downward IR flux to have increased due to the smoke and, in fact, at the base of the smoke layer (10 km) the downward IR increases from 26 to 75 W m$^{-2}$ with the addition of the smoke. However, most of the IR radiation received by the surface originates in the lower troposphere, which has become colder due to the loss of solar radiation; hence, the surface receives less net IR radiation. Overall, the surface experiences a net energy loss (solar plus IR) of 37 W m$^{-2}$. This is balanced by: 1) a decrease in surface temperature that reduces IR emission by about 30 W m$^{-2}$, and 2) a reduction in heat transfer to the atmosphere of about 7 W m$^{-2}$. In our simple model, this reduction in heat transfer produces a consequent reduction in the depth of the troposphere. In a more complex model, the partitioning of the energy reduction between a decrease in temperature and a decrease in latent and sensible heat transfer could be substantially different. This emphasizes the need for more sophisticated treatments of surface and boundary layer processes in order to understand the long-term effects.

In a second series of simulations, we investigated the response of surface temperature to changes in the infrared characteristics of the smoke layer. The curve labeled "Baseline" in Fig. 2 is the same 180-day lifetime case as shown in Fig. 1. It is bracketed by two cases, one in which we assumed the particles had no infrared absorption and the second in which we assumed the particle absorption and extinction were constant with wavelength (*i. e.*, the ratio of optical depth between 0.55 and 10 $\mu$m is 1:1). As expected from the previous work of Turco et al. (1983) and Ramaswamy and Kiehl (1985), the temperature minimum is a weak function of the particle infrared opacity. Even the somewhat extreme case of increasing the infrared optical depth to equal the visible optical depth ($\lambda^0$) produces only a modest reduction of the surface cooling. The small effect of particle infrared opacity is the result of several effects. As noted above, most of the additional thermal energy radiated by the layer never reaches the surface and, in any case, cannot compensate for the solar energy lost to the surface and troposphere by absorption. The solar energy is preferentially absorbed at the top of the smoke layer due to the exponential nature of radiative transfer, while infrared cooling occurs at both boundaries of the layer. This tends to produce a temperature gradient through the smoke layer such that the top radiates more effectively than the bottom, reducing the amount of thermal energy radiated downward. Finally the slower time response of the stratosphere allows the surface to cool beyond the radiative equilibrium point before the stratosphere heats sufficiently to alter the radiation budget (see the fixed smoke case in Fig. 1).

Perhaps the most interesting result in Fig. 2 is the top curve. In this case, we arbitrarily increased the absolute humidity between 9 and 21 km by a factor of 10, and then held it fixed for the duration of the simulation. Additional water vapor is expected in a fire plume due to water of combustion, evaporated surface water, and the entrained water vapor from the relatively moist boundary layer. In the case of the large fires generated by nuclear weapons, we would expect some of the water vapor to be removed immediately by condensation and precipitation (the so-called "black rain" observed at Hiroshima). The amount of water vapor remaining after this initial removal process and its ultimate fate is unknown. However, it seems plausible to assume that the smoke layer would have an enhanced concentration of water vapor. Our assumption of a factor of 10 increase in the smoke layer resulted in an enhancement of the total water vapor column concentration of about 1.2%, with the majority of that increase occurring in the vicinity of 10 km. Since water vapor is a very efficient greenhouse gas, this concentration increase at high altitude reduced the minimum surface temperature by 3-4 K and resulted in a surface temperature increase of about 3 K after 300 model days. This very crude estimation of potential water vapor effects leads to the conclusion that, while additional water vapor carried

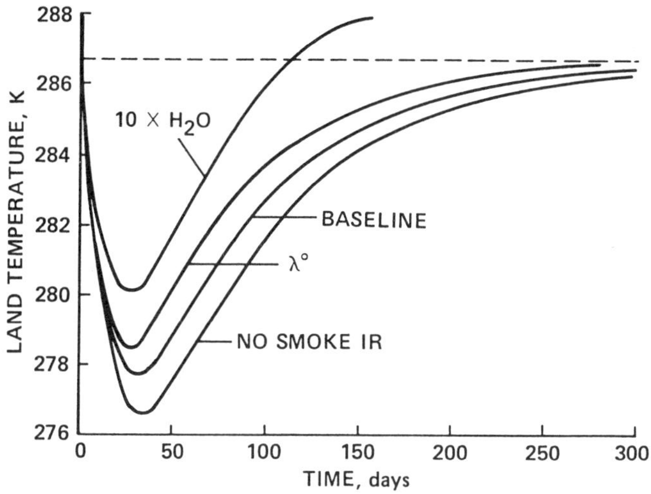


FIGURE 2. Surface land temperatures as a function of model time in days following the smoke injection. The curves (from bottom to top) are for: 1) no smoke infrared optical depth, 2) the baseline case (see text), 3) smoke optical properties constant with wavelength (labelled $\lambda^0$), and 4) a factor of 10 increase in water vapor between 10 and 20 km. All curves are calculated for an assumed e-folding decay time of 180 days.

along in the plume could mitigate the surface cooling, particularly at modest values of the smoke optical depth, it would not prevent surface cooling.

Changes in the water vapor distribution suggest possible changes in cloud amount and cloud optical properties. We have not included any cloud feedback changes here because of the difficulty of incorporating them into a 1-D model in a physically sensible manner. GCM simulations show a general reduction of cloudiness under the smoke cloud due to a suppression of convection (*e. g.*, Malone et al., 1986). Whether this trend would continue on the timescale of 6 months to a year is uncertain. The question of how well the GCMs generate stratiform cloud and fog under cooling conditions also needs to be considered. The issue of changes in cloud properties is particularly critical because the effects of these cloud changes on the radiation budget could dominate surface temperature changes on the longer timescale.

Figures 1 and 2 illustrate the effects of the smoke layer on the surface temperature. However, as shown in Fig. 3 for the base case with a smoke lifetime of 180 days, stratospheric temperatures are also strongly affected. Initially, the model temperatures in the lower stratosphere are about 210 K. Due to the absorption of solar radiation by the smoke, these temperatures increase to about 280 K at day 80 and then decrease slowly. Note that the temperatures at day

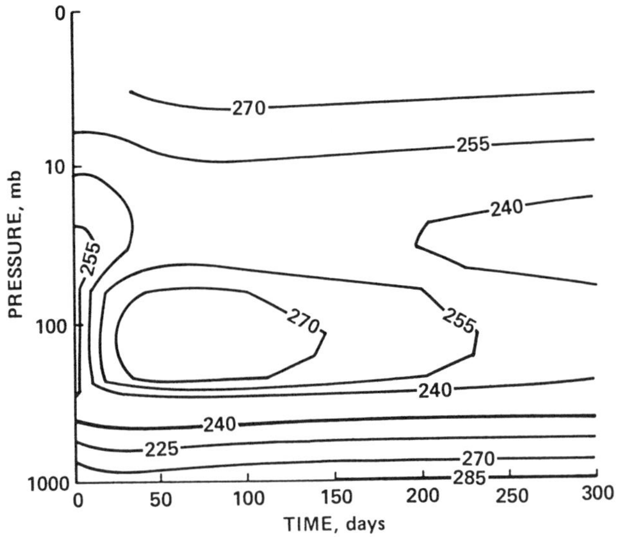


FIGURE 3. Atmospheric temperature as a function of time and pressure for the base case smoke injection and an e-folding decay time of 180 days.

300 are still about 35 K above their equilibrium value due both to the residual smoke still present and the long radiative time constant of the lower stratosphere. Thus one might expect perturbations in stratospheric circulation and chemistry on this longer timescale. How these perturbations might affect the troposphere and surface need to be examined in more detail with 3-D models.

## 4. CONCLUSIONS

We have carried out a sensitivity study of the effects of residual smoke layers following a large-scale nuclear exchange. We estimate that the most plausible e-folding lifetime of this smoke is about 6 months to a year, although there is considerable uncertainty attached to this estimate. The quantity of smoke remaining in the upper atmosphere on a timescale of 1 month to 1 year is sufficient to have a significant climatic impact on the global scale. While it is difficult to assess the actual changes in surface temperature over long periods with a 1-D model, it is clear that surface coolings would persist on timescales exceeding the lifetime of the smoke. The extent of the cooling would depend primarily on the amount of smoke remaining above the convective troposphere. Since the surface energy balance would be substantially altered by the presence of the smoke, it is likely that surface fluxes of latent and sensible heat would also be altered. This in turn would affect atmospheric relative humidity and cloud formation. Determination of the long-term impacts of residual smoke layers will require sophisticated models and careful parameterization of boundary layer processes and of the hydrologic cycle.

The infrared opacity of the smoke layer would act to mitigate the severity of the long-term cooling, but only slightly. Water vapor injected with the smoke is more effective at reducing the cooling but cannot prevent the cooling from occurring.

The stratosphere would experience substantial warming, probably exceeding 50 K. Such warmings would have profound effects on stratospheric dynamics and chemistry that have yet to be explored. These effects could in turn further alter the climate of the troposphere and surface.

## REFERENCES

Crutzen, P., and J. Birks, 1982: The Atmosphere After a Nuclear War: Twilight at Noon, *Ambio*, **11**, 114-125.

Cuzzi, J. N., T. P. Ackerman, and L. C. Helmle, 1982: The Delta-four-stream Approximation for Radiative Flux Transfer, *J. Atmos. Sci.*, **39**, 917-925.

Ghan, S. J., M. C. MacCracken and J. J. Walton, 1987: Chronic Effects of Large Atmospheric Smoke Injections: Interactions With the Ocean Mixed Layer, Sea Ice, and Ground Hydrology. Presented at the *Symposium on Aerosols and Climate*, XIX General Assembly of the IUGG, Vancouver, Canada, August 9-22, 1987.

Hofmann, D. J., 1987: Aerosols From Past and Present Volcanic Emissions. Presented at the *Symposium on Aerosols and Climate*, XIX General Assembly of the IUGG, Vancouver, Canada, August 9-22, 1987.

Hofmann, D. J., and J. M. Rosen, 1987: On the Prolonged Lifetime of the El Chichon Sulfuric Acid Aerosol Cloud, *J. Geophys. Res.*, **92**, in press.

Kent, G. S., C. R. Trepte, U. O. Farrukh, and M. P. McCormick, 1985: Variation in the Stratospheric Aerosol Associated With the North Cyclonic Polar Vortex as Measured by the SAM II Satellite Sensor, *J. Atmos. Sci.*, **89**, 5303-5314.

Malone, R.C., L.H. Auer, G.A. Glatzmaier, M.C. Wood and O.B. Toon, 1985: Influence of Solar Heating and Precipitation Scavenging on the Simulated Lifetime of Post-nuclear War Smoke, *Science*, **230**, 317-319.

Malone, R. C., L. H. Auer, G. A. Glatzmaier, M. C. Wood and O. B. Toon, 1986: Nuclear Winter: Three-dimensional Simulations Including Interactive Transport, Scavenging, and Solar Heating of Smoke, *J. Geophys. Res.*, **91**, 1039-1054.

Malone, R. C., and G. A. Glatzmaier, 1986: Model Validation Studies for Global Transport: Candidate Analogues. In *Technical Papers Presented at the Defense Nuclear Agency Global Effects Review. Volume III.*, DASIAC-TN-86-29-V3, Defense Nuclear Agency, TITL, Washington, D. C., 2-24.

Manabe, S., and R. T. Wetherald, 1967: Thermal Equilibrium of the Atmosphere with a Given Distribution of Relative Humidity, *J. Atmos. Sci.*, **24**, 241-259.

Medalia, A. I., and L. W. Richards, 1972: Tinting Strength of Carbon Black, *J. Colloid Interface Sci.*, **40**, 233-252

Pittock, A. B., T. P. Ackerman, P. J. Crutzen, M. C. MacCracken, C. S. Shapiro and R. P. Turco, 1986: *Environmental Consequences of Nuclear War. Volume I. Physical and Atmospheric Effects.* John Wiley & Sons, Chichester, 342 pp.

Pittock, A. B., J. Frederiksen, J. R. Garratt and K. Walsh, 1987: Climatic Effects of Smoke and Dust Produced From Nuclear Conflagrations. Presented at the *Symposium on Aerosols and Climate*, XIX General Assembly of the IUGG, Vancouver, Canada, August 9-22, 1987.

Pollack, J. B., and T. P. Ackerman, 1983: Possible Effects of the El Chichon Volcanic Cloud on the Radiation Budget of the Northern Tropics, *Geophys. Res. Letts.*, **10**, 1057-1060.

Pruppacher, H. R., and J. D. Klett, 1978: *Microphysics of Clouds and Precipitation.* D. Reidel, Dordrecht, Holland.

Raatz, W. E., R. C. Schnell, M. A. Shapiro, S. J. Oltmans and B. A. Bodhaine, 1985: Intrusions of Stratospheric Air Into Alaska's Troposphere, March 1983, *Atm. Env.*, **19**, 2153-2158.

Ramaswamy, V., and J. T. Kiehl, 1985: Sensitivities of the Radiative Forcing Due to Large Loadings of Smoke and Dust Aerosols, *J. Geophys. Res.*, **90**, 5597-5613.

Roberts, R. E., L. M. Biberman and J. E. A. Selby, 1976: Infrared Continuum Absorption by Atmospheric Water Vapor in the 8-12 $\mu$m Window, *Appl. Opt.*, **15**, 2085-2090.

Rodgers, C. D., and C. D. Walshaw, 1966: The Computation of Infrared Cooling Rate in Planetary Atmospheres, *Q. J. R. Meteorol. Soc.*, **92**, 67-92.

Stephens, S. L., and J. W. Birks, 1987: Ozone as a Sink for Atmospheric Carbon Aerosols: Today and Following Nuclear War. Presented at the *Third International Conference on Carbonaceous Particles in the Atmosphere*, Lawrence Berkeley Laboratory, University of California at Berkeley, October 5-9, 1987.

Stephens, S., M. J. Rossi and D. M. Golden, 1986: The Heterogeneous Reaction of Ozone On Carbonaceous Surfaces, *Intern. J. Chem. Kinetics*, **18**, 1133-1149.

Turco, R. P., O. B. Toon, T. P. Ackerman, J. B. Pollack and C. Sagan, 1983: Nuclear Winter: Global Consequences of Multiple Nuclear Explosions, *Science*, **222**, 1283-1292.

Turco, R. P., O. B. Toon, T. P. Ackerman, J. B. Pollack and C. Sagan, 1987: Climate and Smoke: An Appraisal of Nuclear Winter. Submitted to *Science.*

Vupputuri, R. K. R., 1986: The Effect of Ozone Photochemistry on Atmospheric and Surface Temperature Changes Due to Large Atmospheric Injections of Smoke and $NO_x$ by a Large-scale Nuclear War, *Atm. Env.*, **20**, 665-680.

# THE DEVELOPMENT OF LOW LEVEL COASTAL FLOW FIELDS WHEN SOLAR RADIATION IS BLOCKED BY SMOKE

*C.R. Molenkamp*
University of California, Lawrence Livermore National Laboratory
Livermore, California 94550, USA

## ABSTRACT

Thick layers of smoke, injected into the atmosphere by fires ignited by a nuclear exchange, would block the incident solar radiation and allow continental regions to cool while oceanic areas remain at constant temperature. It has been suggested that the resultant horizontal temperature gradient could initiate precipitation along the coastline that would rapidly scavenge the smoke. This hypothesis is being investigated using an enhanced version of the Colorado State University mesoscale model to simulate the flow field in the vicinity of continental coastlines. The model has been modified to include cloud formation and the effects of these clouds on the transfer of infrared radiation as well as improve the basic longwave radiation parameterization. For moderate westerly winds over the east coast (offshore flow), a layer of fog forms just above the ground that, after about 2 days, evolves into a stratus cloud. Over the ocean a similar stratus cloud forms because of upward mixing of water vapor and radiative cooling. Once these similar cloud layers form, the cooling rates over land and sea become nearly the same eliminating the proposed mechanism for development of convective precipitation and scavenging of mid- or high-level smoke. For the same westerly winds on the west coast (onshore flow), the sequence is similar but evolves more quickly because more moisture is present in the upwind maritime air. In both cases, the formation of ground fog prevents the air temperature in the surface layer from falling below the temperature at which saturation occurs. Before the temperature decreases further, moisture must be removed from the layer.

## 1. INTRODUCTION

Great quantities of smoke, injected into the atmosphere by many large fires ignited by explosions during a major nuclear exchange, could lead to rapid cooling over the continents as the solar radiation is blocked from the surface of the earth. At the same time, the surface temperature over oceanic regions changes only slowly. It has been suggested that large horizontal temperature gradients would develop across the continental coastlines that would drive flow fields causing zones of persistent precipitation along the coastlines and thereby producing an efficient mechanism for removing smoke (Teller, 1984; Singer, 1984; National Research Council, 1985). This hypothesis of anomalous precipitation and scavenging along continental coastlines is examined here using a numerical mesoscale model.

 A. DEEPAK Publishing
ISBN 0-937194-11-5

## 2. MODEL DESCRIPTION

The basic model used for these simulations is the Colorado State University mesoscale model (Mahrer and Pielke; 1976, 1977a,b). It is a hydrostatic, three-dimensional primitive equation model, although only two-dimensional runs are included in this paper. It includes a detailed boundary layer parameterization, a surface energy budget (McCumber and Pielke, 1981), and longwave and shortwave radiative cooling representation. The stable boundary layer diffusion coefficients of McNider and Pielke (1981) are used in regions where nocturnal cooling produces a stable surface layer.

Earlier simulations with the model pointed out some needed modifications. As a result of cooling near the surface the atmosphere became saturated; therefore a representation of the formation of clouds or fog is required. Once clouds are added to the model their effects have to be included in the radiation parameterization. Also, in these earlier simulations, the calculated radiative cooling rates in the lower and middle troposphere were slightly underestimated at less than 1 deg per day indicating the need for an improved longwave radiation calculation.

Two variables are used to represent moisture in the new model: cloud water and total water (sum of cloud water and water vapor). No precipitation mechanism is included. The same conservation equation used for water vapor in the original dry model is used to represent total water in the new model. This is valid since the cloud droplets are assumed to have negligible terminal velocities. The cloud water content is evaluated, at each time step, assuming the atmosphere is either subsaturated (if the actual vapor pressure is less than the saturation vapor pressure) or exactly saturated. If there is either condensation or evaporation, the temperature and liquid water content must be adjusted iteratively, since the saturation vapor pressure is a function of temperature. This iteration is made computationally efficient by taking into account the relationship between changes in saturation vapor pressure and temperature given by the Clausius-Clapeyron equation (Pruppacher and Klett, 1978). The evaluation of the saturation vapor pressure uses the method of Derickson and Cotton (1977).

In consideration of the importance of longwave radiative cooling of the atmosphere and the surface in the current problem, a more detailed and accurate parameterization was developed. The upward and downward fluxes are calculated by vertically integrating the appropriate transfer equations through the atmosphere, taking into account the temperature at each level and the emissivities through the intervening layers. The emissivities are evaluated using the broadband method of Liou and Ou (1983), considering three water vapor bands and the carbon dioxide band as well as an overlap correction. The ozone emissivity is ignored except in a pseudo-layer above the model top. The optical depths in each layer, including pressure and temperature broadening, are estimated using the technique of Liou and Ou (1981). When a layer of cloud is present in the model, it is treated as a boundary for the clear air flux calculation; the boundary flux is calculated assuming the cloud is radiatively gray. The optical thickness of the cloud is evaluated following Stephens (1978). Since relatively small amounts of liquid water produce emissivities very close to 1, most clouds act as blackbodies. Since the radiation parameterization is now fairly complex, it is performed at 5-minute intervals rather than every time step. This procedure works quite well because atmospheric cooling rates change slowly as long as the vertical distribution of clouds is not modified. Since the surface temperature is quite sensitive to the downward longwave flux at the ground, this quantity is evaluated every time step.

## 3. SIMULATION PROCEDURE

A two-dimensional form of the model with the new cloud and longwave radiation parameterization was used for the simulations. The coastline was assumed to lie in a north-south direction with the plane of the model oriented east-west normal to the coastline. Both east and west coast simulations have been run using data from a study along the Oregon coast for a day in August by Mizzi and Pielke (1984) for the input sounding and most other model parameters including the sea surface temperature of 287 K. The one major change in the input is the mean synoptic wind, which is 5 m/s from the west, giving onshore flow for the west coast run and offshore flow for the east coast case. The grid spacing is 4 km in the horizontal for the first 40 km on either side of the coastline; the entire domain, using stretched coordinates beyond 40 km from the coastline, runs from about 400 km west (upwind) to more than 800 km east (downwind) of the coastline.

Smoke is included only as a passive tracer in the version of the model used for these simulations; its effect on the evolution of the flow fields was represented by assuming that it totally obscured the ground from the incident sunlight. The model was run for an initial period of 4 hours with no radiation effects to allow the flow field to approach equilibrium, since different roughness lengths are specified over land and sea. After this adjustment, the model was run through a normal 24-hour diurnal cycle starting at dawn, which occurs at 5:20 a.m. local solar time; this initial diurnal cycle is considered part of the initialization of the model in this study. By dawn of the second day smoke was assumed to cover the domain of the simulation, so no incident solar radiation reached the ground after day 1. The absorption of solar energy at the top of the high altitude smoke layer was ignored. The model was then run for an additional 24 hours in the west coast case and 96 hours for the east coast simulation without any solar input. In effect, the simulation represents one normal daytime heating cycle followed by 36 or 108 hours of night.

## 4. SIMULATION RESULTS

### 4.1 WEST COAST SIMULATION

The United States west coast simulation has been described previously (Molenkamp, 1987) and will be summarized only briefly. When the sun goes down in the evening of the first day the lowest layer of the atmosphere over land cools rapidly. At about 285 K it becomes saturated and a layer of ground fog forms. This fog first occurs at a distance of 12–24 km inland around midnight of the first night and by dawn of the second day covers the entire land surface. The fog alters the radiative properties of the layer so that it experiences a rapid radiative heat loss, but this heat loss is counteracted by the release of latent heat of condensation producing a layer of nearly constant temperature.

Shortly after dawn on the second day, with the rising sun totally obscured from the ground by smoke, the fog begins to move upward and from 6:20 a.m. to 10:20 a.m. changes from a thin ground fog to a layer of cloud between 200 m and 600 m in altitude. Due to radiative cooling at the cloud top, the air below is destabilized and the resultant vertical diffusion produces a well-mixed region from the cloud top to the ground. As a result of continued cooling at the cloud top, the cloud grows to the 850 m layer by the end of the simulation at dawn of the third day.

In the atmosphere over the ocean, radiative cooling, the evaporation of water vapor from the ocean and the constant sea-surface temperature combine to produce a growing well-mixed layer above the surface. About 8:20 a.m. on the second day, at the same time the fog is lifting off the land surface, a cloud forms in this well-mixed layer over the ocean. As the cooling and diffusion of moisture continue, the water content of the cloud increases and the cloud moves upward. By the end of the simulation at dawn on the third day, the cloud over the ocean also extends up to 850 m.

The clouds that form in the simulation are typical of the stratus clouds found off the coast of Oregon in the summertime and that typically spread inland at night. On some days these clouds persist throughout the day near the coast while further inland they dissipate due to solar heating. In an extended simulation with no solar heating, the clouds would just persist and continue to grow vertically. Since they cover both land and sea, the differences in surface radiative characteristics are of minor importance and there is no mechanism to produce the thermal gradients suggested for the development of precipitation and scavenging of high-level smoke.

## 4.2 EAST COAST SIMULATION

Many features of the Unites States east coast simulation are similar to the west coast run. There are also some significant differences; one of the most important is that the upwind boundary is over land and therefore much drier than the marine boundary layer, which is upwind in the west coast simulation.

During the afternoon of the first day, before smoke arrives to block the solar radiation, a weak sea breeze develops near the coastline against the 5 m/s westerly synoptic wind. The potential temperature, specific humidity, and horizontal velocity normal to the coastline at 5:20 p.m. are shown in Fig. 1; the vertical velocity in the vicinity of the coastline is shown in Fig. 2. The well-mixed daytime boundary layer extending to about 1.25 km is clearly evident over land where the potential temperature, specific humidity, and wind speed are nearly constant with height. The weak intrusion of marine air just inland of the coastline is rapidly eroded as it is mixed with thick eastward moving air aloft. At the coastline near the surface, moisture is advected slightly inland and then is rapidly mixed upward. Evidence of eastward transport of heat and moisture can be clearly seen in the layer from 500 to 1500 m above the surface. The mean vertical velocities shown in Fig. 2 are quite small, but they do illustrate the typical lifting at the sea breeze front and the smaller more dispersed downward return flow. They also show that the sea breeze front penetrates no more than 20 km inland. These mean vertical motions, although small, are much larger than any that occur during the extended nocturnal part of the simulation.

Over the ocean during the day a surface-based inversion forms as heated continental air is advected offshore over the much cooler ocean surface. The evidence for this is the packing of the potential temperature contours just above the ocean surface, particularly close to the coastline. This inversion acts as a barrier to the vertical mixing of the moisture that is continuously injected into the atmosphere from the sea; therefore an unusually strong vertical gradient of moisture occurs in the surface layer just offshore. A more typical vertical distribution of moisture occurs over the ocean beyond 200 km from the coastline where continental effects have not yet penetrated.

After sunset on day 1, the air over land begins to cool, rapidly near the surface and more slowly aloft. An inversion and very stable layer develops just above the ground. At a temperature of about 280 K the surface layer becomes saturated and a layer of ground fog forms; this occurs about 11:00 a.m. on day 2 over the entire land surface except within 10 km of the coastline. For the next 48 hours, this layer of fog persists and the surface layer maintains a nearly constant temperature. A sequence of vertical temperature and

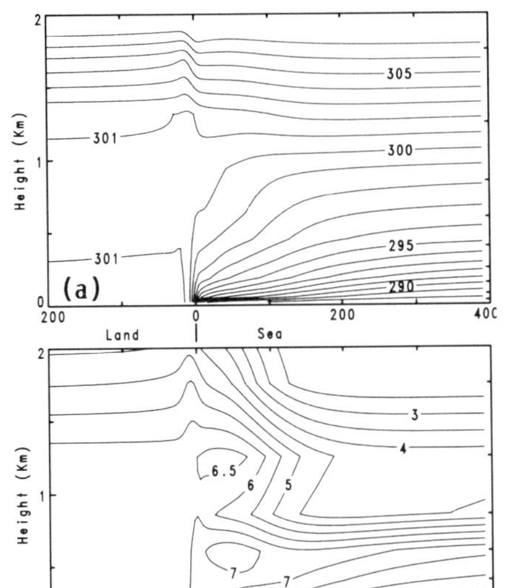


FIGURE 1. Distribution of potential temperature (a), specific humidity (b), and wind normal to the coastline (c) at 5:20 p.m. on day 1. Potential temperature is in degrees K, specific humidity in g/kg, and wind speed in m/s. The tick marks on the right ordinate are the actual grid points for the plotted variable.

potential temperature profiles at 12-hour intervals are shown in Fig. 3 at grid points 50 km inland and 4 and 180 km offshore. Over land the cooling follows an interesting pattern. Initially the most rapid cooling is at the surface, but once the ground fog forms and the surface is held at constant temperature, the layer of most rapid cooling moves constantly upward. This causes the height of the inversion to rise while the lower part of the temperature profile tends to become isothermal. After about 48 hours the second layer above the surface becomes saturated and the top of the fog layer rises. During the next 6 hours the layer of ground fog changes into a cloud layer at an altitude of 200 to 800 m. This transformation, shown in Fig. 4, occurs fairly rapidly because the lower part of the atmosphere is nearly isothermal and has an almost constant specific humidity. Due to the rapid cooling at the cloud top and the associated constant sea surface temperature, the atmosphere below this level becomes well-mixed as shown by the layer of constant potential temperature. During the entire simulation the air temperature never drops below the temperature at which saturation occurs, indicating the important role moisture plays in limiting the cooling of the atmosphere.

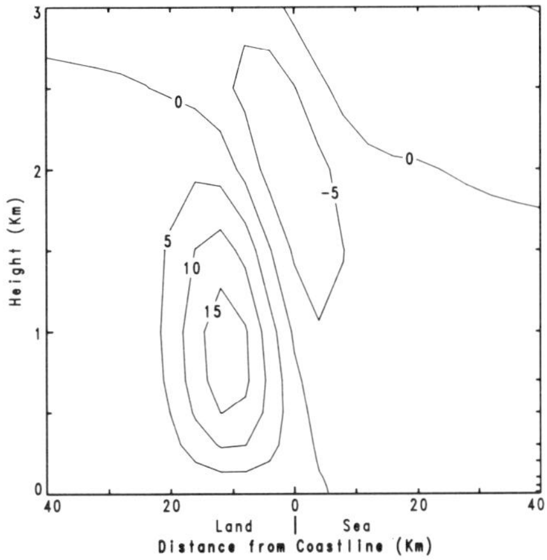


FIGURE 2. Distribution of vertical velocity at 5:20 p.m. on day 1. The vertical velocity is in cm/s.

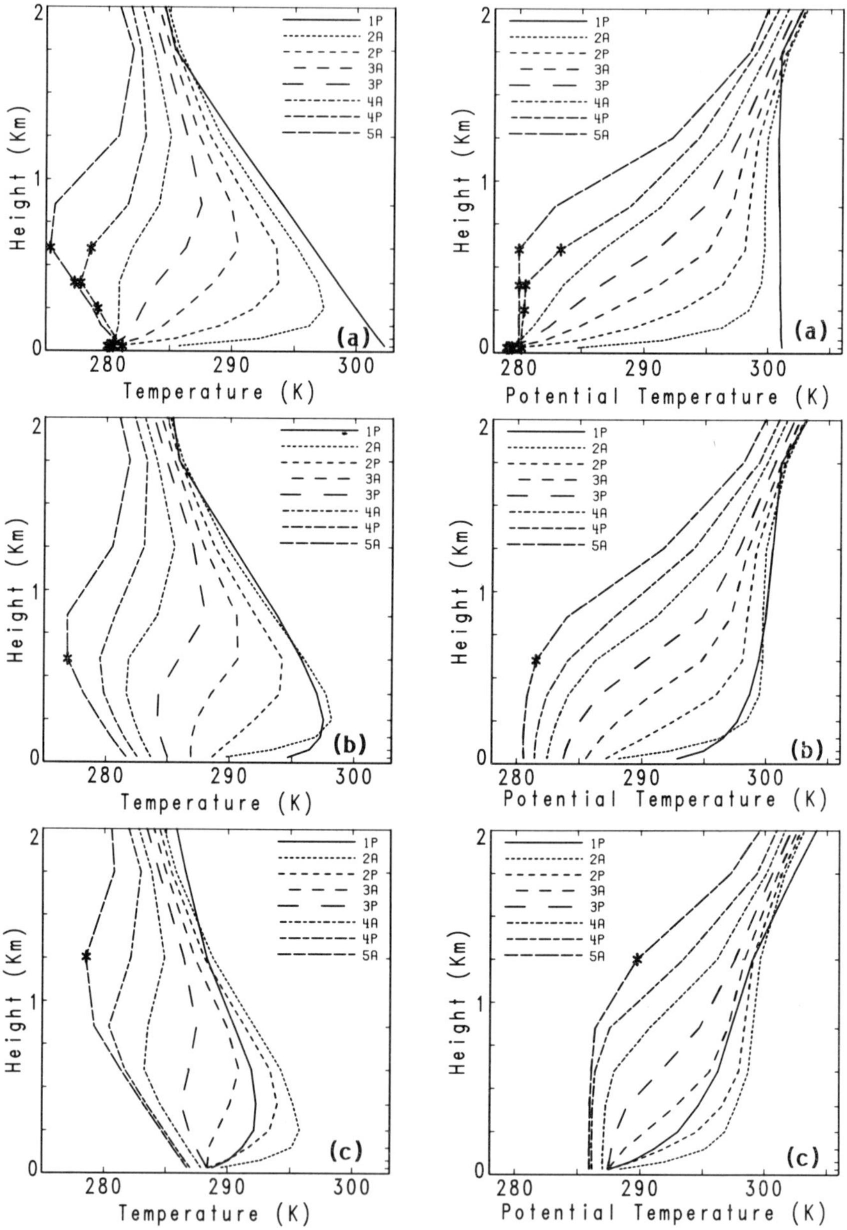


FIGURE 3. Vertical temperature and potential temperature profiles at 12-hour intervals at grid points: (a) 50 km inland, (b) 4 km offshore, and (c) 180 km offshore. The labels of the curves indicate the day (1...5) and time, 5:20 a.m. (A) or 5:20 p.m. (P). The *'s on the profiles indicate levels where clouds are present.

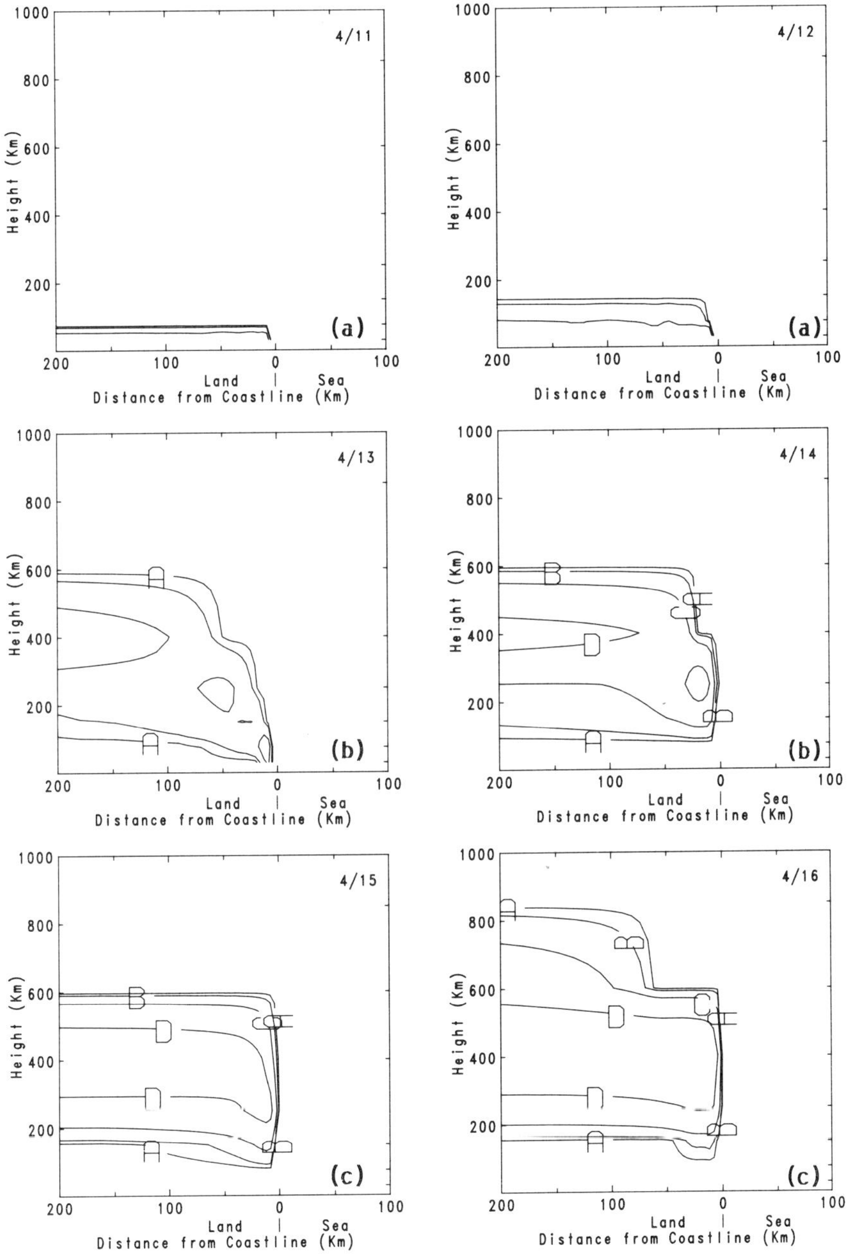


FIGURE 4. Growth of cloud water content over land on day 4 from 11:00 a.m. to 4:00 p.m. The contours, in g/kg, are .01 (A), .03 (B), .1 (C), and .3 (D). The labels in each plot indicate the day/hour.

Over the ocean the evolution of the vertical temperature profile is somewhat different. The earliest profile in Fig. 3 for the grid point 180 km offshore is essentially the initial temperature profile; it has moderate stability throughout the lowest 2 km of the atmosphere with slightly greater stability near the surface. During the next 24 hours the layer below 1 km becomes warmer as air that was heated during its passage over land is advected over this location. Very close to the surface the temperature remains near the original temperature because of its proximity to the sea surface and because there is not much advection of heated air. After the middle of day 2, the air temperature above the ocean starts to decrease because of advection of cooler air from the continent and radiative heat losses. During this temperature decrease aloft, air at the lowest level stays at the constant temperature of the sea surface; therefore, a small layer of neutral stability is found just above the sea surface by evening of day 3. As the cooling continues this unstable layer expands upward. Within the unstable layer, water vapor, evaporated from the sea, is mixed upward leading to formation of a cloud at about 5:00 a.m. on day 5 in the grid layer centered at 1.25 km. This is the first evidence of a cloud over the sea, but within 3 hours the entire ocean is covered by a cloud layer.

The spatial evolution of the structure of the atmosphere from the end of the first night to the end of the simulation at 5:20 a.m. on day 6 can be seen in Figs. 5–7, which show the distribution of potential temperature, specific humidity, wind normal to the coastline, and cloud water content at 5:20 a.m. on each day. An extra set at 5:20 p.m. on day 2 is included in Fig. 5. The very strong stability in the surface layer over land on days 2 and 3 can be seen in the strong packing of the isentropes. As this very cold and stable air moves over the warmer ocean surface, it is quickly heated to a temperature very close to that of the sea surface. Above the very cold surface layer over land, the remnants of the daytime neutral layer can be seen as the region of potential temperatures near 300 K; the advection of this layer over the ocean is responsible for the horizontal gradient of potential temperature beyond 200 km offshore at 5:20 a.m. on day 2. The growth in height of the stable layer over land and the increased cooling of layers between 200 m and 1 km show clearly in Figs. 5 and 6. As the stable layer grows in thickness, however, the stability near the surface decreases, particularly over the ocean, and this allows more upward mixing of water vapor. These effects are evident over the ocean as increased heights of enhanced water vapor amounts in the specific humidity profiles in Figs. 5 and 6. The westerly component of the wind has almost no variation from land to sea or with height, except for the effects of surface roughness that tend to retard motion in the surface layer more over land than over the ocean. There is also an inertial oscillation in the wind field in the simulation that produces changes in the overall magnitude of the normal component of the wind outside of the surface layer. At 5:20 a.m. on day 2, no clouds have formed, but from 11:00 a.m. on day 2 until the middle of day 4 there is a layer of ground fog over land.

A few hours after dawn on day 4, the cooling over land has progressed to the point where layers above the surface are saturated, the fog lifts off the ground and evolves into a layer of cloud between 200 and 800 m. As this cloud forms and produces high rates of cooling near its top, the thickness of the layer over land that is colder than the sea surface grows to over 500 m by dawn of day 4 and over 1 km on day 6. This leads to some significant changes in the evolution of the fields. First, there is a horizontal gradient in potential temperature offshore that is large enough to generate a weak land breeze; the evidence is the enhanced westward wind near the surface and retarded flow above 750 m in Fig 7. This land breeze is quite weak and diffuse; the maximum mean vertical velocities are 6 cm/s downward and 1.5 cm/s upward, far too small to suggest the initiation of

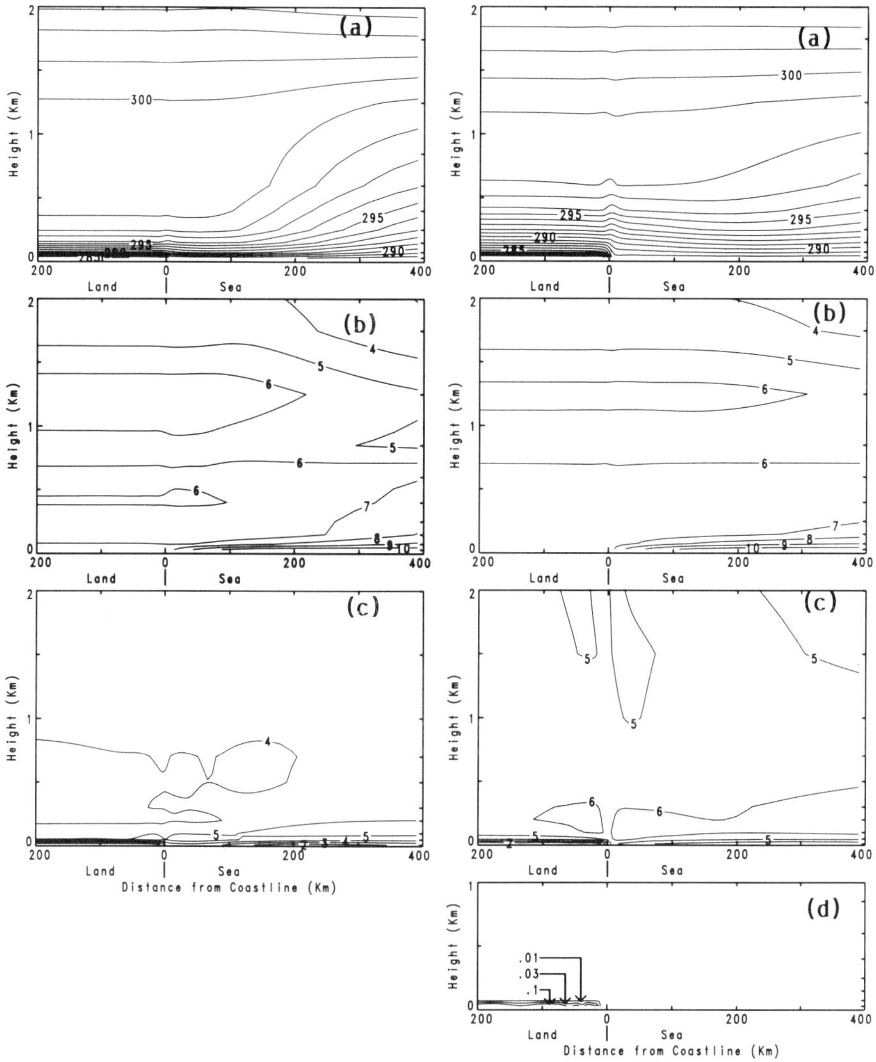


FIGURE 5. Distribution of: (a) potential temperature, (b), specific humidity, (c) wind normal to the coastline, and (d) cloud water content for day 2 at 5:20 a.m. (left) and 5:20 p.m. (right). The potential temperatures are in K, the specific humidity and liquid water content in g/kg and the wind speed in m/s. The tick marks on the right of each panel indicate the actual vertical grid levels for the appropriate variable.

clouds that might significantly scavenge smoke. Second, as this cold air moves over the ocean, it produces an unstable layer. Between 5:20 a.m. and 11:20 a.m. on day 5 upward diffusion of moisture and downward mixing of cold air in this unstable region leads to formation of a cloud. As the cloud top cools, the well-mixed layer grows in height with an associated reduction of the horizontal gradient of potential temperature and a weakening of the land breeze. At the end of the simulation, the only significant deviation in the

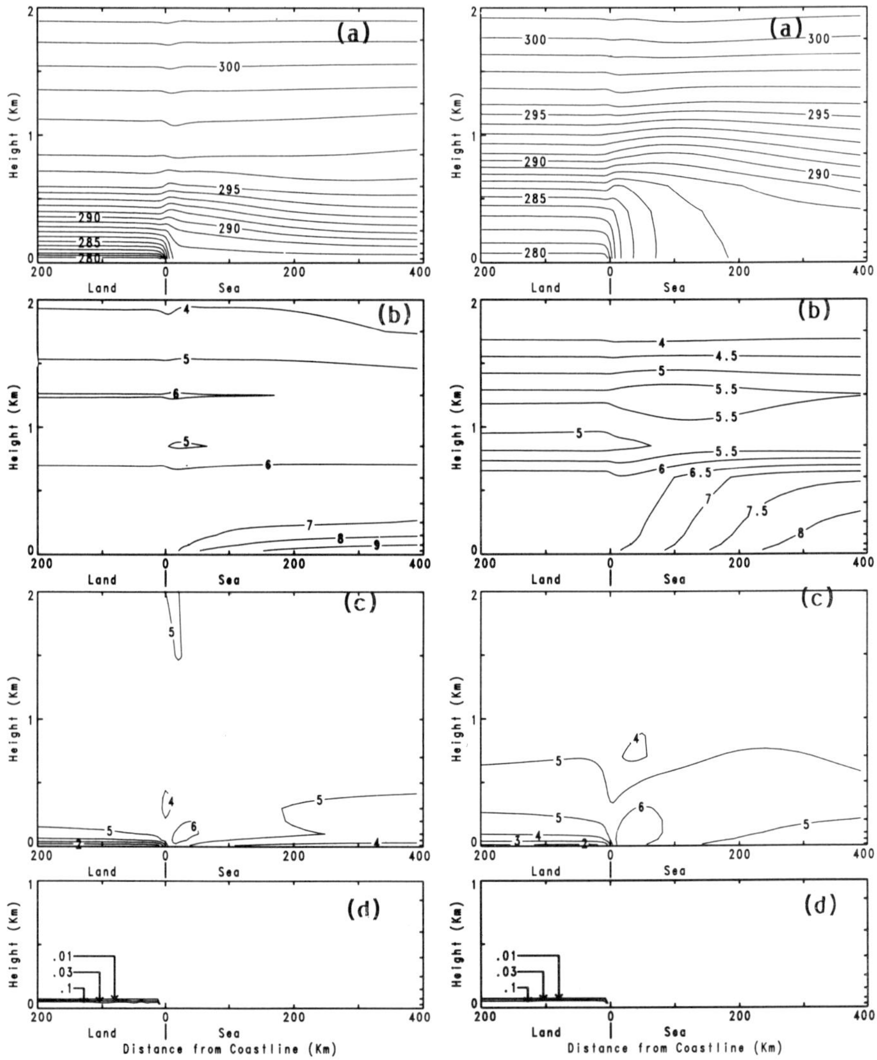

FIGURE 6. Distribution of: (a) potential temperature, (b) specific humidity, (c) wind normal to the coastline, and (d) cloud water content for day 3 at 5:20 a.m. (left) and day 4 at 5:20 a.m. (right). The units for the contours are the same as in Fig. 5.

horizontal wind, from its imposed synoptic component of 5 m/s, is a reduced magnitude near the surface over land due to surface roughness; nowhere is the mean vertical velocity as large as 1 cm/s.

## 5. CONCLUSIONS

Simulations for both onshore and offshore synoptic winds contradict the hypothesis that much of the smoke injected into the atmosphere after a large-scale nuclear exchange

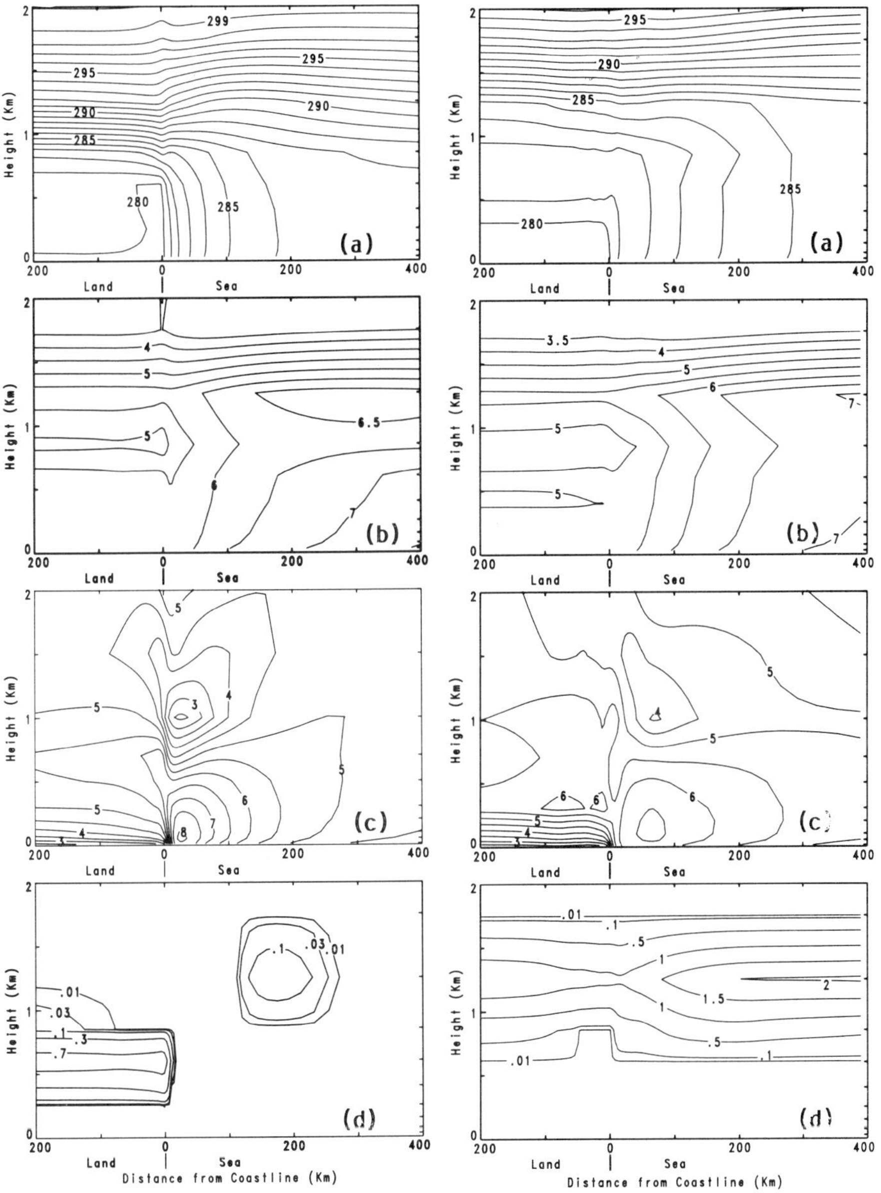


FIGURE 7. Distribution of: (a) potential temperature, (b) specific humidity, (c) wind normal to the coastline, and (d) cloud water content for day 5 at 5:20 a.m. (left) and day 6 at 5:20 a.m. (right). The units for the contours are the same as in Fig. 5.

could be scavenged by precipitation along the continental coastlines by showing that a cloud layer, forming over both land and sea within a few days after solar obscuration prevents the development of thermal gradients with sufficient intensity to induce the formation of convective clouds. The formation of low-level clouds and fog halts the atmospheric temperature decrease through the release of latent heat of condensation and changed radiative properties, and this limit on temperature decrease moderates the degree of continental cooling that occurs when the sun is obscured.

## ACKNOWLEDGMENTS

I am very grateful to Dr. Roger Pielke for giving me a copy of the Colorado State University mesoscale model and for the help that he, Dr. Richard McNider, Dr. Robert Kessler, and Dr. Ray Arritt have given me to understand the various parts of the model.

This work was performed under the auspices of the U.S. Department of Energy by the Lawrence Livermore National Laboratory under contract W-7405-Eng-48.

## REFERENCES

Derickson, R.G., and W.R. Cotton, 1977: On the Use of Finite Taylor's Series Approximations to Certain Exponential and Power Functions Employed in Cloud Models, Atmospheric Sciences Paper No. 268, Colorado State University, Available from NTIS, Springfield, VA, (PB-267 834).

Liou, K.N., and S.S. Ou, 1981: Parameterization of Infrared Transfer in Cloudy Atmospheres, *J. Atmos. Sci., 38,* 2707–2716.

Liou, K.N., and S.S. Ou, 1983: Theory of Equilibrium Temperatures in Radiative-Turbulent Atmospheres, *J. Atmos. Sci., 40,* 214–229.

Mahrer, Y., and R.A. Pielke, 1976: Numerical Simulation of the Airflow Over Barbados, *Mon. Wea. Rev., 104,* 1392–1402.

Mahrer, Y., and R.A. Pielke, 1977a: A Numerical Study of the Airflow Over Irregular Terrain, *Beitr. Phys. Atmos., 50,* 98–113.

Mahrer, Y., and R.A. Pielke, 1977b: The Effects of Topography on Sea and Land Breezes in a Two-Dimensional Numerical Model, *Mon. Wea. Rev., 105,* 1151–1162.

McCumber, M.C., and R.A. Pielke, 1981: Simulation of the Effects of Surface Fluxes of Heat and Moisture in a Mesoscale Numerical Model, 1. Soil Layer, *J. Geophys. Res., 86,* 9929–9938.

McNider, R.T., and R.A. Pielke, 1981: Diurnal Boundary Layer Development Over Sloping Terrain, *J. Atmos. Sci., 38,* 2198–2212.

Mizzi, A.P., and R.A. Pielke, 1984: A Numerical Study of the Mesoscale Atmospheric Circulation Observed During a Coastal Upwelling Event on 23 August 1972, Part I: Sensitivity Studies, *Mon. Wea. Rev., 112,* 76–90.

Molenkamp, C.R., 1987: Simulation of Coastal Flow Fields when the Incident Solar Radiation is Obscured. Proceedings of the International Conference on Energy Transformations and Interaction with Small and Mesoscale Atmospheric Processes, Lausanne, Switzerland, March, 1987, Available as Lawrence Livermore National Laboratory Report, UCRL-94994.

National Research Council, 1985: *The Effects on the Atmosphere of a Major Nuclear Exchange,* National Academy Press, Washington, DC, 193pp.

Pruppacher, H.R., and J.D. Klett, 1978: *Microphysics of Clouds and Precipitation,* D. Reidel Pub. Co., London, 714pp.

Singer, S.F., 1984: Is the "Nuclear Winter" Real? *Nature, 310,* 625.

Stephens, G.L., 1978: Radiation Profiles in Extended Water Clouds, II: Parameterization Schemes, *J. Atmos. Sci., 35,* 2123–2132.

Teller, E., 1984: Widespread After-Effects of Nuclear War, *Nature, 310,* 621–624.

# THE INTERNATIONAL AEROSOL CLIMATOLOGY PROJECT

*Adarsh Deepak*
Science and Technology Corporation
Hampton, Virginia 23666, USA

and

*Gabor Vali*
University of Wyoming
Laramie, Wyoming 82071, USA

ABSTRACT

Under the auspices of the International Commission on Cloud Physics and the International Radiation Commission, a Joint Working Group has been set up to take the leading role in the organization of an International Aerosol Climatology Project. The major goal of this project is to plan and facilitate implementation of international collaboration in the study of the properties and impacts of atmospheric aerosols. Several planning sessions have identified the need for a project of international scope.

## 1. BACKGROUND

A brief background of the International Aerosol Climatology Project (IACP) is provided here. The International Radiation Commission (IRC) Working Group (WG) on Aerosols and Climate was formed in 1978, under the chairmanship of Dr. Joachim Joseph (Tel Aviv University), to address the following three tasks: the compilation of aerosol data for models; the encouragement of sensitivity studies of the effects of aerosols on climate; and the encouragement of contact between climate and aerosol researchers.

The work of this WG continued under the follow-on ad hoc WG on Aerosol Optical Properties, under the cochairmanship of Drs. Hermann Gerber of the Naval Research Laboratory (NRL), Washington, DC, and Joachim Joseph. In March 1983, this ad hoc WG organized a WMO/NASA/NRL Meeting of Experts on Aerosols and Their Climatic Effects in Williamsburg, Virginia. This meeting resulted in WMO Report WCP-55 entitled "Experts Meeting on Aerosols and Their Climatic Effects," edited by A. Deepak and H. Gerber (1983). It was at this meeting that the idea of an IACP was first discussed, with Dr. H.-J. Bolle making suggestions for its implementation.

The details of the proposed IACP plans are given in Chapters 6 and 9 of WCP-55 (Deepak and Gerber, 1983). The establishment of a World Aerosol Data Library (WADL) was given high priority.



ISBN 0-937194-11-5

Subsequently, at the International Radiation Commission's 1984 meeting in Perugia, Italy, it was suggested by Dr. Bolle that the ad hoc WG prepare a firm proposal in one year, detailing how and where an aerosol data base might be set up, and how it might be staffed and funded. This report was prepared by Dr. Joseph and distributed to a few aerosol experts. The highlights of this report were then presented at the Symposium (M-13) on the Variability of Aerosol Optical Properties on 6 August 1985, during the International Association of Meteorology and Atmospheric Physics (IAMAP) General Assembly, in Honolulu, Hawaii. Based on the discussions that ensued, it became apparent that many of the aerosol scientists present were neither aware of the IACP nor of the Joseph report (1985), with the result that the WADL concept was not received as warmly as had been anticipated. A need was felt and a recommendation was made that, as a first step, a directory of aerosol researchers be prepared listing those who have aerosol data, information, instrumentation, models, etc., and/or are engaged in or planning aerosol field programs.

The formation of a Joint Working Group (JWG) of the International Commission on Cloud Physics (ICCP) and the IRC on the International Aerosol Climatology Project was proposed at the 1985 Honolulu meeting, with Drs. Adarsh Deepak (IRC) and Gabor Vali (ICCP) as cochairmen. A proposal seeking official endorsements for the JWG on the IACP was submitted to Dr. J. Lenoble, President of IRC, at the 1986 COSPAR Meeting in Toulouse, France, and to Professor P. V. Hobbs, President of the ICCP. Endorsements were received from them subsequently. The terms of reference of the JWG on the IACP are described below.

## 2. TERMS OF REFERENCE OF THE JOINT IRC-ICCP WORKING GROUP

### 2.1 GENERAL TASKS

The JWG will formulate plans and facilitate implementation of international collaboration in the study of the properties and impacts of atmospheric aerosols within the framework of the terms of reference of the International Radiation Commission and the International Commission of Cloud Physics.

### 2.2 SPECIFIC TASKS

1) The JWG will review the recommendations developed by the ad hoc Working Group on Aerosol Optical Properties and the plans developed by the Committee on Nucleation and Atmospheric Aerosols for a field study of tropical aerosols.

2) Based on the previously formulated plans, and taking into account recent related activities, the JWG will develop recommendations and plans of implementation for the International Aerosol Climatology Project. These plans are to be presented for approval to the IRC and the ICCP.

3) Elements of the project to be considered are:

i) Preparation of a directory of researchers or research groups who can provide data on atmospheric aerosol properties and spatial distribution.
ii) Survey of existing aerosol measurement programs.
iii) Identification and review of existing aerosol measurement techniques.
iv) Survey of existing data bases and mechanisms of data interchange.
v) Development of plans for an automated worldwide aerosol data base.
vi) Organization of quadrennial symposia on atmospheric aerosols.
vii) Encouragement of coordination and implementation of regional experiments, with appropriate standardization and cross-referencing of instrumentation.
viii) Encouragement of collaboration and the formation of national or regional associations among scientists performing atmospheric aerosol-related studies.

## 2.3 PERFORMANCE PERIOD

A performance period of three years is anticipated in order to successfully complete the proposed tasks.

## 2.4 INITIAL TASKS

The first task of the IACP JWG will be to prepare a worldwide directory of scientific workers or groups who can provide data on atmospheric aerosol properties and their spatial and temporal distributions. An informational questionnaire will be prepared for distribution among the aerosol researchers.

## 2.5 MEMBERS

A list of 15 members invited to the JWG follows. For practical reasons, the number of members of the JWG had to be kept to a minimum; however, all of those who want to actively participate are welcome and will be associated with the activity of the JWG. The members are:

Dr. Adarsh Deepak, Science and Technology Corp. (USA), Cochairman
Dr. Gabor Vali, U. of Wyoming (USA), Cochairman
Dr. John DeLuisi, NOAA/ERL (USA)
Dr. Hermann E. Gerber, Naval Research Laboratory (USA)
Dr. Jost Heintzenberg, U. of Stockholm (Sweden)
Prof. Dr. Ruprecht Jaenicke, U. of Mainz (FRG)
Dr. Joachim H. Joseph, Tel Aviv U. (Israel)
Dr. Peter Koepke, U. of Munich (FRG)
Dr. Jacqueline Lenoble, U. of Lille (France)
Dr. E. Meszaros, Inst. of Atmospheric Physics (Hungary)
Dr. M. Patrick McCormick, NASA LaRC (USA)
Dr. Josef Podzimek, U. of Missouri-Rolla (USA)

Dr. Joseph M. Prospero, U. of Miami (USA)
Dr. Ming-xing Wang, Inst. of Atmospheric Physics (China)
Dr. Vladimir E. Zuev, Inst. of Atmospheric Optics (USSR)

## 3. CURRENT STATUS OF PROJECT ELEMENTS

### 3.1 DIRECTORY OF AEROSOL RESEARCHERS

Mailing Lists: In response to our request to the JWG members and other researchers, several mailing lists containing names and addresses of aerosol researchers or research groups have been received. A computerized data base of aerosol researchers is being compiled and continually updated.

Questionnaire: A questionnaire requesting aerosol-related information was prepared shortly after the first meeting of the JWG on the IACP held on the evening of 13 August 1987.

### 3.2 SURVEY OF EXISTING AEROSOL MEASUREMENT PROGRAMS AND DATA BASES

Based on the responses received from JWG members and other aerosol researchers, a summary of the existing or planned aerosol measurement and data base programs, which is to be completed with additional information, is shown as Table 1.

### 3.3 MECHANISMS OF DATA INTERCHANGE

The important mechanisms of aerosol data and information interchange are:

1) a directory listing of aerosol researchers and their data,
2) a periodical newsletter,
3) IACP meetings, and
4) international symposia.

### 3.4 PLANS FOR AN AUTOMATED WORLDWIDE AEROSOL DATA BASE

It was decided that since the stratospheric aerosol data were well taken care of, the JWG should concentrate its initial efforts on developing the tropospheric aerosol data base. The development of an automated worldwide aerosol data base has been discussed by Deepak (Deepak and Koziana, 1984), Derr (1984), Joseph (1985), and d'Almeida and Koepke (1985). Further discussions on this topic will continue to be held within the aerosol community.

# TABLE 1. EXISTING OR PLANNED AEROSOL MEASUREMENT AND DATA BASE PROGRAMS

| Information Source | Program | Description | Organizer |
|---|---|---|---|
| J. DeLuisi (NOAA/ERL, Boulder) | 1) Information Meeting of Stratospheric Aerosol World Lidar Network (Sept/Oct 1987, Italy) | • Data Needs, Applications, and Reporting<br>• Uniform Data Processing Procedures<br>• Reporting of Data, New Findings<br>• Writing a Short Position Paper | G. Fiocco (U. of Rome) and M.P. McCormick (NASA LaRC)<br><br>J. DeLuisi |
| | 2) Submicron (<0.05 $\mu$m) Particle Measurement Intercomparison Workshop Preliminary Planning | | |
| | 3) CCN Counter Design Workshop (Preliminary Planning) | | J. DeLuisi |
| | 4) ISCCP Microphysics of Clouds (Needs To Be Done) | • Microphysical Processes of Cloud Formation and Aerosol Scavenging | |
| J. Heintzenberg (U. of Stockholm) | 1) AEROTEX Project of EUROTRAC (1986–90) | • European Air Chemistry Program (Regional Aerosol Data)<br>• Finland, UK, Norway, Sweden, FRG<br>• Processes Occurring in Transport of Pollution Aerosols From Central Europe to Northeast Sector: (i) Ground and Airborne Experiments, and (ii) Development of Combination of Long-Range Transport and Aerosol Life-Cycle Models, With Validation by Experiments | J. Heintzenberg |
| | 2) WG on Atmospheric Transparency | • To Develop a Generally Acceptable Approach for Atmospheric Correction of Satellite Data<br>• To Design a Major Field Experiment To Help Clarify Uncertainties That Exist in the Assessment of the State of Atmospheric Transparency | R. Rizzi (U. of Bologne) |

## TABLE 1 (Continued)

| Information Source | Program | Description | Organizer |
|---|---|---|---|
| | | **Tasks:**<br>• Intercomparison of Inversion Methods for Retrieval of Aerosol Size Distribution | J. Heintzenberg and V. Cuomo (U. of Naples) |
| | | • Influence of Aerosols on Temperature and Humidity Profile Retrieval | R. Sounder (Oxford) and S.A. Clough (AFGL) |
| | | • Correction of Atmospheric Effects | B. Sturm (Ispra) |
| | | • Intercomparison of Remote Measurements From Ground and Radiation of Space and Ground-Based Remote Sensing Techniques | H.-J. Bolle (Free U., Berlin) and R. Rizzi |
| P. Koepke (Munich, FRG) | Aerosol Optical Climatology | • Radiative Properties of Tropospheric Aerosols (ext. scat. abs. coeffts, w, g, P)<br>• Temporal (1 month)<br>• Spatial (5°×5°, lat.×long.)<br>• Mie Theory; Log-Normal SD<br>• Bulk Radiative Transfer (Standard Radiation Atmosphere Models)<br>• Components: Water Soluble, Dustlike, Soot, Oceanic<br>• Types:Continental, Urban, and Maritime Desert, Polar, and Mixed<br>• 60 Wavelengths (0.3 $\mu$m–40 $\mu$m) | G. A. d'Almeida (U. of Munich) and P. Koepke |
| E.P. Shettle AFGL | 1) Global Aerosol Climatology | • Climatology of Tropospheric Aerosol<br>• Spatial WMO-Data Blocks<br>• Temporal:<br>• Sources, Crustal, Biogenic, Marine, etc.<br>• Radiative Properties: Scattering, Absorption, Asymmetry Parameter<br>• Wavelengths: 0.2 to 40 $\mu$m | R. Husar (Washington U., St. Louis) and E.P. Shettle |
| | 2) Radiative Aerosol Models | • Standard Radiation Atmospheres - Aerosols<br>• Descriptive Aerosol Models: Rural, Maritime, Urban, Stratospheric, Volcanic, Size Dist., Ref. Index | H.-J. Bolle E.P. Shettle and R.W. Fenn (AFGL) |

## TABLE 1 (Continued)

| Information Source | Program | Description | Organizer |
|---|---|---|---|
| | | • Global Aerosols for Climatology Calculations | O.B. Toon and J.B. Pollack (NASA ARC) |
| M.P. McCormick (NASA LaRC) | 1) GOSSA Net (including lidar data) | • Global Observation and Studies of Stratospheric Aerosols<br>• User's Guide<br>• Directory of Lidar Researchers | M.P. McCormick |
| | 2) NASA Aircraft Expedition to Antarctic (DC8&ER2) | • To Measure Antarctic Aerosols and Gases to Investigate Springtime $O_3$ Depletion | |
| | 3) SAGE II Data Archival (Fall 1987) | • Atmospheric and Upper Tropospheric Aerosol Data | |
| J. Podzimek (U. of Missouri–Rolla) | 1) FORMEZ Program | • Tropical Aerosol Measurement Experiment | J. Fontan (Toulouse) |
| | 2) Tropical Aerosol Workshop, Congo (1987–88) | | J. Fontan |
| | 3) Amazon River Basin Tropical Aerosol Mission | | A. Netto (Sao Paulo) |
| J. Prospero (U. of Miami) | 1) AEROCE Program (1987–90) | • Atmospheric–Ocean Chemistry Expt. (NSF)<br>• Daily Measurement of Aerosols, Precipitation, Gases<br>• 4 Stations: Barbados, Bermuda, Mace Head (Ireland), Tenerife (Canary Is.) | J. Prospero |
| | 2) NASA GLOBE/LAWS | • B(180°) Statistics Over Remote Ocean Areas | D. Fitzjarrald (NASA MSFC) |
| G. Vali (U. of Wyoming) | 12th International Conference on Atmospheric Aerosols, Condensation and Ice Nucleation, Vienna, Austria, 21–28 Aug 1988 | | G. Vali |
| V. Derr (NOAA/ERL) | Atmospheric Data Tables | • Atmospheric Handbook: Atmospheric Data Tables Available on Computer Tape | V. Derr |

## TABLE 1 (Concluded)

| Information Source | Program | Description | Organizer |
|---|---|---|---|
| L.L. Stowe, C.R.N. Rao (NOAA/ NESDIS) | Ad Hoc Working Group for Validation of NOAA Aerosol Products | **Requirements:**<br>• Accurate $\delta_{ext}^{aer}$ (0.5 $\mu$m) To Serve as Ground Truth<br>• n(r), m, $m_2$, Sky Conditions, Aerosol Episodes (Dust, Smoke, etc.), N(r), T, and WV Profiles<br>• Island & Coastal Sites/Sea Vessels | C.R.N. Rao and L.L. Stowe |
| W. Boehme (Met Dept., Potsdam) | WMO (CAS) WG on Climatic Problems (Oct 1986) | • Tasked by JSC of ICSU and WMO<br>• To Develop a Methodology Through Which Different Met Services Can Make Long-Term Measurements of Atmospheric Aerosol Properties (for Aerosol Climatology) | W. Boehme |
| E.V. Browell (NASA LaRC) | 1) Amazon B.L. Experiment (ABLE-2A)—Dry Season (1985); (ABLE-2B)—Wet Season (1987)<br>2) Arctic B.L. Experiment (ABLE-3A) | • Lidar Aerosol Data<br>• Entire Amazon Basin<br>• Over Arctic and Subarctic | E.V. Browell |
| P.V. Hobbs (U. of Washington, Seattle) | On-Going Research in Various Locales | • Airborne Measurements of Tropospheric Size Distributions<br>• Aerosol and Optical Properties in Remote and Polluted Environments | P.V. Hobbs |

### 3.5 ORGANIZATION OF QUADRENNIAL SYMPOSIA ON ATMOSPHERIC AEROSOLS

Several suggestions have been made regarding the organization of a quadrennial International Symposium on Atmospheric Aerosols (ISAA), with emphasis on the optical properties and the climatic effects of aerosols. The thrust of the upcoming 12th International Conference on Atmospheric Aerosols, Condensation and Ice Nucleation in Vienna, Austria, 21-28 August 1988, does not adequately take into consideration these aspects. A workshop focused on implementation of the IACP to be held in May 1988 is another option. Such a workshop could then set the planning in motion for organizing the first quadrennial symposium on atmospheric aerosols, perhaps as early as 1989.

### 3.6 THE INTERNATIONAL UNION OF GEODESY AND GEOPHYSICS SUPPORT FOR IACP PROPOSAL TO THE INTERNATIONAL GEOSPHERE BIOSPHERE PROGRAM (IGBP)

To support the IACP, the International Union of Geodesy and Geophysics (IUGG) has allocated a small amount of funding to bring together a small group of scientists in order to write an IACP proposal as an input to the IAMAP proposal to the International Geosphere Biosphere Program (IGBP). Dr. Bolle, President of IAMAP, has suggested that a report of the group could result in more detailed plans than currently exist on:

1) How to extract global aerosol (atmospheric turbidity) information from operational satellites.
2) How aerosol information can be gained from special space missions.
3) How the lidar network should be used to deduce aerosol profiles.
4) What information can be deduced from ground-based photometric measurements.
5) How the data could be managed.
6) What (minimal) organizational structure is necessary to conduct the project.

The introduction of the report or proposal should contain information on the envisaged changes in the aerosol regime of the globe taking into account volcanoes, desertification, and anthropogenic pollution.

Prior to convening a workshop to write such a proposal, Dr. Lenoble proposed that the task would be made much easier if we allocated the work in several areas (such as satellite observations, ground-based observations, and microphysics) by forming sub-working groups. At the first meeting of the IACP working group, which took place in Vancouver, Canada, on 13 August 1987, the following sub-WGs were formed:

Sub-WG 1: Aerosol Physics and Chemistry
Will address questions related to optical properties, composition, microphysics, and chemistry of atmospheric aerosols.

Sub-WG 2: Interaction between Aerosols and Clouds
Will address questions related to composition, growth, evaporation, and scavenging of aerosols.

Sub-WG 3: Ground-Based and In Situ Aerosol Measurements
Will address questions such as what network stations exist; where should validation measurements be performed; what is the quality of observations; and on the intercomparison of instruments, such as lidars, radiometers, and photometers.

Sub-WG 4: Space Tropospheric Aerosol Observations
Will address questions on the comparison of algorithms, validation experiments, and nadir/limb experiments.

Sub-WG 5: Global Aerosol System Requirements
Will address data users' needs for different applications, such as requirements for climatic effects, studies, instrumentation design, etc.

Input from the aerosol scientific community, and its active cooperation and participation in this project, will be the key to the IACP's effectiveness and success.

## REFERENCES

d'Almeida, G. A., and P. Koepke, 1985: Aerosol Optical Climatology (AOC). A letter report submitted to the IRC. (Available from the authors: Meteorologisches Institut der Universität München, D-8000 München 2, Fed. Rep. of Germany.)

Deepak, A., and J. Koziana, 1984: The World Aerosol Data Library (WADL) Concept, in Aerosols and Their Climatic Effects, H. Gerber and A. Deepak (Eds.), A. Deepak Publishing, Hampton, VA, USA, 105-115.

Deepak, A., and H. Gerber (Eds.), 1983: Report of the Experts Meeting on Aerosols and Their Climatic Effects, WCP-55, World Meteorological Organization, Geneva, Switzerland.

Derr, V. E., 1984: Atmospheric Handbook: Atmospheric Data Tables, Report UAG-89, National Geophysical Data Center, NOAA/NESDISI E/GC2, 325 Broadway, Boulder, CO 80303, USA, 1-55 (available on computer tapes).

Joseph, J. H., 1985: The International Aerosol Climatology Program (IACP) and Its World Aerosol Data Library (WADL). A letter report to members of the ad hoc Working Group on Aerosols and Climate. (Available from the author: Dept. of Geophysics and Planetary Sciences, Tel-Aviv University, Ramat-Aviv, Tel-Aviv 69978, Israel.)

# SUBJECT INDEX*


A

Absorption of radiation, 271
Aerosol(s)
  anthropogenic impact, 241
  chemistry, 471
  climatology, 125
  concentrations, 117
  data library, 471
  effects on climate, 335, 349, 385, 395, 471
  layer, 117
  mass, 215
  modeling, 261
  number density, 21
  optical properties, 81, 325
  over oceans, 69
  particle lifetimes, 443
  particle removal, 443
  particles, 241
  physics, 471
  remote sensing of, 39, 51
  scattering function, 117
  simulation, 233
  size distribution, 101, 117
  sizes, 175
Aerosols (types)
  anthropogenic, 373
  arctic, 271
  biogenic, 31
  interstitial, 423
  manmade, 253
  marine, 1
  moist, 373
  stratospheric, 215, 277, 287, 305, 313, 325
  volcanic, 335
Agriculture, 385
Antarctica, 287
Antarctic aerosol and ozone, 195
Arctic, 271
Arctic climate, 373
Arctic haze, 271, 335, 373
Atmospheric
  aerosol(s), see under Aerosols
  chemistry, 9
  models, 147, 459
  optical thickness measurements, 89
  turbidity, 39
AVHRR data, 69

B

Background stratospheric aerosols, 195
Backscattering, 215
Balloon-borne particle center, 215
Balloon data, comparison, 305
Balloon measurements, 117
Biogenic aerosol, 31
Breaking waves, 1
Bubbles, 1

C

$CFCl_3$, 313
Chinese forest fires, 195
Chlorine-catalyzed ozone destruction, 287
Chlorofluorocarbons, 313
Climate, 139, 223, 253
  change, 335, 459
  effects, 349, 423
  effects of smoke on, 395
  variations of, 195
  instabilities, 385
Clouds, 471
Cloud
  cover, 139
  physics, 9
  properties, 241, 253
Clouds and fog, 459
Coastal precipitation, 459
Condensation growth, 423
Condensation nuclei, 195

D

Desert aerosol, 261
Desert dust, 147
Diurnal cycle, 459
Dusts, 395

E

Elemental carbon, 373
Extinction of light, 165, 215
Extinction coefficient, 185
Evaporation rates, 21


*Number refers to first page of appropriate paper.

F

Film drops, 1
Fires, 411, 423, 435
Forecast errors, 435
Forest fires, 411, 435

G

Gas-to-particle conversion, 195
General circulation model (GCM), 373
Global loading of aerosol, 51

H

Harmattan, 175
Heat balance, 21
Heating rate, 271
Hemispheric cooling, 385
Heterogeneous chemistry, 287
High Altitude Sampling Program (HASP), 313
$HNO_3$, 287

I

Ice nuclei, 31
Infrared radiation, 165, 271
Infrared radiative transfer, 459
International Aerosol Climatology Project (IACP), 471
Interstitial aerosols, 423

J

Jet drops, 1

L

Land breeze, 459
Lidar, 195, 215, 325, 471
  in space, 81
  measurements, 385
Lower stratosphere, 287

M

Marine aerosol, 1
Mauna Loa Observatory, 325
Mie theory, 165
Multiwavelength extinction coefficients, 305

N

Nighttime cooling rates, 175
Nitric acid haze, 287
Northern Hemisphere stratosphere, 313
Nuclear war, 423
Nuclear winter, 335, 395, 411, 435, 443, 459
Nucleation, 423
Numerical models, 147, 233, 261, 313, 335, 443

O

Oceanic whitecaps, 1
Optical depth, 241, 271
Optical properties, 125, 233, 261
Ozone, 443
Ozone-aerosol relations, 195
Ozone hole, 287

P

Particle(s)
  from fires. 411
  lifetime, 443
  removal, 443
  size distribution, 233
Particulates, 185
Planetary albedo, 241
Planetary boundary layer, 459
Polar stratospheric clouds, 287
Polar stratospheric haze, 287

R

Radiation, 271
  terrestrial, 241
  total-diffuse, 271
Radiative
  characteristics, 125
  cooling, 459
  convective model, 443
  forcing, 349
  perturbations, 349
  transfer, 51, 261
Radiometers, 471
Remote sensing, 39, 51, 69, 81, 89, 215
Rocket measurements, 117

S

SAGE II, 305
Satellite
  data, 101
  imagery, 51
  observations, 253
Scattering of light, 165
Sea breeze, 459
Sea salt, 1
Single scattering albedo, 261

Size distribution of particles, 153, 241, 277, 305
Smoke, 443
  effects on surface temperature, 435
  injection, 459
  particles, 423
Solar, 271
  radiation measurements, 89, 101, 241
  spectrum measurements, 89
Soot, 373
Space, remote sensing of aerosols from, 471
Space shuttle, 81
Stratosphere, 215, 277
Stratospheric
  aerosols, 195, 305, 325, 385
  layers, 233
  mean concentration trend, 313
  trace gas measurements, 313
Stratus clouds, 459
Sulfur
  chemistry, 195
  dioxide, 195
Sulfuric acid, 195, 373
Surface-active materials, 1
Surface temperature, 435

T

Terrestrial radiation, 241
Total-diffuse radiation, 271
Tracer transport, 147
Turbidity, 139, 175

U

Ultraviolet, 165

V

Vertical mixing, 459
Visibility, 185
Visual range, 21, 185
Volcanic, 277
  aerosols, 195, 325
  eruptions, 195, 385
    El Chichon, 117, 195, 335
    Nevado del Ruiz, 117, 195
    Sierra Negra, 117
Volcanism, 215
Volcanoes, 117, 139, 223

W

Weather modification, 459

# AUTHOR INDEX

Ackerman, T.P., 271, 335, 443
Amato, U., 101
Arnold, F., 287
Bernstein, R.L., 253
Blanchet, J.-P., 373
Brock, C.A., 411
Brogniez, C., 305
Brühl, C., 287
Chuan, R.L., 277
Coakley, J.A., Jr., 253
Cohen, A., 21
Crutzen, P.J., 287
Cuomo, V., 101
d'Almeida, G.A., 125
Deepak, A., 471
DeFoor, T.E., 325
Durkee, P.A., 253
Edwards, L.L., 423
Endemann, M., 81
Fontana, F., 101
Frederiksen, J.S., 395
Garratt, J.R., 395
Gore, W.J.Y., 271
Grassl, H., 241
Hegg, D.A., 411
Helmes, L., 139
Hobbs, P.V., 411
Hofmann, D.J., 195, 215
Hoppel, W.A., 9
Hummel, J.R., 261
Jaenicke, R., 31, 139
Jäger, H., 215
Jayaraman, A., 117
Joussaume, S., 147
Kästner, M., 81
Kaufman, Y.J., 51
Koepke, P., 125
Krekov, G.M., 233
Leifer, R., 313
Lenoble, J., 305
Littfass, M., 215
Litvak-Neumann, M., 21
Longtin, D.R., 261
Lyons, J.H., 411
Macchiato, M., 101
Mahrer, Y., 21
Mao Jietai, 89
Masuda, K., 39
Matthias, S., 31
McClain, E.P., 69
Molenkamp, C.R., 459
Monahan, E.C., 1
Oluwafemi, C.O., 175
Penner, J.E., 423
Pittock, A.B., 395
Podzimek, J., 153
Pryce, J.D., 261
Quenzel, H., 81
Radke, L.F., 411
Rakhimov, R.F., 233
Ramaswamy, V., 349
Rao, C.R.N., 39, 69
Rasmussen, R., 411
Renger, W., 81
Reiter, R., 385
Robinson, E., 325
Robock, A., 435
Rosen, J.M., 215
Sadourny, R., 147
Sapper, J., 69
Saxena, V.K., 185
Schmailzl, U., 287
Sendra, C., 51
Serio, C., 101
Shettle, E.P., 165, 261
Stowe, L.L., 69
Subbaraya, B.H., 117
Takashima, T., 39
Toon, O.B., 443
Trakhovsky, E., 165
Turco, R.P., 443
Valero, F.P.J., 271
Vali, G., 471
Walsh, K., 395
Wang Qiang, 89
Weil, M.L., 271
Weintraub, D., 185
Weiss, M.A., 21

Weiss, R., 411
Wiegner, M., 81
Woods, D.C., 277
Woolf, D.K., 1
Xu Qun, 223
Zhang Chengxiang, 89
Zhao Bolin, 89
Zuev, V.E., 233